Advances in Control

Springer
London
Berlin
Heidelberg
New York
Barcelona
Hong Kong
Milan
Paris
Santa Clara
Singapore
Tokyo

Paul M. Frank (Ed)

Advances in Control

Highlights of ECC '99

Springer

Paul M. Frank
Gerhard-Mercator Universität, GH Duisburg, FB9/MRT, Bismarckstr. 81,
D-47048 Duisburg, Germany

British Library Cataloguing in Publication Data
Advances in control : highlights of ECC '99
1.Automatic control - Congresses
I.Frank, Paul, 1934-
629.8

Library of Congress Cataloging-in-Publication Data
European Control Conference (5th : 1999 : Karlsruhe, Germany)
Advances in control : highlights of ECC '99 / Paul M. Frank (ed.)
P. cm.
"Plenary papers and mini-courses of the ECC '99"--CIP table of contents p.
Includes bibliographical references (p.).

ISBN-13: 978-1-4471-1216-7 e-ISBN-13: 978-1-4471-0853-5
DOI: 10.1007/978-1-4471-0853-5

1. Automatic control--Congresses. I. Frank, Paul M. II. Title.
TJ212.2.E97 1999 98-49851
629.8--dc21 CIP

Softcover reprint of the hardcover 1st edition 1999

Typesetting: Camera ready by contributors

69/3830-543210 Printed on acid-free paper SPIN 10693279

Preface

This book contains the text of the plenary lecture, the semi-plenary lectures and the mini-courses of the 5th European Control Conference (ECC '99) held in Karlsruhe, Germany, from August 31 to September 3, 1999. On the threshold of the new millennium, 9 prominent authorities express their views on current developments and perspectives in the theory and practice of automatic control and systems engineering. In addition, a number of experts in the field present self-contained sets of lecture notes giving a broad overview of special topics of current interest in modern control theory and its applications.

The first essay is by the plenary speaker of the conference, K. J. Åström. The author gives some personal reflections on the development of automatic control - theory and applications - over the past 5 decades with an assessment of the current status and some speculations about future developments. The contribution by I. Mareels, S. Van Gils, J. W. Polderman and A. Ilchmann analyses the complex dynamics in adaptive control with a view to providing a more theoretical insight to be used for the design of supervisory adaptive control systems. Challenging current control problems in multi-modal tele-presence and tele-action systems which typically consist of a human operator, the possibly distant tele-operator and various types of barriers between them, are discussed in the article by M. Buss and G. Schmidt.

According to the increasing endeavours of the organisers of the ECC to address more issues of control applications, there are four application oriented contributions. The article from D. E. Seborg provides a perspective on the opportunities for advanced process control technology placing emphasis on developments that have occurred during the past 5 years. The paper by M. J. Grimble and G. Hearns deals with the advanced control of hot rolling mills, addressing performance and robustness benefits through the use of H-infinity controllers with a decentralised structure and advanced simulation techniques. The essay by M. Blanke outlines current techniques of fault tolerant control of technical plants including both fault detection and isolation and system re-configuration. The article from J. Lévine presents 3 industrial nonlinear control applications in the field of mechanical systems combining the physical model-based approach with singular perturbations and flatness-based motion planning and tracking controller design.

The next 2 articles address selected topics of modern control theory. The paper by B. H. Krogh and A. Chutinan outlines the modelling and supervisory control of hybrid systems, briefly reviewing recently developed tools for computer-aided modelling, simulation and verification of hybrid systems. The paper by S. L. Campbell, R. Nikoukhah and F. Delebecque is dedicated to nonlinear descriptor systems. It surveys some of the nonlinear theory and then relates it to several control problems from a mathematical point of view, emphasising nonlinear and linear time varying systems.

The following texts refer to the 3 four-hour mini-courses presented at the ECC '99. These tutorial exposés describe recent developments in modern control theory and its applications.

The contribution from T. Kaczorek is composed of 6 papers. Bose and Yang investigate the consequences of imposing various symmetry conditions on the bounds of wavenumber response of interval finite impulse response (FIR) multidimensional digital filters. D. Franke outlines the 2D-analysis of a special class of hybrid dynamical systems. J. Gregor reports on attempts to establish a general theory of discrete systems by successive specialisations of basic structures. T. Kaczorek, one of the fathers of multidimensionl systems theory, introduces new classes of positive 2D discrete and continuous-discrete linear systems. J. Klamka surveys the recent results concerning controllability of different types of linear 2D systems with constant coefficients. D. A. Owens, E. Rogers and K. Galkowski present an introduction to repetitive processes and they survey the work to-date on the development of a control theory for these processes and its application to classes of iterative learning control schemes. P. E. Wellstead, M. B. Zarrop and S. R. Duncan discuss 2D system concepts for industrial web processes.

The exposé by S. Kowalewski, S. Engell and O. Stursberg discusses the formal verification of logic controllers for processes with switched continuous dynamics. It combines available analysis tools for real-time and hybrid systems from computer science and a signal-flow oriented modular modelling framework for mixed discrete-continuous systems.

The multi-authored lecture notes by F. Allgöwer, T. A. Badgwell, J. S. Qin, J. B. Rawlings and S. J. Wright deal with nonlinear predictive control. They give a tutorial introduction and an overview of the active areas of model predictive control (MPC) and moving horizon estimation (MHE). A wide spectrum of topics is discussed ranging from the history of MPC and MHE, the underlying control and estimation theory, to computational aspects and issues of industrial applications. The focus of this article is on the relatively new areas of nonlinear MPC and nonlinear MHE, but the linear problem is also covered. This contribution concludes the text of the book.

I want to thank all those who have contributed to this book for their excellent work and their great cooperation in the performance of the text. Mrs. Barbara Frischemeier deserves my special gratitude for her great efforts and perfect work concerning the formatting of the text.

Duisburg, April 10, 1999

Paul M. Frank, ECC '99 chairman

Table of Contents

1

Automatic Control - The Hidden Technology

K. J. Åström

Department of Automatic Control, Lund Institute of Technology, Lund, Sweden, e-mail: kja@control.lth.se

Summary. The field of automatic control is about 50 years old. This paper presents some reflections on the dynamic development. Automatic control represented a paradigm shift from the traditional split of engineering into civil, mechanical, electrical, and chemical. Key ideas in the development are presented. The interplay of theory and applications are discussed as well as relations to other fields such as mathematics and computer science. It is attempted to make an assessment of the current status. The lecture ends with some speculations about future development.

1 Introduction

Automatic control systems are today pervasive. They appear practically everywhere in our homes, in industry, in communications systems, in all types of vehicles and in scientific instruments. Control systems are increasingly becoming mission critical, a failure of the control system will thus lead to a system failure. In spite of this automatic control is not very much talked about. It is therefore appropriate to label the technology the hidden technology. Automatic control is now (1999) about 50 years old. Books about history of automatic control are appearing, for example [49], [9], [10]. Sessions on history of control have also been introduced in many of the conferences. The purpose of this paper is not to give an historical account but rather to give some personal reflections on the development of the field.

Education and research in engineering grew out of specific technologies such as mining, building of roads and dams, construction of mechanical and machines, generation and transmission of electricity, and industrial use of chemistry. This led to an organization of engineering schools based on departments of mining, civil engineering, mechanical engineering, electrical engineering, and chemical engineering etc. This served very well in the en of the 19th century and the beginning of the 20th century. The situation changed significantly with the advent of fields like automatic control, that cut cross traditional department boundaries. Industry has adapted quickly to the new demands but academia has not.

Automatic control is crucial for practically all engineering activity. Typical examples are generation and transmission of energy, process control, manufacturing, communication, transportation and entertainment. Automatic control has been a key element in the design of experimental equipment and instrumentation used in basic sciences. Principles of automatic control also have impact on such diverse fields as economics, biology, and medicine.

Automatic control, like many other branches of engineering science, has developed in the same pattern as natural science. Although there are strong similarities between natural science, and engineering science it is important to realize that there are some fundamental differences. The inspiration for natural science is to understand phenomena in nature. This has led to a strong emphasis on analysis and isolation of specific phenomena, an extreme case being particle physics. A key goal is to find basic laws that describe nature. The inspiration of engineering science is to understand, invent, and design man-made technical systems. This places much more emphasis on design. It is natural to consider interaction instead of isolation. A key goal is to find system principles that makes it possible to effectively deal with complex systems. Feedback, which is at the heart of automatic control, is an example of such a principle.

The wide applicability of automatic control has many advantages, but it has also created some difficulties. Because automatic control can be used in so many different fields, it is a very good vehicle for technology transfer. There are of course also difficulties, one of them is to keep a proper balance and a good perspective.

2 The Roots

Although we can find very early examples of the use of feedback, the development of automatic control is strongly connected to the industrial developments associated with the industrial revolution. Whenever new sources of power were discovered the need to control them immediately arose. When new production techniques were developed there were needs to keep them operating smoothly with high quality. In this section we will follow some of the roots of automatic control. We will start with the consequences of the development of steam power and industrial production. We will then discuss some implications regarding the development of ships and airplanes and we will end with some consequences for the emergence of telecommunications.

2.1 Governors

The need for control devices appeared already in the operation of windmills. The famous centrifugal governor, Figure 1, first appeared in this context, see [49]. The first major application, however, was in connection with the development of steam power. The desire to keep textile mills running at constant speed was a strong motivation. The result was an interesting development of governors for machines over a 100 year period stretching from late 1700. This development if very nicely described in [9]. Theoretical understanding of the devices also developed, starting with the papers [48] and [74]. They provided basic understanding and posed interesting stability problems, which were later solved independently by [63] and [29]. There was little interaction between the scientists. Routh and Hurwitz were not aware of each other's contributions and neither knew about the fundamental work on stability by Lyapunov [46]. Vyshnegradskii's paper had a very strong influence on engineering practice and was widely used in control systems for hydroelectric power. The book [69] remained standard material for control of machines for a long time. This book presented of the state of the art of governors at the turn of the century. Methods of analysis based on linearization and analysis of the roots of the characteristic equation based on Hurwitz theorem are presented. The presentation is, however, strongly focused on the application.

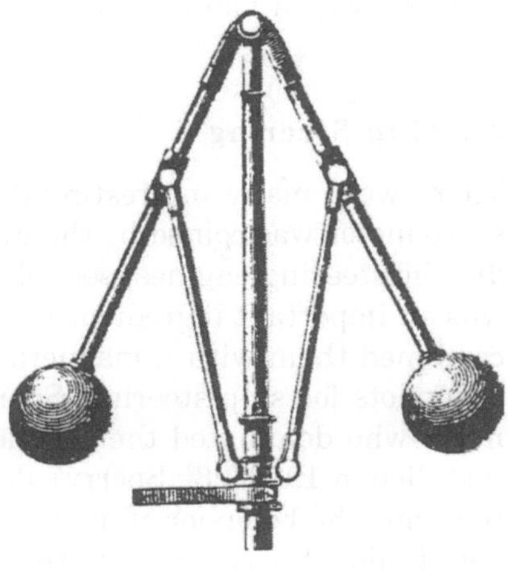

Fig. 1 The centrifugal governor which has become a symbol of the field of automatic control.

2.2 Industrial Process Control

Automation of process and manufacturing industries evolved from the late 1800. The development of control in that field was largely parallel to the development of governors. It resulted in the standard PID controller now commonly used in industry. Proportional feedback as well as integral and derivative action was rediscovered because the workers in the field failed to see the similarity with the already existing work on governors. Emergence of sensors, instruments and controllers resulted in the creation of new companies. By mid 1930 there were more than 600 control companies with Foxboro, Brown, Taylor Instruments, Honeywell, Leeds & Northrup and Fisher & Porter among the leading ones, see [68].

There were very modest theoretical developments. One reason for this is may be a lack of organizations with a critical mass of researchers. ASME created the Industrial Instruments and Regulators Committee in 1936, which provide a forum for discussion. The state of the art is illustrated by the following quote from [30]:

> "In spite of the wide and ever-increasing application of automatic supervision in engineering, the science of automatic regulation of temperature is at present in the anomalous position of having erected a vast practical edifice on negligible theoretical foundation."

One explanation is that analysis was difficult because of the fact that most process control systems had time delays, another is that the industry structure was such that it permitted few luxuries and that hardware problems were still predominant.

One lasting contribution is an empirical method for tuning the standard PID controller [78]. It is interesting to observe that this work made extensive use of simulation of closed loop dynamics using the differential analyzer which had just been developed by Vannevar Bush, see [19].

2.3 Ship Steering

There were many interesting developments in ship steering. The word servo motor was coined by the French engineer Farcot who developed hydraulic steering engines, see [9]. These devices which provided actuation was an important ingredient towards automation of ship steering. Sperry combined them with a magnetic compass and gyroscopes and developed autopilots for ship steering. Sperry was one of the most successful engineers who dominated the commercial field with more than 400 systems installed in 1932 [28]. Sperry's design was largely intuitive where he tried to mimic the behavior of an experienced helmsman. By observing experienced pilots Sperry came to the conclusion that

> "An experienced helms man should 'meet' the helm, that is, back off the helm and put it over the other way to prevent the angular momentum of the ship carrying it past its desired heading."

This effect can obtained by derivative action. Sperry succeeded very well and his autopilot, which was nicknamed the "Metal-Mike" was very successful because it captured much of the behavior of a skilled pilot.

There were also interesting theoretical development of ship steering. A major contribution was given by Minorsky in the paper [52] which presented a taxonomy of many different controllers and a recommendation to use a PID controller. The results were also verified in full scale experiments. In [9] there is an interesting discussion of the contributions of Sperry and Minorsky and their impact on actual auto-pilot design.

2.4 Flight Control

There were many experiments with manned flight in the 18th century. The Wright brothers were the first to be successful. One major reason was that they clearly understood the relation between dynamics and control. This was expressed very clearly by Wilbur Wright when lecturing to the Western Society of Engineers in 1901 [50], we quote:

> " Men already know how to construct wings or airplanes, which when driven through the air at sufficient speed, will not only sustain the weight of the wings themselves, but also that of the engine, and of the engineer as well. Men also know how to build engines and screws of sufficient lightness and power to drive these planes at sustaining speed ... Inability to balance and steer still confronts students of the flying problem. ... When this one feature has been worked out, the age of flying will have arrived, for all other difficulties are of minor importance."

By using their insight and skillful experiments the Wright brothers made the first successful flight with Kitty Hawk in 1905. The fact that this plane was unstable was a strong impetus for the development of autopilots based on feedback, [28]. Sperry made a very spectacular demonstration of his autopilot as early as 1912, see Figure 2. His son Lawrence flew the plane close to the ground with his hands in the air with a mechanic walking on the wing!

Although a good theoretical understanding of flight dynamics was based on linearized equations and analysis of the characteristic equation based on Routh's stability criterion was available as early as 1911, the theoretical work did not have any impact on practical autopilot design until the mid-1950s. One possible reason is that tools for computation were lacking.

Fig. 2 The dramatic demonstration of Sperry's airplane stabilizer at the Paris competition for safety of airplanes in 1914. (The picture is drawn from a fuzzy photo.)

The success of the Wright brothers is a very early example of what we today call integrated process and control design. The key idea is that automatic control gives the designer extra degrees of freedom. It is interesting that Minorsky was well aware of the issues. He captured it in the phrase: "It is an old adage that a stable ship is difficult to steer."

It is interesting to observe that modern high performance fighters are designed to be unstable. They rely on a control system to maintain stability.

2.5 Telecommunication

Feedback was an enabling factor in the development of telecommunication. The key contribution was the invention of the feedback amplifier by Black. An interesting perspective on the invention is given in the paper [13]. In his efforts to obtain a highly linear amplifier that was very insensitive to variations in the characteristics of the electron tubes Black had made the observation that:

> "By building an amplifier whose gain is deliberately made, say 40 decibels higher than necessary and then feeding the output back on the input in such a way as to throw away the excess gain, it had been found possible to effect extraordinary improvement in constancy of amplification and freedom from non-linearity."

A circuit diagram of the feedback amplifier is shown in Figure 3. If the gain of the amplifier is A the input-output behavior is given by

$$\frac{V_2}{V_1} = -\frac{R_2}{R_1}\frac{1}{1+\frac{1}{A}\left(1+\frac{R_2}{R_1}\right)}$$

This equation shows clearly that input output ration V_2/V_1 is determined by the passive components R_1 and R_2. If the gain A is sufficiently large,

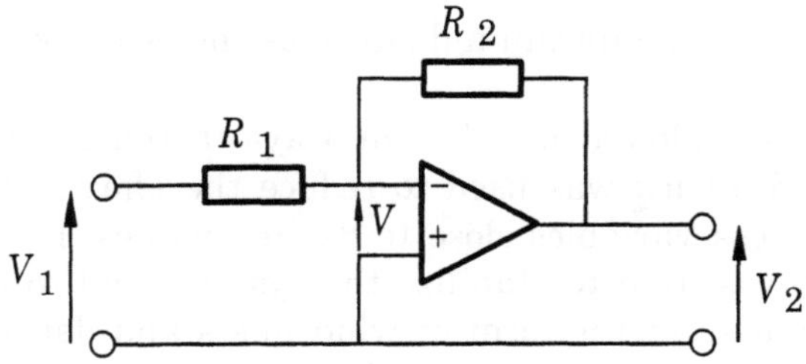

Fig. 3 Circuit diagram of an amplifier with negative feedback.

the ratio is also practically constant even for large variations of A. This is sometimes expressed by saying that gain is the hard currency that can be traded off to reduce the sensitivity to parameter variations. Similar ideas were used in the development of pneumatic controllers. Controller with stable linear behavior were constructed using an amplifier in the form of the flapper valve which is highly nonlinear and variable.

The importance of Blacks invention is illustrated by the following quote by Mervin B. Kelley of Bell Labs at the presentation of the IEEE Lamme Medal in 1957.

> "Although may of Harold's inventions have made great impact, that of the negative feedback amplifier is indeed the most outstanding. It easily ranks coordinate with De Forests invention of the audion as one of the two inventions of broadest scope and significance in electronics and communications of the past 50 years ... it is no exaggeration to say that without Blacks invention, the present long-distance telephone and television networks which cover our entire country and the transoceanic telephone cables would not exist. The application of Blacks principle of negative feedback has not been limited to telecommunications. Many of the industrial and military amplifiers would not be possible except for its use."

An interesting fact about Blacks invention was that nine years was used in the patent office because the officers refused for a long time to believe that the amplifier would work. Blacks work also inspired much theoretical work, for example the stability analysis [54] and [14] which are all became pillars of automatic control.

2.6 Reflections

It is interesting to reflect a little about the differences in the development of the feedback in ship steering, process control and telecommunications. In the mid 1930s feedback was used extensively in all fields. Feedback had made it possible to develop system that could not been realized without feedback, there was a significant industrial use of feedback in all the three fields.

Feedback was exploited in different ways in the different areas. In process control feedback was used to reduce the effects of disturbances and to keep process variables close to desired values. In ships and aircrafts feedback was used to stabilize the systems and to reduce the effects of disturbances in the form of wind gusts and density variations. In telecommunications feedback was thus to overcome variations in the process in the form of nonlinearity and variability of electron tube characteristics.

There were considerable differences in implementation. In a feedback amplifier the controller and the amplifier was an integrated design. It was relatively easy to make a compensator by adding resistors and capacitors. The pneumatic controllers used in process control was much more primitive. The hardware used in the autopilots also imposed severe restrictions.

The complexity of the models differed significantly. The following quote from Bode, [16] is illuminating.

> "The two fields also differ radically in their mathematical flavor. The typical regulator system can frequently be described, in essentials, by differential equations of no more than perhaps the second, third or fourth order. On the other hand, the system is usually highly nonlinear, so that even at this level of complexity the difficulties of analysis may be very great. In contrast, the order of the set of differential equations describing the typical negative feedback amplifier used in telephony is likely to be very much greater. As a matter of idle curiosity, I once counted to find what the order of the set of equations in an amplifier I had just designed would have been, if I had worked with the differential equations directly. It turned out to be 55."

Unquestionably the most advanced theoretical advances had occurred in telecommunications where frequency response methods had been developed. These methods had strong impact on the design of feedback amplifiers. A particular advantage was that it was possible to deal with systems of high order. This was necessary because the feedback amplifiers had many capacitors and inductors. Some theory had been developed by Minorsky for ship steering. The theory, which was based on mathematical models of low order, had very little impact on the actual design of autopilots. A little theory was also available in process control but much of the controller tuning was done empirically. The presence of time delays was a particular feature of process control systems which made analysis difficult. Telecommunication had the strong advantage of having Bell Laboratories with many strong researchers. Similar organizations did not exist for ship steering or process control. It is interesting to speculate what could have happened if there had been strong research groups in these fields. Judging from wide claims of Blacks patent it was clear that the patent layers at Bell Labs had a vision of the wide applicability of feedback.

3 The Field Emerges

A large number of control systems based on feedback were in operation in a wide range of fields around 1940. Although some ideas were the same for example linearization and analysis of the closed loop characteristic equation, the commonality of the systems were not appreciated. One of the most striking illustrations of this was that feedback was reinvented to many times. Another is that the nice properties of integral and derivative action was also rediscovered.

By the late 1930s the attractive properties of feedback were reasonably well understood in many disciplines. For example that feedback could give systems with highly linear responses that were insensitive to disturbances and parameter variations. This is expressed very well in the seminal paper [27].

> "For a simple vacuum tube amplifier the linear response was due to to the constancy of the parameters within the amplifier. Any departure from constancy of these parameters affects the relation between input and output directly. For a servomechanism, however, the only function of the servo-amplifier element is to apply sufficient force to the servo output to bring it rapidly to the correspondence of the servo input. Such an amplifier can be a relatively crude affair."

In Blacks patent on feedback there are also claims that feedback can be applied not only to electronic amplifiers but to a wide range of other systems, see [12] and [13]. Similar ideas are expressed very clearly in the following quote from [58].

> "But there was no common approach to control problems in the different engineering disciplines. For example in 1937 I gave a conference paper on flight control, which was later published in an aeronautical journal. This was the first systematic study of the topic, and at the end of the paper I also compared flight control with the control of other variables such as pressure and temperature, giving the equivalent mechanical system in each case and comparing systems equations and typical dynamic behavior. This idea of a common ground between entirely different types of control systems was quite new but the paper was not known outside aeronautics."

3.1 War Pressures

Although there were glimpses of a pattern there was still not enough structure to call it a discipline. Although there were techniques to analyze stability of a system the existing methods did not indicate how the system should be modified to be stable. Neither were there any systematic methods to design a control system with specified properties.

It is reasonable to claim that the field of automatic control emerged as a discipline during the Second World War when it was realized that science could have a drastic impact on the war effort. The development was strongly stimulated by the cross disciplinary teams that were formed. Similar things happened in many countries, most notably in the US where the efforts were very well documented. The strongest driving force came from the development of radar and fire control. In the US there was a National Defense Research Committee (NDRC) directly under the President Roosevelt. This committee organized and funded several laboratories at MIT among them the Radiation Laboratory which at one time had about 4000 researchers most of them came from outside MIT. Fire Control was one of the themes where much work on automatic control was done. A characteristic feature of the laboratory was a strong interdisciplinary staff with a wide range of backgrounds. A major contribution of the Radiation laboratory was the development of systematic design methods based on the techniques that were developed at Bell labs by Bode and Nyquist.

Major contributions were also given by other Laboratories at MIT notably the Instrumentation Laboratory and the Servomechanisms Laboratory. Both laboratories had very strong leaders Charles Stark Draper and Gordon S. Brown respectively. The Instrumentation Laboratory is now called the Draper Lab and the Instrumentation Laboratory is now called LIDS. There were also major contributions from AT& T's Bell labs.

There were also strong pressures to improve process control for the military needs. The stringent demands for the advanced systems used for uranium enrichment in Los Alamos stimulated development of better controllers for the process industries.

There were also very significant developments in many other countries. In the UK a servo panel was formed in 1942 to exchange ideas on weapons control. There was also substantial activities in the USSR and in Germany.

3.2 Servomechanism Theory

Since so many applications centered around servo problems the resulting theory became known as servomechanism theory although there were major applications in wide range of fields. The first publication on servomechanism is probably the work by Hazen [27]. The Radiation Laboratory was dissolved immediately after the war. The research work at the laboratory was published in an impressive series of 28 volumes. The book [31] dealt with the theory of servomechanisms. The interdisciplinary nature of automatic control is reflected by the fact that the book was written by the physicist Hubert James, the control engineer Nathaniel Nichols and a mathematician Ralph Phillips. This book was followed by others written by authors from the Servomechanism Laboratory [17] and Bell Labs [15], [47]. To get a perspective on the development it is also enlightening to read some of the original publications, for example [26]. This also gives a clear indication that there was a very wide network of researchers working on the problems.

Servomechanism theory used a graphical representation of systems in terms of block diagrams that is independent of the technological details of the systems. An important consequence of the block diagram notation was that the similarity of many different control system was obvious. Block diagrams is an early example of information hiding. The behavior of the individual blocks was described by transfer functions which were introduced by Harris, [26]. The mathematical foundation was based on linearization, complex variables, and Laplace transforms. An important factor that significantly contributed to the success of servomechanism theory was that the transfer function of a system could be determined experimentally by investigating the propagation of sine waves through the system. In this way it was possible to deal with systems where physical modeling was difficult. A systematic method for designing a controller that gave a closed loop system with specified properties was a key element. The particular method developed was based on graphical methods,using Bode and Nichols diagrams.

The books by Smith [64] and Eckman [22] show that there was an analog development in process control with concepts and ideas very similar to the servomechanism theory. Major contributions was also given in other countries. Oldenbourg and Sartorius of Siemens published the book [55] in 1944. It became available in English in 1948, [56]. Oppelt [59], [60] also made major contributions. Comprehensive presentations of British contributions appeared later.

The servomechanisms theory was used extensively in the industries involved in the war time efforts and it spread rapidly. The availability of good books contributed to this. Among the early books we can mention [71], [21], [70] and [25]. In view of the wide application range it is perhaps more adequate to call the approach *frequency response*. This captures the idea that much of the analysis and design focused on the notion of frequency response. The name is also not so strongly tied to a specific application.

The book by Tsien [71] called Engineering Cybernetics was very advanced for its time. It expressed the fact that something new had really happened in the following way:

> "A distinguishing feature of this new science is the total absence of considerations of energy, heat, and efficiency, which are so important in other natural sciences. In fact, the primary concern of cybernetics is on the qualitative aspects of the interrelations among the various components of a system and the synthetic behavior of the complete mechanisms."

Later he goes on to say:

> "An engineering science aims to organize the design principles used in practice into a discipline and thus to exhibit the similarities between different areas of engineering practice and to emphasize the power of fundamental concepts."

Another very interesting feature of the book is that it anticipated much of the future development. The word cybernetics had been used earlier in the very influential book by Wiener [75].

The book [70] summarized many of the results and it also presented a systematic method for controller design.

3.3 Education

Automatic control was recognized as a very powerful technique that could be applied to many problems in diverse fields. It quickly received world wide acceptance in industry and academia. Control groups were created in many companies, and new industrial enterprises specializing in control were established. Courses in the field were introduced at practically all engineering schools and a large number of textbooks were written.

The academic positioning of the discipline is difficult because the traditional organization of the universities. One approach that is commonly used is to have separate control groups within traditional departments of electrical, mechanical, and chemical engineering. This is easy to fit into the existing structure but it may leads to fragmented groups which lack critical mass. Another is to have a central control group with global responsibility. This makes it possible to have groups of reasonable size. It is also possible to organize courses very efficiently. It is mandatory that a central group maintains strong contact with a variety of application areas to make sure that it can acquire the knowledge about the specific systems that are indispensable for successful use of automatic control.

3.4 Applications

The growing knowledge and awareness of automatic control led to a virtual explosion of applications in a wide range of fields, for example.

- Generation and distribution of electricity and other forms of energy
- Process control in a wide range of industries
- Manufacturing and robotics

Robot-Piloted Plane Makes Safe Crossing of Atlantic

No Hand on Controls From Newfoundland to Oxfordshire—Take-Off, Flight and Landing Are Fully Automatic

By ANTHONY LEVIERO
Special to THE NEW YORK TIMES.

WASHINGTON, Sept. 22 — A Douglas C-54 Skymaster with a mechanical brain landed without human aid near London today after a robot directed hop from Newfoundland.

The revolutionary flight across the Atlantic, effected by the push of the button, was hailed by Air Force spokesmen said. Two ships somewhere in the Atlantic furnished bearings to the Skymaster's brain. She had 3,700 gallons of fuel aboard.

Great Britain several weeks ago had asked as a favor that the Air Force send the Skymaster there to make demonstration flights for

Fig. 4 Article in New York Times on September 23, 1947, describing the first fully automatic transatlantic flight.

- Transportation, space vehicles, airplanes, ships, trains, cars
- Communication
- Buildings, structural stabilization, heating, ventilation and air conditioning
- Materials
- Instruments
- Entertainment

There were also a number of spectacular applications. One of them is a fully automatic flight across the Atlantic was made by the Robert E. Lee on September 23, 1947, see Figure 4. The complete flight including take off and landing was fully automatic. It is described in the insightful article [51] which also gives an interesting perspective on the development of flight control.

3.5 Organization

Much of the early work in automatic control was classified because of its military connection. Many ideas emerged independently in many countries. Many researchers felt that there were strong benefits from an open exchange of ideas, which resulted in a strong drive for internationalization. This need became very strong after the war. The International Conference on Automatic Control held in July 1951 in Cranfield, UK [73] was one of the first international meetings. This was probably the first manifestation of Automatic Control as an established field. It was followed by the a symposium in New York in 1953, see [57]. In 1956 there were plans for no less than eight national meetings on automatic control in Europe. Good leadership resulted in the formation of an international body IFAC

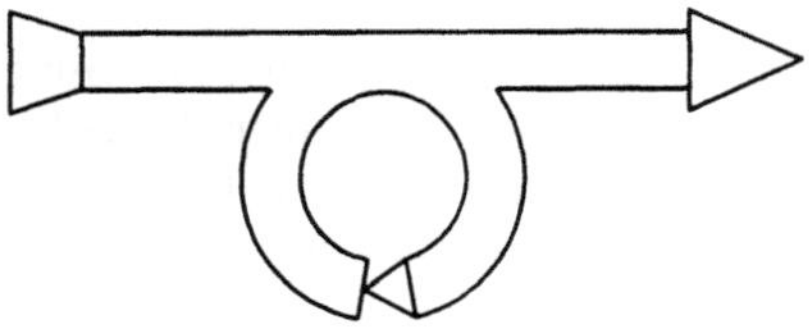

Fig. 5 Slight variations of this symbol has been used as a logotype for the World Congresses of IFAC, see [34].

(International Federation of Automatic Control), which became the international forum for our field, see [57], [20] and [44]. An organizational structure was set up, tri-annual World Congresses, symposia, and workshops were organized, and a journal was started [45]. Many issues were settled in a meeting in Heidelberg in 1956 with participants from 19 countries. A key meeting in IFAC is the tri-annual World Congress [34] with the first World Congress held in Moscow in 1960. This conference had a very large impact because it offered an opportunity for researchers who had been working in isolation to meet with colleagues who had worked on similar problems. IFAC's activities have expanded substantially and today there are IFAC meeting almost every week. IFAC also gave the field of automatic control a new symbol, see Figure 5. Although IFAC is the global international organization for automatic control there are also other important actors. The Automatic Control Council in USA, which is the national member organization of IFAC in USA, organizes the yearly American Control Conference in collaboration with many engineering societies. The European Control Conference, which meets every other year, initiated with a meeting in Grenoble in 1991. An Asian Control Conference had its first meeting in 1994. Theoretical aspects of the field are covered in the

Automatic control also has an important place in the meetings of many engineering societies, AIAA, AIChE, AISE, ASME, IEEE, ISA, and SCS often in cooperation with IFAC. The organization MTNS focuses on theoretical issues in system theory and organize biannual conferences.

The establishment of journals is an important aspect of a scientific field. A number of high quality journals appeared simultaneously with the emergence of organizations.

3.6 The Second Wave

Automatic control in the form of servomechanism theory or frequency response was very well established in the beginning of the 1960's in the form of servomechanism theory with a wide range of applications. One factor that strongly contributed to the excitement of the field was that shortly after its establishment there was a second wave. There was a substantial injection of ideas from several different sources. The space race which began with Sputnik in 1957 was a strong driving force. Computers started to be used for implementation of control systems, and there was an influx of ideas from mathematics.

The development of space travel posed many challenges. There were many optimization problems, for example: How to make the most effective use of moderate sized rocket to put the largest payload in orbit? How to find efficient trajectories for interplanetary travel? How to minimize heating at reentry to the earths atmosphere? Attempts to solve problems of this type led to the development of optimal control theory. Major contributions were made by mathematicians and control engineers. The development led to a strong revitalization of the classical calculus of variations. Pontryagin and his coworkers in Moscow followed the tradition of Euler and Lagrange when they developed the maximum principle, see [61]. Bellman instead followed the ideas of Hamilton and Jacobi when he developed dynamic programming, see [7], [8]. The special case of linear system with quadratic criteria was solved by Bellman in special cases and in [35]. The books [6] [18] presented the results in a form that was easily accessible to engineers. These books also dealt with computational issues.

The idea to control industrial processes using digital computers emerged in the mid 1950s. Louisiana Power and Light Company installed a Daystron computer for plant monitoring in September of 1958 and Texaco installed a RW-300 at their Port Arthur refinery. This system did closed loop control for the first time on March 15, 1959. The use of digital computers then expanded very rapidly and computer control is today the standard technique for controlling industrial processes. The theory of sampled system had already started to be developed in connection with air defense systems. Early work was don at Columbia University in the US, in the UK and at the Institute of Control Problems in Moscow. The early contributions are well described by the key contributors in their textbooks [62], [33] and [72].

Disturbances is a key ingredient in a control problem, without disturbances and process uncertainties there is no need for feedback. Modeling of disturbances is therefore important. In [31] and [66] it was proposed to use stochastic processes as models for disturbances. A key problem in fire control was to predict the future motion of an aircraft. Very elegant solutions to this problem were given independently by Wiener and Kolmogorov, see [76] and [41]. A reformulation by Kalman and Bucy let to the Kalman filter as a solution which is very attractive computationally, see [38] and [36].

Stochastic control theory emerged from a combination of optimal control with the theory of stochastic processes. In then case of linear systems with quadratic criteria it led to a very elegant separation of the complete problem into a combination of a Kalman filter and a state feedback from the estimated states, see [1]. This separation does not hold for general systems. The solution then becomes very complicated. It leads to dual controllers where the controller has the dual task of driving the system towards its goal but also to inject signals so that we obtain better information about the state of the system, see [23].

The theoretical approaches to control were originally based on differential equations. The servomechanism theory was a radical change because the systems were now characterized by their frequency responses. The influx of ideas during the 1960s led to a return to differential equation models which at the time was called state space theory, see [77]. Many fundamental questions were also raised such as controllability and observability, [40]. Algebra was also used heavily to develop the theory, see [37] and [39].

Frequency response was a very powerful method to determine a transfer function experimentally. With the emergence of state-space theory came need for experimental methods to determine appropriate systems models. This coincided with the availability of digital computers. The result was a dynamic development of the field of system identification, see [2].

Initially the emergence of all new ideas led to quite a bit of controversy between "classical" and "modern" control but the viewpoints have now been unified and the advantages of having several different views of a problem are widely appreciated. The development led to a very dynamic and rapid development of the field.

4 Current Status

One way to explore a field is to investigate the contents of the leading conferences in the field over a number of years. A perspective on the current status and the development of automatic control can be obtained by analyzing the contents of the IFAC World Congresses, the American, European and Asian Control Conferences and the IEEE Decision and Control Conference. Such a reading shows how subspecialities are emerging blossoming and decaying. It also gives an interesting view of how the field is expanding, see [42].

4.1 System Theory

In practically all conferences there are sessions on linear systems (time invariant as well as time-varying), nonlinear systems, stochastic systems, discrete time systems, time delay systems, distributed parameter systems, and decentralized systems. These sessions naturally have a strong flavor of applied mathematics. The issues discussed are representations, metrics, stability, observability, controllability, robustness, sensitivity, system structure etc. The tools used are typically ordinary and partial differential equations, but many other mathematical tools such as operator theory and algebra have had a strong impact. The theory of nonlinear systems has for example been strongly influenced by differential geometry. Lately there has been a significant interest in systems described by differential algebraic equations (behavioral systems). This is also a natural approach when modeling physical systems. The field of linear systems has been declared uninteresting at many instances but interest has often been renewed due to new viewpoints and introduction of new theories. Lately there has been a significant interest in hybrid systems and in discrete event dynamical systems. There are many challenging problems because of the combined continuous and discrete nature of such systems.

4.2 Modeling and Identification

Models are a very important element of automatic control. It is a significant portion of the development of all control systems. It is also an area where specific process knowledge plays a major role. There are two approaches to modeling, one is based on physics (white-box modeling) the other on plant experiments (black-box modeling). It is also possible to combine the approaches (grey-box modeling). The early efforts on experimental modeling were focused on frequency response which was an important ingredient in the servomechanism theory. The emergence of state space theory gave a strong stimulation to develop other experimental modeling techniques geared to that theory. This gave birth to the field of system identification which had strong connections to many branches of applied mathematics such as statistics and numerical mathematics. Lately there has been a significant interest in identification for control. The area of model reduction is a closely related field, which pops up regularly.

4.3 Design

It is difficult to capture design problems formally because there are so many factors to consider. A common approach has been to formalize one or a few aspects of the problem so that it can be dealt with mathematically. A lot of creativity has been demonstrated in this area. There is a reasonably good understanding of the role of feedback and there are good tools to deal with servo and regulation problems. Very few techniques are available for task oriented control, most contributions in this important area comes from robotics. Early work on control system design had a strong engineering flavor with emphasis on graphical methods like loop shaping and root locus. Analytical design methods such as pole placement were introduced later.

Optimization is a natural tool that has stimulated much research on optimal control. It comes in many flavors, LQG, LTR, dynamic programming, and the maximum principle. Because of the nature of the problem and the caliber of the people involved much interesting mathematics has also appeared. A focus on disturbances led to the development of stochastic control theory. Robustness and sensitivity were essential elements of frequency response. These issues were largely forgotten in the euphoria of the development of state space theory. In the last decade there has been a significant effort to consider model uncertainty. This resulted in design methods such as QFT, H_∞, L_1-methods, μ, and LMI.

Finding fundamental limitations for achievable performance is a basic aspect of control system design. The early work by Bode gave considerable insight into this problem, but it is an important area that deserves much more research. It is particularly important for integrated control and process design.

Control system design is a difficult area. If all engineering aspects are considered the problem becomes very messy. If the problem is simplified so that it can be solved elegantly many aspects have to be neglected. This has caused much controversy. It will probably take a long time before we have good design tools that will deal with the real issues of control system design.

An important requirement is that the methods also should be developed and packaged so that they can be used by an engineer with a training at the master's level. A consequence of this is that special design methods have appeared in application areas with very little initial contact with the main stream control theory. Typical examples are dynamic matrix control (DMC), in chemical process control, and the computed torque method in robotics. A dynamic matrix is nothing but a truncated impulse response for a multi-variable system. Since the chemical engineers were ignorant of this, they solved the problem and invented a completely new vocabulary (DMC, QDMC, IDCOM). There is a good lesson to be learned from this. We must package our results so that they are easily accessible. The following quotation from a speech by the chairman of IBM, Louis V Gerstner Jr., gives an interesting perspective:

> "The information revolution will happen, but only when the industry stops worshiping technology for its own sake and starts focusing on *real value* for the customers."

4.4 Learning and Adaptation

To have systems that automatically learn about a process and redesign the controller when necessary has been a long standing vision for control engineers. There have been sessions on this from the 1960 Moscow Conference and onwards. This has led to the model reference adaptive controller and the self-tuning regulator. It is, however, only lately in the 1980s that the simpler problem of providing control systems with devices for automatic tuning of the controller has been approached. This has had a significant impact particularly on the field of process control. Much work, however, remains before the longstanding goal is fulfilled. There are connections to the problem of machine learning—an area that is currently developing in a very interesting way.

4.5 Computing and Simulation

The development of computational tools has been very important for the development of automatic control. Computers are used to analyze systems, they are also an integral part of practically all controllers. Much of the early development was hampered by poor equipment. This is very noticeable in the development of autopilots and industrial PID controllers. The following quote by Vannevar Bush from 1927 gives an interesting perspective.

> "Engineering can proceed no faster than the mathematical analysis on which it is based. Formal mathematics is frequently inadequate for numerous problems pressing for solution, and in the absence of radically new mathematics, a mechanical solution offers the most promising and powerful attack ... ."

The mechanical differential analyzer developed by Bush 1928-1931 was a very influential contribution and there was literally a revolution when the microprocessor and the personal computer arrived. Automatic control has been strongly linked to computing throughout its history. Computing is important for analysis, simulation, and implementation. Digital computing replaced analog computing that was used in the beginning.

The development of the computer is probably the factor that has changed the most in the short history of the field. An illustration of this is that at the First IFAC World Congress in Moscow computing appeared under the sub-heading "Pneumatic Components and Computing Devices" in the components category! The emergence of computers have also put much more emphasis on numerics, which is important both for analysis and implementation. Lately there have been both special sessions and special symposia on Computer Aided Control Engineering. The emergence of software for matrix calculations such as Matlab and Matrix$_X$ and computer algebra software like Maple and Mathematica has also had a significant influence. In recent conferences we can find several special sessions on this. Simulation is a very important area that now has its own conferences and publications. It is also interesting to observe that there are numerical analysts that have taken a real interest in problems that arise from analysis and design of control systems.

4.6 Implementation

Implementation is an important issue. Because of the rapid development of micro electronics it is also one of the factors that has changed the most over the past 50 years. At the Moscow Congress there were three major groups: theory, components, and applications. Implementation issues came under components. Some aspects on implementation, such as sampled data systems, are very well represented but many equally important aspects are not dealt with at all. There has, only been a few sessions on integrator windup. Selector control is rarely dealt with although it is widely used in practice. Many important aspects on implementation are not covered in textbooks. A good implementation require knowledge of control systems as well as certain aspects of computer science.

It is necessary that we have engineers from both fields with enough skills to bridge the gap between the disciplines. Typical issues that have to be understood are windup, real-time kernels, computational and communication delays, numerics, and man-machine interfaces. Implementation of control systems is far too important to be delegated to a code generator. Lack of understanding of implementation issues is in my opinion one of the factors that has contributed most to the notorious GAP between theory and practice.

4.7 Commissioning and Operation

There are a wide range of interesting problems that appear after a control system is designed and implemented. First, the system has to be commissioned. Then it is necessary to continuously supervise and assert that the system is running properly. Many of the problems occurring during this phase have only recently been addressed. Typical issues are fault detection and diagnosis, but there are also many other interesting problems, such as loop assessment and performance assessment. Development in this area are strongly motivated by the drive for higher quality. It is interesting to observe that there is a growing interest in this area. It is also interested that commissioning can be influenced substantially by a proper control design. Automatic tuners can drastically simplify the commissioning procedure.

4.8 Applications

The broad application areas have not changed much over the years. At the Moscow Conference there were sessions on metalworking, electrical power systems, electrical drives, transport, industrial processes, chemical and oil, thermal and nuclear power, and metallurgical processes. In later years we can also find applications in aerospace, automotive, micro electronics, and consumer electronics (CD, Video etc). A long experience with journals and conferences has shown that it is very difficult to get good applications papers. The engineers who really know about the applications do not have the time or the permission to publish. Many of those who do write have only a superficial knowledge about the applications. This sends distorted signals in all directions. There are occasional efforts with special issues of journals, where really good applications papers sometimes appears. We need those badly for better education of the next generation of control engineers.

5 The Future

Automatic control has emerged as an interesting and successful discipline in a period of about 50 years time. The development of the field has been very dynamic. Automatic control it the first systems discipline that transcends the traditional boundaries of mechanical, electrical and chemical engineering. Automatic control has a firm base in applied mathematics and an unusually wide range of applications. Automatic control systems are increasingly becoming mission critical in the sense that the system fails if the controller fails. Examples are found in cars, airplanes, industrial systems and CD-players. Successful development have contained ingredients such as demanding problems, strong mathematical theory and good engineering. A natural question to ask is if the development will continue or if it has reached is apex. Having a positive attitude I strongly believe in a continuation, but what happens will depend on ourselves, see [67]. Below I will discuss some factors that may influence the development.

5.1 Our Knowledge Base

A large amount of knowledge in automatic control has been generated since the field emerged in the mid 1940s. There is a vast amount of intellectually stimulating material that is useful for a wide range of engineering problems. The knowledge is available in a very large number of books, articles and courses. The books typically deal with a narrow specialization in great detail. Such books are not easily accessible to the engineers who are designing control systems. Courses have developed randomly and there is much diversity. The material is not very well structured and the textbooks have not kept up with the development. Introductory courses in control are often very similar to courses given twenty or thirty years ago even if the field itself has developed substantially. The only difference may be a sprinkling of Matlab exercises. We need to take a careful look at our knowledge base and explore how it can be weeded and streamlined. We should probably also pay more attention to the academic positioning of our field.

Experiments with automatic control is a very good way to enhance the students engineering ability. It provides a complex system that includes sensors, actuators, electronics, computers, algorithms. Automatic control is thus a good base for training the engineers of the future.

5.2 Intellectual Drivers

The health of any discipline depends strongly on its ability to attract bright students. Stimulating and challenging problems is a strong attractor as are spectacular demonstrations and good applications. Automatic control is well endowed from these points of view but it is essential that it is brought out in the courses. The introductory courses are particularly important. If they are not sufficiently interesting to capture the bright students we do not have a great future.

5.3 Relations to Other Fields

Automatic control has a strong base in mathematics. The interaction goes in both directions, a wide range of mathematics has found its use in control, and control has occasionally stimulated the development of mathematics. We are also fortunate to have good mathematicians in our community. Typical examples are optimal control, which revitalized the field of calculus of variations, stochastic control, and nonlinear control. A more detailed discussion on this is found in [24].

There are many links to computer science. Computing is used extensively for analysis and design, computers are key elements in implementation of control systems. There are also many areas where there is cross fertilization in methodology, see [11].

Ideas of automatic control are also used in other fields. There are many uses of feedback and feedforward in economics. Simple program trading schemes can be viewed as applying decentralized control to a complex coupled systems. Therefore it is not a big surprise that instabilities can be created. Regulatory agencies are responding by switching off program trading but do they understand how the system behaves?

There are many interactions between automatic control and specialized disciplines that have knowledge about the processes we control. It is, of course, very important to have a good understanding of the systems to be controlled. This is particularly important for integrated process and control design.

There are many feedback mechanisms in biology and medicine. Feedback is a central feature of life. All organisms have the ability to sense how they are doing and to make necessary modifications. The process of feedback governs how we grow, respond to stress and challenge, and regulate factors such as bode temperature, blood pressure and cholesterol level. Feedback allows us to stand upright. Some of these functions are simple functions such as stabilization and regulation but others are more complicated task oriented functions.

5.4 Balance of Theory and Applications

The balance between theory and practice is another key issue. Throughout the short history of our field there has been a continuous and sometimes heated debate about the GAP between theory and practice. I personally believe that it is important to strike a balance. It is important to look inwards to improve our basic understanding, our theory and our design methods. But this is not enough; we must also look at the applications.

I feel very strongly that automatic control is a systems discipline, which means that we should take full systems responsibility. This means that we should educate students who are capable of solving the whole problem from conceptual design to implementation and commissioning. Control will fade away if we consider the job done when a mathematical description of the control law is obtained. This means that we need to consider implementation, commissioning, and operation.

5.5 Complex Systems

There is a general tendency for engineering systems to become more and more complex. There are many mechanisms that introduces complexity; size and interaction are two factors. It is not uncommon to have chemical processes with many thousand feedback loops. Astronomical telescopes with adaptive optics may have a very large number of reflecting surfaces whose orientation are controlled individually by feedback systems. In the process industries it is common to use recirculation to save energy and to reduce pollution. Recirculation introduces a coupling from the output stream to the input stream, which can generate very complex behavior. Another factor that introduces complexity is that continuous systems are mixed with logic and sequencing. Such systems are very difficult to analyze and design. The emerging field of hybrid systems is one attempt to address these challenges. This is an interesting meeting place of control and computer science, see [11]. Systems with a high degree of autonomy is another.

5.6 Integrated Process and Control Design

A cardinal sin of automatic control is to believe that the system to be controlled is given a priori. Very good arguments for an integrated design of a process and its control system were put forward very eloquently already in the 1943 paper by Ziegler et al. [79]. In that paper it is emphasized that much too often the control engineer is faced with a process that is very difficult to control properly. It is suggested that a study of the wide sense controllability of a process should always be done at the design stage because it is very costly to make changes afterwards

Flight control is a good example of integrated process and control design. The Wright brothers succeeded where others failed, because they made an unstable airplane that was maneuverable. Later one learned to build stable and maneuverable aircraft. There are still substantial advantages in having unstable aircrafts that rely on a control system for stabilization. Modern fighters obtain their performance in this way, see Figure 6.

When the automatic control system becomes a critical part of the process it may also become mission critical which means that the system will

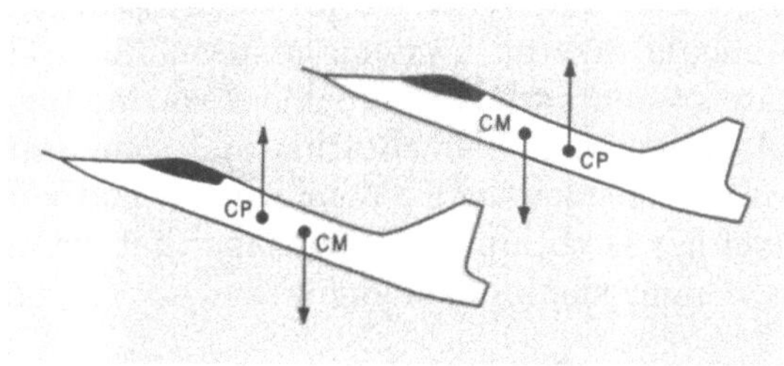

Fig. 6 The center of pressure of an aircraft shifts backwards when a plane goes supersonic. To be stable the center of pressure must be behind of the center of mass. If the plane is stable at sub-sonic speeds it becomes even more stable at supersonic speeds. Very large forces and large control surfaces are then required to maneuver the airplane. A more balanced design is obtained by placing the center of pressure in front of the center of mass at sub-sonic speeds. Such a plane will however by unstable at sub-sonic speeds, i.e. at take off and landing. This imposes severe constraints on the safety and robustness of the control system.

fail if the control system fails. This induces strong demands on the reliability of the control system. Interesting discussion of the consequences of this are found in [67].

Other examples are found in development of instrumentation where precision and bandwidth could be improved by several orders of magnitude by a proper combination of feedback control and instrument design. In robotics, substantial gains in performance, precision, and price are obtained by a light flexible mechanical structure that obtains its stiffness by an automatic control system. Similar examples are found in mechatronics and power systems.

6 Conclusions

Automatic control, which emerged as a discipline in the mid 1940s has had a very dynamic development. It is one of the first systems disciplines that transcends the boundaries of the traditional engineering fields. A wide range of control systems appear in practically all technological systems. Automatic control has also had a strong impact on all fields where feedback plays a role. To continue to be a strong field it is necessary that we pay attention to our knowledge base and that we make a strong effort to continue to attract bright students. We should perhaps also pay some attention to our trade marks. The importance of trademarks is obvious to anyone who sees how carefully the major companies watches them. In automatic control we use too many names for our discipline: control engineering, feedback control, systems science, and cybernetics, to name a few. I think it is important to pick one word for our discipline and stick to it. We should probably also consider the academic positioning of the field more carefully.

References

1. K. J. Åström. *Introduction to Stochastic Control Theory*. Academic Press, New York, 1970. Translated into Russian, Japanese and Chinese.
2. K. J. Åström and T. Bohlin. Numerical identification of linear dynamic systems from normal operating records. In *Proc. IFAC Conference on Self-Adaptive Control Systems*, Teddington, UK, 1965.
3. K. J. Åström and T. Hägglund. *PID Controllers: Theory, Design, and Tuning*. Instrument Society of America, Research Triangle Park, NC, second edition, 1995.
4. K. J. Åström and B. Wittenmark. *Computer Controlled Systems—Theory and Design*. Prentice-Hall, Englewood Cliffs, New Jersey, second edition, 1990.
5. K. J. Åström and B. Wittenmark. *Adaptive Control*. Addison-Wesley, Reading, Massachusetts, second edition, 1995.
6. M. Athans and P. L. Falb. *Optimal Control*. McGraw-Hill, New York, 1966.
7. R. Bellman. *Dynamic Programming*. Princeton University Press, New Jersey, 1957.
8. R. Bellman, I. Glicksberg, and O. A. Gross. Some aspects of the mathematical theory of control processes. Technical Report R-313, The RAND Corporation, Santa Monica, Calif., 1958.
9. S. Bennet. *A History of Control Engineering 1800-1930*. Peter Peregrinus, London, 1979.
10. S. Bennet. *A History of Control Engineering 1930-1955*. Peter Peregrinus, London, 1993.
11. A. Benveniste and K. J. Åström. Meeting the challenge of computer science in the industrial applications of control: An introductory discussion to the Special Issue. *Automatica*, 29:1169–1175, 1993.
12. H. S. Black. Stabilized feedback amplifiers. *Bell System Technical Journal*, 13:1–18, 1934.
13. H. S. Black. Inventing the negative feedback amplifier. *IEEE spectrum*, pages 55–60, December 1977.
14. H. W. Bode. Relations between attenuation and phase in feedback amplifier design. *Bell System Technical Journal*, 19:421–454, July 1940.
15. H. W. Bode. *Network Analysis and Feedback Amplifier Design*. Van Nostrand, New York, 1945.
16. H. W. Bode. Feedback – The history of an idea. In *Proceedings of the Symposium on Active Networks and Feedback Systems*. Polytechnic Institute of Brooklyn, 1960.
17. G. S. Brown and D. P. Campbell. *Principles of Servomechanisms*. Wiley & Sons, New York, 1948.

18. A. E. Bryson and Y. C. Ho. *Applied Optimal Control Optimization, Estimation and Control*. Blaisdell Publishing Company, 1969.
19. V. Bush. The differential analyzer. *JFI*, 212(4):447–488, 1931.
20. H. Chestnut, editor. *Impact of Automatic Control—Present and Future*. VDI/VDE-Gesellschaft, Duesseldorf, 1982.
21. Harold Chestnut and Robert W. Mayer. *Servomechanisms and Regulating System Design*. Wiley, New York, 1959.
22. D. P. Eckman. *Principles of industrial process control*. Wiley, New York, 1945.
23. A. A. Feldbaum. *Optimal Control Theory*. Academic Press, New York, 1965.
24. W. H. Fleming, editor. *Future Directions in Control Theory—A Mathematical Perspective*. Society for Industrial and Applied Mathematics, Philadelphia, 1988.
25. J. C. Gille, M. J. Pelegrin, and P. Decaulne. *Feedback Control Systems*. McGraw-Hill, New York, 1959.
26. H. J. Harris. The analysis and design of servomechanisms. *Printed under Auspices of the Fire Control Committee (Sec. D-2) of NDRC, OSRD*, 1942.
27. H. L. Hazen. Theory of servomechanisms. *JFI*, 218:283–331, 1934.
28. T. P. Hughes. *Elmer Sperry Inventor and Engineer*. The John Hopkins University Press, London, 1993.
29. A. Hurwitz. On the conditions under which an equation has only roots with negative real parts. *Mathematische Annalen*, 46:273–284, 1895.
30. A. Ivanoff. Theoretical foundations of the automatic regulation of temperature. *J. Inst. Fuel*, 7:117–130, 1934. discussion pp. 130–138.
31. H. M. James, N. B. Nichols, and R. S. Phillips. *Theory of Servomechanisms*. Mc-Graw-Hill, New York, 1947.
32. Rolf Johansson. *System Modeling and Identification*. Prentice Hall, Englewood Cliffs, New Jersey, 1993.
33. E. I. Jury. *Sampled-Data Control Systems*. John Wiley, New York, 1958.
34. S. Kahne. A history of the ifac congress. *IEEE CSM*, 16(2):10–12, 78–83, 1996.
35. R. E. Kalman. Contributions to the theory of optimal control. *Boletin de la Sociedad Matématica Mexicana*, 5:102–119, 1960.
36. R. E. Kalman. New methods and results in linear prediction and filtering theory. Technical Report 61-1, RIAS, February 1961. 135 pp.
37. R. E. Kalman. On the general theory of control systems. In *Proceedings first IFAC Congress on Automatic Control, Moscow, 1960*, volume 1, pages 481–492, London, 1961. Butterworths.
38. R. E. Kalman and R. S. Bucy. New results in linear filtering and prediction theory. *Trans ASME (J. Basic Engineering)*, 83 D:95–108, 1961.
39. R. E. Kalman, P. L. Falb, and M. A. Arbib. *Topics in Mathematical System Theory*. McGraw-Hill, New York, 1969.
40. R. E. Kalman, Y. Ho, and K. S. Narendra. *Controllability of Linear Dynamical Systems*, volume 1 of *Contributions to Differential Equations*. John Wiley & Sons, Inc., New York, 1963.
41. A. N. Kalmogorov. *Interpolation and Extrapolation of Stationary Random Sequences*. Math. 5. Bull. Moscow Univ., USA, 1941.

42. A. H. Levis. Challenges to control—a collective view report of the workshop held at university of santa clara on september 18-19, 1986. *IEEE Trans. Automat. Contr.*, AC-32:274–285, 1987.

43. L. Ljung. *System Identification—Theory for the User*. Prentice Hall, Englewood Cliffs, New Jersey, 1987.

44. U. Luoto. *20 Years Old; 20 Years Young. An Anniversary Publication 1957–1977*. Pergamon Press, 1978. IFAC International Federation of Automatic Control.

45. U. Luoto. 20 years old, 20 years young. an anniversary publication 1957-1977. *Pergamon Press*, 1978.

46. A. M. Lyapunov. *The General Problem of the Stability of Motion (in Russian)*. Kharkov Mathematical Society (250 pp.), 1892. Collected Works II, 7. Republished by the University of Toulouse 1908 and Princeton University Press 1949 (in French, republished in English by Int. J. Control 1992.

47. L. A. MacColl. *Fundamental Theory of Servomechanisms*. D. Van Nostrand Company, Inc., New York, N. Y., 1945.

48. J. C. Maxwell. On governors. *Proceedings of the Royal Society of London*, 16:270–283, 1868. Also published in "Mathematical Trends in Control Theory" edited by R. Bellman and R. Kalaba, Dover Publications, New York 1964, pp. 3–17.

49. O. Mayr. *Zur Frühgeschichte der Technischen Regelungen*. R. Oldenbourg, München, 1969.

50. M. W. ed. McFarland. *The Papers of Wilbur and Orville Wright*. Mc Graw Hill, New York, 1953.

51. D. McRuer and D. Graham. Eighty years of flight control: Triumphs and pitfalls of the systems approach. *J. Guidance and Control*, 4(4):353–362, July-August 1981.

52. N. Minorsky. Directional stability of automatically steered bodies. *J. Amer. Soc. of Naval Engineers*, 34(2):280–309, 1922.

53. George C. Newton, Jr, Leonard A. Gould, and James F. Kaiser. *Analytical Design of Linear Feedback Controls*. John Wiley & Sons, 1957.

54. Harry Nyquist. Regeneration theory. *Bell System Technical Journal*, 11:126–147, 1932. Also published in "Mathematical Trends in Control Theory" edited by R. Bellman and R. Kalaba, Dover Publications, New York 1964, pp. 83–105.

55. R. C. Oldenbourg and H. Sartorius. *Dynamik der Selbstättigen Regelungen*. Oldebbourg-Verlag, München, 1944.

56. R. C. Oldenbourg and H. Sartorius. *Dynamics of Automatic Controls*. ASME, New York, N. Y., 1948.

57. R. Oldenburger. *Frequency Response*. The MacMillan Company, New York, 1955.

58. W. Oppelt. Die Flugzeugkurssteuerung im Geradeausflug. *Luftfahrtforschung*, 14:270–282, 1937.

59. W. Oppelt. *Kleines Handbuch technischer Regelvorgænge*. Verlag Chemie, 1947.

60. W. Oppelt. *Kleines Handbuch technischer Regelvorgänge*. Verlag Chemie, Weinheim, 1964.
61. L. S. Pontryagin, V. G. Boltyanskii, R. V. Gamkrelidze, and E. F. Mischenko. *The Mathematical Theory of Optimal Processes*. John Wiley, New York, 1962.
62. J. R. Ragazzini and G. F. Franklin. *Sampled-Data Control Systems*. McGraw-Hill, New York, 1958.
63. E. J. Routh. *A Treatise on the Stability of a Given State of Motion*. Macmillan, London, 1877. Reprinted in Fuller (1975).
64. E. S. Smith. *Automatic Control Engineering*. McGraw Hill, New York, 1944.
65. Torsten Söderström and Petre Stoica. *System Identification*. Prentice-Hall, London, UK, 1989.
66. V. V. Solodovnikov. The frequency-response method in the theory of regulation. *Automatica i Telemekhanika*, 8:65–88, 1947.
67. G. Stein. Respect the unstable. In *The 1989 Bode Lecture*, IEEE, Piscataway, NJ, 1989. IEEE Control Systems Society.
68. J.. Strothman. More than a century of measuring. *InTech*, pages 52–78, June 1995.
69. M. Tolle. *Die Regelung der Kraftmaschinen*. Springer-Verlag, Berlin, 1905. 1909, 2nd edition; 1922, 3rd edition.
70. J. Truxal. *Automatic Feedback Control System Synthesis*. McGraw-Hill, New York, 1955.
71. H. F. Tsien. *Engineering Cybernetics*. McGraw-Hill Book Company, Inc, New York, 1954.
72. Ya. Z. Tsypkin. *Theorie der Relais Systeme der Automatischen Regelung*. R. Oldenburg, Munich, Germany, 1958.
73. A. Tustin, editor. *Automatic and Manual Control: Proceedings of the 1951 Cranfield Conference*, New York, N. Y., 1952. Academic Press.
74. J. Vyshnegradskii. Sur la théorie générale des régulateurs. *Compt. Rend. Acid. Sci. Paris*, 83:318–321, 1876.
75. N.. Wiener. *Cybernetics or control and communication in the animal and the machine*. MIT Press, Cambridge, MA, 1948.
76. N. Wiener. *The Extrapolation, Interpolation, and Smoothing of Stationary Time Series with Engineering Applications*. Wiley, New York, 1949. Originally issued as a classified MIT Rad. Lab. Report in February, 1942.
77. L. A. Zadeh and C. A. Desoer. *Series in Systems Science*. McGraw-Hill, Inc., USA, 1963.
78. J. G. Ziegler and N. B. Nichols. Optimum settings for automatic controllers. *Trans. ASME*, 64:759–768, 1942.
79. J. G. Ziegler, N. B. Nichols, and N. Y. Rochester. Process lags in automatic-control circuits. *Trans. ASME*, 65:433–444, July 1943.

2

Asymptotic Dynamics in Adaptive Gain Control

I. Mareels, S. Van Gils, J. W. Polderman and A. Ilchmann

Department of Electrical and Electronic Engineering, The University of Melbourne, Vic 3052, Australia, i.mareels@unimelb.edu.au
Faculty of Mathematical Sciences, The University of Twente, Enschede, The Netherlands, s.a.vangils@math.utwente.nl
Faculty of Mathematical Sciences, The University of Twente, Enschede, The Netherlands, j.w.polderman@math.utwente.nl
School of Mathematical Sciences, University of Exeter, Exeter EX4 4QE, UK, ilchmann@maths.exeter.uk

Summary. It is well known that linear SISO systems that can be rendered passive through constant output feedback can be adaptively stabilised through a single gain adaptation law. We revisit the dynamical behaviour of such systems and exhibit through a bifurcation analysis a rich variety of potential asymptotic dynamics, for which we provide a control theoretic interpretation. This in turn leads us to question the actual adaptive control question and solution approach.

1 Introduction

Adaptive output gain control for the purposes of regulation and without reliance on identification or probing signals constitutes about the simplest adaptive control objective one can consider. Research into this idea was initiated by Mareels [3], Morse [12], and further developed by Willems and Byrnes [10]. For an overview and extensive reference list see Ilchmann [4]. The work of

Willems and Byrnes lead to the whole topic of universal controllers see e.g. [7] and [2, Chapter 6]. The topic of adaptive stabilisation via output feedback is also studied in great detail by [11]. In this paper we revisit this adaptive scheme from a dynamical systems perspective. Our aim is to understand the adaptive control question better and to be in a position to recast adaptive control questions in a more appropriate context.

In order to apply adaptive output gain control one starts from the assumption that the plant to be controlled is a finite dimensional SISO linear system that can be stabilised by static high gain output feedback. Adaptation is required because the amount of gain in order to achieve regulation is unknown. A large class of linear systems that can be stabilised by output gain feedback is characterised by: the plant's transfer function has stable zeroes and relative degree one. Such systems can be conveniently represented as follows [2, Section 6.4] (see also the Appendix I):

$$\begin{aligned} \dot{x} &= Ax + by, \\ \dot{y} &= -cx - dy + gu. \end{aligned} \tag{1}$$

Here u is the input, y the output and $(x, y) \in \mathbb{R}^n \times \mathbb{R}$ is the state of the linear system (1). $A \in \mathbb{R}^{n\times n}$ is a matrix whose eigenvalues have negative real part. The eigenvalues of the matrix A are precisely the zeroes of the transfer function of the system (1). We assume that $g > 0$, g is the high frequency gain from input to output. We assume that the original system (1) is controllable. As a consequence, see Appendix I, the pair (A, b) is controllable. It is clear that for all sufficiently large $k > 0$, the output feedback control law $u = -ky$ stabilises the system (1), see e.g. [3, 5].

In this paper we are interested in characterising the qualitative behaviour (the phase portrait) when the adaptive feedback law

$$\begin{aligned} u &= -ky + e, \\ \dot{k} &= -\sigma k + y^2, \quad k(0) > 0. \end{aligned} \tag{2}$$

is applied to (1). Here $\sigma \geq 0$ is a small positive constant representing the so called sigma modification [8] and $e \in \mathbb{R}$ represents a control offset error. The control offset error is introduced to unearth some of the robustness issues of concern in adaptive control. One could object that a constant offset error is not a realistic disturbance as it is easily overcome via integral control action. As is shown in Appendix V, this is false, and the discussion we present remains relevant even if integral action is explicitly taken into account.

For $\sigma = 0$ and $e = 0$ the intuition behind the adaptation law is simple. As long as $y \neq 0$ the feedback gain k increases, eventually rendering the (x, y) subsystem asymptotically stable. The output will decrease to zero in an exponential manner which limits the feedback gain. Unfortunately, any nonzero measurement error or disturbance such as e will force the feedback gain to grow indefinitely.

This observation of lack of robustness lead to the so called sigma modification. This idea was first suggested by Ioannou and Kokotovic [8] in the context of adaptive model reference control. It was adopted for systems of the form (1) by Bar-Kana and co-workers, see [11] for a detailed overview of this work.

Another possible modification to overcome this lack of robustness problem whilst preserving the simplicity of the controller is the so called λ-tracking approach (first suggested in [3] and introduced by Ilchmann and Ryan [6], see also Ilchmann [5]). In this line of work a dead-zone in the gain adaptation is invoked. The dead-zone reduces the performance of regulation in the disturbance free case in that the output converges to a prespecified, λ-neighbourhood of the reference signal. This is a small price to pay for the added robustness with respect to small disturbances. The λ-tracking modification will not be considered in the present paper.

Compared to the above simple intuitive picture, the sigma modification alters the phase portrait in a non trivial way. This is the topic studied in this paper. The prime purpose of the sigma modification is to obtain robustness with respect to measurement errors. This is indeed achieved, but at a price, as the asymptotic dynamics are no longer as desired. We investigate this phenomenon in some detail, suggesting some alternatives.

Despite the rather large literature on the topic it is our contention that our understanding of the dynamics at large of these simple adaptive control systems is not complete. In this paper we collect a number of results describing what can be expected both for the transients as well asymptotic behaviour. Moreover describing these dynamics leads us to an understanding of relevant adaptive control questions.

The paper is organised as follows, in the preamble following the introduction, we gather a number of known results and introduce some notation. The next section is devoted to the ultimate boundedness that the adaptively controlled system dynamics enjoy. The results established here serve to underscore the importance of the sigma modification. Then we discuss the simplest possible scenario, when the linear system to be controlled has no zeroes. This situation is actually simpler than what (1) suggests, but nevertheless informative. This leads to a planar adaptive system, whose dynamics can be well understood. It provides a good pointer for the complications we may expect. Next follows an incomplete analysis for the case when the linear system to be controlled has a single stable zero and two poles. Finally we conjecture what a phase portrait in the general case may look like. The paper concludes with suggestions on how we may have to modify the adaptive law based on the understanding of the exposed dynamical behaviour in order to obtain a robust adaptive system.

2 Preamble

For convenience sake and without loss of generality (as far as the dynamical system analysis is concerned) we study the closed-loop system (1), (2) for

- $g = 1$, if $g \neq 1$ consider the state space transformation $(x, y, k) \to (x\sqrt{g}, y\sqrt{g}, k/g)$,
- (A, b) in controllable canonical form.

The closed-loop adaptive system can thus be represented as:

$$\begin{aligned} \dot{x} &= Ax + by, & x(0) \in \mathbb{R}^n, \quad & b \in \mathbb{R}^n,\ A \in \mathbb{R}^{n \times n} \text{ stable}, \\ \dot{y} &= -cx - dy - ky + e, & y(0) \in \mathbb{R}, \quad & c^T \in \mathbb{R}^n,\ d, e \in \mathbb{R}, \\ \dot{k} &= -\sigma k + y^2, & k(0) \geq 0, \quad & \sigma \geq 0. \end{aligned} \tag{3}$$

We refer to the subsystem with state (x, y) as the plant. We consider the parameters e, d and the vector c as bifurcation parameters.
We use the notation:

$$\begin{aligned} z(s) &= \det(sI - A), \\ p(s) &= \det \begin{pmatrix} sI - A & -b \\ c & s \end{pmatrix}. \end{aligned} \tag{4}$$

Notice that with the above notation:

$$\det \begin{pmatrix} sI - A & -b \\ c & s + \beta \end{pmatrix} = p(s) + \beta z(s).$$

We conclude this section with some simple observations about the phase portrait of the system (3):

- **Fact (i)**
  $k(t) \geq 0$ for all $t \geq 0$ if $k(0) \geq 0$.
- **Fact (ii)**
  If $e = 0$, $(x(t), y(t), k(t))$ is a solution of (3) if $(-x(t), -y(t), k(t))$ is one.
- **Fact (iii)**
  For $\sigma > 0$, $e = 0$ and $(x(0), y(0)) = (0, 0)$, it follows that $k(t) = e^{-\sigma t} k(0)$ and $(x(t), y(t)) = (0, 0)$ for all $t \geq 0$.
- **Fact (iv)**
  For $\sigma = 0$ and $e = 0$, the trajectories converge to a point in state space of the form $(0, 0, k_\infty)$, for some $k_\infty \geq 0$ which depends on the initial condition. (See [2]).
- **Fact (v)**
  For $\sigma = 0$ and $e \neq 0$, $(x(t), y(t))$ converge to zero and $k(t)$ diverges.
- **Fact (vi)**
  The transformation $(x, y, k) = \sigma(x_1, y_1, k_1)$ and time scaling $\sigma t = \tau$,

leads to a system description of the form:

$$\begin{aligned} x_1^{'} &= A_1 x_1 + b_1 y_1, \\ y_1^{'} &= -c_1 x_1 - d_1 y - k_1 y_1 + e_1, \\ k_1^{'} &= -k_1 + y_1^2. \end{aligned} \tag{5}$$

Here $\sigma A_1 = A$, $\sigma b_1 = b$, $\sigma c_1 = c$, $\sigma d_1 = d$ and $\sigma^2 e_1 = e$. The $'$ denotes derivative with respect to the new time variable τ. Clearly (5) is of the same form as (3). Hence, when discussing the dynamics we can limit ourselves to the case $\sigma = 1$. Yet, from a control perspective, we prefer (3), as σ is a design variable and A, b, c, d are not.

3 Global stability

It is clear from **Fact (iv)** and **Fact (v)** that the adaptive system with $\sigma = 0$ is not robust with respect to input offsets. The sigma modification has the property that all trajectories are ultimately bounded[1], regardless of the disturbance e. This is probably the main motivation for its introduction. The ultimate boundedness result can be established using a Lyapunov argument.

Theorem 3.1 *Consider (3) with $\sigma > 0$. Let $P = P^T > 0$ be the unique positive definite solution of the Lyapunov equation $A^T P + PA = -I$. For all $\sigma < \bar{\sigma} = 1/\lambda_{\max}(P)$ the compact set*

$$\mathbb{A} = \left\{(x, y, k) \ : \ x^T P x + y^2 + (k - \mu)^2 \leq \mu^2 + |e|\right\}, \tag{6}$$

with

$$\mu = \max\{0, -d + \frac{1}{2}\left(\frac{|e|}{2\sigma} + \sigma + (Pb - c^T)^T (I - \sigma P)^{-1} (Pb - c^T)\right)\} \tag{7}$$

is globally attractive and positively invariant.

Proof Consider the comparison function

$$V(x, y, k) = x^T P x + y^2 + (k - \mu)^2. \tag{8}$$

Its derivative along the solutions of system (3) is given by:

$$\begin{aligned} \dot{V}(x(t), y(t), k(t)) &= -x^T x + 2x^T P b y - 2x^T c^T y - 2dy^2 - 2ky^2 + 2ye \\ &\quad -2\sigma k(k - \mu) + 2(k - \mu) y^2, \\ &= -x^T x + 2x^T P b y - 2x^T c^T y - 2dy^2 + 2ye - 2\mu y^2 \\ &\quad -2\sigma (k - \mu)^2 + 2\sigma \mu^2. \end{aligned}$$

[1] The solutions of $\dot{x} = f(x)$ are said to be *ultimately bounded*, if there exists a $B > 0$ such that for all initial conditions $x(0) = x_0$ there exists a $T(x_0) > 0$ such that the solutions satisfy $||x(t, x_0)|| \leq B$ for all $t \geq T$.

Selecting μ as indicated in (7) implies that

$$-x^T x + 2x^T Pby - 2x^T c^T y - 2dy^2 + 2ye - 2\mu y^2 \leq -\sigma(x^T Px + y^2) + 2\sigma|e|,$$

which leads to:

$$\dot{V}(x(t), y(t), k(t)) \leq -2\sigma V(x(t), y(t), k(t)) + 2\sigma\mu^2 + 2\sigma|e|$$

This establishes our claim. □

Remark 3.2 The theorem statement is particularly useful for the case that σ is small. It is clear from the proof that ultimate boundedness can be established for all $\sigma > 0$, but we prefer the above formulation as it provides us with a reasonably tight estimate of the form $V(x, y, k) \leq \mu^2 + |e|$ for a positively invariant set. In case $e = 0$ our estimate cannot be improved as the set $\mathbb{A}$ has the origin (an equilibrium in this case) on its boundary. Moreover, "small" σ is in line with practice. □

For $e = 0$ we can glean from the theorem statement that for sufficiently large, positive d the origin of the state space is globally asymptotically stable. It can be established that this is the case for all systems (1) that have strict positive real transfer function, (i.e. $\mathrm{Re}\{z(j\omega)/[p(j\omega) + dz(j\omega)]\} > 0$ for all $\omega \in \mathbb{R}$).

Theorem 3.3 *Consider system (3) with $\sigma > 0$ and $e = 0$. Assume furthermore that the linear system (1) has a strictly positive real transfer function, i.e. there exists a positive definite solution $P = P^T$ such that $A^T P + PA < 0$ and $d > ||c^T - Pb||^2$. Then the origin of the state space is globally uniformly asymptotically stable.*

Proof The result is derived using the comparison function V introduced in equation (8) with $\mu = 0$. The conditions of the Theorem 3.3 lead to $\dot{V}(x(t), y(t), k(t)) < -2\sigma k^2(t) - W(x(t), y(t))$, with positive definite $W(x, y)$. □

The gap between the two results stated in Theorem 3.3 and 3.1 does not appear to be large. One could wonder whether the same result as stated in Theorem 3.3 could hold true for the larger class of stable minimum phase systems. This is not the case, as will become clear. Here is a simulation example to illustrate what can happen.
Consider a system of the form (3) with the parameters selected as:

$$A = \begin{pmatrix} 0 & 1 \\ -1 & -10 \end{pmatrix} \quad b = \begin{pmatrix} 0 \\ 1 \end{pmatrix} \quad c = \begin{pmatrix} -9 & -10 \end{pmatrix} \quad d = -1 \quad \sigma = 0.1$$

The linear system (1), with the above parameter values, has a transfer function with stable zeroes and stable poles. The transfer function is given by $z(s)/(p(s) + dz(s)) = (s^2 + s + 10)/(s^3 + 2s^2 + s + 1)$. This transfer function is not strictly positive real. A typical trajectory for the above system with initial conditions (0.1, 0.1, 1, 0) is presented in Figure 1. Both the plant output y and the adaptive gain k are displayed. Notice the asymptotically periodic behaviour. A root locus plot for that root of $p(s) + (d + k)z(s) = s^3 + 3s^2 + 2s + 11 + (-1 + k)(s^2 + s + 10)$ with the largest real and largest imaginary part is presented in Figure 2. Observe that the adaptive gain hovers asymptotically around that value of the gain above which the root locus remains in the left half plane, clearly the plant is but marginally stabilised through the adaptive control law. Observe that no control would have been better in this case, as the open loop system is stable.

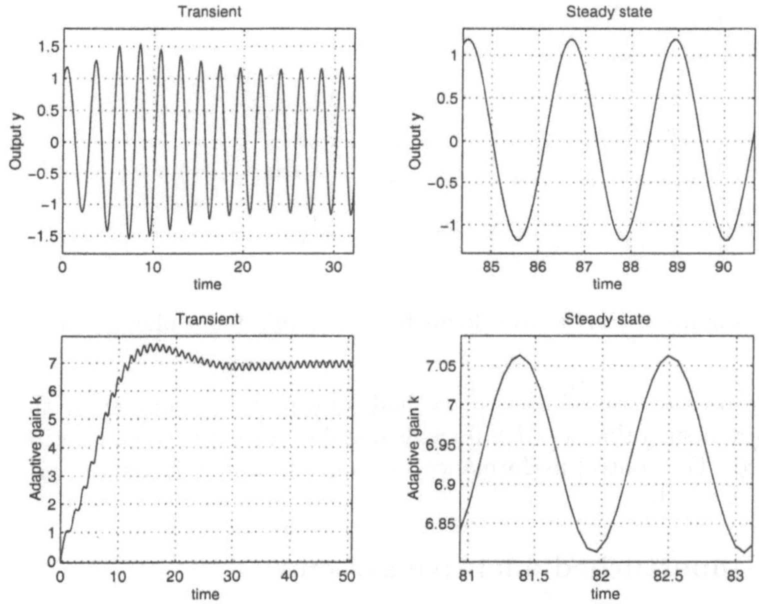

Figure 1: Limit cycle behaviour for adaptively controlled stable minimum phase plant

4 Planar adaptive system

Let us first consider the adaptive stabilisation through adaptive gain output feedback of a one dimensional linear plant:

$$\begin{aligned} \dot{y} &= -dy - ky + e, \quad y(0), \\ \dot{k} &= -\sigma k + y^2, \qquad k(0) \geq 0. \end{aligned} \tag{9}$$

Here d is the open loop pole of the plant to be stabilised, y is the measured output, k the adaptive feedback gain and e represents the input disturbance.

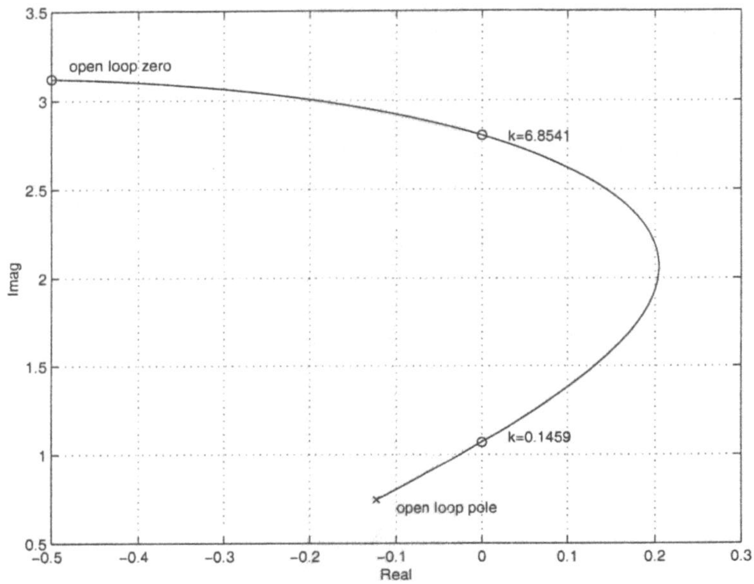

Figure 2: Partial root locus for stable minimum phase plant

For purposes of comparison, if we had known the parameter d, then any control gain such that $k + d > 0$ yields a stable system with equilibrium $y_e = e/(k+d)$. For control performance we would presumably require $|y_e| < |e|$, hence $k + d > 1$.

4.1 Unperturbed adaptive system

For the adaptive system (9) with $e = 0$, we can make the following observations:

1. **An open loop stable plant $d \geq 0$ implies that the trivial solution is globally asymptotically stable.**

   Clearly the solution $(y(t), k(t)) = (0, 0)$ is uniformly asymptotically stable as seen from considering the Lyapunov function $V(y, k) = y^2 + k^2$. For the derivative of V along the solutions of (9) we obtain $\dot{V}(t) = -2dy^2(t) - 2\sigma k^2(t)$. It follows that the origin is globally asymptotically stable. This observation is also a corollary of Theorem 3.3, as any linear stable first order system has a strictly positive real transfer function.

2. **An open loop unstable plant $d < 0$ leads to solutions that are ultimately bounded**

   Consider the Lyapunov function $V(y, k) = y^2 + (k + d - \sigma)^2$. We obtain for the derivative along the solutions of (9) that $\dot{V}(y(t), k(t)) =$

$-2\sigma y^2(t) - 2\sigma k^2(t) + 2\sigma(-d+\sigma)k(t)$. Considering that $k(t) \geq 0$, we obtain the estimate $\dot{V}(y(t), k(t)) \leq -2\sigma V(y(t), k(t)) + 2\sigma(-d+\sigma)^2$. From which it follows that the set characterised by $V(y,k) \leq (-d+\sigma)^2$ is positively invariant. This observation is also a corollary of Theorem 3.1.

3. **For an open loop unstable plant $d < 0$ the solutions either converge to $(\sqrt{-\sigma d}, -d)$ or $(-\sqrt{-\sigma d}, -d)$, unless $y(0) = 0$ in which case the solutions converge to $(0,0)$.**

   If $y(0) = 0$ it follows that $y(t) = 0$ for all $t \geq 0$ and furthermore that $k(t) = \exp(-\sigma t)k(0)$. If $y(0) > 0$ then obviously $y(t) > 0$ for all $t \geq 0$. Consider the Lyapunov function $V(y,k) = y^2 - 2\sigma d \ln y + (k+d)^2$ on the domain $D = \{y > 0, k > 0\}$. V achieves its minimum in $(\sqrt{-\sigma d}, -d)$. For its derivative along the solutions of (9) we find $\dot{V}(y(t), k(t)) = -2\sigma(k(t)+d)^2$. Applying LaSalle's invariance principle, it follows that all solutions starting in D converge to the point $(\sqrt{-\sigma d}, -d)$. This establishes our claim, as the situation $y(0) < 0$ can be dealt with in an analogous manner.

The above describes a single pitch fork bifurcation for the origin attained at $d = 0$. The typical local stability picture associated with this bifurcation has global validity in this particular case. For $d > 0$ the origin is globally asymptotically stable, for $d < 0$ the origin becomes a saddle and two new locally stable equilibria come into existence. In this particular situation the stable manifold of the saddle (the origin) acts as a separatrix between the domains of attraction for the other equilibria.
The control interpretation is as follows. In the absence of any disturbances, the adaptive sigma modified control law adjust the feedback gain k to the minimum feedback action required to stabilise the open loop plant. Indeed, if the open loop plant was stable, asymptotically no feedback is applied. If the open loop plant was unstable, the minimum feedback gain to render the linear plant marginally stable is applied. In doing so the adaptive control law sacrifices the control objective of achieving regulation of the output y, instead $|y|$ converges to $\sqrt{-\sigma d}$. For small values of σ (meaning $\sigma << 1/d$) this loss of asymptotic performance may be acceptable. It is nevertheless disappointing. In this particular case of a first order system we know a priori that arbitrary transient performance can be achieved for correspondingly large gains. The adaptive law however limits the gain to the minimum gain required to just achieve stability.

4.2 Perturbed adaptive system

It is disturbing to notice that the adaptive law leads to a marginally stabilised linear plant in the case the latter was open loop unstable. This is the more annoying when we realise that this is the only case in which some control is

required and that good performance can be achieved via high feedback gain. What happens if measurement errors are involved, i.e. $e \neq 0$?
Importantly, $e \neq 0$ breaks the symmetry inherent in (9) when $e = 0$, hence the pitchfork bifurcation pattern breaks as $e \neq 0$. Moreover in view of the fact we now have a two parameter family of planar vector fields (e, d are the important parameters, σ is not) we expect that we can organise the local dynamics around a Takens-Bogdanov bifurcation point. This would lead to the presence of a homoclinic bifurcation and a Hopf bifurcation in the bifurcation diagram. A bifurcation diagram is presented which describes the local dynamics. For the transients we offer some analytic results, but mainly rely on simulation results.

1. **For $d \geq 0$, $e \neq 0$ there is a unique globally asymptotically stable equilibrium.**

   The equilibrium is defined via $\sigma k_e = y_e^2$ and $y_e^3 + \sigma d y_e = \sigma e$. With $d \geq 0$ the latter equation implies that the equilibrium exists and is unique. Local stability follows at once from the Jacobian:

$$J = \begin{pmatrix} -d-k & -y \\ 2y & -\sigma \end{pmatrix}. \tag{10}$$

   The characteristic polynomial of J at an equilibrium (y_e, k_e) is given by $\lambda^2 + (d + k_e + \sigma)\lambda + \sigma d + 3y_e^2$. Using the fact that $y_e^2 = \sigma k_e$ at equilibrium, this can be rewritten in the more informative format

$$\det(\lambda I - J) = (\lambda + d)(\lambda + \sigma) + k_e(\lambda + 3\sigma). \tag{11}$$

   Global stability can be established using a Lyapunov argument. Consider the case $e > 0$. Let $V(y, k) = (y - y_e)^2 + (k - k_e)^2$. Observe that the set $\mathbb{D} = \{(y, k) : y \geq 0, k \geq 0\}$ is attractive and positively invariant. On $\mathbb{D}$ we have $\dot{V}(y(t), k(t)) \leq -\sigma(k - k_e)^2 - (d + k_e)(y - y_e)^2$.

2. **For $d < 0$ and $|e| > e_{SN}$, with $e_{SN} = \frac{2}{3}\sqrt{\frac{\sigma}{3}}(-d)^{3/2}$ there is a unique, locally stable equilibrium.**

   Equilibria satisfy $\sigma k_e = y_e^2$ and $y_e^3 + \sigma d y_e = \sigma e$. With $d < 0$ and $|e| > e_{SN}$ the latter equation implies that the equilibrium exists and is unique.

   Local stability follows at once from the Jacobian (10). As under the given circumstances $y_e^2 > -\sigma d$ or $k_e > -d$ it follows that $d + k_e + \sigma > \sigma > 0$ and $\sigma(d + 3k_e) > -2\sigma d > 0$, which by inspection of (11) implies that (y_e, k_e) is locally asymptotically stable.

3. **For $d < 0$ and $|e| \leq e_{SN}$, there are three equilibria, $(y_{ei}, \sigma y_{ei}^2)$ $i = 1, 2, 3$ with $-y_{SN} \leq y_{e1} \leq -\sqrt{\frac{-\sigma d}{3}} \leq y_{e2} \leq \sqrt{\frac{-\sigma d}{3}} \leq y_{e3} \leq y_{SN}$, where y_{SN} is the positive root of $y^3 + \sigma d y = \sigma e_{SN}$.**

The locus

$$e^2 = \frac{4}{27}\sigma(-d)^3 \tag{12}$$

describes the Saddle-Node bifurcation in the bifurcation diagram.

The equilibrium $(y_{e2}, \sigma y_{e2}^2)$ is an unstable saddle. Indeed the product of the eigenvalues of the Jacobian evaluated at this equilibrium is given by $\sigma d + 3y_{e2}^2 \leq 0$. The equilibria for $i = 1, 3$ may be stable or unstable depending on the values of d, e. If $\sigma + 2d/3 > 0$ they are locally stable, regardless of the value of e. They change their stability status through a Hopf bifurcation if $\sigma + 2d/3 < 0$. When e is less than but near e_{SN} the equilibrium $i = 3$ is locally stable, whilst the equilibrium $i = 1$ is unstable.

The Hopf bifurcation locus in the bifurcation diagram is described by:

$$e^2 = -\sigma^3 d - \sigma^4 \quad \text{with } d < -3/2\sigma. \tag{13}$$

The equilibria undergoing the Hopf bifurcation are described by:

$$\begin{aligned} y_e^2 &= \sigma(d+\sigma)^2, \quad y_e e < 0, \\ k_e &= -d - \sigma. \end{aligned} \tag{14}$$

4. Takens-Bogdanov bifurcation point

In order to organise the bifurcation diagram more clearly, we investigate the presence of a Takens-Bogdanov bifurcation point and verify its local dynamics. In view of the fact that we are dealing with a two-parameter family of vector fields we expect to find some isolated Takens-Bogdanov points. The calculations used to obtain the normal form for the unfolding are summarized in Appendix II.

The Takens-Bogdanov points for (9) are given by:

$$y_{TB1} = \frac{\sigma}{\sqrt{2}}, \quad k_{TB} = \frac{\sigma}{2}, \quad d_{TB} = -\frac{3\sigma}{2}, \quad e_{TB1} = -\frac{\sigma^2}{\sqrt{2}}, \tag{15}$$

and

$$y_{TB2} = -\frac{\sigma}{\sqrt{2}}, \quad k_{TB} = \frac{\sigma}{2}, \quad d_{TB} = -\frac{3\sigma}{2}, \quad e_{TB2} = \frac{\sigma^2}{\sqrt{2}}. \tag{16}$$

As is easily verified, the Takens-Bogdanov points are at the extremities of the Hopf-bifurcation locus described in equation (13) and of course on the Saddle-Node bifurcation locus (12). In Appendix II it is shown that the planar system undergoes a regular Takens-Bogdanov bifurcation at either of these points. The corresponding normal form is given by:

$$\begin{aligned} \dot{\eta_1} &= \eta_2, \\ \dot{\eta_2} &= \beta_1 + \beta_2\eta_1 + \eta_1^2 + \eta_1\eta_2. \end{aligned} \tag{17}$$

It follows from Theorem 8.4 [13] that in the bifurcation diagram of the system (9) we encounter a line corresponding to a Hopf bifurcation creating an unstable limit cycle together with a stable focus and a homoclinic bifurcation of the saddle through which the unstable limit cycle is annihilated. The loci in the bifurcation diagram corresponding to the homoclinic bifurcations are obtained numerically and displayed in Figure 3. The Hopf-bifurcation locus is described in equation (13).

5. Bifurcation diagram

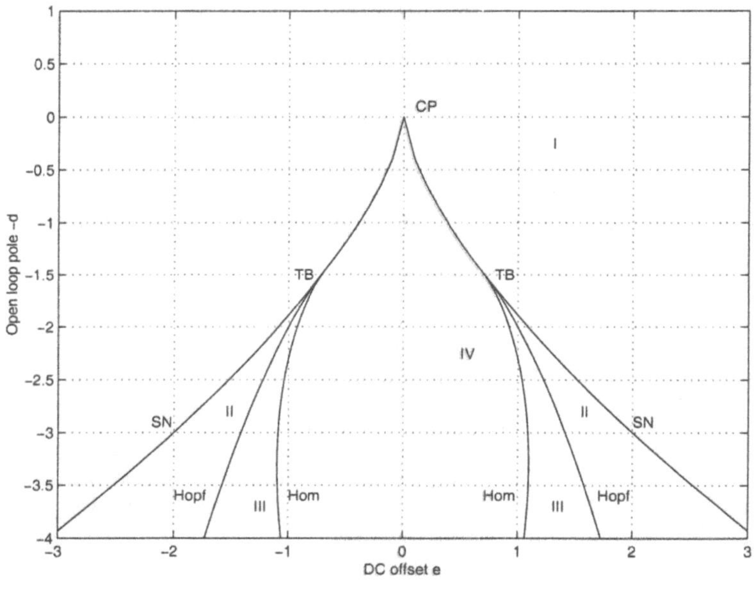

Figure 3: Bifurcation diagram for the planar adaptive system

We can now piece together a rather complete picture of the asymptotic dynamics as function of the bifurcation parameters d, e. The bifurcation diagram is depicted in Figure 3, whilst Figures 5 to 7 represent a number of typical phase portraits corresponding to the various possibilities suggested by the bifurcation diagram. In all simulations we used $\sigma = 1$. The bifurcation diagram was generated using Dtool. The phase portraits were simulated using Matlab, with the aid of the suite pplane5.m.

The line **SN** represents the Saddle-Node bifurcation locus, equation (12). The cusp point CP is the origin in the bifurcation diagram $(e, d) = (0, 0)$ with corresponding equilibrium $(y_e, k_e) = (0, 0)$. The points **TB** identify the Takens-Bogdanov points (15) and (16), which here clearly are the points around which the bifurcation diagram can be organised. The lines **Hom** identify the homoclinic cycle bifurcation, terminating at a TB point. The lines **Hopf** identify the Hopf bifurcation, terminating at a TB point, see (13).

Let us follow a counterclockwise direction through the bifurcation diagram starting in the region marked **I**. In region **I** there is a unique locally stable equilibrium, seemingly globally attractive (only confirmed analytically for $d > 0$). Figure 5 depicts a representative phase portrait.

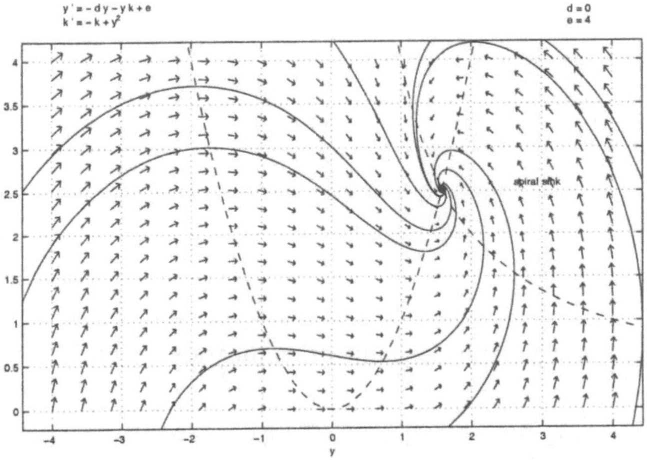

Figure 4: Phase portrait for bifurcation diagram region I in Fig. 3

In the regions **II**, after the saddle-node bifurcation, three equilibria coexist. A saddle and unstable focus come into existing through the Saddle-Node bifurcation. It appears from the phase portrait in Figure 5 that apart from the stable manifold of the saddle, all solutions converge to the stable focus.

In the regions **III**, just after the Hopf bifurcation, the spiral source equilibrium sheds through the Hopf bifurcation an unstable limit cycle and becomes a stable focus. The interior of the limit cycle is the domain of attraction of this focus. The other stable equilibrium attracts the other solutions, apart from the stable manifold of the saddle. This is shown in Figure 6, which represents a typical phase portrait for Region **III**.

The limit cycle continues to grow (in size and period) as we further move towards the line Hom. There a homoclinic bifurcation occurs, which destroys the limit cycle. After the homoclinic bifurcation in Region **IV**

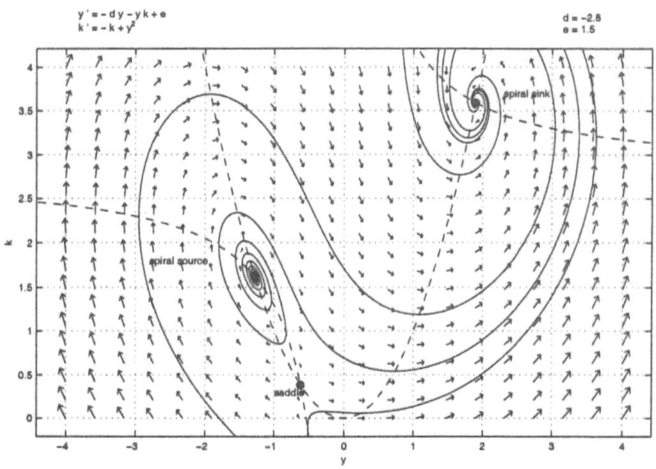

Figure 5: Phase portrait for bifurcation diagram region II in Fig 3.

it appears from Figure 7, that the stable manifold of the saddle acts as a separatrix for the domains of attraction of the stable foci. (On the line $e = 0$ this was analytically demonstrated above.)

4.3 Control interpretation

Despite that the disturbance e complicates the phase portrait in a significant way, its influence from a control perspective is not so important. Both transient behaviour (although we have not presented a complete proof for this) and asymptotic dynamics appear entirely acceptable. Certainly, the asymptotic dynamics are bounded. More importantly the output of the plant becomes of the same order of magnitude as the disturbance e. This is in complete accordance with linear control intuition. One simply can not expect any better from a static gain output feedback. The main difference with static output feedback is of course that through the adaptation law there are multiple coexisting possibilities for the asymptotic dynamics, but all of similar control performance.

The main drawback of the adaptive law is that there is no real control over the actually achieved control performance and that it only achieves marginal stabilization of the plant.

It could nevertheless be argued that the σ modification with adaptive gain does work, and is robust with respect to input disturbances. This follows because over the class of systems controlled by the adaptive system it is not possible to estimate the achievable performance. Also no actual requirements

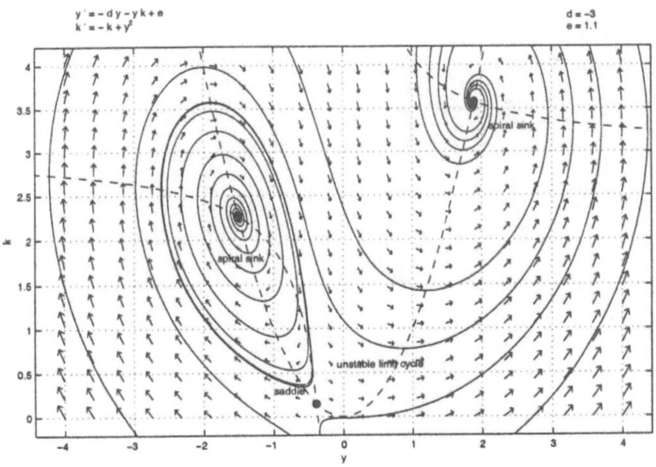

Figure 6: Phase portrait for bifurcation diagram region III in Fig 3.

are incorporated in the adaptation law, apart from the fact that we would like y to be regulated. In the presence of disturbances this becomes impossible, and how much we want to reject the disturbances has not been specified at all. Hence, we could not really expect any better behaviour than what is achieved. This identifies in a sense where the suggested adaptation mechanism falls short: the achievable control performance is to be identified in order to realise a true adaptive control strategy.
The above picture is however too optimistic as will transpire from the next example.

5 Three dimensional adaptive system

In this section we investigate the possible asymptotic dynamics when the plant has a single stable zero, here fixed at $-a < 0$. The adaptive system (1) takes the form:

$$\begin{aligned} \dot{x} &= -ax + y, \\ \dot{y} &= -cx - dy - ky + e, \\ \dot{k} &= -\sigma k + y^2. \end{aligned} \tag{18}$$

We treat the parameters c, d and e as bifurcation parameters. A complete bifurcation diagram will not be presented.
Before we start our discussion of the bifurcation diagram, let us observe that the phase portrait for $c = 0$ is already known by virtue of our discussion in

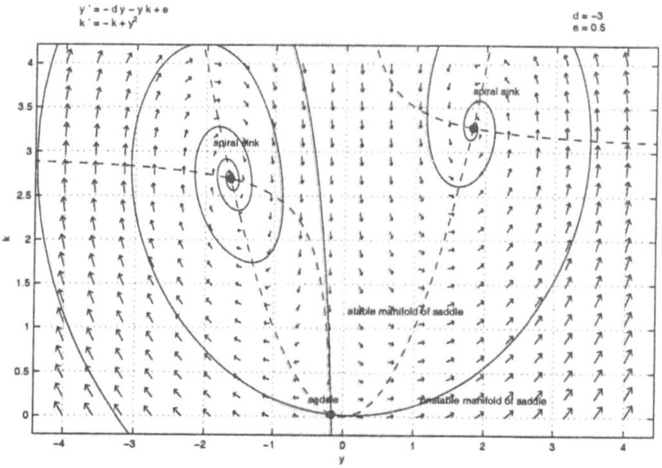

Figure 7: Phase portrait for bifurcation diagram region IV in Fig 3.

the previous section. Indeed for $c = 0$ the x-state does not affect the (y, k) dynamics, which coincide with the planar system discussed before, see equation (9). Hence the phase portrait will contain for the appropriate parameters unstable limit cycles, homoclinic orbits and multiple equilibria.

5.1 Equilibria and Jacobian

The equilibria are characterised by:

$$ax_e = y_e, \quad \sigma k_e = y_e^2, \quad \sigma(c/a + d)y_e + y_e^3 = \sigma e. \tag{19}$$

The equilibrium is unique if

$$c + da \geq 0, \text{ or } c + da < 0 \text{ and } e^2 > -\frac{2\sigma}{27}(d + c/a)^3. \tag{20}$$

There are three distinct equilibria if:

$$c + da < 0 \text{ and } e^2 < \frac{2\sigma}{27}(-d - c/a)^3. \tag{21}$$

At an equilibrium the Jacobian is given by:

$$J_e = \begin{pmatrix} -a & 1 & 0 \\ -c & -d - k_e & -y_e \\ 0 & 2y_e & -\sigma \end{pmatrix}. \tag{22}$$

The characteristic polynomial of the Jacobian, evaluated at an equilibrium, is given by;

$$det(\lambda I - J_e) = (\lambda + \sigma)(\lambda^2 + (d+a)\lambda + c + da) + k_e(\lambda + a)(\lambda + 3\sigma). \quad (23)$$

5.2 Unperturbed adaptive system $e = 0$

1. **An open loop stable plant $d + a > 0$ and $ad + c > 0$ implies that the trivial solution is globally asymptotically stable**

   This observation is a corollary to Theorem 3.3, as any stable second order linear system with a stable zero has indeed a strictly positive real transfer function.

2. **Unstable system with unstable zero leads to unbounded response**

   The condition $a > 0$ is necessary for stability whenever the open loop system is unstable. Consider for example the adaptive system:

$$\begin{aligned} \dot{x} &= -ax + y, \\ \dot{y} &= -dy - ky, \\ \dot{k} &= -k + y^2. \end{aligned} \quad (24)$$

   Obviously, for all $a \leq 0$ we have that x grows without bound (for almost all initial conditions, and all d). The adaptive feedback is only meaningful when the system has stable zeroes.

3. **Pitchfork bifurcation**

   The pitchfork bifurcation locus in the bifurcation diagram is given by $c + da = 0$, see equation (21) with $e = 0$. For $c + da < 0$ the origin is a saddle. The other two equilibria are described by $(ay_e, y_e, y_e^2/\sigma)$ with $y_e^2 = -\sigma(c/a + d)$. Their local stability is determined from the location of the roots of (23):

$$\lambda^3 + \frac{1}{a}(a\sigma + a^2 - c)\lambda^2 + \frac{\sigma}{a}(-2(ad + c) - c + a^2)\lambda - 2\sigma(ad + c) \quad (25)$$

   These equilibria are stable when $a^2 - c > \max(a\sigma, -2(ad + c))$, and $(a^2 - c)(a^2 + a\sigma - c) - 2(ad + c)(a\sigma - c) > 0$ unstable otherwise. This means that immediately after the pitchfork bifurcation, i.e. for $ad+c < 0$ but close to 0, the new equilibria are locally stable if $d + a > 0$, and saddles otherwise.

4. **Takens-Bogdanov point: open loop system is double integrator**

   The origin is a Takens-Bogdanov point when $d + a = 0$ and $ad + c = 0$. The corresponding Jacobian has two zero eigenvalues with geometric multiplicity one, and one eigenvalue $-\sigma$. Using the comparison function

$V = x^2 + y^2 + k^2 - 4k$, it can be verified that the origin is globally asymptotically stable in this situation.

The bifurcation locus for the homoclinic orbit bifurcation is contained in the parameter domain $0 < c < a^2$ and $d < -a$ (see next item Hopf bifurcation).

5. **Hopf bifurcation**

The origin undergoes a regular Hopf bifurcation for $d + a = 0$ and $ad + c > 0$. The origin remains the unique equilibrium, but becomes an unstable focus as $d + a$ becomes negative. A stable periodic orbit comes into existence which persists for all $d < -a$ and $c > a^2$, at least for sufficiently small σ. This can be demonstrated using averaging ideas. The calculations are summarized in Appendix III and IV. In Appendix III we establish a domain of attraction for an invariant torus, which we demonstrate to contain a unique periodic orbit in Appendix IV. Appendix IV provides existence, uniqueness and local stability. It is demonstrated that the periodic orbit is for sufficiently small σ well approximated by:

$$\begin{aligned} x(t,\sigma) &= -\sqrt{\frac{-2\sigma(d+a)}{1+a^2}}\sin(\omega t) + O(\sigma), \\ y(t,\sigma) &= \sqrt{\frac{-2\sigma(d+a)}{1+a^2}}(-a\sin(\omega t) + \omega\cos(\omega t)) + O(\sigma), \\ k(t,\sigma) &= -(d+a) + O(\sqrt{\sigma}), \\ \omega^2 &= c - a^2. \end{aligned} \tag{26}$$

In Appendix III it is demonstrated that the domain of attraction is virtually all of $\mathbb{R}^3$ apart from the axis $\{x = y = 0\}$ (more precisely, any bounded domain in $\mathbb{R}^3/\{x = y = 0\}$ is contained in a domain of attraction for sufficiently small σ).

The above observations complete the local bifurcation picture for the trivial equilibrium.
Following the non-zero equilibria it can be observed that they too undergo a Hopf bifurcation, creating locally attractive limit cycles. This takes place for $ad + c < 0$ and $(\sigma a + a^2 - c)(a^2 - c - 2(ad + c)) = -a^2(ad + c)$. The limit cycles themselves undergo a sequence of saddle-node bifurcations, eventually evolving into complex attractors. These limit cycles and attractors do not scale with σ as the limit cycles originating from the Hopf bifurcating origin (26). An example is represented in the Figure 8. The aperiodic response in Figure 8 is generated from the initial condition $(x_0, y_0, k_0) = (0, 1, 1)$ for the system (18) with $a = 1$, $d = -2.751$, $c = -5.301$, $\sigma = 0.001$. It is reminiscent of drift-burst behaviour observed in other direct adaptive control schemes [2]. A partial, numerically obtained bifurcation diagram is represented in Figure 9. The bifurcation parameter is α. The system (18) has parameters $a = 1$, $d =$

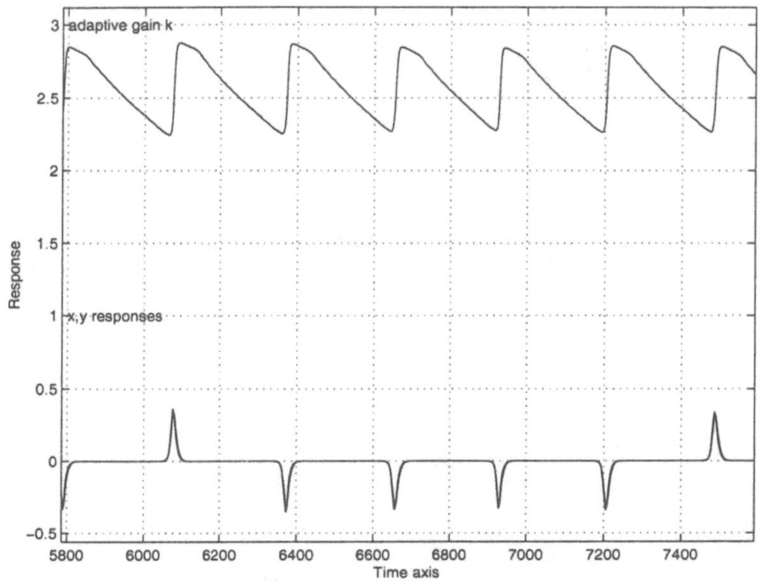

Figure 8: Aperiodic behaviour after pitch fork bifurcation followed by Hopf bifurcations

$-\alpha$, $c = -\alpha/4 + 3/4$ and $\sigma = -0.1$. The bifurcation parameter ranges from 0 to 2.5. A phase space portrait displaying some of the possible competing asymptotic dynamics for $\alpha = 0.82$ is displayed in Figure 10. The numerical bifurcation diagram reveals that the origin undergoes a pitch fork bifurcation $\alpha = 0.6$. The origin remains unstable for all $\alpha > 0.6$. The new equilibria are unstable in the interval $0.831 < \alpha < 1.86$, and locally asymptotically stable otherwise. In the same parameter region $0.831 < \alpha < 1.86$ we identify a strange attractor, with a structure not dissimilar from the Lorenz attractor. The non zero equilibria undergo regular Hopf bifurcations for $\alpha = 0.831$ and 1.86. The Hopf bifurcations are supercritical. Regions of hysteresis are clearly identifiable, indicative of saddle node bifurcations for the limit cycles.

5.3 Perturbed adaptive system

We now consider concisely the influence of the input offset e.

1. **Stable plant:** $c + ad > 0$ **and** $d + a > 0$

   Under these condition there is a unique equilibrium, see equation (19), which is at least locally uniformly stable, see equation (23). The output is no longer regulated, at equilibrium $|y_e| < |e|$.

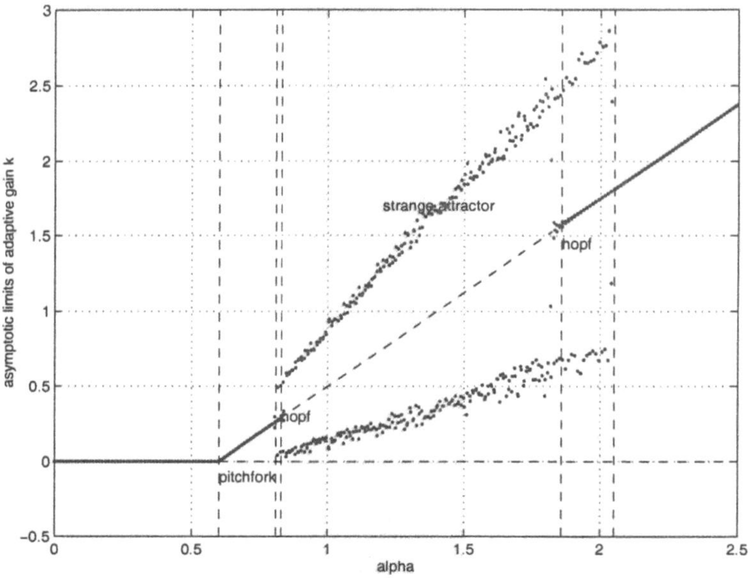

Figure 9: Partial bifurcation diagram for system (18)

2. **Unstable plant, real poles: $c + da < 0$**

   In this situation there may be up to three equilibria, (see equation (19), depending on the magnitude of e. The main event in this part of the bifurcation diagram is the saddle-node bifurcation. This situation corresponds very much to the planar system discussed above. It suffices to consider the local dynamics in the centre manifold after eliminating a one dimensional stable manifold.

   (a) The saddle-node bifurcation locus is given by

$$e^2 = \frac{2\sigma}{27}(-d - c/a)^3. \tag{27}$$

   For $e^2 > \frac{2\sigma}{27}(-d-c/a)^3$ there is but one locally stable equilibrium, for $e^2 \leq \frac{2\sigma}{27}(-d-c/a)^3$ there are three equilibria.

   The control interpretation is that a larger DC offset enforces a larger control gain, which in turn provides more stability.

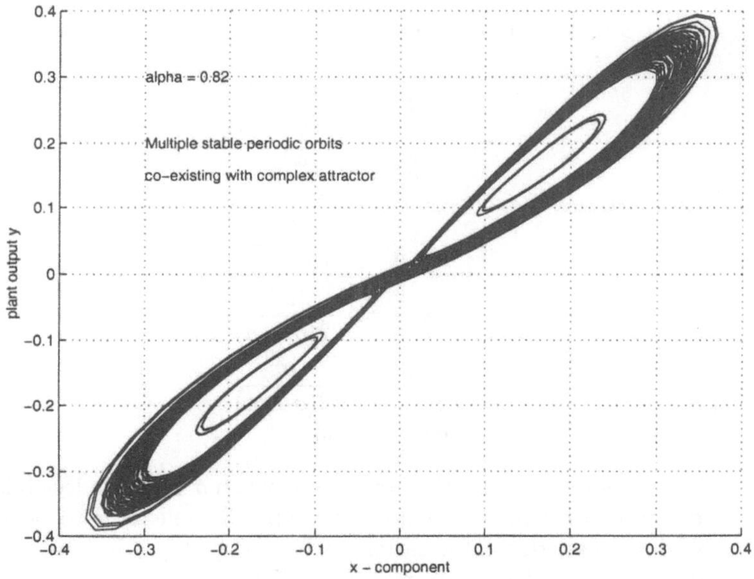

Figure 10: Co-existing asymptotically stable dynamics for (18)

(b) **Takens-Bogdanov points** There is a variety of Takens-Bogdanov points characterised by:

$$\begin{aligned}(\sigma + a - \sigma a - 1/3a^2)d + (1 + \sigma - a/3)c &= -\sigma a, \\ -4\sigma(c + ad)^3 + 27e^2 &= 0, \\ k_{TB} &= -\tfrac{c+ad}{3}, \\ y_{TB}^2 &= \sigma k_{TB}, \\ ax_{TB} &= y_{TB}.\end{aligned} \tag{28}$$

(c) **Organising centres**

The most degenerate bifurcation points, also the most likely organising centres in the bifurcation diagram are those equilibria with a zero eigenvalue of algebraic multiplicity three and with a single eigenvector. One expects to find complex asymptotic attractors in the neigbourhood of such points. These points only exist if

$\sigma > a > 0$, and are characterised by:

$$
\begin{aligned}
d &= \frac{2a^2 - 3\sigma^2}{2(\sigma - a)}, \\
c &= \frac{a^3}{a - \sigma)}, \\
e^2 &= \frac{2\sigma^7}{(\sigma - a)^3}, \\
k_e &= \frac{\sigma^2}{2(\sigma - a)}, \\
y_e^2 &= \frac{\sigma^3}{2(\sigma - a)}, \\
ax_e &= y_e.
\end{aligned}
\tag{29}
$$

These points, because they only exist when $\sigma > a$, are less relevant from a control perspective as one typically selects σ to be small. Nevertheless, this information serves to qualify what we have to understand under small σ, and furthermore serves to underscore the difficulty in selecting an appropriate σ value. Indeed, a proper value of σ, if one wants to avoid the complex dynamics associated with such points, does require non trivial prior knowledge about the location of the plant transfer function zeroes.

3. Hopf bifurcations

In case $c + ad > 0$ there is unique equilibrium, that is locally asymptotically stable for $d + a > 0$, and may undergo a Hopf bifurcation for $d + a \leq 0$. The Hopf bifurcation locus corresponding to this bifurcation situation is described by:

$$
\begin{aligned}
a^2 e^2 &= \sigma k(c + d + k)^2, \\
(c + ad) + \sigma(a + d) + (a + 3\sigma)k &> 0, \\
(\sigma + d + a + k)(a\sigma + c + (a + \sigma)d + k(a + 3\sigma)) &= \sigma((c + ad) + 3ak).
\end{aligned}
\tag{30}
$$

(The elimination of k from the above expression (30)) is feasible, but leads to a rather cumbersome and uninteresting semi-algebraic expression.) Given that the constant term in (23) is always positive, we observe that a stable focus looses stability and a locally stable limit cycle is created at the Hopf bifurcation. In case $e = 0$, the origin is the only equilibrium, and the Hopf bifurcation takes places at $d = -a$. The averaging results established in Appendix III and IV can be repeated for this situation as well.

The equilibria generated through the saddle node bifurcation also undergo a Hopf bifurcation.

5.4 Control interpretation

The adaptive control law fails to achieve regulation in all but the rather non-interesting case of a stable open loop system. The presence of a DC offset has two main effects, first it acts stabilising in that it enforces a larger feedback gain than would have been the case without it, but it also induces further complex dynamics in the behaviour depending on the actual plant parameters. We conclude that the σ modification limits the adaptive gain too severely to achieve the desired control objective. Moreover we observe that the selection of σ is not trivial. A proper selection, one that avoids as much as feasible complicated dynamics and thus assists in predicting the typical asymptotic dynamics, requires a σ value which is small compared to the dominant time constants and zeroes of the plant's transfer function.

6 Generalisations

In higher dimensions, the system (3) with $n > 1$, all the above discussed bifurcation phenomena re-occur and more complicated ones emerge.
We limit the discussion to the generic bifurcations corresponding to equilibria. The saddle-node bifurcation phenomenon we observed in the planar system and the Hopf bifurcation we observed in the three dimensional example are the main events to consider.
The emphasis is on a control theoretic interpretation.

6.1 Equilibria

For the general case, see equation (3), the equilibria (x_e, y_e, k_e) satisfy the equations:

$$\begin{aligned} x_e &= -A^{-1} b y_e, \\ k_e &= \frac{y_e^2}{\sigma}, \\ \sigma e &= \sigma(-cA^{-1}b + d) y_e + y_e^3. \end{aligned} \tag{31}$$

Provided $d - cA^{-1}b = p(0) + dz(0) \geq 0$ there is but one equilibrium. Notice also that in this case $|y_e| \leq |e|$ and y_e has the same sign as e.
If $d - cA^{-1}b < 0$ there are possibly three equilibria, depending on the magnitude of e. At least one of the equilibria is larger in magnitude than

$$y^* = \sqrt{-\sigma(d - cA^{-1}b)}; \text{ whenever } d - cA^{-1}b < 0. \tag{32}$$

For future reference we denote

$$k^* = {y^*}^2/\sigma = cA^{-1}b - d. \tag{33}$$

There are three equilibria only if

$$|e| \leq e^* = \frac{2(cA^{-1}b - d)^{3/2}}{3\sqrt{3}}. \tag{34}$$

The equation (34) represents the saddle-node bifurcation locus in the bifurcation diagram. When the condition (34) is satisfied, for one equilibrium the output value is larger in magnitude than y^*, this equilibrium we refer to as y_{e1}. Another equilibrium has an output value contained in the interval $-y^*/\sqrt{3} \leq y_{e2} \leq y^*/\sqrt{3}$. The last equilibrium is such that the output value has the opposite sign of y_{e1} and satisfies $y^{*2} \geq y_{e3} \geq y^{*2}/3$.

6.2 Local stability of equilibria

The characteristic equation of the Jacobian evaluated at the equilibria is given by:

$$\begin{aligned} J(\lambda) &= \det\begin{pmatrix} \lambda I - A & -b & 0 \\ c & \lambda + d + k_e & y_e \\ 0 & -2y_e & \lambda + \sigma \end{pmatrix}, \\ &= (\lambda + \sigma)(p(\lambda) + (d + k_e)z(\lambda)) + 2y_e^2 z(\lambda) \\ &= (\lambda + \sigma)(p(\lambda) + dz(\lambda)) + k_e(\lambda + 3\sigma)z(\lambda). \end{aligned} \tag{35}$$

From this it follows that for sufficiently small σ the root loci of the plant $p(\lambda) + (d + k_e)z(\lambda)$ as a function of k_e is a good indicator for what to expect in the phase portrait of the adaptive system. (Again sufficiently small σ means small with comparison to the dominant time constants and zeroes of the plant's transfer function.)

6.2.1 Case 1: Open loop stable plant

The origin is a locally asymptotically stable equilibrium in this case (see equation (35) with $k_e = 0$. If moreover the plant has a strictly positive real transfer function, then the equilibrium is globally asymptotically stable. The local analysis conforms of course. Indeed, in this case $p(\lambda) + (d + k)z(\lambda)$ is a Hurwitz polynomial for all $k \geq 0$. Hence $p(0) + dz(0) > 0$ which confirms that the equilibrium is unique and from equation (35) we observe that it is locally stable.

In general, for stable and minimum phase but not strictly positive real plants, there will be multiple asymptotically stable attractors in the phase portrait, the origin being one of them. See the example (Figure 1) at the end of Section 3 for an illustration. (It can be verified that the domain of attraction of the origin is rather small.)

6.2.2 Case 2: Open loop unstable plant, single positive pole

In this scenario, the open loop plant is unstable with a single positive eigenvalue; the root loci of $p(\lambda) + (d+k)z(\lambda)$ with k viewed as a parameter, has a single (real axis) branch cutting the imaginary axis for $k = k^* = -p(0)-dz(0)$. In this case we have obviously $p(0) + dz(0) < 0$. There may be as many as three equilibria, see equation (34). The equilibrium corresponding to y_{e1} for which $|y_{e1}| \geq y^*$ is also locally asymptotically stable. Indeed we can rewrite the Jacobian (35) in this case as: $J(\lambda) = (\lambda+\sigma)(p(\lambda)+(d+k_e)z(\lambda))+2y_e^2 z(\lambda)$. Because $p(\lambda) + (d+k^*)z(\lambda) = \lambda h(\lambda)$ for some Hurwitz polynomial $h(\lambda)$. It follows that under this scenario with $k_e \geq k^*$, $J(\lambda)$ is Hurwitz.
The equilibrium with y_{e2} is always unstable, because $J(0) = \sigma(p(0)+dz(0)) + 3y_{e2}^2 z(0) \leq 0$.
The equilibrium corresponding to y_{e3} may be either stable or unstable.
Clearly the transition from Case 1 to Case 2 is effectuated via a saddle-node bifurcation, e.g. associated with a variation of the parameter d. The bifurcation value being $d = d_p$ for which $p(0) + d_p z(0) = 0$. This phenomenon is completely captured by the planar example.
But as indicated in the three dimensional example, depending on the plant, the global dynamics may be rather more complicated with possibly co-existing attractors, either multiple locally stable equilibria, but possibly also periodic orbits and even complex attractors, see Figure 10.

6.2.3 Case 3: a single pair of complex conjugate unstable poles

When the open loop plant is unstable with a single pair of complex conjugate eigenvalues, we have that $p(0) + dz(0) > 0$, and there is but one equilibrium. The local stability now depends in a crucial manner on the magnitude of the disturbance e. The transition from Case 1 to Case 3 corresponds to a Hopf bifurcation.
In particular for $e = 0$, the equilibrium is the origin $(x_e, y_e, k_e) = (0,0,0)$ and clearly this equilibrium is unstable, the Jacobian being equal to $J(\lambda) = (\lambda + \sigma)(p(\lambda) + dz(\lambda))$, which by assumption has a single pair of complex conjugate roots with positive real part. In this situation the adaptive system adjust k such that a stable limit cycle comes into existence. Using averaging ideas as in the Appendix IV we may show that the limit cycle, for sufficiently small σ, is well approximated by:

$$\begin{aligned} x(t) &= O(\sqrt{\sigma}), \\ y(t) &= \sqrt{2\sigma k_0}\sin(\omega t + \phi) + O(\sigma), \\ k(t) &= k_0 + O(\sqrt{\sigma}). \end{aligned} \tag{36}$$

Where k_0 and ω are such that:

$$\det\left(j\omega I - \begin{pmatrix} -A & -b \\ c & d+k_0 \end{pmatrix}\right) = 0.$$

As y_e and hence the corresponding k_e increases monotonically with e it follows that under this scenario at some critical level the equilibrium undergoes a Hopf bifurcation. For $|e| > e_H$ (such that $k_e = k_0$) the equilibrium is locally asymptotically stable, for $|e| \leq e_H$ it is unstable and a stable limit cycle exists. Again it is a direct consequence of the center manifold theorem that the three dimensional example captures the essence of the local dynamics. There exists an asymptotically stable periodic orbit for all $|e| < e_H$ and the equilibrium is asymptotically stable otherwise.

6.3 Other scenarios

The transition from a stable plant to an unstable plant via a Saddle-Node or Hopf bifurcation is but the simplest scenario. Obviously, much more complicated bifurcation phenomena are possible. Nevertheless, within the class of linear systems considered, these are the generic bifurcations. The different asymptotic dynamics corresponding to these transitions being explored in the above two and three dimensional examples.

Typically the least structurally stable (and hence most interesting) plants to consider are those linear plants that have all their poles in the origin. In the bifurcation diagram for such adaptive systems (order > 2) we expect to encounter chaotic dynamics.

Moreover as for a typical plant, there may be multiple values of the feedback gain k for which the plant is marginally stabilised, there will be in the phase portrait of the adaptively controlled system a multiplicity of different and co-existing asymptotic dynamics, each corresponding to a different level of adaptive feedback gain.

7 Alternative adaptive gain laws

Clearly the adaptive systems we discussed may exhibit a rather rich variety of asymptotic dynamics. One may wonder, could we design an adaptive gain controller of the form $\dot{k} = f(k, y)$ with smooth f that achieves regulation in a structurally stable manner, i.e. without displaying periodic behaviour and or multiple co-existing asymptotic dynamics for all sufficiently small perturbations of the vector field. (Allowing for perturbations other than the constant input offset considered here.) Unless we are given more prior information about the class of systems which need to be controlled, the answer appears to be negative.

Indeed, the class of systems to be controlled, linear systems with a transfer function characterised by stable zeroes and pole-zero excess of one, is simply too rich. Regardless of the adaptive gain law $f(k, y)$ we use to control the linear system at hand, the closed loop dynamics may exhibit the richness we discussed above.

In order to find a structurally stable adaptive gain law we need to consider adaptation laws of the form $\dot{z} = f(z, y)$ with $k = g(z, y)$ or perhaps $\dot{k} =$

$f(k, y, t)$. These options are presently under consideration.
Alternatively, we may want to invoke further prior information on the systems to be controlled in order to achieve robust adaptive regulation.
Let us observe that the bifurcation analysis has provided us with a lot of information about the behaviour of these systems, and this knowledge can be exploited to design a supervisory adaptive control law to decide on an appropriate feedback gain level. A supervisor could observe the asymptotic dynamics, from a catalog decide in which regime the adaptive system is operating and reset the feedback gain accordingly. An adaptive feedback law building on this principle could take on the form:

$$\dot{k} = -\sigma(k - k_r) + y^2. \tag{37}$$

Where both the value of σ and $k_r > 0$ are set by the supervisor. Appropriate selection rules for these parameters are under investigation.

8 Conclusions

Through a bifurcation analysis, we explored the possible asymptotic dynamics in adaptive high gain feedback applied to linear almost passive systems, when the adaptive law has been modified with the so called σ-modification, $\dot{k} = -\sigma k + y^2$. In the disturbance free case, when the uncontrolled plant does not possess a strictly positive real transfer function, even when the plant is stable and minimum phase, asymptotically periodic and complex dynamics are to be expected. For the less interesting case of plants with strict positive real transfer function the system is regulated.
It became apparent that even in this simple situation the design of the adaptive law, the selection of σ is not trivial. A proper selection requiring some information about the size of the time constants and zeroes of the plant's transfer function.
Despite the fact that the adaptive control law always leads to a bounded system response regardless of the plant, the main alarming observation is that the adaptation leads at best to a control gain which only achieves marginal stability. Hence the adaptation law can never be switched off, or stability may be lost. We conjecture that no smooth time-invariant adaptive gain law of the form $\dot{k} = f(k, y)$ can be constructed which possesses the property of regulation for the class of linear systems with stable zeroes and pole-zero excess of one, in a structurally stable manner.
The results here should be seen as a first step in the re-exploration of the adaptive control question with the aim of understanding the dynamics of adaptive systems as to develop a supervisory control approach.
Finally let us observe that we have not really dealt with the issue of transient performance in gain adaptive systems.

I System representation

For the sake of completeness we explain here how one can arrive at the (1) representation for systems that possess a transfer function of relative degree one and with stable zeroes.
Let us start from the input/output description of the system:

$$P(\frac{d}{dt})y = Q(\frac{d}{dt})u, \tag{38}$$

where $P(\xi)$ is a monic polynomial of degree $n+1$ and $Q(\xi)$ is a polynomial of degree n. We assume that the system is controllable. This corresponds to $P(\xi)$ and $Q(\xi)$ being coprime. Since the degree of $P(\xi)$ is exactly one larger than the degree of $Q(\xi)$, we can write $P(\xi) = \frac{1}{q_n}(\xi - d)Q(\xi) + r(\xi)$ with $\deg r(\xi) < \deg Q(\xi)$. Equation (38) can then be rewritten as:

$$(\frac{1}{q_n}(\frac{d}{dt} + d) + r(\frac{d}{dt}))y = Q(\frac{d}{dt})u. \tag{39}$$

Using elimination theory, [1, Chapter 6] it can be seen that (39) is input/output equivalent to

$$\frac{d}{dt}y = -dy + q_n u_1, \quad Q(\frac{d}{dt})u_2 = r(\frac{d}{dt})y, \quad u = u_1 + u_2. \tag{40}$$

Notice that since $\deg Q(\xi) > \deg r(\xi)$, the second equation in (40) defines a proper input/output relation between y and u_2. Notice also that since $P(\xi)$ and $Q(\xi)$ are coprime, so are $Q(\xi)$ and $r(\xi)$. Therefore the (y, u_2) system is controllable. We can therefore consider the controller canonical state space representation of that system, which is given by

$$\frac{d}{dt}x = Ax + by \quad u_2 = hx, \tag{41}$$

where the eigenvalues of A are the zeroes of $Q(\xi)$ and (A, b) is controller canonical form. Combining (41) with the first and the third equation of (40), we obtain

$$\begin{aligned} \frac{d}{dt}x &= Ax + bu, \\ \frac{d}{dt}y &= -q_n hx - dy + q_n u. \end{aligned} \tag{42}$$

Putting $c = q_n h$ and $g = q_n$ yields (1).

II Normal form representation for the Takens-Bogdanov point in (9)

Without loss of generality we start with the system description (9) with $\sigma = 1$, see **Fact (vi)**. The Takens-Bogdanov points are equilibria with a nil potent Jacobian.

For the system (9) (with $\sigma = 1$) these are given by:

$$y_{TB} = \frac{1}{\sqrt{2}}, \quad k_{TB} = \frac{1}{2}, \quad d_{TB} = -\frac{3}{2}, \quad e_{TB} = -\frac{1}{\sqrt{2}}, \tag{43}$$

and

$$y_{TB} = -\frac{1}{\sqrt{2}}, \quad k_{TB} = \frac{1}{2}, \quad d_{TB} = -\frac{3}{2}, \quad e_{TB} = \frac{1}{\sqrt{2}}. \tag{44}$$

We consider the local behaviour around the latter. The former, due to symmetry will have mutatis mutandis a completely similar normal form. Introduce new variables $(x_1, x_2, \lambda_1, \lambda_2)$ through:

$$\begin{aligned} y &= -\tfrac{1}{\sqrt{2}} + x_1, \\ k &= \tfrac{1}{2} + x_2, \\ d &= -\tfrac{3}{2} - \lambda_1, \\ e &= \tfrac{1}{\sqrt{2}} + \lambda_2 + \tfrac{1}{\sqrt{2}}\lambda_1. \end{aligned} \tag{45}$$

The system (9) described in the new variables takes on the form:

$$\begin{aligned} \dot{x_1} &= \lambda_2 + (1+\lambda_1)x_1 + \tfrac{1}{\sqrt{2}}x_2 - x_2 x_1, \\ \dot{x_2} &= -\sqrt{2}x_1 - x_2 + x_1^2. \end{aligned} \tag{46}$$

We now transform the variables such as to reduce the first equation to a pure integrator. (This also ensures that the linear part for $(\lambda_1, \lambda_2) = (0, 0)$ is in Jordan form.) We use the transformation:

$$\begin{aligned} y_1 &= x_1, \\ y_2 &= \lambda_2 + (1+\lambda_1)x_1 + \tfrac{1}{\sqrt{2}}x_2 - x_2 x_1. \end{aligned} \tag{47}$$

This transformation is invertible in a neighbourhood of the origin contained in the domain $y_1 < \frac{1}{\sqrt{2}}$. The inverse is given by:

$$\begin{aligned} x_1 &= y_1, \\ x_2 &= \frac{-\lambda_2 + y_2 - (1+\lambda_1)y_1}{\frac{1}{\sqrt{2}} - y_1}. \end{aligned} \tag{48}$$

This leads to a system description of the form:

$$\begin{aligned} \dot{y_1} &= y_2, \\ \dot{y_2} &= \lambda_2 + \lambda_1 y_1 + \lambda_1 y_2 + \tfrac{3}{\sqrt{2}}y_1^2 - y_1^3 + \sqrt{2}y_2 \frac{\lambda_2 - (1+\lambda_1)y_1 - y_2}{1 - \sqrt{2}y_1}. \end{aligned} \tag{49}$$

Only retaining terms up to second order in (y_1, y_2) we arrive at a locally equivalent vector field:

$$\begin{aligned} \dot{z_1} &= z_2, \\ \dot{z_2} &= \lambda_2 + \lambda_1 z_1 + (\lambda_1 + \sqrt{2}\lambda_2) z_2 - \frac{3}{\sqrt{2}} z_1^2 + \\ &\quad (2\lambda_2 - \sqrt{2}(1+\lambda_1)) z_1 z_2 - \sqrt{2} z_2^2. \end{aligned} \tag{50}$$

Following Section 8.4.1 in [13], the term linear in z_2 in the z_2-equation can be eliminated by a parameter dependent shift of z_1. The z_2^2 term in the z_2-equation can be eliminated through a time scaling transformation. Finally observe that the $z_1 z_2$ term and the z_1^2 term have both negative sign, which can be made positive through the linear transformation $(z_1, z_2) \to (-z_1, -z_2)$. It follows that the normal form for the Takens-Bogdanov bifurcation in the system (9) is given by:

$$\begin{aligned} \dot{\eta_1} &= \eta_2, \\ \dot{\eta_2} &= \beta_1 + \beta_2 \eta_1 + \eta_1^2 + \eta_1 \eta_2. \end{aligned} \tag{51}$$

III Averaging computations

Consider the system (18) and assume that $\sigma << 1$. Given **Fact (vi)**, this is actually without much loss of generality, as we can always scale the vector field as to satisfy this condition.
Consider the following time varying state space transformation:

$$\begin{aligned} R &= \begin{pmatrix} -a & 1 \\ -c & a \end{pmatrix}, \\ z &= \frac{1}{\sqrt{\sigma}} \exp(-Rt) \begin{pmatrix} x \\ y \end{pmatrix}, \\ \xi &= \frac{k+a+d}{\sqrt{\sigma}}. \end{aligned} \tag{52}$$

Notice that $\exp(Rt)$ is a rotation matrix. Let $\omega = \sqrt{c - a^2}$. Let $b^T = (0\ 1)$. We have that (z, ξ) satisfy the following differential equation:

$$\begin{aligned} \dot{z} &= \sqrt{\sigma}\xi \exp(-Rt) b b^T \exp(Rt) z, \\ \dot{\xi} &= -\sigma\xi - \sqrt{\sigma}(-a - d - z^T \exp(R^T t) b^T b \exp(Rt) z). \end{aligned} \tag{53}$$

This is in standard form to apply averaging techniques ([9]), using $\sqrt{\sigma}$ as a small parameter. The averaged system of equations is given by:

$$\begin{aligned} \dot{z^a} &= \sqrt{\sigma}\xi^a \Gamma z^a, \\ \dot{\xi^a} &= -\sigma\xi^a - \sqrt{\sigma}(-a - d - z^{aT} \Delta z^a). \end{aligned} \tag{54}$$

For a first order averaging approximation one could neglect the term $-\sigma\xi^a$. Nevertheless, because we are interested in an approximation valid over the entire time axis, and because this term provides stability, we carry it along in the averaged equation. For a discussion of the validity of this approach see e.g.[9] Chapter 4.
Here Γ and Δ are given by:

$$\Gamma = \frac{1}{2\omega^2}\begin{pmatrix} \omega^2+1 & -2 \\ c & \omega^2-1 \end{pmatrix} \quad \Delta = \frac{1}{2\omega^2}\begin{pmatrix} c^2 & -c \\ -c & \omega^2+1 \end{pmatrix}. \tag{55}$$

Importantly observe that $\Delta\Gamma + \Gamma^T\Delta = \Delta = \Delta^T > 0$.
Let us consider the following comparison function for the averaged system (54).

$$V(z^a,\xi^a) = \frac{1}{2}(z^{aT}\Delta z^a + d + a) - \frac{\sqrt{\sigma}}{2}\log\frac{z^{aT}\Delta z^a}{-a-d} + \frac{1}{2}\xi^{a2}. \tag{56}$$

Notice that V is positive on the domain $D = \{(z^a,\xi^a) : z^a \neq 0\}$, and achieves a minimum $V = 0$ on the ellipse $E = \{(z^a,\xi^a) : \xi^a = 0;\ z^{aT}\Delta z^a = -a - d\}$. Moreover V grows unbounded whenever its argument approaches the boundary of the domain D.
Assuming that $z^a(0) \neq 0$, we have that along the solutions of the averaged system equations (54) $V(z^a(t),\xi^a(t)) = -\sigma\xi^a(t)^2 \leq 0$. Using LaSalle's invariance principle we deduce that all solutions starting inside D converge to a fixed point in E. Locally this convergence is exponentially fast.
From the results in Chapter 4 [9] it follows that the original system of equations (53) contains an attractive invariant tubular neighbourhood of radius $O(\sqrt{\sigma})$ centred on the ellipse E with a domain of attraction arbitrarily large inside the interior of D. The convergence is exponential. (Considering a larger domain of attraction, within D, may require one to consider smaller values of σ.)
Re-interpreting the result in terms of the original coordinates (18) we obtain:

Theorem III.1 *Consider the system (18) under the conditions $e = 0$, $d+a < 0$ and $c + ad > 0$. Let $x(0) \neq 0$ and $k(0) \geq 0$. For sufficiently small $\sigma > 0$, all solutions satisfy:*

$$\begin{aligned} \limsup_{t\to\infty}|k(t) + a + d| &\leq C\sigma, \\ \limsup_{t\to\infty}|\,(x(t)\ y(t))\exp(-R^T t)\Delta\exp(-Rt)\begin{pmatrix} x(t) \\ y(t)\end{pmatrix} + (a+d)\sigma| &\leq C\sigma^{3/2}. \end{aligned} \tag{57}$$

Here C is some positive constant independent of σ. The matrix R is defined in (52).

Remark III.2 In view of the fact we can rescale the equations such that $\sigma = 1$, see **Fact (vi)**, it is clear that the above Theorem III.1 remains valid for σ not necessarily small. However, obviously the theorem's conclusions are strongest for small values of σ. □

Remark III.3 Consider d as bifurcation parameter. In conjunction with the observation that a regular Hopf bifurcation takes place for $c + ad > 0$ and d passing through $-a$, we conclude from the Theorem III.1 that the Hopf bifurcation is such that as d decreases through $-a$ a stable limit cycle forms whilst the equilibrium becomes unstable. Appendix IV establishes this fact rigorously. □

Finally let us provide the details of the calculations for Δ and Γ. Recall (52), we have

$$\exp(Rt) = \begin{pmatrix} \cos(\omega t) - \frac{1}{\omega}\sin(\omega t) & \frac{1}{\omega}\sin(\omega t) \\ -\frac{c}{\omega}\sin(\omega t) & \cos(\omega t) + \frac{1}{\omega}\sin(\omega t) \end{pmatrix}. \tag{58}$$

It follows that

$$\exp(-Rt)bb^T\exp(Rt) = \begin{pmatrix} -\frac{c}{\omega^2}\sin^2(\omega t) & -\frac{1}{\omega^2}\sin(\omega t)(\omega\cos(\omega t + \sin(\omega t)) \\ -\frac{c}{\omega^2}(\omega\cos(\omega t) - \sin(\omega t))\sin(\omega t) & \cos^2(\omega t) - \frac{1}{\omega^2}\sin^2(\omega t) \end{pmatrix}, \tag{59}$$

from which we readily deduce the expression for Γ. Similarly from

$$\exp(R^T t)bb^T\exp(Rt) = \begin{pmatrix} \frac{c^2}{\omega^2}\sin^2(\omega t) & -\frac{c}{\omega^2}\sin(\omega t)(\omega\cos(\omega t + \sin(\omega t)) \\ -\frac{c}{\omega^2}(\omega\cos(\omega t) + \sin(\omega t))\sin(\omega t) & (\cos(\omega t) + \frac{1}{\omega}\sin(\omega t))^2 \end{pmatrix}, \tag{60}$$

we compute Δ as the average over one period of the above expression.

IV Existence and uniqueness of limit cycle

Consider the adaptive system (18) with $e = 0$ and parameters in the range $c > a^2$ and $d < -a$. Let $\omega^2 = c - a^2 > 0$ as in Appendix III. Introduce the variables $z_1 = x$, $z_2 = -ax + y$, $z_3 = k - d - a$. The system description becomes:

$$\begin{aligned} \dot{z_1} &= z_2, \\ \dot{z_2} &= -\omega^2 z_1 - z_3(z_2 + az_1), \\ \dot{z_3} &= -\sigma z_3 + (z_2 + az_1)^2 + \sigma(d + a). \end{aligned} \tag{61}$$

Introduce now cylinder co-ordinates, $z_1 = \sqrt{\sigma} r \sin(\omega\theta)$, $z_2 = \sqrt{\sigma}\omega r \cos(\omega\theta)$ and $z_3 = \sqrt{\sigma} z$. Denote $\sqrt{\sigma} = \varepsilon$. This leads to the system description:

$$\begin{aligned}
\dot{r} &= -\frac{\varepsilon}{\omega} z r(\cos(\omega\theta) + a\sin(\omega\theta))\sin(\omega\theta), \\
\dot{\theta} &= -1 + \frac{\varepsilon}{\omega^2} z(\cos(\omega\theta) + a\sin(\omega\theta))\cos(\omega\theta), \\
\dot{z} &= -\sigma z + \varepsilon\left((d+a) + r^2(\cos(\omega\theta) + a\sin(\omega\theta))^2\right).
\end{aligned} \tag{62}$$

Utilizing Theorem 1.1 pp261 [14], we deduce that there exists $\varepsilon^* > 0$ such that for all $0 < \varepsilon < \varepsilon^*$ the system (62) possesses an isolated periodic orbit $(r(\theta,\varepsilon), z(\theta,\varepsilon) = (\sqrt{\frac{-2(d+a)}{1+a^2}}, 0) + O(\varepsilon)$. The periodic orbit $(r(\theta,\varepsilon), z(\theta,\varepsilon)$ is continuously differentiable in ε.
The conditions of Theorem 1.1 pp261 [14] are that the equations:

$$\begin{aligned}
0 &= z_0 r_0, \\
0 &= (d+a) + r_0^2\frac{(1+a^2)}{2},
\end{aligned} \tag{63}$$

have a solution for which

$$\det\begin{pmatrix} 0 & -a\sqrt{\dfrac{-(d+a)}{2(1+a^2)}} \\ \sqrt{-2(d+a)(1+a^2)} & 0 \end{pmatrix} = -a(d+a) \neq 0. \tag{64}$$

The latter is ensured by assumption as $a(d+a) < 0$.
For the local asymptotic stability of the periodic orbit (in (r,z)-space), it suffices to observe that the trace of the Jacobian is $-\sigma < 0$. Alternatively, we can exploit the averaging ideas of Appendix III.
From the above we deduce that the system (18) with $a + d < 0$ and $c > a^2$ possesses a locally asymptotically stable periodic orbit for all sufficiently small σ:

$$\begin{aligned}
x(t) &= -\sqrt{\frac{-2\sigma(d+a)}{1+a^2}}\sin(\omega t) + O(\sigma) \\
y(t) &= \sqrt{\frac{-2\sigma(d+a)}{1+a^2}}(-a\sin(\omega t) + \omega\cos(\omega t)) + O(\sigma) \\
k(t) &= -(d+a) + O(\sqrt{\sigma}) \\
\omega^2 &= c - a^2 > 0
\end{aligned} \tag{65}$$

V How to remove DC offset error

As is well known, the DC offset error can be removed from the output through integral action. It is probably well less known that the resulting dynamical system remains in the same class of systems studied in this paper. Hence our analysis also applies to these systems where integral action is present. Starting from the system description:

$$\begin{aligned} P(\frac{d}{dt})y &= Q(\frac{d}{dt})u, \\ \frac{d}{dt}y_I &= y, \\ u &= -k(y+\alpha y_I)+e. \end{aligned} \tag{66}$$

Here $\alpha > 0$ is the relative integral action applied in the control and y_I is the integral of y. Considering a new output $\eta = y + \alpha y_I$ we can equivalently represent the control system as:

$$\begin{aligned} \frac{d}{dt}P(\frac{d}{dt})\eta &= (\frac{d}{dt}+\alpha)Q(\frac{d}{dt})u, \\ u &= -k\eta + e, \\ (\frac{d}{dt}+\alpha)y &= \frac{d}{dt}\eta. \end{aligned} \tag{67}$$

This system (67) is clearly of the same form as the system (1). We can now proceed as in the rest of the paper, with the added flexibility of using y and/or η in the adaptive law that defines k. The benefit of the integral action is that whenever the adaptive system regulates to an equilibrium that then the output y is regulated to zero. Clearly this is important from a control perspective, but it does not alter in any significant sense the overall dynamics of the adaptive system. Moreover as regulation to an equilibrium is by no means guaranteed, nor the most likely event, integral action looses its attraction.

Acknowledgments
Iven Mareels acknowledges the School of Mathematical Sciences, Exeter for its support for his visit in 1997, during which this paper was initiated and the support of NWO for his visit in January 1999 to the University of Twente, during which period this paper was finalised. Achim Ilchmann acknowledges the support received by the University of Melbourne for his visit to the Department of Electrical and Electronic Engineering in 1998.

References

[1] J.W.Polderman, J. C. Willems, *An Introduction to Mathematical Systems Theory: A Behavioral Approach*, Springer, Berlin 1998.

[2] I. Mareels and J.W. Polderman, *Adaptive Systems: An Introduction*, Birkhäuser, Boston, 1996.

[3] I. Mareels, A simple selftuning controller for stably invertible systems *Systems & Control Letters*, Vol 4, pp5-16, Feb.1984.

[4] A. Ilchmann *Non-Identifier-Based High-Gain Adaptive Control*, Springer-Verlag, London, 1993.

[5] A. Ilchmann, Adaptive λ-tracking for polynomial minimum phase systems; *Dynamics and Stability of Systems* Vol 13, pp341-371, 1998.

[6] A. Ilchmann, E.P. Ryan, Universal λ-tracking for nonlinearly-perturbed systems in the presence of noise, *Automatica* Vol 30, pp 337-346, 1994.

[7] B. Mårtensson, J.W. Polderman, Correction and simplification to: The order of a stabilizing regulator is sufficient a priori information for adaptive stabilization, *Systems & Control Letters*, Vol 20, pp. 465-470, 1993.

[8] P. Ioannou, P.V. Kokotović, Instability analysis and improvement for robustness of adaptive control, *Automatica* Vol 20, pp 583-594, 1984.

[9] J.A. Sanders, F. Verhulst, *Averaging methods in nonlinear dynamics*, Springer verlag, Applied Mathematical Sciences, Vol 59, New York 1985.

[10] C.I. Byrnes, J.C. Willems, Global Adaptive Stabilization in the Absence of Information on the Sign of the High Frequency Gain. Part I: Analysis and Optimization of Systems. In: *Proc. 6th Intern. Conf. on Anal. and Optimiz. of Syst.*, A. Bensoussan and J.L. Lions Eds. Springer Verlag, pp 49-57, New York, 1984.

[11] H. Kaufman H., I. Bar-Kana, K. Sobel, *Direct Adaptive Control Algorithms:Theory and Applications*, Springer-Verlag, New York, 1994.

[12] A. S. Morse, Recent problems in parameter adaptive control, in *Outils et Modèles Mathématiques pour l'Automatique, l'Analyse de Systèmes et le Traitment du Signal*, I.D. Landau, Ed., pp.733-740 (Editions du CNRS 3, Paris)

[13] Y. Kuznetsov, *Elements of applied bifurcation theory*, 2nd ed, Applied Mathematical Sciences, Vol 112, Springer, New York 1998.

[14] J.K. Hale, *Ordinary Differential Equations*, 2nd ed, Krieger, Malabar Florida, 1980

3

Control Problems in Multi-Modal Telepresence Systems

M. Buss and G. Schmidt

Institute of Automatic Control Engineering, Technische Universität München, D-80290 München, Germany, E-mail: {M.Buss, G.K.Schmidt}@ieee.org

Summary. In this chapter we discuss challenging control problems in multi-modal telepresence and teleaction systems. The typical system structure consists of a human operator, the possibly distant (and differently scaled) teleoperator and various types of barriers between them. Fundamental control modes to organize and distribute control between human and system are discussed. Potential applications and enabling technologies are sampled. A subjective listing of central control issues is given including multi-modal control, impact of time-delays, local control measures, real world, virtual environment (VE) and augmented environment interaction, sensory substitution, etc. An experimental prototype system for multi-modal telepresence and teleaction developed in our laboratory is presented and some experiments further illustrating typical control challenges are highlighted.

1 Introduction

This chapter reviews a subjective selection of topics in the area of telepresence systems with an emphasis on multi-modal control problems. Target systems also include classical teleoperation systems, which recently most often cover multi-modal aspects. Of particular interest is not only passive telepresence, but the more important combined issue of telepresence and teleaction. The term teleaction is to express the importance of active manipulation by the human operator in the remote environment. Here, remote environments include possibly distant and/or scaled physical environments, virtual environments—VEs[1] and augmented realities.

[1] The term "virtual reality—VR" is often used instead of virtual environment, which is a contradiction of words and therefore preferably avoided.

One of the central issues in modern telepresence systems is multi-modality in the *human-system interface—HSI*[2] accompanied by appropriate sensing techniques at the teleoperator site comprising theoretically all the human senses. In current technical applications most important and only partly realized are the visual, auditory, and haptic — i.e. kinesthetic, tactile and temperature — senses. Here, the haptic modality comprises kinesthetic/force *and* tactile/temperature feedback and therefore refers to the combination of all senses stimulated by direct mechanical interaction.

Recently, systems with multi-modal human interaction in VEs or augmented realities are of increasing interest to the research community. One of the most important enabling factors for this development are immense advances in multimedia technology. Multi-modality is the extension of multimedia in that the kinesthetic and tactile senses are included as well. Feedback data to the human operator in multi-modal telepresence systems is often artificially computed in a *virtual environment engine* and may be augmented or fused with data received from the teleoperator sensors. The VE engine performs real-time *multi-modal rendering*, which is a generalization of the term "rendering" often used in the area of computer graphics. It is to express that the goal is to render the virtual environment not only in graphics, but to include kinesthesia, tactility, temperature, audio, and eventually smell in the rendering process.

In telepresence and even more so in teleaction systems closed-loop control problems arise in all the implemented modalities. The human operator acts as the controller and uses the multi-modal HSI to control the possibly distant teleoperator or a virtual implementation thereof. Physical interaction between the teleoperator and objects in the remote environment is measured by sensors and fed back to the human operator using modality dependent feedback devices in the HSI. In case of VE interaction the feedback signals are computed by a multi-modal VE engine in real-time.

For the organization of the chapter: After this introduction a prototypical multi-modal telepresence and teleaction system — abbreviated TPTA system hereafter — is presented and major control modes for interaction control of the human operator and teleoperator are discussed. Typical barriers and possible application areas are mentioned. Section 3 lists and briefly discusses control challenges in TPTA systems. In Section 4 the focus is on a prototype experimental multi-modal TPTA implementation in our laboratory. By discussing features of different experiments some of the control problems in multi-modal teleaction are highlighted from an experimental point of view and first novel solutions are presented.

---

[2] The term human-system interface (HSI) is to generalize all other forms of e.g. human-machine, human-robot interface etc.

2 Multi-Modal Telepresence and Teleaction Systems

2.1 General System Structure

The *general structure of a multi-modal telepresence and teleaction (TPTA) system* is depicted in Figure 2.1. On the operator-site the human operator gives multi-modal command inputs to the HSI using motion, voice or symbolic input devices. The commands are transmitted to the executing teleoperator on the remote-site across barriers. The teleoperator — often an executing robotic system such as a mobile service robot — is controlled according to the commands received from the human operator. Sensors mounted on the teleoperator gather data about the interaction between the teleoperator and the environment. Typically visual, acoustic, force and tactile sensors are employed. Measurement data is transmitted back to the human operator and displayed[3] using modality dependent hardware in the HSI. Clearly, this control loop is closed via the human operator.

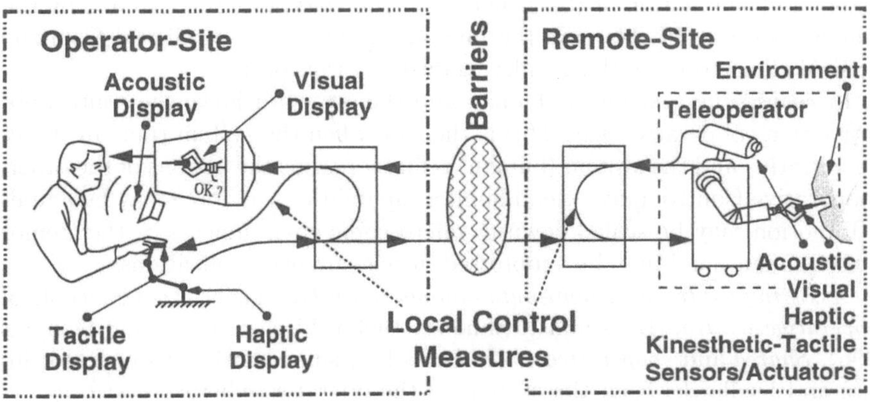

Fig. 2.1. Multi-Modal Telepresence and Teleaction (TPTA) System.

A central issue is the degree of coupling between the human operator and the teleoperator. If the operator gives symbolic commands to the teleoperator by pushing buttons and watching the resulting action in the remote environment the coupling is *weak*. The coupling is comparably *strong* for the kinesthetic modality in a bilateral teleoperation scenario. Here, most often the motion (force) of the human operator is measured, communicated and imposed on the teleoperator. On the other hand the resulting forces (motion) of the teleoperator in the remote environment are sensed, communicated and fed back to the human operator.

In what follows, common definitions of important terms are given and principle control modes for TPTA systems are listed. We will also focus on

[3] Again the term *multi-modal display* includes the conventional terms of *graphic display* and *haptic display*.

what kind of barriers may be encountered in TPTA systems, on potential application areas and enabling technologies for the current and to be expected rapid development in this field.

2.2 Control Modes

The literature on TPTA systems distinguishes between *remote*, *shared*, *co-operative*, *assisting*, *semi-autonomous*, *symbolic* and *trading control*, see e.g. (Anderson 1996; Buss 1994; Buss and Hashimoto 1993b; Fischer, Buss, and Schmidt 1996; Johannsen 1993; Sheridan 1992).

Remote control as widely applied in consumer electronics or modern toys is a special case of Figure 2.1 with straightforward continuous control of some of the teleoperator actuators without feedback to the human, except by direct visual observation of the teleoperator without any barrier.

Shared control means the sharing of *different* control aspects between the human operator and the technical system (Sheridan 1992). An example is a tele-grinding task, where the grinding force is best controlled locally by the system and the motion of the grinder over the workpiece is controlled by the human operator. In this case the human operator is also relieved from the physical effort to press the grinder against the workpiece.

In *cooperative control* the human and the system achieve the control objective in a cooperative way. This is the case when the system (human) takes the initiative and the system (human) refines or trims the actions of the other. Cooperative Control may extend human capabilities, e.g. force may be amplified, motion may be scaled down to microscopic environments or the tremor of the human hand may be suppressed in tele-surgery applications.

Assisting control and *semi-autonomous control* are generalized concepts of *cooperative control* (Buss and Hashimoto 1993b; Fischer, Buss, and Schmidt 1996). *Shared* and *cooperative control* mostly assume continuous operator inputs and feedbacks from the system to the operator, whereas *assisting* and *semi-autonomous control* rely also on the communication of symbols between human and system. Teleoperator task goal objectives as which object to pick up, may be specified by voice input and simultaneously cooperative control may be employed to actually pick up the desired object.

Semi-autonomous control means that some of the subtasks are executed completely autonomously and the human operator only interacts with the teleoperator from time to time to control the overall task execution.

Symbolic control or symbolic teleoperation uses symbols or a language definition to specify task goals to a more or less autonomous teleoperator (Rest, Fleureau, and Marcé 1997; Stein and Paul 1994; Zelek 1997).

Trading control refers to situations, where the system backs the human operator up, i.e. normally the human operator has all control, while in case of unreasonable operator action the system takes over. A computer completely taking over control with no human interaction is a special case of trading control. To realize *trading control* the central issue is to decide *when* the system should take over the control initiative, whereas for *shared control* the question is *what* control aspects to share.

It is important to remember that *shared control* and *cooperative control* are *not* necessarily hierarchical control architectures (Sheridan 1992); as opposed to supervisory control. This means that sometimes the human operator may be lower in the control hierarchy than the teleoperator and vice versa. The other control modes are more hierarchical with the initiative being with the human operator mostly; except for *trading control*, where the computer decides when to take over.

Local control refers to local control measures taken on the operator *and/or* remote sites. *Semi-autonomous* local control is usually used for the teleoperator. To stabilize kinesthetic teleaction systems local measures are required on both sites as will be explained in greater detail in Section 3.4. Clearly, *local control* reduces the degree of coupling between operator and teleoperator over the barrier.

Under *remote control*, the degree of coupling between operator and teleoperator is *extremely weak*. *Shared control* is *weakly* coupled also, since control aspects are different between operator and teleoperator, i.e. there is no direct feedback signal from the task aspect the teleoperator performs automatically. *Cooperative control* and *assisting control* naturally require *strong* coupling, whereas *symbolic* or *semi-autonomous control* are *weakly* coupled control modes. Important questions to be answered by future research are: What amount of coupling is best to achieve the task or if a time-varying degree of coupling is useful and how it may be implemented. The possible degree of coupling also strongly depends on the type of barrier. In a time-delay environment a strong kinesthetic coupling between operator and teleoperator may be difficult to achieve, cf. Sections 3.2 and 3.4.

A more general question is how to manage and adjust the control mode (possibly depending on the task) of the TPTA system to one of the above modes of remote, shared, cooperative, assisting, trading, symbolic, semi-autonomous control and how to choose the degree of coupling appropriately. Such *dynamic control mode management* may be important in case of time-varying barriers.

2.3 Types of Barriers

Telepresence technology is an absolute requirement to overcome the *barrier* between the human operator and teleoperator in TPTA systems, see Figure 2.1. For many types of barriers (e.g. hazardous environments) direct human operator action in the remote environment is difficult or impossible to realize by other means. One obvious barrier is distance, i.e. the operator and teleoperator environments are separated by up to several minutes or hours of travel. Telepresence can reduce the need for expensive and time consuming transport of human expert operators.

Restricted accessibility e.g. requirements to wear special clothing and to cross air locks to access clean-rooms are another type of barrier. Certain remote environments are hazardous (e.g. radioactive) to humans and therefore telepresence and teleaction technology is absolutely necessary. Ergonomic considerations can be another motivation to employ telepresence technology, e.g. some surgery procedures require the surgeon to maintain a stressful and even painful body posture sometimes for several hours, eventually causing fatigue and back problems in the long term.

Scaled remote environments also require telepresence. An application example is the assembly of micro or miniature mechatronic components. Currently the human operator assembles the devices mostly under the microscope without adequate force feedback. Here a scaled telepresence system would largely simplify the assembly process while also increasing accuracy by force feedback.

In some machining tasks it would be a great advantage to have a "look under the hood", meaning that the casing of the machine is virtually taken off and the operator or maintenance person can actually see — and feel if multi-modal — the interior of the running machine in a VE. The casing of the machine in this case is another type of physical barrier disabling operator access. A similar situation arises in *minimal invasive surgery.*

2.4 Applications

Application areas of telepresence and teleaction systems are countless, to name only a few: tele-programming, tele-installation, tele-diagnosis, tele-service, tele-maintenance, tele-manufacturing, miniature or micro mechatronics, inner and outer space operations, tele-teaching, tele-medicine, tele-surgery, tele-shopping, etc.

2.5 Enabling Technologies

First prototypes and applications of telemanipulation systems were located in adjacent rooms only separated by a few meters. Mostly the remote environment was directly visible for the operator; in some applications through protective glass if required. The master and slave manipulators were mechanically coupled e.g. by tendons. Later the mechanical coupling was replaced by computer control and visual communication enabled the environments to be widely separated like in the early space teleoperation experiments.

In recent years the situation has changed drastically. The cost for computers capable of performing high-speed computations more complex than ever before is comparably negligible. Computer graphics capabilities to visualize 3D (virtual) environments in real-time are becoming available for steeply decreasing cost. Worldwide communication is practically guaranteed and taken for granted by everybody. Various communication channels, e.g. telephone, ISDN, ATM, Internet, global live television via communication satellites are available. The bandwidth of global and national communication backbones is growing at an exponential rate. Software development for graphical user interfaces as well as control systems has been largely standardized and simplified.

In summary, we can formulate the conjecture that the enabling technologies — high-speed computational, graphical capabilities and high-bandwidth communication — for TPTA systems have become or soon are to be available at reasonable cost.

2.6 Hardware Devices

In recent years various hardware devices for the design of TPTA and VE systems have been proposed and some devices are now commercially available, see `http://www.3dsite.com` and (Barfield, Hendrix, Bjorneseth, Kaczmarek, and Lotens 1995; Burdea and Zhuang 1991; Burdea 1996; Brooks 1990; Shimoga 1993a; Shimoga 1993b).

Three dimensional *visual feedback* to the human operator is achieved by displaying separate images to both eyes. A simple technique uses shutter glasses with coordinated image display on the computer monitor and alternating transparency of the glasses. Head-Mounted-Displays (HMDs) are capable of displaying both images simultaneously directly in front of the eyes with a resolution of up to 1280 $\times$ 1024 pixels. The operator feels highly immersed, however, there are a few drawbacks in that the operator cannot see the surrounding environment when wearing an HMD or a wearable computer device; this may lead to dangerous operator behavior or operator sickness (Mann 1998). The heavy weight of commercially available HMD devices is uncomfortable when wearing them for long periods of time. Polarized glasses together with video beaming devices displaying two polarized images is another possibility to generate 3D visual impressions.

Immersive acoustic 3D feedback systems and methodologies are less developed (Doel and Pai 1998; Hendrix and Barfield 1996; Hollander 1994). One commercially available surround sound product is the Convolvotron by Crystal River Engineering (Wenzel, Wightman, Kistler, and Foster 1988).

Devices for *kinesthetic feedback* to the operator are numerous. More common are sensing devices without force feedback like e.g. the data-glove or data-suit. Force feedback is usually limited to the human hand and arm using an exoskeleton to produce forces and moments. A more popular form of force feedback devices are multi-DOF robotic manipulators held by the human operator's hand; a popular example is the PHANToM device (Massie and Salisbury 1994). Some research groups also use standard robotic manipulators equipped with a force/torque sensor, e.g. (Kosuge, Fujisawa, and Fukuda 1993). The DeKiFeD3 developed in our laboratory (see Section 4.2) closes a gap between smaller and weaker devices like the PHANToM and an industrial robot, also reducing the possibility of operator injury. Probably the only commercial product to generate force feedback at several human fingers is the CyberGrasp by Virtual Technologies (`http://www.virtex.com`). As an example of low-cost products with reduced functionality and therefore very simplistic force feedback capabilities we mention force feedback joysticks for computer games becoming available recently.

Prototypes of devices for *tactile feedback including temperature* have been proposed (Caldwell, Lawther, and Wardle 1996; Kontarinis and Howe 1995). A commercially available product is the CyberTouch by Virtual Technologies using vibrating voice coils.

2.7 Methodologies

In this rather young field of multi-modal telepresence generally applicable methodologies and technological design paradigms are still lacking. In our opinion advances in the field should be in four major directions: 1. technological advances in sensing and actuation devices for teleoperator and the HSI; 2. computational software and hardware architectures; 3. real-time applicable algorithms; 4. control concepts including variable time-delay and variable teleoperator-environment contact situations.

In the previous section we have mentioned some of the *multi-modal feedback hardware devices*. Most of the development is very recent and there is ample room for technological improvement of the available device proposals. The most important advances are to be expected in the mechatronic and miniature mechatronic sensing and actuation technologies considered crucial for the implementation of more sophisticated feedback devices.

Computational hardware and software is developing at a rapid pace in general. Advances to be expected soon are mainly due to continually increasing computational and graphical rendering capabilities. However, what is missing are custom computational architectures combined with efficient software implementations specially designed for virtual environment applications. To mention a specific but hypothetical example: an abstract virtual world modeling language could be combined with a target compiler for a specially designed computing architecture to perform *multi-modal rendering* of virtual and augmented environments.

The third important methodology is the development of *real-time capable algorithms* for multi-modal operator presence in virtual and augmented environments. A very important but little developed area is haptic rendering (Kunii, Buss, and Hashimoto 1995; Kunii and Hashimoto 1996; Basdogan, Ho, and Srinivasan 1997). Haptic rendering has to be performed at high sampling rates of several 100Hz, which makes real-time computation of non-simplistic VEs a challenging undertaking (Nahvi, Nelson, Hollerbach, and Johnson 1998; Thompson, Johnson, and Cohen 1997; Thompson, Nelson, Cohen, and Hollerbach 1997). Haptic and graphical rendering of deformable objects is an even more complex implementation goal, cf. Section 4.4.4.

After a summary of the overall TPTA system architecture, the discussion of principle control modes, technological challenges and major research directions, we will next focus on the fourth key methodology of control concepts. The following section discusses in greater detail some of the control and control design issues encountered in the multi-modal control loop including the human operator.

3 Control Challenges

The multi-modal control loop of TPTA systems is shown in Figure 3.1. From the human operator the multi-modal command consisting of voice, haptic (force/motion) and symbolic input is transmitted across the barrier to the teleoperator. Simultaneously the command is also input to a model of the teleoperator and remote environment implemented in a virtual environment for local feedback. Data measured by teleoperator sensors results in multi-modal feedback to the human operator across the barrier. Multi-modal feedback consists of 2D or 3D visual, mono/stereo acoustic, haptic (force, motion, tactile, temperature) and symbolic information. The remote local control loop using a human operator model is also shown.

Feedback to the operator can also be augmented, i.e. remote data is augmented and fused with data from a remote environment model computed in the VE engine. Augmentation on the remote site uses human control expertise from a human operator model to support local control of the teleoperator. Augmentation is possible for all mentioned human modalities, but most established are kinesthetic and (predictive) visual augmentation.

In the following the central problems in this multi-modal control loop are briefly outlined. For details the reader is referred to the references and the experiments in Section 4, where we present more details on some of the basic control issues. A more specialized survey of control problems focusing on the haptic modality is given by (Salcudean 1998).

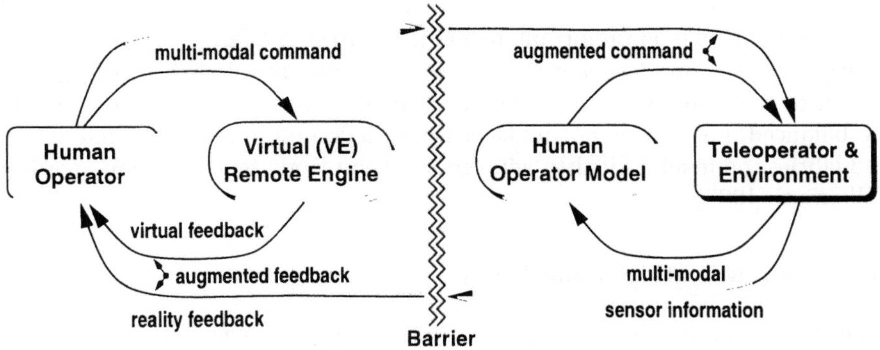

Fig. 3.1. Multi-modal control loop in TPTA systems.

3.1 Multi-Modal Control

Telepresence systems emphasizing the technical realization of one of the human sensing modalities have been thoroughly investigated. The feedback choice to the human operator in distant teleoperation with significant time-delay (e.g. space telerobotics) usually is the *visual modality.* Commands to the teleoperator are taken from continuous motion measurement input devices without force feedback from the teleoperator. Alternatively, the commands are symbolic by keyboard or voice input. In both cases the control loop is weakly coupled with respect to haptic presence. A much better telepresence and teleaction quality can be achieved with strong coupling and inclusion of the kinesthetic and tactile modalities.

The challenge is to realize real multi-modal control with a strongly coupled control-loop using an appropriate mix of modalities. The basic and most important modalities are kinesthetic, tactile, each in closed-loop, and auditory as well as 3D visual feedback from the teleoperator site. Very little has been reported about experimental realizations of all these modalities together in one multi-modal control loop and in integrated systems.

It is clear that the required and tolerable closed-loop characteristics strongly depend on the type of modality. It is widely accepted that for bilateral teleoperation high sampling rates with a time-delay < 50ms are advantageous. It is known from space applications that a time-delay of 0.5-2s in the visual modality is acceptable if a local predictive visual display is implemented and an increased task completion time is not crucial.

One immediate problem is that the modalities require different frequency bandwidths in which the human operator is able to process modal sensations, e.g. kinesthetic up to 10-60Hz, tactile up to 100-800Hz, acoustic $>$5000Hz, visual 30Hz for motion and approximately 3-10Hz for perception (Brooks 1990; Card, Moran, and Newell 1983; Shimoga 1993a). A comparison of human sensory capabilities with the feedback quality achieved by available hardware is given in (Barfield, Hendrix, Bjorneseth, Kaczmarek, and Lotens 1995). Hence, the amount of communication data across the barrier also strongly depends on the modality.

When establishing a communication link between the human operator and the teleoperator it is important to implement the modality control-loops in a way that intermodal causal consistency is guaranteed thus avoiding confusion of the human operator. Further, the quality of the modalities has to be balanced, i.e. it does not make sense to degrade the quality of one of the modalities too much as it then gets ignored completely by the human operator (Wermuth 1998).

3.2 Communication Time-Delay

The crucial communication aspect is *time-delay* often also termed *communication latency*, which is usually time-varying in the currently available communication networks. A delayed visual modality (delay < 1s) decreases the performance of the human operator in teleaction systems a little only, especially when predictive displays are implemented on the operator site. It takes much longer to complete a given task because the human operator waits for most recent images from the teleoperator site. The problems coming along with time-delays in the acoustic modality are worse. Sound is heavily distorted even by very short pauses and the understanding of speech becomes nearly impossible.

The problems resulting from time-delay in kinesthetic control-loops are particularly serious. Without extra measures in the control-loop it is very likely to become unstable. Furthermore, operator task performance decreases rapidly as time-delay increases in the kinesthetic feedback.

When implementing telepresence and teleaction systems with available communication technology, some limitations in communication bandwidth and latency are unavoidable. Typical time-delays of inner space applications are in the order of 0.5-2s. In terrestrial applications the problem is somewhat relaxed but still exists. Especially, when choosing widely available communication infrastructures like e.g. the Internet, intercontinental time-delay varies between 100ms up to several minutes (Denk 1997). Bandwidth limitations are a crucial issue too. Even for commercial communication technologies like ISDN digital phone lines the service providers cannot guarantee communication channel characteristics such as a fixed and guaranteed time-delay.

It needs to be mentioned that this area is currently developing and changing at an extremely high pace. New communication technologies with Quality-of-Service (QoS) are becoming available and are under current development. The intention of QoS is to be able to guarantee bandwidth and latency on demand. This is important if telephone, video-conferencing and digital packet oriented communication for electronic mail and the web are to be implemented on the same digital communication network. QoS may turn out to be an essential requirement for high-quality TPTA systems.

3.3 Bilateral Kinesthetic Teleoperation

Specializing the general multi-modal control loop of Figure 3.1 to *bilateral (kinesthetic) teleoperation* results in the simplified two-port model shown in Figure 3.2. The (possibly nonlinear) mechanical environment impedance Z_e relates the teleoperator velocity v_e to the force f_e by[4]

$$f_e = Z_e(v_e) \ . \tag{3.1}$$

The *teleoperator interface* is used to transmit this environment impedance Z_e to the operator site, where it is felt by the human as the transmitted impedance Z_t. The force f_h and velocities v_h of the human operator's hand are related by this impedance as $f_h = Z_t(v_h)$. Transparency — meaning that the teleoperator interface of Figure 3.2 is totally transparent as if it were not existent — is defined by impedance matching, i.e. the human operator feels exactly the mechanical impedance of the teleoperator environment. Formally, transparency is written as

$$Z_t = Z_e \ . \tag{3.2}$$

It has been shown that transparency requires the teleoperator interface to be of the so-called four-channel architecture depicted in Figure 3.3. The forces f_h^*, f_e^* are external to the human, teleoperator, respectively. The blocks are

[4] Here the more common convention of the force-voltage, velocity-current (force-effort, velocity-flow) analogy is preferred from its dual.

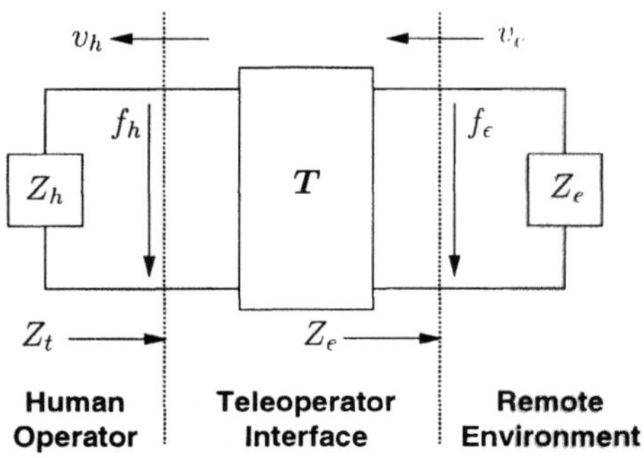

Fig. 3.2. Two-port model of bilateral teleoperation.

the HSI (master) device impedance Z_m, its velocity v_m equal to the human hand velocity v_h, the local master controller C_m, the human operator's hand impedance Z_h, the teleoperator (slave) impedance Z_s, its velocity v_s equal to the contacted environment velocity v_e, the local teleoperator (slave) controller C_s, the remote environment impedance Z_e. The four blocks C_1, C_2, C_3, C_4 denote the four communication channels for velocity forward, force backward, force forward, velocity backward, respectively; note that all blocks in Figure 3.3 can be nonlinear and may involve dynamics. It has been shown that all four channels are needed to achieve transparent teleoperation (Lawrence 1993), which requires that the motions *and* forces need to be measured on both the operator and teleoperator site.

In the linearized case (with $f_e^* = 0$) the teleoperation system considered as a two-port is characterized by the general hybrid matrix formulation in the Laplace domain as

$$\begin{bmatrix} f_h(s) \\ v_h(s) \end{bmatrix} = \begin{bmatrix} H_{11}(s) & H_{12}(s) \\ H_{21}(s) & H_{22}(s) \end{bmatrix} \begin{bmatrix} v_e(s) \\ -f_e(s) \end{bmatrix} . \tag{3.3}$$

Solving for f_h and v_h in terms of v_e and f_e, using (3.1) in the linearized version $f_e(s) = Z_e(s)v_e(s)$, eliminating v_e and dropping the Laplace arguments for convenience, yields

$$f_h = Z_t v_h = (H_{11} - H_{12}Z_e)(H_{21} - H_{22}Z_e)^{-1} v_h . \tag{3.4}$$

Some insights about the block parameters H_{ij} are given in (Lawrence 1993). The general teleoperator architecture shown in Figure 3.3 is described in the form (3.3) with

$$H_{11} = (Z_m + C_m)D(Z_s + C_s - C_3C_4) + C_4 \tag{3.5}$$
$$H_{12} = -(Z_m + C_m)D(I - C_3C_2) - C_2 \tag{3.6}$$
$$H_{21} = D(Z_s + C_s - C_3C_4) \tag{3.7}$$
$$H_{22} = -D(I - C_3C_2) , \tag{3.8}$$

where $D = (C_1 + C_3Z_m + C_3C_m)^{-1}$. Substituting (3.5) through (3.8) into (3.4) one can obtain a general expression for the transmitted impedance Z_t, which can then be compared to the desired impedance or to alternative control algorithm implementations.

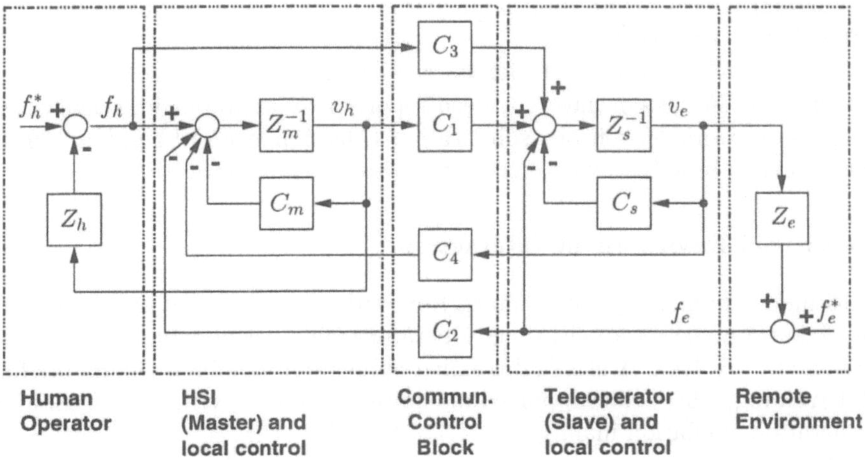

Fig. 3.3. Block diagram of a four-channel bilateral teleoperation system.

3.4 Local Control Measures

Local control uses models of the human operator, teleoperator and the environment to implement local feedback on the operator *and/or* remote-site, see Figure 3.1. Such local feedback loops may help to avoid instability caused e.g. by a communication time-delay and are often used to extract parameters of task goals and the environment. Another already mentioned possibility is to augment data from local models to substitute delayed teleoperator data. A predictive visual display is one such approach. The more general term of *modality augmentation* comprises this feature and the proposal to substitute missing data for one modality by other sensations.

In a bilateral kinesthetic teleoperation scenario local feedback has been used successfully to achieve stability (in fact passivity) of the communication link by interpreting it as a lossless two-port (Anderson and Spong 1989; Baier, Buss, and Schmidt 1997; Kosuge, Murayama, and Takeo 1996; Niemeyer and Slotine 1998), cf. Section 3.3. Still there remain problems in this approach, e.g. the requirement of knowledge about the generalized mechanical environment impedance Z_e, which can be a-priori or by online estimation techniques. Nevertheless, the proposed solution improves performance in time-delay environments significantly. Another idea employs the concept of a virtual tool with the operator and teleoperator exerting forces on either end of the tool (Kosuge, Itoh, Fukuda, and Otsuka 1995).

Local feedback for other modalities than kinesthesia is a topic remaining mainly open for research. A visual predictive display has been proposed using first a wire-frame and later a solid model image generated by computer graphics hardware and then blended or overlayed with the camera image from the remote site (Bejczy, Kim, and Venema 1990; Hirzinger 1993; Kim 1996). Others have proposed a predictive information display (Mitsuishi, Hori, and Nagao 1994). Very recent and exciting yet to be published work by our colleagues in Munich aims at realistic texture overlay techniques (`http://www.lpr.ei.tum.de`).

Local model-based feedback for the tactile and auditory senses in the context of a TPTA system have not been reported so far to our knowledge; an exception is work on sensory substitution (Massimino and Sheridan 1994), where force is replaced by artificially generated sound, cf. Section 3.6.

3.5 Virtual Environment Interaction

Multi-modal human operator interaction with purely virtual environments has various important applications. Operator training is possible without risk for humans involved. A classical training application are flight simulators for pilot training, where the supported modalities have been visual feedback and acceleration emulation mainly.

Medical applications like e.g. practice runs of complicated surgical procedures are being developed for surgeon training (Satava and Jones 1997). Clearly, multi-modal and realistic feedback to the surgeon will be important to make the training useful. In particular the realization of high fidelity and highly realistic haptic feedback is an important but still widely open problem in this area.

Tele-shopping relying on multi-modal presence in virtual catalogues may become an important application area in the near future, provided that the multi-modal hardware devices in the HSI become available at prices affordable for customers.

Another application of VE interaction is the extraction of operator expertise, its transfer and implementation for semi-autonomous local teleoperator control, cf. Figure 3.1 and Section 3.10.

3.6 Sensory and Modality Substitution

The basic idea of sensory or modality substitution is to provide a proband with information, that is hard or impossible to convey with the available devices in the HSI, by stimulating alternative modalities. In medical terms stimuli substitution is equivalent to inadequate stimulation because of the selectivity of senses. A frequently used example of an inadequate stimulus is the excitation of the rods and cones of the retina by rubbing the eyeball strongly with a finger, thereby producing visual impressions (Handwerker 1989). A method of stimuli substitution applicable in feedback systems is electrical (and therefore inadequate) stimulation of the mechanoreceptors on the human hand (Kaczmarek, Tyler, and Bach-y-Rita 1994).

The success of symbolic sensory substitution with the example of force feedback assumed not to be possible by the hardware and therefore substituted by acoustic feedback has been reported (Massimino and Sheridan 1994). In the experiments probands were provided information on the contact force magnitude by the volume of a sound with 1000Hz constant frequency or by the amplitude of vibrations on the fingertip with a constant frequency of

250Hz. Although the information was conveyed by stimulation of different modalities, the probands were able to fulfill tasks requiring information on contact forces. Other researchers have tried to convey contact conditions and interaction forces of virtual objects by changing their color.

3.7 Auditory Control

Transmission of the acoustic modality from the remote site to the human operator is readily realizable. The acoustic impression can be used as augmented information about the teleoperator actions in the remote environment. However, there is only little work using the acoustic modality in a closed-loop setting in the spirit of Figure 3.1.

A real-time rendering method to produce realistic sounds of colliding physical objects in a virtual environment has been proposed recently (Doel and Pai 1998; Hendrix and Barfield 1996; Hollander 1994). A similar approach for a virtual machining tool (drilling machine) by analyzing dominant parameters of real machining sound and approximate implementation in VE has been developed in our laboratory (Wermuth 1998). The use of acoustic feedback in sensory substitution has already been mentioned in the previous section.

3.8 Telepresence Quality and Performance Evaluation

Sheridan suggests three approaches to evaluate the quality of TPTA systems (Sheridan 1996): 1. measurement of neuromuscular and acoustic reaction times; 2. measurement of subjective presence quality by a questionnaire; 3. measurement of the ability to distinguish between presence in the real and the virtual environment. Evaluation is therefore objective or subjective. The third possibility is not very useful with the rather low quality of currently available multi-modal displays and virtual environments.

A quantitative measure of telepresence including both objective and subjective measurement is suggested in (Schloerb 1995). The degree of subjective telepresence is defined as the probability that the human operator claims to be physically present using the TPTA system. Likewise objective presence is said to be the probability to complete given tasks successfully. Measurements of the subjective sense of presence in a virtual driving task depending on pictorial realism, visual delay and observer interactivity are due to (Welch, Blackmon, Liu, Mellers, and Stark 1996).

With the exception of the above mentioned measurement methods, agreed upon and widely applicable, multi-modal, complete and possibly parametric measures for quantitative task performance and presence quality evaluation are currently not available. It has to be the goal of any such measure to evaluate the human task performance as well as the stress factors and subjective presence quality perceived by the operator. Along this line of research we have achieved first results in establishing objective and subjective measures for the simple task of teledrilling, which depend on tunable system parameters (time-delay, visual frame rate). Section 4.4.2 gives a few more details on this subject.

3.9 Optimal and Ergonomic Multi-Modal Control

Suppose now, that such task performance and presence quality measures are established and available. The next striking challenge is a control design problem in that these measures can be used to design *optimal ergonomic control* strategies for TPTA systems. From given control system parameters such as available communication bandwidth, sampling frequency and time-delay between operator and remote site, an optimal controller would be designed as the optimal solution to the performance/quality measures. From our experience it is possible to give a few heuristic control design guidelines (Wermuth 1998), however, the challenge of a systematic ergonomic control design based on quantitative operability or presence quality measures remains an open and very important area for further research.

3.10 Human Model-Based Control

The use of human models, skill and expertise to supplement local teleoperator control as discussed above and shown in Figure 3.1 has been proposed by several investigators, e.g. (Buss and Hashimoto 1995; Buss 1994; Buss and Hashimoto 1993a; Buss and Hashimoto 1994; Delson and West 1993; Friedrich, Holle, and Dillmann 1998; Kunii and Hashimoto 1997; Yang and Asada 1990; Yang, Xu, and Chen 1993). More closely related to learning rather than TPTA systems is (Arimoto 1991). The basic idea is to have a human expert operator perform tasks by demonstration, extract expert controller knowledge (skill) from the measured data and apply the task skill for semi-autonomous teleoperator control. Though application areas of the mentioned works are plentiful this basic idea applies to most of them. The approaches may be classified into control algorithm acquisition, reference path/trajectory learning and a combination of both.

A recently started approach followed in our laboratory aims at providing mathematical models for human sensory perception capabilities (Kammermeier 1997). From receptor stimulation to the actual symbolic perception in the human brain a sequence of mathematical mappings is defined. This mathematical framework for human perception is quite useful when discussing sensory, modality or stimuli substitution (cf. Section 3.6) and can be applied to haptic devices by including inverse functional maps. As an example, the touch sensation of an object edge was produced using simple piezoelectric vibrotactile actuators attached to the human finger tip in a touch feedback glove adapted from (Caldwell, Lawther, and Wardle 1996). The same experiments have been extended by attaching tactile actuators on the desktop kinesthetic feedback device DeKiFeD3 (see Section 4.2).

3.11 Other Fundamental Control Related Issues

One important and highly desirable feature of teleaction systems is *graceful performance degradation*. This means that the measures of subjective degree of telepresence and objective performance measures should smoothly depend on the quality of the communication network in terms of bandwidth and time-delay. Current systems do not gracefully degrade, i.e. up to some limit of time-delay the performance decreases and then suddenly drops to undesirable system/operator behavior when exceeding this limit. The challenge is a control design procedure to achieve graceful degradation much in the spirit of optimal multi-modal control outlined in Section 3.9.

As mentioned before *performance, operability and ease-of-use* assessment and optimization of multi-modal control designs thereafter are important issues not adequately addressed so far.

A related problem is *loop-shaping* of the frequency characteristics in each modality of the telepresence and teleaction system. Intermodal dependencies, communication bandwidth and time-delay make this a challenging control design task. An overview of loop shaping techniques for bilateral teleoperation systems can be found in (Salcudean 1998). An approach using a H^{∞}-optimal control design to minimize the weighted error between actual and desired transfer function for positions and forces has been presented in (Kazerooni, Tasy, and Hollerbach 1993). A related general framework for H^{∞}-design of teleoperation controllers is given in (Yan and Salcudean 1993) also including constant time-delays using Padé all-pass approximations. Within the μ-synthesis framework a control design approach to optimize performance characteristics with guaranteed stability for a-priori known constant time-delays has been proposed (Leung, Francis, and Apkarian 1993). What we are trying to advocate here, is to consider the development of approaches in the spirit of loop shaping to be used for control design of the other modalities.

The most important requirement for multi-modal TPTA systems is *operator safety, teleoperator system safety* and of course *control system stability.* Stability is certainly an important feature of the system also connected to operator and system safety, but under some operational conditions it is known that human expert operators can stabilize inherently unstable teleoperation systems. Very often the operator gets confused in these cases and it is of course more desirable to guarantee stable behavior of the system at all times and under all conditions. Guaranteed passivity of bilateral teleoperation systems has been mentioned before and certainly is a viable approach to stabilization of the haptic modalities.

A more restrictive requirement to multi-modal control design is *operator and teleoperator safety.* Highest priority in all cases has operator safety. Usually, extra sensors to monitor if the human operator has the hand firmly attached to a haptic display device are implemented. Alternatively, one could implement detection algorithms in the spirit of trading control to avoid unreasonable operator actions possibly destroying the teleoperator or objects in the remote environment. It is worthwhile to protect teleoperators from destruction because these systems can be expensive and time consuming to repair or rebuild. In tele-medicine the highest priority regarding to safety are the patient and the doctor involved.

An interesting area for further research is *scaled telepresence*. An extremely scaled telepresence application toward nano-mechatronic applications has been recently proposed (Sitti and Hashimoto 1998). The goal of this work is to haptically render data gathered by an atomic force microscope (AFM) for haptic display to the human operator. Forces and motions have to be scaled with a factor of about 10^4 to 10^8.

Four basic schemes of how to connect the operator device and the teleoperator in classical bilateral teleoperation systems have been suggested: 1. F-F, 2. F-P, 3. P-F, 4. P-P, where "F-P" means force control of the operator feedback device and position control of the teleoperator, cf. Section 3.3 and e.g. (Hannaford 1989; Salcudean 1998). A fundamentally different approach is a force-velocity (F-V) or velocity-force (V-F) control scheme, where the operator device position corresponds to the teleoperator velocity or vice versa with a similar definition of transparency and notion of impedance. An advantage of F-V control with respect to safety is that if the operator lets go of the kinesthetic input device the teleoperator will also stop with zero velocity. The obvious disadvantage is that the absolute positions of the operator and remote devices do not correspond. F-V control seems to be useful for scaled environments such as remotely controlled excavators; refer to (Salcudean 1998) and the references therein for a more detailed discussion of this topic.

In the multi-modal control context another fundamental question to be answered by further investigations is which *multi-modal variables* of the teleoperator are best connected to which *multi-modal variables* of the human sense and corresponding display device in the HSI. This would include also questions asked in the area of sensory substitution, i.e. one would try to connect measured teleoperator force to color and/or velocity of objects in a virtual environment visible to the human operator.

Other important goals are to increase the number of DOFs of haptic display devices, the quality of tactile feedback, to solve scaled presence and to increase the number of modalities used simultaneously. *Collaborative multipresence* by several operators collaborating in the same virtual or augmented environment poses additional control problems not completely understood to date.

4 Experimental Studies

Experimental works in the area of multi-modal virtual presence and remote telepresence are so numerous that it is impossible to give a complete overview here. A specific and broadly covering survey of significant *multi-modal experimental work* is not available to our knowledge. As a starting point it is recommended to refer to the references at the end of this chapter, the MIT-Press Journal *Presence–Teleoperators and Virtual Environments* committed to this field, recent conference proceedings of the IEEE International Workshop on Robot and Human Communication–ROMAN, IEEE International Conference on Robotics and Automation–ICRA, and the IEEE/RSJ International Conference on Intelligent Robots and Systems–IROS, etc. Given this situation and the specific aim to discuss control problems in telepresence, our focus is mainly on work performed in our laboratory.

The experiments shown subsequently are fundamental results with practical implications for the application areas mentioned in Section 2.4; e.g. deformable objects are important in VE surgery, tactile displays are needed for virtual shopping catalogues, temperature feedback is advantageous for telediagnosis.

We have been developing an experimental environment to perform a variety of TPTA experiments. Emphasis is on multi-modal teleaction in a real (physical) remote environment. In this section we briefly describe the structure and possibilities of the developed experimental TPTA system, see (Baier, Buss, Freyberger, Hoogen, Kammermeier, and Schmidt 1999) for details. Multi-modal presence — with emphasis on the visual, auditory, kinesthetic, tactile and temperature senses — is realizable in a remote environment located in another laboratory, in virtual environments and as augmented telepresence, see Figure 4.1.

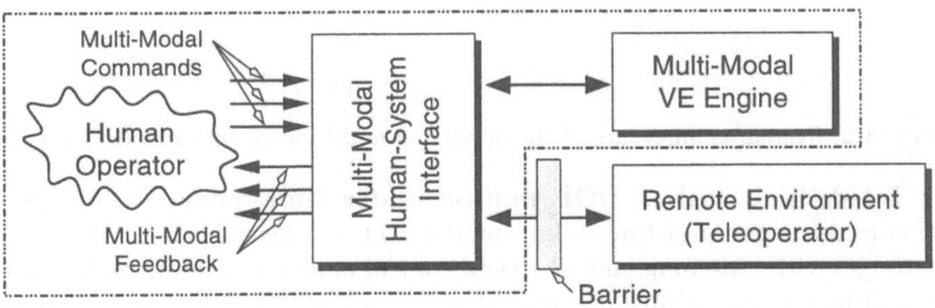

Fig. 4.1. System operation modes of multi-modal virtual presence and telepresence in real (physical) remote environments.

Features of a high-performance haptic display including kinesthetic, tactile and temperature feedback are presented. To discuss some of the control challenges in detail three experiments are described with application examples in multi-modal VE interaction, long-distance tele-drilling between Munich and Tokyo (as a representative laboratory scale experiment for telemanufacturing) and multi-modal teleaction in real remote environments.

4.1 Experimental Environment

The overall system architecture of the experimental TPTA environment is shown in Figure 4.2. In the following we briefly discuss some details of the mechanical design of the subsystems, the computational environment, controller implementation, communication infrastructure and graphics capabilities of the developed mainly PC-based system. Multi-modal presence of the operator can be realized both in a VE and a real (physical) remote environment, see Figure 4.1.

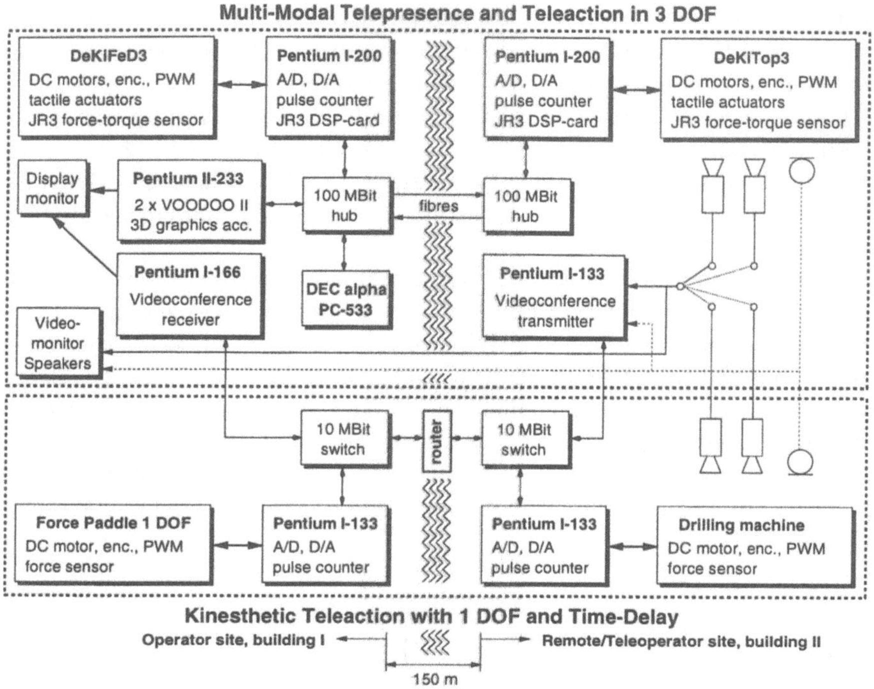

Fig. 4.2. System architecture of the developed multi-modal telepresence system.

Teledrilling with 1 DOF motion under time-delay: For teleoperation experiments in a time-delay environment and investigations of control strategies with switching among several control modes depending on the contact situation, the teledrilling subsystem is used (Baier, Buss, and Schmidt 1997; Baier and Schmidt 1998). The force reflecting paddle with 1 rotational DOF, see Figure 4.3(a), controls the motion of the linear axis in the remote environment, where the drilling machine is mounted on, see Figure 4.3(b). The control strategy uses control mode switching techniques in the framework of a hybrid (discrete-continuous) dynamical system approach.

Multi-modal Presence in 3 DOF: The second subsystem is a multi-modal telepresence setup including a newly developed Desktop-Kinesthetic-Feedback-Device (DeKiFeD3) with 3 active DOF and 1 passive rotation coupled to the Desktop-Kinesthetic-Teleoperator (DeKiTop3) with a similar hardware structure. Actual performance characteristics and the mechanical design of the haptic display DeKiFeD3 and the teleoperator DeKiTop3 are discussed in greater detail in Section 4.2; see also (Baier, Buss, Freyberger, Hoogen, Kammermeier, and Schmidt 1999).

The kinesthetic display capabilities of the DeKiFeD3 are augmented by a vibrotactile display and a Peltier element for temperature feedback to the human operator. Corresponding tactile and temperature sensors are included in the DeKiTop3 setup. Characteristics of the tactile and temperature feedback are discussed in Section 4.3.

Information Processing: Figure 4.2 depicts the computing architecture of the developed telepresence system. The system currently consists of 7 Pentium or Pentium-II PCs and 1 DEC alpha PC. Distribution is logically by placing subprocesses on several PCs and physically because the operator is situated in one building separated from the remote site in a neighboring building 150m apart. For multi-modal communication several types of connections are available (see Figure 4.2): 1. two analog connections by coaxial cable; 2. two optic fibers used for a dedicated 100 MBit/s connection via 2 hubs in the laboratories; 3. via the campus LAN (shared with 8 other laboratories) using 10 Mbit/s.

The four mechatronic subsystems — 1. force paddle, 2. linear axis/drilling machine, 3. DeKiFeD3, 4. DeKiTop3— are each controlled by a PC with appropriate analog interface boards, counters for the optical encoders and DSP-cards for processing of the JR3 force/torque sensor data. The DC motors are current (torque) controlled by standard PWM amplifiers with 25 kHz modulation frequency. All of the force and motion control algorithms (implemented in software) for the haptic devices run at 1 to 2ms sampling time.

The VE part of the system is located on the operator site, see Figure 4.2 and compare Figure 3.1. One PC (P-II) equipped with a dedicated 3D graphic (game) accelerator assures high graphical performance for implementation of fairly complex virtual environments.

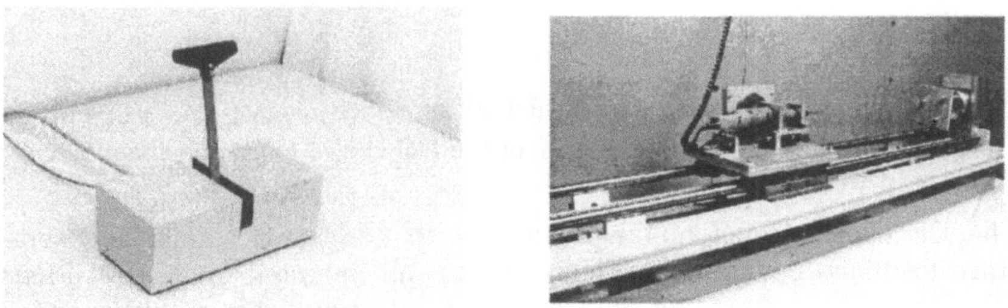

(a) (b)

Fig. 4.3. Force paddle (a); linear axis with drilling machine (b)

4.2 Desktop Kinesthetic Feedback Device DeKiFeD3

Figure 4.4(a) and (b) shows photos of the Desktop-Kinesthetic-Feedback-Device (DeKiFeD3) with 3 active degrees-of-freedom (DOF). The kinematic structure shown in Figure 4.4(c) allows the end-effector (EE) to be positioned in three Cartesian directions. The mechanical design of the third joint is such that the EE orientation remains fixed. The rotation around the vertical axis resulting from the SCARA joints is eliminated using a passive rotational joint.

Each active joint is actuated by a high-performance DC motor (Faulhaber) with planetary gears, optical encoders and standard PWM power amplifiers. The chosen motors have exceptionally good maximum holding torque (1 Nm without gear) to weight (less than 10N including gear and encoder) ratios with a maximum output power of 220W. The gear ratio is 66, 43, 66 in the 1st, 2nd, 3rd joint, respectively. Forces up to 60N can be generated at the EE.

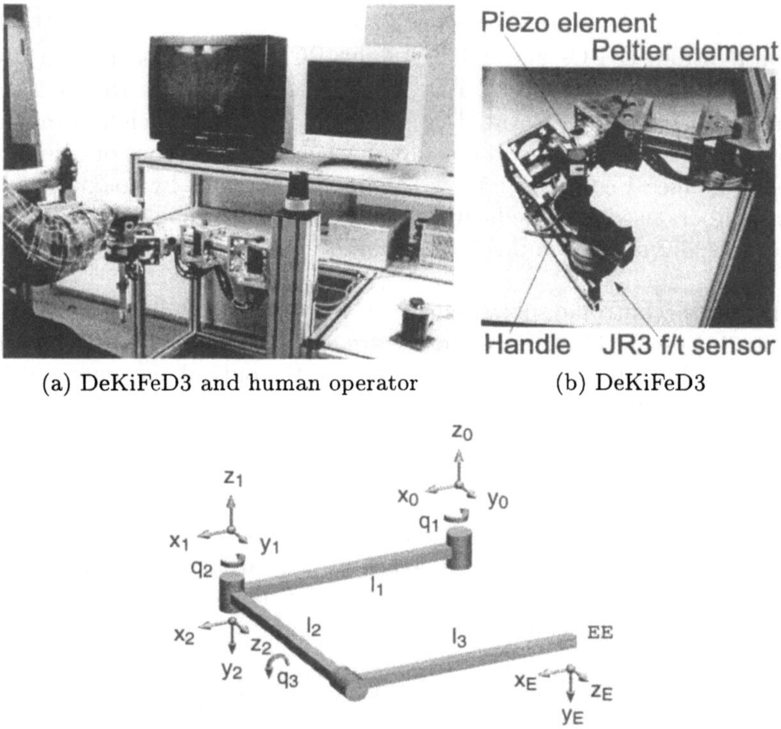

(a) DeKiFeD3 and human operator (b) DeKiFeD3

(c) Kinematics

Fig. 4.4. Photos and kinematic structure of the DeKiFeD3 haptic feedback device.

The EE consists of a holding handle with additional tactile and temperature feedback devices, see Figure 4.4(b), all mounted on a JR3 6-axis force/torque sensor operating at a maximum sampling rate of 8kHz and a measurement range of 100N/6.3Nm. Using the high-quality JR3 force/torque sensor in the DeKiFeD3 gives the advantage of high fidelity force control (possible by sampling times as low as 1ms) of the actual forces between the device and the human operator's hand.

To evaluate the performance and the desired high fidelity kinesthetic feedback capabilities of the DeKiFeD3 we have conducted experiments using a virtual wall. Virtual walls have been reported and used as benchmark experiments by others as well (Colgate, Grafing, Stanley, and Schenkel 1993; Salcudean and Vlaar 1994; Rosenberg and Adelstein 1993).

The DeKiFeD3 is force controlled to zero-force in free space. When the y-position of the EE enters the virtual wall at $y < y_w = 0.22$m, a repelling force F is generated as the set-point for the force controller abiding the virtual wall model of

$$F = K(y_w - y) + B\dot{y} \qquad \text{for} \qquad y < y_w \tag{4.1}$$

with $K = 10000$N/m and $B = 50$Ns/m.

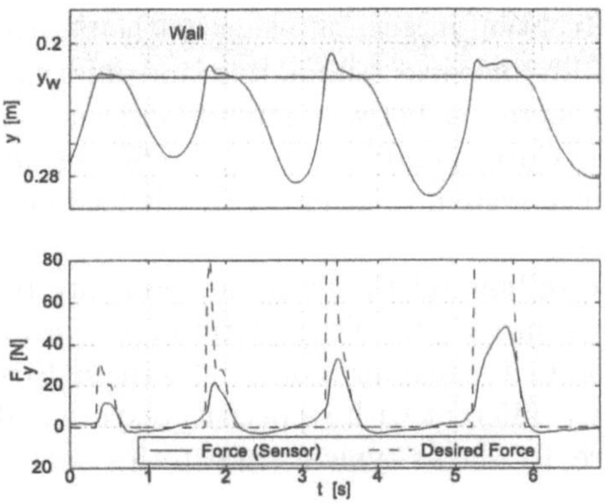

Fig. 4.5. Experimental result when repeatedly hitting the virtual wall.

Figure 4.5 shows a sample experimental result. The repeated entering of the virtual wall in y-direction can be seen while the desired forces and actual forces generated at the human operator site are shown below. Due to the limited actuator bandwidth (13Hz) and actuator saturation (maximum exertable force 60N) the desired force cannot be achieved exactly. However, concerning the exerted forces up to $50 - 60$N and the low penetration depth of the virtual wall ($\leq$ 1cm) the DeKiFeD3 is able to generate a rather stiff subjective impression of hard surfaces with the human operator. This is not an exact psychophysical measurement, but the result of interviewing external expert users.

4.3 Tactile Feedback

Vibrotactile and temperature actuators in the DeKiFeD3 augment the kinesthetic capabilities realizing a haptic display. The tactile sensors mounted on the DeKiTop3 and the actuators on the DeKiFeD3 are shown in Figure 4.6(a).

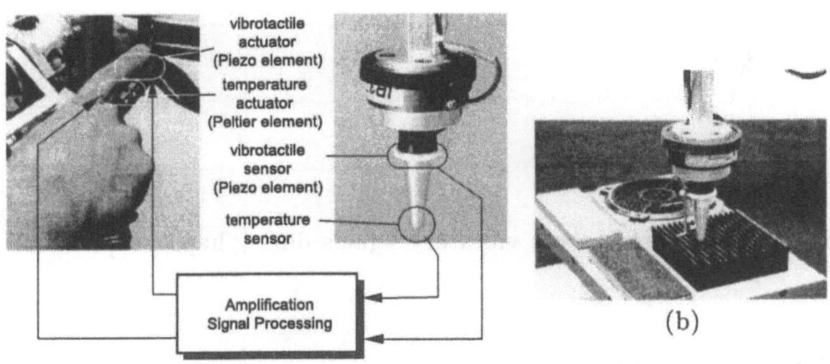

(a) (b)

Fig. 4.6. Tactile sensors and actuators for haptic exploration.

We are investigating modeling paradigms for the human tactile sense from a systems theoretic point of view (Kammermeier 1997) aiming at a framework to describe the processes of human perception in general using mathematical mappings. The human sense of touch, its physical stimulation and information processing, and the deduction of coherences between stimuli and sensory perceptions is one part of this model. Corresponding inverse *tactile sensation maps* can be used in the telepresence system to generate tactile feedback signals (Kammermeier 1997).

Here we can only briefly demonstrate how teletactile impressions are realized in the presented system by an exemplary tactile exploration scenario. A prototypical experimental setup for tactile exploration is shown in Figure 4.6(b) with four objects significantly different with respect to texture and the resulting tactile impression. Figure 4.7 shows typical trajectories during tactile object exploration.

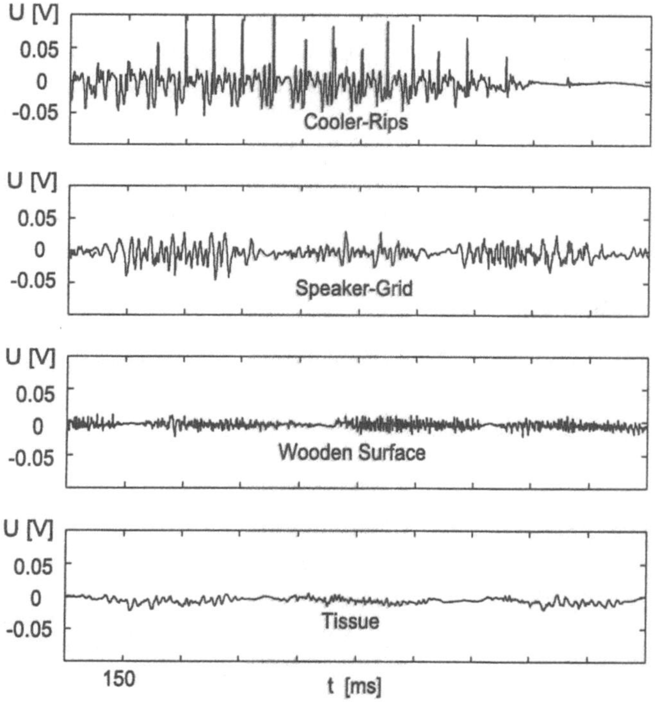

Fig. 4.7. Measured tactile vibratory signals during haptic exploration.

4.4 VE Experiments

As typical scenarios of multi-modal VE presence this section outlines the applications of haptic exploration, virtual drilling, entertainment and force feedback as a result of deforming soft objects.

4.4.1 Haptic World Exploration. The DeKiFeD3 can be used to haptically explore a VE and the objects therein. An example world is shown in Figure 4.8. The relatively flat object on the left is hot, with a soft but rough skin and substantial friction, whereas the box type object on the right is cold, rigid, smooth and frictionless. The half sphere on the ceiling is rigid and frictionless like all the walls of the shown virtual world. The borders of the room are inside the workspace of the DeKiFeD3.

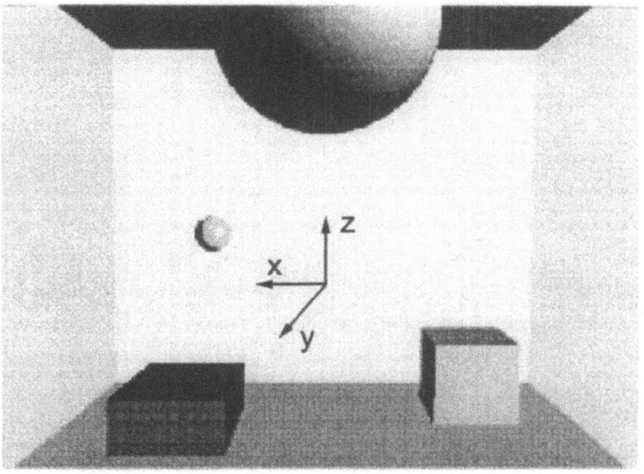

Fig. 4.8. Haptic exploration in a VE.

The small sphere in the middle of the picture represents the virtual probe, which can be moved around using the DeKiFeD3. It can interact with the other objects in the VE and in case of contact with one of the other objects, appropriate feedback forces, tactile feedback stimuli and temperature set-points are generated to display the characteristics of the touched object. The underlying algorithms for modeling and haptic rendering of the VE are for non-deformable objects similar to (Basdogan, Ho, and Srinivasan 1997; Zilles and Salisbury 1995), but extended by an approach to model and display effects like friction forces, roughness and temperature, see (Hoogen 1998) for details. Due to neglecting deformation and simple object shapes this rendering algorithm is very fast ($\leq$ 100 μs) which makes a controller sampling time of 1ms possible.

Tactile feedback is generated depending on parameters representing object characteristics as surface texture, edge type, temperature and thermal conductivity. For VE-experiments the tactile data can either be taken from physical measurements, see Figure 4.7(b), or generated by simulations using functional models for touched shape primitives, fingertip deformation and skin receptors. Alternatively, for ease of implementation deterministic (e.g. peaks, square-wave) or stochastic signals (white noise) can be used.

Figure 4.9 shows typical trajectories when a human operator explores the VE shown in Figure 4.8 moving in negative x-direction from the left, making contact and following the shape of the left object, gliding over the ground to the right, and finally moving across the right object. The height of the virtual fingertip corresponding to the flat and the higher box can be easily seen in the first plot (z-position) of Figure 4.9. F_z and F_x indicate the detected impedance and friction forces when moving across the vertical and horizontal surfaces. U shows the activity of the vibrotactile actuator while T is the output for the temperature set-point.

The fact that haptic, vibrotactile and temperature feedback is generated in this exploration scenario adds a dimension to operator immersion quality. The physical object characteristics like e.g. shape, corners, edges, friction, roughness (texture) and temperature can be felt very realistically.

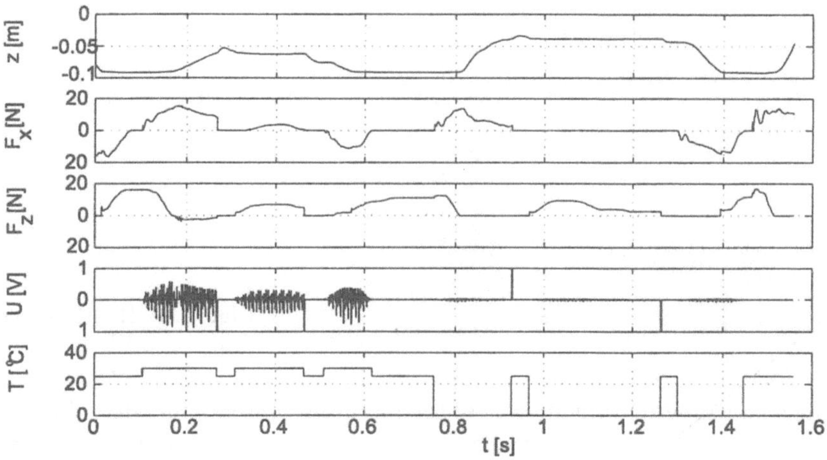

Fig. 4.9. Typical trajectories of a human operator exploring the VE shown in Figure 4.8.

4.4.2 Virtual Drilling. For investigations with respect to evaluation of multi-modal presence quality in visual-haptic telepresence systems we have set up a virtual teledrilling experiment to emulate the real teledrilling application, see Figure 4.10(a). We conducted a subjective and objective measurement of presence quality depending on visual frame rate, intermodal time-delay and consistency as well as importance of each modality by putting up a questionnaire for 25 probands (Wermuth 1998). Main motivation to realize the drilling system in VE was that we had to perform several hundred evaluation experiments. Another reason is that the tuning of modal quality characteristics like frame rate or time-delay is much easier to achieve in the VE realization.

By augmenting synthesized sound to the virtual drilling machine the presence quality could be improved significantly. The idea to generate synthetic auditory feedback is similar to the work reported in (Doel and Pai 1998). We analyzed the sound of the real drilling machine by a Fourier analysis for the two cases of free motion and when drilling an object. The characteristic features of these Fourier transforms were used to synthesize sounds similar to the natural sound, see (Wermuth 1998) for details and detailed quantitative results of the questionnaire. Just to mention one result, which is obvious in retrospect: intermodal consistency is extremely important, because if one modality is realized with particularly low quality (e.g. time-delay larger than for other modalities), human operators will completely ignore it and rely on the higher quality modalities.

(a) virtual drill (b) ping pong

Fig. 4.10. VE applications.

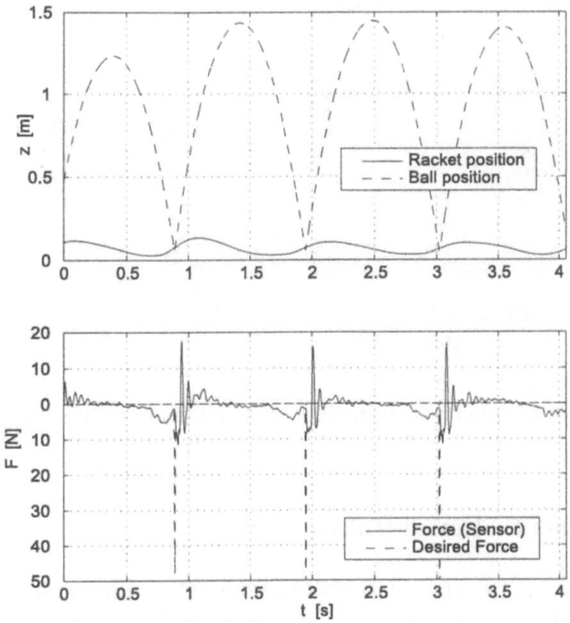

Fig. 4.11. Trajectories during the ping pong game.

4.4.3 Ping Pong. As an exemplary entertainment application we have implemented a simple ping pong game, see Figure 4.10(b). The human player uses the DeKiFeD3 to move the racket, hit the ball and make it bounce in the vertical direction. Typical trajectories of the ball, the racket, the desired and measured DeKiFeD3 feedback forces to the player are plotted in Figure 4.11.

4.4.4 Deformable Objects. In addition to the non-deformable virtual world in Section 4.4.1 we implemented an environment with deformable objects, see Figure 4.12. These objects consist of 100-300 mass nodes which are connected to each other by spring-damper systems (Deussen and Kuhn 1995). This approach allows a highly realistic simulation of soft objects with deformable shape. The drawback of nodal models lies in the required numerical integration time. In our implementation the integration time is $\leq$ 10ms for haptic rendering. Details about the implementation of the deformable VE can be found in (Hall, Kolssouzidis, and Schickinger 1998). This virtual world can also be explored with the DeKiFeD3. Some results of object deformations are shown in Figure 4.12.

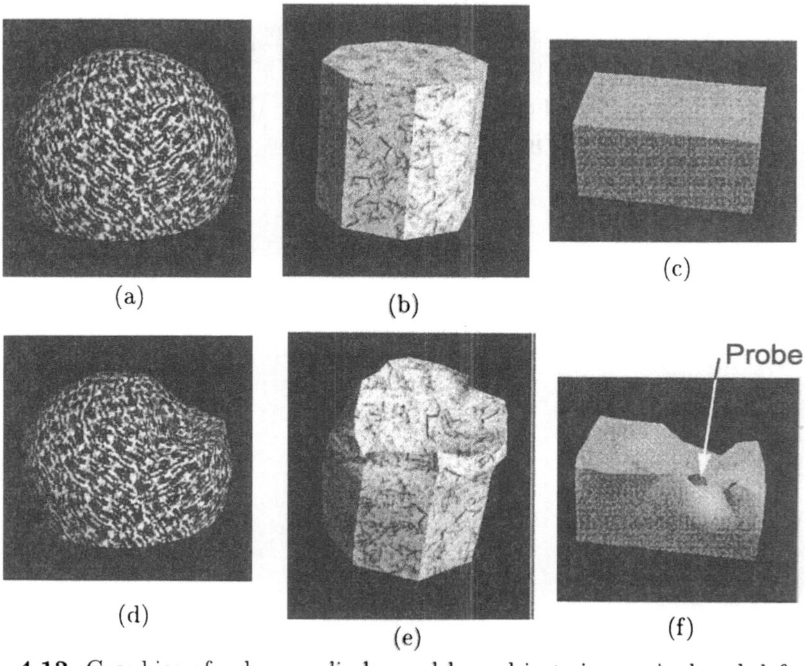

Fig. 4.12. Graphics of sphere, cylinder and box objects in nominal and deformed configuration.

4.5 Teledrilling Experiment Munich-Tokyo

The example 1 DOF teleaction system is a laboratory scale teledrilling system, which is controlled via the Internet (Baier, Buss, and Schmidt 1997; Baier and Schmidt 1998). The control system is of bilateral force-position F-P type using task dependent control mode switching to P-F. We distinguish between three types of experiments, one with the operator and teleoperator in our laboratory and communication running via the local network, the second experiment with a relay station in Tokyo and the third with the operator site in Tokyo and the teleoperator in Munich. In all cases the communication network (Internet) introduces unknown and variable time-delays of 0.2 to 2s into the teleaction system; sometimes complete data packets are lost.

In teledrilling we apply — in addition to guaranteeing the communication two-port (teleoperator interface) to be lossless — different controllers for different contact situations of the drilling machine in the remote environment, e.g. a position controller for free motion of the drilling machine and a force controller during drilling with environment contact. This scenario results in a continuous variable control system depending on the discrete states of a finite state machine. Hence we have a *hybrid discrete-continuous dynamical system.* One important and challenging problem in this realization is the synchronization of discrete state transitions in the corresponding operator and teleoperator state machines.

This can be illustrated by the following example: assume that the drilling machine is position controlled locally and the operator gets force feedback. If environment contact of the drilling machine is detected by a sensor or estimation algorithm on the remote site, the teleoperator state machine switches to local force control. However, because of time-delay the operator site is still in position control. Therefore, communicated position reference values are meaningless for the force controller. This problem can be solved by introducing safe and stable transition states to synchronize the control modes of the HSI and the teleoperator.

Figure 4.13 depicts a simplified version of the hybrid automaton with the states of `APPROACH` and `DRILLING` for both the operator and teleoperator mode-switched local controllers. The simplifications from the original automaton (Baier and Schmidt 1998) are that the `APPROACH`-state additionally includes a velocity restrictive substate when the drill is near the workpiece. Another state is inserted between `DRILL` and `APPROACH` to move the drilling machine away from the workpiece back to a starting position. The waiting states to achieve discrete state coincidence between operator and teleoperator device are also omitted in Figure 4.13 for briefness, for details and completeness refer to (Baier and Schmidt 1998).

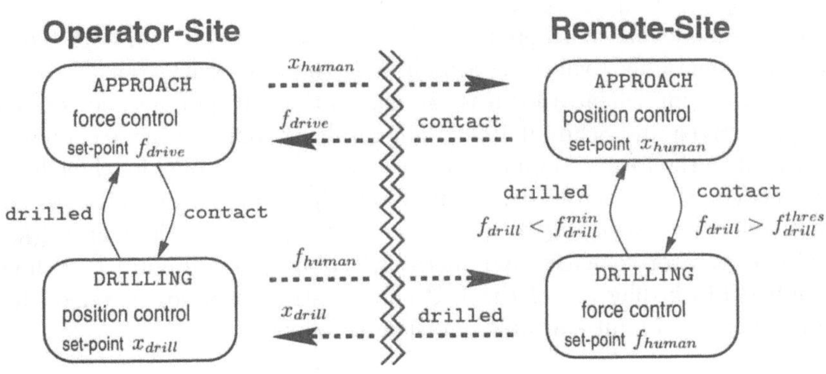

Fig. 4.13. Simplified hybrid automaton used for teledrilling control mode switching.

The teledrilling machine is assumed to be equipped with a drilling force sensor f_{drill}, a linear axis driving force sensor f_{drive} and a drill position sensor x_{drill}. The (simplified) drilling sequence starts with the APPROACH phase, where the position x_{human} of the operator device is communicated across the barrier, used as the desired position for the local teleoperator position (motion) controller. The force f_{drive} required to drive the teledrilling machine towards the workpiece is measured and communicated to the force-controlled operator device as the desired force set-point. Hence, in the APPROACH phase the teleaction control-loop is of type F-P.

In addition to the continuous variables f_{drive} and x_{human} the discrete state APPROACH is sent from the operator to the teleoperator (for mode synchronization) and the discrete event signal `contact` is communicated from the teleoperator to the operator. The communication between operator and teleoperator is indicated by dashed arrows in Figure 4.13; note that we assume significant time-delay. If the event `contact` — detected by a sudden increase in the drilling force $f_{drill} > f_{drill}^{thres}$ — occurs the desired position x_{human} for the drilling machine position is frozen and the teleoperator enters a safe waiting state. The event `contact` is communicated and the teleoperator waits in this safe state until the operator device has received and acknowledged the event signal. If both sites have been synchronized by this handshake procedure, the state transition to DRILLING is executed.

During the DRILLING phase the teleoperation scheme is P-F, i.e. the operator device is position controlled to the teleoperator position sensor x_{drill} and the teleoperator is force controlled to the force input f_{human} by the human operator. Once the hole has been drilled — detected by a sudden decrease of the drilling force $f_{drill} < f_{drill}^{min}$ and a simultaneous fast motion of the drill behind the hole — the system again enters a safe synchronization state followed by a move away state. It is then ready to drill the next hole.

When viewing e.g. the F-P control mode of this control scheme in the context of the four-channel architecture, cf. Figure 3.3, the force forward block C_3 and velocity backward block C_4 are not used, see Figure 4.14. The velocity forward block $C_1 = B_s + K_s/s$ is of integrator type and the force backward block is $C_2 = K_f$ with the scaling factors B_s, K_s, K_f.

If a time-delay T is considered this can be included as e^{-sT} in both blocks C_1, C_2. As mentioned before in Section 3.4, the scattering transformation can be used to implement a lossless communication two-port (Anderson and Spong 1989; Baier and Schmidt 1998). The resulting control scheme cannot be transformed to the form of Figure 3.3, but results in additional local feedback blocks between the velocities v_h, v_e^d and forces f_h^d, f_e on the operator (h) and remote environment (e), respectively. Note, that v_e^d, f_h^d are the transmitted, desired teleoperator velocities and operator feedback force. Figure 4.14 shows a simplified block diagram of the F-P control strategy used for teledrilling using a generalized multi-variable notation as follows

$$\begin{bmatrix} v_e^d(s) \\ f_h^d(s) \end{bmatrix} = \begin{bmatrix} G_{11}(s) & G_{12}(s) \\ G_{21}(s) & G_{22}(s) \end{bmatrix} \begin{bmatrix} f_e(s) \\ v_h(s) \end{bmatrix} , \tag{4.2}$$

where the inputs and outputs of the of the communication block have been redrawn from left to right in Figure 4.14. Clearly, the diagonal components G_{11}, G_{22} correspond to blocks C_1, C_2 of Figure 3.3 and the non-diagonal components G_{12}, G_{21} to the local feedback measures to be taken for lossless implementation of the communication two-port. The detailled structure of $G(s)$ in our teledrilling implementation are lengthy and therefore omitted here; for details refer to (Baier, Buss, and Schmidt 1997; Baier and Schmidt 1998).

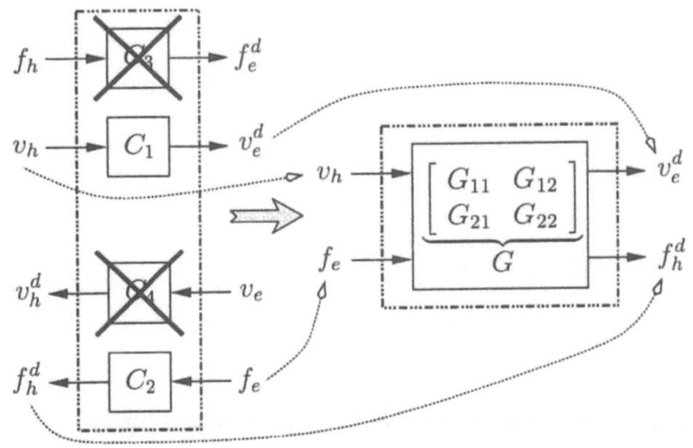

Fig. 4.14. Structure of the communication block for the bilateral F-P control scheme under time-delay, cf. Figure 3.3.

In the second control mode when drilling is performed, the control mode is switched to P-F control, which means that the velocity forward block C_1 and the force backward block C_2 are not used. This also emphasises the hybrid nature of the control system in that in different modes a different block diagram is drawn to represent the corresponding control algorithm.

Extensive experimental and simulation studies of this hybrid teledrilling control algorithm have been performed, but cannot be presented here because of space limitations. The essence of the hybrid approach has been described and visualized with the simplified hybrid automaton in Figure 4.13, see (Baier, Buss, and Schmidt 1997; Baier and Schmidt 1998) for details and experimental results.

4.6 Teleaction in 3 DOF

An example of a telemanipulation task is shown in Figure 4.15, where the 3 triangular parts have to be assembled such that they fit together forming a square. Figure 4.15(a) shows the DeKiTop3 pushing a smaller triangle. The corresponding force measurements during this telepushing operation are shown in Figure 4.15(c). The teleoperator may also lift the part by putting the hook into the ring. This can be seen in the photo of Figure 4.15(b) with the corresponding measured force trajectory in Figure 4.15(d).

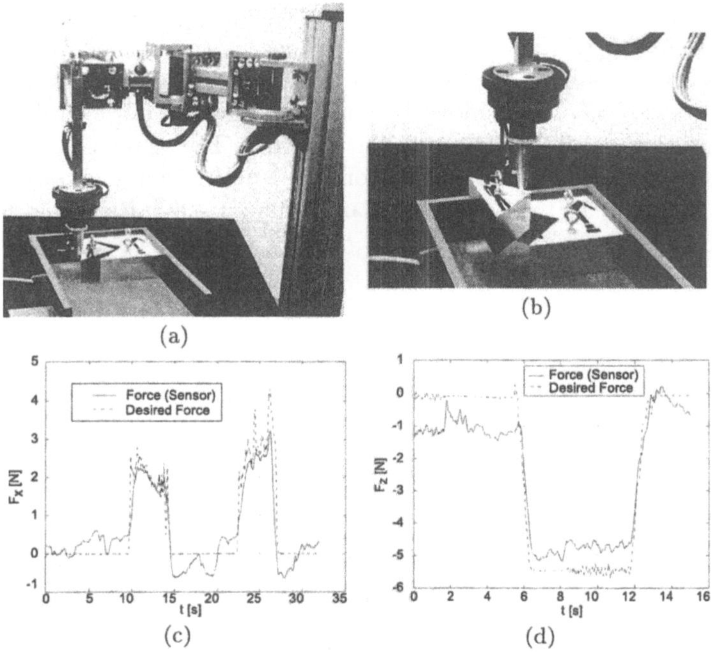

Fig. 4.15. Telemanipulation of triangular objects and measured forces.

5 Conclusions

This chapter presented the general system and control architecture of multi-modal telepresence and teleaction systems. Typical control modes such as remote, shared, cooperative, assisting, trading, symbolic, semi-autonomous control were discussed in detail. One of the most critical issues in TPTA systems is variable time-delay mainly introduced by the communication infrastructure. Another type of barrier between human operator and teleoperator is environment scaling such as encountered for telepresence in micro- or nano-scaled environments. Enabling technologies and available hardware devices to realize truly multi-modal human-system interfaces–HSIs have been briefly summarized. The conclusion is that all the required ingredients from technology, communication infrastructures, computational power, etc. are available and a rapid development of this exciting area of multi-modal telepresence is to be expected. Major directions of research need to include technological advances of (micro-)mechatronic sensor/actuator hardware, computational advances in hardware, software and algorithmic concepts, and of course the development of advanced multi-modal control concepts.

The second part of this chapter has discussed the TPTA control challenges of multi-modal closed-loop control, strategies to deal with unavoidable communication time-delay, bilateral kinesthetic teleoperation, local control measures implementable on the operator and remote sites locally, virtual environment interactive control, sensory substitution and auditory control. Performance evaluation of multi-modal TPTA systems is a very important issue, but remains a topic widely open for further research. The inverse problem of optimal ergonomic multi-modal control design is the consequential challenge once adequate multi-modal performance measures become available. Human model-based control exploiting human operator expert knowledge to implement local control strategies in semi-autonomous TPTA systems has been discussed. Some fundamental issues including multi-modal loop shaping, operator safety, environment scaling and modal interconnection methodology have been mentioned.

To further illustrate some of the mentioned control challenges and discuss problems in greater detail, we have given a brief overview of experiments performed with the multi-modal TPTA system being developed in our laboratory. This TPTA system includes the visual, acoustic, kinesthetic, tactile and temperature senses for telepresence and teleaction as well as augmented presence. It is again emphasized that significant contributions to this exciting field of multi-modal telepresence and teleaction are to be expected by developing integrated systems and performing truly multi-modal experiments.

Acknowledgments

Development and integration of the experimental multi-modal TPTA system is due to joint efforts by the authors together with H. Baier, F. Freyberger, J. Hoogen, P. Kammermeier. The implementation of the on-line deformable object simulation and interaction with the DeKiFeD3 is due to A. Hall, S. Kolssouzidis and T. Schickinger. The authors greatly appreciate technical support by J. Gradl, H. Kubick, T. Lowitz and T. Stoeber of the laboratory workshop during construction of the force paddle, teledrilling machine and DeKiFeD3/DeKiTop3 experimental hardware and control systems. The assistance of W. Jaschik during the integration of the presented system into the laboratory local area network is also appreciated.

This research was partly supported by the German Research Foundation DFG within the Collaborative research centre SFB 453 on "High-Fidelity Telepresence and Teleaction".

References

Anderson, R. (1996). Autonomous, Teleoperated, and Shared Control of Robot Systems. In *Proceedings of the IEEE International Conference on Robotics and Automation*, Minneapolis, Minnesota, pp. 2025–2032.

Anderson, R. and M. Spong (1989). Bilateral Control of Teleoperators with Time Delay. *IEEE Transaction on Automatic Control 34*, 494–501.

Arimoto, S. (1991). Learning for Skill Refinement. In *Proceedings of the International Workshop on Intelligent Robots and Systems IROS*, pp. 10–13.

Baier, H., M. Buss, F. Freyberger, J. Hoogen, P. Kammermeier, and G. Schmidt (1999). Distributed PC-Based Haptic, Visual and Acoustic Telepresence System—Experiments in Virtual and Remote Environments. In *Proceedings of the IEEE Virtual Reality Conference VR'99*, Houston, TX, pp. 118–125.

Baier, H., M. Buss, and G. Schmidt (1997). Control Mode Switching for Teledrilling Based on a Hybrid System Model. In *Proceedings of the IEEE/ASME International Conference on Advanced Intelligent Mechatronics AIM'97*, Tokyo, Japan, Paper No. 50.

Baier, H. and G. Schmidt (1998). Control Aspects of Closed-Loop Haptic Teleoperation. In *Proceedings of the 7th International Workshop on Robotics in Alpe-Adria-Danube Region (RAAD'98)*, Smolenice, Slovakia, pp. 151–156.

Barfield, W., C. Hendrix, O. Bjorneseth, K. Kaczmarek, and W. Lotens (1995, Fall). Comparison of Human Sensory Capabilities with Technical Specifications of Virtual Environment Equipment. *Presence 4*(4), 329–356.

Basdogan, C., C. Ho, and M. Srinivasan (1997). Haptic Rendering: Point- and Ray-Based Interactions. In *Proceedings of the Second PHANToM Users Group Workshop*, Dedham, MA.

Bejczy, A., W. Kim, and S. Venema (1990). The phantom robot: Predictive displays for teleoperation with time delay. In *Proceedings of the IEEE International Conference on Robotics and Automation*, Cincinnati, Ohio, pp. 546–551.

Brooks, T. (1990). Telerobotic response requirements. In *Proceedings of the IEEE International Conference on Systems, Man, and Cybernetics*, pp. 113–120.

Burdea, G. (1996). *Force and Touch Feedback for Virtual Reality.* New York: John Wiley & Sons.

Burdea, G. and J. Zhuang (1991). Dexterous telerobotics with force feedback — an overview. part 1: Human factors, part 2: Control and implementation. *Robotica 9*, 171–178,291–298.

Buss, M. (1994, June). *Study on Intelligent Cooperative Manipulation.* Ph. D. thesis, University of Tokyo, Tokyo.

Buss, M. and H. Hashimoto (1993a). Human manipulative skill analysis and acquisition for the intelligent assisting system. In *Proceedings 12th IFAC World Congress, International Federation of Automatic Control*, Volume 9, Sydney, Australia, pp. 321–326.

Buss, M. and H. Hashimoto (1993b). Skill acquisition and transfer system as approach to the intelligent assisting system—ias. In *Proceedings of the 2nd IEEE Conference on Control Applications*, Vancouver, British Columbia, Canada, pp. 451–456.

Buss, M. and H. Hashimoto (1994). Manipulation skill modeling for dexterous hands. In *Proceedings of the IEEE International Conference on Robotics and Automation*, San Diego, California, pp. 3403–3408.

Buss, M. and H. Hashimoto (1995). Motion scheme for dextrous manipulation in the intelligent cooperative manipulation system—icms. In V. Graefe (Ed.), *Intelligent Robots and Systems*, pp. 279–294. Amsterdam: Elsevier Science.

Caldwell, D., S. Lawther, and A. Wardle (1996). Tactile Perception and its Application to the Design of Multi-modal Cutaneous Feedback Systems. In *Proceedings of the IEEE International Conference on Robotics and Automation*, Minneapolis, Minnesota, pp. 3215–3221.

Card, S., T. Moran, and A. Newell (1983). *The Psychology of Human-Computer Interaction.* Lawrence Erlbaum Associates.

Colgate, J., P. Grafing, M. Stanley, and G. Schenkel (1993). Implementation of stiff virtual walls in force-reflecting interfaces. In *Proceedings of the IEEE Virtual Reality Annual International Symposium VRAIS*, pp. 20–26.

Delson, N. and H. West (1993). Robot programming by human demonstration: Subtask compliance controller identification. In *Proceedings of the IEEE/RSJ International Conference on Intelligent Robots and Systems IROS*, pp. 33–41.

Denk, J. (1997). Teleoperation of the ROMAN-Gripper over the Internet. Internal Report, Institute of Automatic Control Engineering, Technical University of Munich.

Deussen, O. and Kuhn (1995). Echtzeitsimulation deformierbarer Objekte über nodale Modelle. In *Proc. Intergration von Bild, Modell und Text, ASIM Mitteilungen No. 46*, Universität Magdeburg.

Doel, K. and D. Pai (1998, August). The Sounds of Physical Shapes. *Presence 7*(4), 382–395.

Fischer, C., M. Buss, and G. Schmidt (1996). HuMan-Robot-Interface for Intelligent Service Robot Assistance. In *Proceedings of the IEEE International Workshop on Robot and Human Communication (ROMAN)*, Tsukuba, Japan, pp. 177–182.

Friedrich, H., J. Holle, and R. Dillmann (1998). Interactive Generation of Flexible Robot Programs. In *Proceedings of the IEEE International Conference on Robotics and Automation*, Leuven, Belgium, pp. 538–543.

Hall, A., S. Kolssouzidis, and T. Schickinger (1998, June). Modeling of Deformable Objects in Virtual Worlds. Internal Report, Institute of Automatic Control Engineering, Technical University of Munich.

Handwerker, H. (1989). General Sensory Physiology. In R. Schmidt and G. Thews (Eds.), *Human Physiology*, pp. 176–195. Berlin Heidelberg New York London Paris Tokyo Hong Kong: Springer-Verlag.

Hannaford, B. (1989, August). A design framework for teleoperators with kinesthetic feedback. *IEEE Transactions on Robotics and Automation 5*(4), 426–434.

Hendrix, C. and W. Barfield (1996, Summer). The Sense of Presence within Auditory Virtual Environments. *Presence 5*(3), 290–301.

Hirzinger, G. (1993). Rotex–the first robot in space. In *Proceedings of the ICAR International Conference on Advanced Robotics*, pp. 9–33.

Hollander, A. (1994). An exploration of virtual auditory shape perception. Master's thesis, Human Interface Technology Laboratory, University of Washington, Seattle, WA.

Hoogen, J. (1998). Development of an Experimental System for Kinesthetic Feedback in Telepresence Systems. Internal Report, Institute of Automatic Control Engineering, Technical University of Munich.

Johannsen, G. (1993). *Mensch-Maschine-Systeme.* Springer.

Kaczmarek, K., M. Tyler, and P. Bach-y-Rita (1994). Electrotactile Haptic Displays on the Fingertips. In *Proceedings of the IEEE International Conference on Engineering in Medicine and Biology*, New York.

Kammermeier, P. (1997). Development of a Haptic Display for Tactile Feedback in Telepresence Systems. Internal Report, Institute of Automatic Control Engineering, Technical University of Munich.

Kazerooni, H., T.-I. Tasy, and K. Hollerbach (1993). A Controller Design Framework for Telerobotic Systems. *IEEE Transaction Control Systems Technology 11*, 105–116.

Kim, W. (1996, Spring). Virtual Reality Calibration and Preview/Predictive Displays for Telerobotics. *Presence 5*(2), 173–190.

Kontarinis, D. and R. Howe (1995, Fall). Tactile Display of Vibratory Information in Teleoperation and Virtual Environments. *Presence 4*(4), 387–402.

Kosuge, K., Y. Fujisawa, and T. Fukuda (1993). Control of robot directly maneuvered by operator. In *Proceedings of the IEEE/RSJ International Conference on Intelligent Robots and Systems IROS*, pp. 49–54.

Kosuge, K., T. Itoh, T. Fukuda, and M. Otsuka (1995). Scaled Telemanipulation System Using Semi-Autonomous Task-Oriented Virtual Tool. In *Proceedings of the IEEE/RSJ International Conference on Intelligent Robots and Systems IROS*, Pittsburgh, PA, pp. 124–129.

Kosuge, K., H. Murayama, and K. Takeo (1996). Bilateral Feedback Control of Telemanipulators via Computer Network. In *Proceedings of the IEEE/RSJ International Conference on Intelligent Robots and Systems IROS*, Osaka, Japan, pp. 1380–1385.

Kunii, Y., M. Buss, and H. Hashimoto (1995). Force feedback using dynamic force simulator. In *Video Proceedings of the IEEE International Conference on Robotics and Automation*, Nagoya, Japan, pp. 15.

Kunii, Y. and H. Hashimoto (1996). Dynamic force simulator for haptic interface. In J. Gertler, J. Cruz, and M. Peshkin (Eds.), *Preprints of the 13th World Congress*, Volume L, pp. 403–408. San Francisco: International Federation of Automatic Control—IFAC.

Kunii, Y. and H. Hashimoto (1997). Tele-teaching by human demonstration in virtual environment for robotic network system. In *Proceedings of the IEEE International Conference on Robotics and Automation*, Albuquerque, New Mexiko, pp. 405–410.

Lawrence, D. (1993, October). Stability and Transparency in Bilateral Teleoperation. *IEEE Transactions on Robotics and Automation 9*(5), 624–637.

Leung, G., B. Francis, and A. Apkarian (1993). Bilateral controller for teleoperators with time delay via μ-synthesis. *IEEE Transactions on Robotics and Automation 10*, 105–116.

Mann, S. (1998, November). Humanistic Computing: "WearComp" as a New Framework and Application for Intelligent Signal Processing. *Proceedings of the IEEE 86*(11), 2123–2151.

Massie, T. and J. Salisbury (1994). The PHANToM Haptic Interface: A Device for Probing Virtual Objects. In *Proc. of the ASME Winter Annual Meeting, Symposium on Haptic Interfaces for Virtual Environment and Teleoperator Systems*, Chicago.

Massimino, M. and T. Sheridan (1994). Sensory Substitution for Force Feedback in Teleoperation. *Presence 2*(4).

Mitsuishi, M., T. Hori, and T. Nagao (1994). Predictive Information Display for Tele-Handling/Machining System. In *Proceedings of the IEEE/RSJ/GI International Conference on Intelligent Robots and Systems IROS*, Munich, Germany, pp. 260–267.

Nahvi, A., D. Nelson, J. Hollerbach, and D. Johnson (1998). Haptic Manipulation of Virtual Mechanisms from Mechanical CAD Designs. In *Proceedings of the IEEE International Conference on Robotics and Automation*, Leuven, Belgium, pp. 375–380.

Niemeyer, G. and J. E. Slotine (1998). Towards Force-Reflecting Teleoperation Over the Internet. In *Proceedings of the IEEE International Conference on Robotics and Automation*, Leuven, Belgium, pp. 1909–1915.

Rest, E. L., J.-L. Fleureau, and L. Marcé (1997). Semantics and implementation of a language for telerobotics. In *Proceedings of the IEEE/RSJ International Conference on Intelligent Robots and Systems IROS*, Grenoble, France, pp. 1807–1812.

Rosenberg, L. and B. Adelstein (1993). Perceptual decomposition of virtual haptic surfaces. In *Proceedings of the IEEE Symposium on Research Frontiers in Virtual Reality*, San Jose, CA, pp. 46–53.

Salcudean, S. (1998). Control for Teleoperation and Haptic Interfaces. In B. Siciliano and K. Valavanis (Eds.), *Lecture Notes in Control and Information Sciences 230: Control Problems in Robotics and Automation*, pp. 51–66. London: Springer.

Salcudean, S. and T. Vlaar (1994). On the emulation of stiff walls and static friction with a magnetically levitated input/output device. In *Proceedings of ASME WAM*, New York, pp. 303–309.

Satava, R. and S. Jones (1997). Virtual environments for medical training and education. *Presence 6*, 139–146.

Schloerb, D. (1995, Winter). A Quantitative Measure of Telepresence. *Presence 4*(1), 64–80.

Sheridan, T. (1992). *Telerobotics, Automation, and Human Supervisory Control.* Cambridge, Massachusetts: MIT Press.

Sheridan, T. (1996, Spring). Further Musings on the Psychophysics of Presence. *Presence 5*(2), 241–246.

Shimoga, K. (1993a). A survey of perceptual feedback issues in dexterous telemanipulation: Part i. In *Proceedings of the IEEE Virtual Reality Annual International Symposium VRAIS*, pp. 263–270.

Shimoga, K. (1993b). A survey of perceptual feedback issues in dexterous telemanipulation: Part ii. In *Proceedings of the IEEE Virtual Reality Annual International Symposium VRAIS*, pp. 271–279.

Sitti, M. and H. Hashimoto (1998). Tele-Nanorobotics using Atomic Force Microscope. In *Proceedings of the IEEE/RSJ International Conference on Intelligent Robots and Systems IROS*, Vancouver, Canada, pp. 1739–1746.

Stein, M. and R. Paul (1994). Operator interaction, for time-delayed teleoperation, with a behavior-based controller. In *Proceedings of the IEEE International Conference on Robotics and Automation*, San Diego, California, pp. 231–236.

Thompson, T., D. Johnson, and E. Cohen (1997). Direct haptic rendering of sculptured models. In *Proceedings Symposium on Interactive 3D Graphics*, Providence, pp. 167–176.

Thompson, T., D. Nelson, E. Cohen, and J. Hollerbach (1997). Maneuverable NURBS models within a haptic virtual environment. In *Proceedings ASME Dynamic Systems and Control Division*, Dallas, pp. 37–44.

Welch, R., T. Blackmon, A. Liu, B. Mellers, and L. Stark (1996, Summer). The Effects of Pictorial Realism, Delay of Visual Feedback, and Observer Interactivity on the Subjective Sense of Presence. *Presence 5*(3), 263–273.

Wenzel, E., F. Wightman, D. Kistler, and S. Foster (1988). The Convolvotron: Realtime synthesis of out-of-head localization. In *Second Joint Meeting Acoustic Societies of America and Japan*, Honolulu, HI.

Wermuth, S. (1998, Mai). Quantitative Investigations of Multi-Modal Telepresence Quality in Visual-Haptic Telepresence Systems. Internal Report, Institute of Automatic Control Engineering, Technical University of Munich.

Yan, J. and S. Salcudean (1993). Teleoperation controller design using h^∞ optimization. *IEEE Transaction Control Systems Technology 4*, 244–258.

Yang, B. and H. Asada (1990). Skill acquisition for robot grinding: Learning of high-level feedback laws from teaching data. In *Proceedings of the 1990 JAPAN-U.S.A Symposium on Flexible Automation*, pp. 649–655.

Yang, J., Y. Xu, and C. Chen (1993). Hidden markov model approach to skill learning and its application in telerobotics. In *Proceedings of the IEEE International Conference on Robotics and Automation*, Atlanta, Georgia, pp. 396–402.

Zelek, J. (1997). Human-robot interaction with a minimal spanning natural language template for autonomous & tele-operated control. In *Proceedings of the IEEE/RSJ International Conference on Intelligent Robots and Systems IROS*, Grenoble, France, pp. 299–305.

Zilles, C. and J. Salisbury (1995). A Constraint-based God-object Method For Haptic Display. In *Proceedings of the IEEE/RSJ International Conference on Intelligent Robots and Systems IROS*.

4

A Perspective on Advanced Strategies for Process Control (Revisited)

D. E. Seborg

Department of Chemical Engineering, University of California,
Santa Barbara, CA 93106
U.S.A, e-mail: seborg@engineering.ucsb.edu

Summary. This paper provides a personal perspective on the current status of advanced process control. First, process control strategies are classified according to the extent to which they have been applied in industry. Then important new developments in information technology and plant automation are summarized. Prominent advanced process control methods are critiqued with emphasis placed on key issues and unresolved research problems. Finally, recent developments in an important related field, process monitoring, are reviewed.

1 Introduction

As the end of the millennium approaches, it is appropriate to reflect on the current status of advanced process control, with regard to both state of the art and industrial practice. Since digital computers were first used for process control 40 years ago, it is readily apparent that dramatic improvements have occurred in computer control hardware, software and price/performance. During this period, the fields of control theory and process control have synergistically evolved relying on model-based approaches to consider an ever broader range of control problems and issues. Because excellent overviews of these historical development are available [245], [154], [155], [156], [66], [62], [23], [177] and [15], it is unnecessary to include a historical review here. Instead, the objective of this paper is to present a broad overview of the current state of advanced process control with emphasis placed on developments that have taken place since 1994, when a previous review was published by the author [203].

But first a few disclaimers. This paper is intended to provide a broad overview of advanced process control rather than an in-depth treatment of specific topics. Admittedly, this survey is somewhat subjective and undoubtedly reflects a North American perspective (or perhaps, bias). As a partial remedy, 252 references are included to serve as starting points for locating additional information about individual topics.

The term "advanced process control" is subjective, meaning different things to different people, depending on their background and experience. One way of classifying process control strategies is shown in Table 1.

Table 1. Classification of process control strategies according to the degree of use in industry.

Category I: Conventional Control Strategies
- Manual Control
- PID Control
- Ratio Control
- Cascade Control
- Feedforward Control

Category II: Advanced Control: Classical Techniques
- Gain Scheduling
- Time Delay Compensation
- Decoupling Control
- Selective/Override Controllers

Category III: Advanced Control: Widely Used Techniques
- Model Predictive Control
- Statistical Quality Control
- Internal Model Control
- Adaptive Control

Category IV: Advanced Control: Newer Techniques with Some Industrial Applications
- Optimal Control (LQG)
- Nonlinear Control
- Robust Control
- Neural Network Controllers
- Fuzzy Control
- Expert Systems

Category V: Advanced Control: Proposed Strategies with Few (if any) Industrial Applications

Category I consists of standard control strategies that have been widely used for several decades. The vast majority of automatic control loops in the process industries (~ 90%) still rely on various forms of the ubiquitous PID controller which has been commercially available for over 60 years. For many batch processing operations, process control is achieved via infrequent manual adjustments by plant operators. Consequently, it can be argued that the substantial capabilities of modern control theory and computer control systems are greatly underutilized. However, a compelling counter argument is that many industrial control problems are quite simple and consequently, one or more conservatively tuned PI or PID controllers provide a satisfactory solution. (In fact, it is doubtful whether derivative control action is widely used.) As noted by Bialkowski [28] in 1986, "Add some cascaded loops, ratio control, possibly some feedforward compensation, and you are essentially at the state of the art as practiced in the process industries". This statement is still valid for many plants. However, actual implementation of these conventional techniques can be quite complex for various configurations of selectors, ratio control, and overrides (e.g., furnace and boiler controls).

For the vast majority of process control problems, there is little incentive to apply more complicated "advanced control" methods. Instead, these strategies should be reserved for difficult control problems where they can provide significant improvements over conventional single loop controllers. Although the number of difficult problems is small compared to the total number of control loops, they typically involve critical process variables which strongly affect key control objectives such as safety, product quality, process operability, and compliance with environmental standards. Thus the economic impact can be significant [214].

The advanced control strategies in Category II are referred to as *classical* because they have been used in industry for over 30 years and were described in textbooks written by industrial control engineers in the 1960's (e.g., books by Buckley [33] and Shinskey [209]). Although these control techniques are not used in every plant or even in most plants, they do provide cost effective solutions for important classes of problems. Furthermore, the basic concepts of time delay compensation and gain scheduling are common features of many modern model-based control strategies. The process control strategies in Category III have been widely used in industry and are described in process control textbooks (e.g., Seborg et al [204]).

Category IV contains both old and new advanced control strategies that have apparently not been widely used in industry, even though successful industrial applications of each of these methods have been reported. For example, Linear Quadratic Gaussian optimal control (LQG)is a powerful control strategy that has been successfully applied to process control problems for many years (e.g., [67], [17], [248] and [225]). However, LQG has not enjoyed widespread application in the process industries for a variety of reasons that include the lack of accurate linear state-space models. (This situation could change in the future due to the new generation of state-space identification methods that have become commercially available [130], [111] and [65].) Furthermore, LQG was an important precursor to the development of the Model Predictive Control (MPC) techniques that are now widely used. Because LQG is well known and has been available for over 35 years, it will not be considered further in this paper.

Category V is a default category that includes published control strategies with few, if any, industrial applications. However, some of these techniques could very well have a significant impact on industrial practice in the future.

To what extent are the advanced control strategies in Categories III and IV actually being used in industry? It is difficult to obtain definitive answers to the question because comprehensive surveys are not available. However, specific surveys of industrial applications have been reported for Japanese industry [248], [225] and model predictive control [186] and [187]. As one specific example, consider the Mitsubishi Chemical Corporation's ethylene plant in Mizushima, Japan. Shoda [213] has reported that for the 5500 control loops, the ratio of PID control, classical advanced control, and multivariable model predictive control is 100:10:1. Obstacles that impede the application of advanced control strategies have been discussed by Pearson [171].

Specific advanced control strategies in Categories III and IV are discussed in Sections 3-7. An important related topic, process monitoring, is considered in Section 8. But first, important new developments in information technology and plant automation will be considered because they are expected to have a major impact on the practice of process control.

2 Information Technology and Plant Automation

In order for companies in the process industries to remain competitive in the international marketplace, there are two key requirements related to process control. First, there is a need for accurate, complete and timely data to be made available not only to plant operating personnel, but to other groups and even to other sites within the company. Second, virtually continuous improvements in plant automation technology are required. This technology is changing rapidly with evolutionary advances occurring every 18-30 months [77].

The current revolution in information technology is having a major impact on various process control activities. Although a detailed discussion of this important topic is beyond the scope of this paper, a few key developments will be briefly described:

2.1 Computer Networks for Plant Automation

During the past 10-15 years, a variety of digital communication networks have become commercially available and widely used in the process industries at several different levels: device, sensor, and field levels. Field level networks are of special interest in process control because they offer a number of important advantages: reduced wiring and installation costs, flexibility, and peer-to-peer communication among intelligent devices (e.g., between sensors and actuators). They also facilitate the further distribution of control functions from the computer level to the sensor and actuator level. The digital network capability and smart instruments open many new opportunities for improved plant monitoring including better diagnostic information and faster response times.

While many of these communication networks are proprietary, two prominent networks, Fieldbus and Profibus, rely on an open architecture based on published, international standards [77] and [98]. These networks have been developed by two industrial consortia (with over 100 companies each) that include virtually all of the leading suppliers of instrumentation and process control equipment. In principle, field instruments and control systems manufactured by different suppliers which are Fieldbus (or Profibus) compliant will work together seamlessly in a network of Fieldbus (or Profibus) devices. If this objective is achieved, it will allow customers to choose the "best of breed" in different instrumentation categories instead of being captive to a single supplier in order to ensure compatibility.

In recent years *smart instruments*, that is, instruments which incorporate a microcomputer, have become available and offer a number of important advantages [90], [105], [202] and [52]. Smart transmitters and actuators have the ability to condition data, perform self-diagnostic tests, and respond to network signals. Thus the use of smart instruments and digital networks such as Profibus and Fieldbus enable a wider range of process monitoring strategies. Smart instruments allow much of the computations required for routine data acquisition and control calculations to be performed at a lower level.

2.2 Object Technology and Open Software

The development and maintenance of software for tasks such as process modeling, simulation, control and optimization has been a major problem for the process industries [63]. Key issues include (i) the incompatibility of software from different vendors and (ii) the difficulty of converting legacy software to new computers or new operating systems. For plant automation it is not sufficient to merely generate data and store it using a data historian. Checking the integrity of the data and making the data available to a variety of groups within the company in a timely fashion is becoming increasingly important. Different corporate constituencies such as engineering, maintenance, plant management and corporate management, have traditionally used different software packages which include commercial packages as well proprietary software. Consequently, until recently it has been very difficult to establish and maintain an efficient flow of information through the "enterprise".

There are encouraging signs that emerging software technology such as object technology and open architectures may offer significant improvements. For over 20 years, the predominant computer control configuration has been the hierarchial distributed control system (DCS) that consists of a network of microcomputers performing the low level data acquisition and control functions, and workstations performing higher level functions. But in recent years there has been strong trend in North America toward using personal computers (primarily Windows NT) in a client-server configuration [116] and [108]. The combination of object technology, open architecture, component software, and standard protocols facilitates information flow to interested constituencies via intranets and the Internet [109], [129], [108], [63] and [97]. For example, DuPont is using object-oriented software construction methodology to develop modeling and simulation software tools [63]. The Java language and technology has also been advocated as a means to integrate real-time and enterprise data [108] and [12].

A key requirement for the desired information flows is the establishment of standard software protocols. For example, the CAPE-OPEN industrial consortium in Europe is attempting to achieve a consensus for an open architecture that will allow seamless integration of simulation and modeling software from different vendors. The ISA SP95 Enterprise/Control System Integration Committee is working to remove obstacles to integrating plant and business data.

Finally, the pervasive influence of the Internet on process control activities is already strong and growing rapidly. Clearly, the Internet provides a rapid, inexpensive, and widely available source of information that can be used to purchase instrumentation and computer hardware & software. It also has the potential for a broad range of new process control activities, providing that computer security issues can be resolved. Remote process monitoring and control loop tuning via the Internet are already a reality. Commercial products such as Honeywell's Loop Scout™ are starting to appear that facilitate remote control loop performance monitoring via Internet data transfer [96]. Balchen [16] has proposed commissioning process control research experiments at distant locations via the Internet (CYBERLAB) and a coordinating organization.

3 Selection of Control Structure

The selection of the control system structure is a major decision in control system design. Two key issues are:

1. Selection of controlled, measured and manipulated variables.
2. The choice of the control strategy (e.g., multivariable vs. multi-loop, linear vs. nonlinear).

These design issues have been recognized for decades but have become more critical in recent years, as new processes involve more recycles and greater energy integration, in response to economic pressures. In view of its central importance and widespread occurrence, it is surprising that the subject of control structure selection has received much less attention than other topics such as specific control methods (e.g., MPC, adaptive control) and the much narrower topic of PID controller tuning. However, in recent years there has been a renewal of interest among the research community and some promising new results have been reported.

Before choosing which process variables to control and manipulate, we need to determine the number of variables that can be controlled. Thus we must determine the *degrees of freedom*, i.e., the maximum number of variables that can be independently specified without generating an over-determined system. The traditional approach [204] is to determine the degrees of freedom N_F from the number of process variables N_V and the number of independent equations which relate these variables N_E,

$$N_F = N_V - N_E \tag{1}$$

In general, the number of controlled variables N_C will be less than or equal to N_F. For an interesting discussion of the determination of N_C and the relevant literature, see Luyben [141].

Van der Wal and Jager [234] have reviewed the literature on control structure design. Design methods have been based on controllability and operability considerations as well as economic considerations 160], [239] and [216]. Frequency domain analysis and robustness analysis can also provide useful insight [216] and [231].

Some interesting new concepts in control structure selection and design will be briefly mentioned. Shinnar and co-workers [4], [5] and [208] have introduced the concept of *partial control* to describe the common situation where a number of control objectives can be realized by controlling a smaller number of process variables via a decentralized control scheme. If general synthesis techniques can be developed for partial control strategies, this approach will be of considerable value. Georgakis [74] has proposed using extensive thermodynamic variables in control structure selection and to interpret slow and fast dynamic modes.

While determination of the best control structure is a challenging task, it is particularly daunting when the scope of the project is the entire plant itself, rather than a specific unit. Although research on the synthesis of *plantwide control systems* has been considered intermittently for over 30 years, a resurgence of interest has occurred in recent years. Several surveys [195], [221], [215] and a textbook [144] are recommended for their historical perspectives and evaluations of existing synthesis methods.

The plantwide control synthesis problem is especially difficult for a number of reasons that include:

1. The number of process variables that are potential candidates to be controlled or manipulated variables is very large (hundreds to thousands of variables).
2. The selection of controlled and manipulated variables is less obvious for an entire plant than it is for a single process unit.
3. The decomposition of the plantwide control problem into smaller, interconnected control problems may not be obvious due to process and control system interactions.

In order to analyze plantwide controls problem, mathematical formulations have been proposed which pose the problem as a constrained vector optimization problem. (See [221] and its references.) Existing design methods typically involve decomposition of the large plantwide control problem into a set of smaller control problems and/or the application of heuristic rules. For example, the problem decomposition can be done on the basis of process units, control structures, control objectives, or time & frequency domain scales [181], [182], [18], [194], [143], [144] and [148]. Typically, the decomposition results in a hierarchical approach with two or more levels.

It is important that various aspects of the plantwide control problem be considered during the process design itself [68], [175], [239] and [251]. The relation between process design and control can be succinctly summarized by the following quotation:

> "In the application of automatic controllers, it is important to realize that controller and process form a unit; credit or discredit for results obtained are attributable to one as much as the other. A poor controller is often able to perform acceptably on a process which is easily controlled. The finest controller made, when applied to a miserably designed process, may not deliver the desired performance. True, on badly designed processes, advanced controllers are able to eke out better results than older models, but on these processes there is a definite end-point which can be approached by the instrumentation and it falls short of perfection".

While this commentary is still relevant today, the quotation is from a 1943 paper by Ziegler and Nichols [252].

In recent years, there has been renewed interest in the the interaction between process design and process control as reflected by the larger emphasis on dynamics and control issues in process design textbooks [55] and [207].

4 Multivariable Control Strategies

When strong interactions exist between the controlled and manipulated variables, conventional multiloop PID control configurations may not be adequate. In these situations, the more general *multivariable control strategies* offer potential for significant improvements. For the purposes of this paper, the term *multivariable control* will refer to control strategies in which at least one of the manipulated variables is adjusted based on measured values of more than one controlled variable. Multivariable control techniques tend to be model-based and can provide significantly better control because the effect of each manipulated variable on each controlled variable is captured by the dynamic model of the process. However, model accuracy is obviously a key concern for all types of model-based control strategies, including multivariable control techniques.

Early multivariable control techniques consisted of using additional *decoupling controllers* to augment a multiloop control scheme. The purpose of the decouplers was to compensate for undesirable process interactions between manipulated and controlled variables. Typically, decoupling control systems have been used for problems with a small number of controlled variables (e.g., 2-4). But static decouplers have been successfully used for much larger problems such as temperature control in reforming furnaces. Although multivariable control strategies have been available for over 30 years, widespread application did not take place until the 1980's. For example, a 1976 survey article reported only 29 industrial applications of multivariable control techniques, with 10 of these involving decoupling control systems [203]. By contrast, a 1995 vendor survey by Qin and Badgwell [186] reported over 2200 applications of the most widely used multivariable technique, *model predictive control (MPC)*, which is considered in the next section.

Because multivariable control strategies are usually based on a dynamic model of the process, the availability and accuracy of the model are key issues. Although first principles models are preferred, in many practical control problems they are not readily available. Consequently, there is considerable interest in *process identification* (or *system identification*), namely, developing empirical dynamic models from input-output data. Typically, process identification is the most demanding and time consuming step in the industrial implementation of advanced control strategies [196].

A comprehensive discussion of the state of the art of process identification is beyond the scope of this paper. Instead, a few recent developments will be mentioned and references cited. The identification of linear dynamic models is a mature field but important new results continue to appear and excellent software packages are available. *Subspace identification technique*s can be used to determine state-space model structure (including time delays) as well as model parameters from input-output data [201], [65], [130] and [235]. Simulation studies for MIMO systems have demonstrated that subspace identification techniques can provide more accurate models from closed-loop identification even when correlated disturbances are present [111].

In addition to the process model itself, it is desirable to have an estimate of its accuracy. The accuracy is commonly expressed as frequency domain uncertainty bounds that are calculated in some commercial software packages. These uncertainty descriptions play a key role in robustness analysis and the design of robust control systems [216], [159], [41] and [46].

The identification of nonlinear dynamic models is a much more difficult task for several reasons which include: (i) the complex static and dynamic behavior that can occur, and (ii) the lack of a comprehensive theoretical framework for nonlinear identification, in contrast to the elegant results that are available for linear system identification [137]. These issues are discussed in a recent survey paper [136], a book chapter [173], and a new book [172] which analyzes the relation between model structure and the qualitative behavior of discrete-time, nonlinear systems. Neural network models of dynamic systems have received considerable attention in the process control literature [223], [233] and excellent commercial software is available.

Next, two multivariable control strategies that have been widely applied in industry will be considered.

4.1 Predictive Control

Model predictive control (MPC) is the multivariable control strategy that has been most widely applied in the process industries. The first MPC techniques were developed independently by two industrial groups in the 1970's. In 1976 Richalet *et al* [192] published an MPC technique which became commercialized as *IDCOM*. In 1979 Cutler and Ramaker [45] published a related technique, *Dynamic Matrix Control (DMC)*, which was developed independently. During the next 20 years, these and related MPC techniques have become the method of choice for difficult multivariable control problems in oil refineries and petrochemical plants around the world. MPC techniques based on linear dynamic models are considered in this section while nonlinear MPC methods are considered in Section 5.

The MPC literature is voluminous but fortunately, a number of informative survey articles are available [190], [73], [193], [191], [132], [157], [186] and [158]. Surveys of industrial applications [39], [71], [165], [186] and textbooks [180], [218], [34] are also available. The definitive survey paper by Qin and Badgwell [186] is especially recommended because it provides a perceptive introduction to both the theory and practice of MPC. In view of this wealth of information, the present paper will provide only a brief description of MPC followed by a discussion of several key issues.

The basis concept in MPC is that a dynamic model of the process and available measurements are used to predict future process behavior. The control calculations minimize the difference between the predicted process response and the desired trajectory. The MPC control approach was developed for difficult multi-input, multi-output control (MIMO) problems where there are significant interactions between the manipulated inputs and the controlled outputs. A key advantage of MPC is that it can accommodate inequality constraints on process variables.

In constrained versions of MPC, the on-line control calculations consist of solving a linear or quadratic programming problem at each sampling instant, as new measurements become available. Although these calculations are complex and time consuming, the sampling periods are long enough (e.g., 5-30 minutes) so that the calculations can be performed on-line. An advantage of the MPC approach is that the loss of a sensor or an actuator can be accommodated by simply changing the corresponding inequality constraints. Experience has indicated that the model identification is a key step in the successful implementation of MPC techniques. Typically, the model is developed from a series of open-loop experimental tests. In the petrochemical industry, these tests can be very time consuming (e.g., several weeks to several months of around-the-clock plant tests).

In principle the MPC approach can be utilized with a wide variety of process models, which could be physically-based or empirical, linear or nonlinear, static or dynamic, etc. However, the vast majority of the reported industrial applications [186] utilize linear, empirical dynamic models in the form of discrete step response or impulse response models. However, transfer function models [40], [34] and [218] and state-space models [134], [140], [132] can also be employed. In particular an MPC approach based on transfer function models, *Generalized Predictive Control*, has received considerable attention [34], [40], [218] and been applied in industry (e.g., [39], [34] and [31]). Also, a variety of adaptive controllers based on multi-step predictions have been published [205]. Because the various linear models are all interrelated, the choice of MPC model form is largely a matter of convenience and the preference of the control system designer or end user.

In a 1995 vendor survey by Badgwell and Qin [186], over 2200 applications of MPC were reported, largely for oil refineries and petrochemical plants around the world. Since then, the number of MPC applications has grown rapidly so that the current total could be over 3000. The largest application in the 1995 survey involved 40 controlled variables and 80 manipulated variables. According to this survey, MPC has had much less impact in the other process industries. This lack of activity is somewhat surprising because MPC is a general technique that does not rely on specific attributes of oil refineries and petrochemical plants. A 1993 estimate suggested that over 20,000 applications could benefit from MPC [191].

One reason why MPC has become a major commercial success is that there are over 15 vendors worldwide who are licensed to market MPC products and who install them on a turnkey basis. Consequently, even medium-sized companies have been able to take advantage of this new technology. Payout times of 3-12 months have been widely reported. Unconstrained, single-input/single-output (SISO) versions of MPC have been available at the DCS level for many years but appear to have had only limited commercial success. There is no fundamental reason why DCS products cannot include constrained versions of MPC for at least small problems (e.g., 2-10 controlled variables). Generic MPC software at the DCS level would offer users the advantage of freeing them from having to rely on outside vendors for installation and any subsequent modification of proprietary products. New DCS systems are starting to include MPC at the function block level (e.g., Fisher-Rosemount's *Delta V* system).

4.2 Cross Direction Control of Sheet Making Processes

There is a second class of MIMO model-based control technique that has had widespread application in the process industries. Several vendors market software packages which provide *cross direction control (CDC)* of paper machines and other sheet & film making processes. For paper making, typical a control objective is to minimize the variations of paper thickness, moisture content etc. across the width of the moving sheet of paper. The control strategy is based on an empirical MIMO model whose dynamics are dominated by a large time delay.

A 1985 article (*Pulp & Paper*, February, 1985) reported that 960 CDC systems had already been installed at a cost of over $500 million. The typical payout time was six months. Because the 960 systems represented 20% of the total number of paper machine computer systems, it is clear that this model-based control strategy received rapid acceptance. The CDC strategy is so widespread that it would be unusual for a new paper machine to not have at least one CDC application as part of its control strategy.

5 Nonlinear Control Techniques

Because many important industrial control problems exhibit inherently nonlinear behavior, conventional PID controllers and advanced controllers based on linear models (e.g., MPC) must be tuned conservatively to ensure stability over the entire range of operating conditions. This traditional approach can result in very sluggish responses for much of the operating region. Consequently, there are significant incentives for the development of control strategies based on nonlinear process models, especially physically-based models. But the implementation of nonlinear model-based control strategies typically requires a much greater investment of time and effort, compared to linear control strategies such as MPC [164] and [43]. Thus, the size of the potential market for nonlinear model based control strategies is not readily apparent. Cott of Shell Canada [43] has suggested that, "... the typical number of nonlinear loops in a plant will be another order of magnitude less than the number of linear MPC loops. But these nonlinear loops will certainly be extremely high value loops with large economic driving forces".

For purposes of control system design, it would be desirable to have an effective screening technique to determine if a nonlinear control strategy will be required, based on a limited amount of *a priori* information. A number of analysis techniques have recently been proposed which attempt to characterize static and dynamic nonlinearities and their effect on closed-loop control [79], [80], [219] and [89]. However, the existing assessment methods have not yet produced practical tools suitable for control system design.

Nonlinear process control, and the more general topic of nonlinear control theory, are very active research areas as indicated by recent survey articles [3], [164], [26], [27], [29], [91], [187] and textbooks [217], [133], [106], [126] and [94]. But first, we consider some of the more traditional approaches which have been used in industry for decades.

5.1 Traditional Nonlinear Control Strategies

Early nonlinear control strategies favored three types of approaches:

1. nonlinear modifications of standard PID control algorithms
2. nonlinear transformations of input or output variables
3. gain scheduling.

Shinskey [211] provides a good introduction to these methods and related techniques while Bequette [27] has recently published a comprehensive literature survey.

Standard PID control laws have been modified by making the controller gain a function of the control error. For example, the controller gain can be piecewise constant or vary with the absolute value of the control error,

$$K_c = K_{c0}(1 + a\,|\,e(t)|\,) \qquad (2)$$

where K_{c0} and a are constants. This type of modification has proven to be useful, even for problems where the process behaves linearly such as level control in surge vessels [211] and [27].

Many classical and modern nonlinear control strategies have the design objective of making the closed-loop operation linear or much less nonlinear. If successful, this general approach allows the process to be controlled over a wider range of operating conditions and in a more predictable manner. Method 2 consists of using simple linear transformations of input or output variables; it can be remarkably effective. Common examples include using the logarithm of a product composition as the controlled variable for high purity distillation columns, or adjusting the ratio of feed flow rates in blending problems. The major limitation of this approach is that it is difficult to generalize because the appropriate variable transformations are application specific. However, the use of thermodynamic extensive variables provides a more general framework [74].

In the third traditional approach, *gain scheduling*, one or more controller settings are adjusted automatically based on the measured value of a *scheduling variable*. This variable is usually the controlled variable or setpoint, but it could be the manipulated variable or some other process variable (an *auxiliary variable*). Usually, only the controller gain is adjusted because many important industrial processes have variable steady-state gains but relatively constant dynamics (e.g., pH neutralization). This type *of nonlinear gain compensation* has been widely used in industry and has been an option in commercial control systems for decades. Consequently, it is surprising that very little theoretical analysis has been available until recent years [131] and [226].

5.2 Input-Output Linearization Techniques

In recent years, general model-based nonlinear control strategies have been developed using both physical models and a wide range of empirical models. A popular synthesis method for classes of nonlinear models is based on *exact linearization* or an *input-output linearization* approach. By incorporating an (approximate) inverse of the process model into the control strategy, the closed-loop system will exhibit linear behavior under ideal conditions. Analytical solutions can be obtained for general classes of nonlinear dynamic such as *affine* models that are linear with respect to each manipulated input. Either first principles models or empirical models can be employed. Specific versions of the general input-output linearization approach include Reference System Synthesis [20], Nonlinear IMC [93], Nonlinear Decoupling [14] and Generic Model Control [133].

The extensive literature on the input-output linearization approach has been surveyed in review articles [125], [149], [92], [26], [124], [164], textbooks [106], [94], and a NATO Workshop [27]. Successful experimental applications of input-output linearization techniques have been reported by a number of university groups (e.g., [93]). The number of industrial applications appears to be small but growing [3], [43], [133] and [164].

The vast majority of these simulation and experimental studies involve control problems with only a few input and output variables. Thus there is little evidence to indicate that the input-output linearization approach will be beneficial for larger control problems (e.g., the Tennessee Eastman Challenge Problem [56]). A major shortcoming of most current input-output linearization methods is that they do not consider inequality constraints which are common in practice. However, extensions to constrained problems are being developed [127].

5.3 Nonlinear Model Predictive Control

Nonlinear model predictive controllers (NMPC) can be designed by extending the popular MPC approach for linear models to classes of problems where the process model is nonlinear. The relevant literature is large and growing rapidly, as indicated by recent survey articles [29], [147], [188], [91], [187], and a book chapter [150].

As indicated in these surveys, a wide variety of nonlinear models and problem formulations can be employed with no single model type being dominant at this time. If a general nonlinear dynamic model is considered with hard inequality constraints, then a nonlinear programming problem results which must be solved on-line at each sampling period. This is a much more difficult and time-consuming task than the LP and QP calculations employed in linear MPC. For large NLP problems (>10,000 variables), interior points methods are very promising approach [78]. An alternative approach is to update linear models as conditions change or to use a set of a set of local models (either linear or nonlinear) to accommodate different operating regimes [110], [230] and [232].

Qin and Badgwell [187] have recently summarized industrial applications of NMPC based on a 1998 survey of five NMPC vendors. This survey located a total of 86 industrial applications. The vast majority of these applications involved two commercial packages: Pavilion Technologies' *Process Perfecter* (43 applications) and Continental Controls' *Multivariable Control* (*MVC)* product (36 applications). Both products divide the control calculations into a steady-state optimization followed by a dynamic optimization; both can accommodate hard constraints. Process Perfecter control calculations are based on a nonlinear steady-state model and a dynamic model that used gains calculated from the steady-state model. Significant reductions in plant testing time have been reported using this approach [115].

5.4. Summary of Nonlinear Process Control

For specific classes of control problems, both traditional and newer model-based control strategies can provide significant improvements over linear control techniques. A number of successful applications of both input-output linearization and NMPC techniques have been reported. Although the potential advantages of nonlinear model-based control strategies are readily apparent, there are a number of major unresolved issues:

1. It would be desirable to have practical screening tool for nonlinear control problems. At the present time we do not have a general and practical method for deciding whether a nonlinear control strategy will offer a significant improvement over linear control alternatives, based on a limited amount of *a priori* information.

2. The amount of time and effort to develop and maintain a nonlinear model is a key issue, perhaps the overriding issue in the implementation of nonlinear model-based control. Typically, the effort required to develop a nonlinear model is much greater than for a linear model. For example, Cott [43] of Shell Canada reports that,

"Nonlinear model development tools are scarce making the modeling procedure time-consuming and imprecise. For example, my experience with nonlinear model development indicates that the time required to obtain accurate nonlinear models is over two orders of magnitude greater than that required for a linear model using time series techniques".

Major advances in nonlinear model development, for purposes of on-line control, would pave the way for more widespread application of nonlinear control techniques.

3. Theoretical analysis of closed-loop performance properties such as stability and robustness is very difficult, especially if inequality constraints are present. Thus a case study approach is often employed with no guarantee that the reported results can be generalized, or are even representative.

4. For the NMPC approach, the computational effort required to solve the required NLP problem at each sampling instant can be substantial with convergence of the optimization algorithm difficult to guarantee. A satisfactory resolution of these computational issues could facilitate the widespread use of NMPC methods.

6 Controller Tuning and Adaptive Control

After a control system is installed in the plant, controller tuning is often required to determine suitable controller settings. This typically involves plant tests which are both time consuming and intrusive. Furthermore, after process conditions change, the old controller settings may not be satisfactory and thus the controller should be re-tuned. If the process characteristics change in significant and unanticipated ways, then continuous re-tuning (or *adaptive control*) may be required. The development of simple, effective methods for updating controller setting to compensate for changing process conditions would be beneficial for both model-based controllers and conventional PID controllers.

The topic of tuning single-loop PID controllers has received an inordinate amount of attention in the process control literature. Arguably, there may be more publications on this topic than any other in process control. Most of the academic publications have been concerned with tuning relations for the ideal PID control algorithm. But since this form is seldom used in practice [7], the recommended controller settings may have to be modified. Many commercial DCS control packages contain automatic tuning features. See Åström and Hägglund [7] for a detailed comparison of commercial auto-tuning techniques.

Despite the extensive literature, there have been a number of interesting new developments in PID controller tuning in recent years. For example, modifications of the popular relay auto-tuning method [204] and [7] have been reported (e.g., [139]) including extensions to multi-loop control [83], [167] and [240]. Refinements to the ZN rules and tuning relations for unstable processes have been published [142], [8] and [9]. A new pattern recognition, adaptive PI controller has been developed and widely used in HVAC applications [206].

In contrast to the voluminous literature on tuning PID controllers, techniques for re-tuning model-based controllers have received relatively little attention in the open literature. Two key issues include: (i) How do we decide that the controller is not performing well? (ii) How do we decide whether to re-tune the controller or to update (or re-identify) the process model?

During the past 25 years, *adaptive control* is one of the advanced control techniques that has received the most attention. The basis concept is very appealing, namely, to have a control system that can automatically adjust its settings to accommodate changing process conditions. Thus the controller "adapts" to changes in the process and to other unforeseen conditions. It is convenient to classify adaptive control strategies as either being *direct methods or indirect methods*. In a direct method the controllers parameters are adjusted based on input-output data collected under closed-loop conditions. By contrast, in the indirect approach, model parameters in an assumed model structure are estimated via a recursive parameter estimation technique from the same closed-loop data. Detailed descriptions of adaptive control techniques for both linear and nonlinear control systems are available in textbooks [11], [84], [102], [126], survey articles [205], [81], [59], [8] and their respective bibliographies.

Adaptive control research has been conducted for over 40 years while commercial adaptive control systems have been available since the early 1980's [244] and [6]. Probably the most widely used STC is the NOVATUNE controller, which was originally marketed by ASEA in Sweden about 1982. A different type of adaptive controller, the EXACT controller, was introduced by the Foxboro Company in 1984. The EXACT is an adaptive PID controller which is based on a proprietary pattern recognition approach.

It is difficult to make accurate assessments of industrial utilization of adaptive control methods. Åström [6] has reported that there were 100,000 adaptive control loops running in 1988 but this estimate includes "auto-tuners" which provide PID controller tuning on an "on-demand" basis.

It is important to note that adaptive control can also be implemented as a custom version of a specific controller, rather than as a general purpose adaptive controller. For example, in a recent application of a nonlinear control strategy to a pH neutralization process, an adaptive version had to be employed in order to obtain satisfactory control [93].

7 Artificial Intelligence Techniques

During the past decade there has been intense interest in developing artificial intelligence (AI) techniques for a wide variety of scientific and engineering applications. A comprehensive survey paper by Stephanopoulos and Han [220] provides a thorough review of intelligent systems in process engineering and contains 385 references. The process control research in this area has largely been concerned with three AI methods: knowledge-based systems, neural networks, fuzzy logic, and various combinations of these techniques. But software based on genetic algorithms and "chaos systems" has also been commercialized [19]. A number of overviews of AI control research are available [237], [242], [103], [85], [49] and [135]. Surveys of industrial applications and commercial AI software for control have also been reported [249], [37] and [19].

7.1 Knowledge-Based Systems

Knowledge-based systems (KBS), also referred to as *expert systems*, use a set of *rules* to perform logical inferences about the state of a process operation or some other activity of interest. Stephanopoulos and Han [220] note that industrial applications of KBS systems have largely been concerned with either diagnostic & monitoring activities or supervisory control. Supervisory control applications have included the following problems: complex control schemes; recovery from extreme conditions; and emergency shutdowns.

A 1992 trade journal article [199] provides an overview of KBS applications in the process industries. They report that "... Du Pont has well quantified the benefits it is reaping from the thousand or so KBSs it has in place". In the same issue, Samdani [198] reports that, " ... [Du Pont] says that there are about 20,000 more areas of applications yet to be tapped. For every dollar spent in implementing a KBS for process control, the payoff is from six to ten dollars per year". A first-hand perspective of Dupont's KBS experience has been reported by Rowan [197].

Applications of expert systems are facilitated by real-time KBS systems that enable the user to integrate plant data and process models in an expert system shell which has a sophisticated graphical interface. This combination provides a powerful vehicle for on-line process monitoring, especially diagnostics and fault detection. At the present time, the most widely used system is the G2 product from the GENSYM Corporation (Cambridge, MA). Over 4000 G2 systems have been installed worldwide [153].

The early enthusiasm for KBS has been tempered by the realization that a considerable effort is required to codify the available expertise. Furthermore, if each potential application has a significant number of unique features, it is less feasible to spread the development costs over a large number of projects. Despite this inherent problem, the industrial employment of KBS for applications such as process diagnosis and supervisory control is significant and growing at an impressive rate [220] and [225].

7.2 Neural Networks

Neural networks have been widely used to develop empirical nonlinear models for a wide variety of scientific, engineering, and business applications [243], [103] and [223]. In the area of control engineering, they have been used for a variety of activities including developing dynamic models, gain scheduling, control system design, process monitoring, and batch-to-batch optimization [101], [118], [233], [223], [2], [226] and [53].

Standard neural network models consist of three layer networks with sigmoidal functions used as the *activation function* for each neuron in the hidden layer. However, networks which consist of linear combinations of *radial basis functions* offer significant theoretical and computational advantages for small to medium sized problems [178], [220] and [88]. There is also considerable interest in incorporating *a priori* physical information (e.g., steady-state relations, knowledge of constraints) into otherwise empirical neural net models [183], [222], [227], [1], [42] and [179].

The commercial availability of neural network software for use by non-specialists should increase the current widespread interest in neural network applications for process control. For example, a recent survey of nonlinear predictive control [187] indicated that most of the industrial applications of nonlinear model predictive control have employed commercially available neural network control software.

7.3 Fuzzy Control Systems

Fuzzy logic provides a conceptual framework for practical problems where some process variables are represented as *linguistic variables* which have only a few possible values (e.g., very large, large, normal, small, etc.). The linguistic variables can then be processed using a set of rules. Thus applications of fuzzy logic and fuzzy control can be viewed as special cases of KBS that have fuzzy boundaries for the rules. Alternatively, fuzzy control can be viewed as a formal methodology for incorporating heuristic knowledge into control system design or as a heuristic design method for nonlinear controllers [250].

There has been considerable controversy concerning fuzzy control and its relative merits viz. conventional control and model-based control [189], [212] and [250]. One of the reasons for this controversy is that there is no theoretical framework for analyzing the closed-loop properties of fuzzy control systems, although some progress has been made and is summarized in recent books [174], [85], [168] and [166]. As Stephanopoulos and Han [220] aptly note, "Using fuzzy controllers takes a lot of testing and/or faith."

In recent years, fuzzy control has been widely used in consumer products such as washing machines, vacuum cleaners, and camcorders where a high level of control system performance is not required [250]. Industrial applications of fuzzy control to process control problems have begun to appear, perhaps more frequently in Japan than in the United States or Europe. For example a 1995 industrial survey in Japan indicated that fuzzy control has been used in 38% of the plants that were surveyed while MPC applications were reported in 41% of the plants [225].

Much of the fuzzy control literature, and some commercial software, employ fuzzy rules in combination with PID control. For example, various combinations of fuzzy logic and auto-tuning have been commercialized [200]. Fuzzy logic has also been effectively used for high-level control and monitoring functions in supervisory control [38] and [70]. As Yurkovich [250] has recently noted, "it is difficult to access what the prevailing attitude in U.S. industry is today toward fuzzy control". A 1993 trade journal article [200] states, "Although FL has yet to prove its worth to most [process] engineers, particularly in the U.S., vendors believe that it is here to stay, and are steadily commercializing FL-based products. They report that more than 10,000 FL-embedded PID controllers have been sold by the Yokogawa Corp. of America. Specific industrial applications of fuzzy control are available in survey papers [57], [220] and [250].

Regrettably, a high degree of "hype" was associated with the initial introduction of these three AI technologies and consequently, early expectations were not always fulfilled. But it is important to keep in mind that these new approaches for process control (and the software) still tends to be "first generation". As the technology and available software continues to improve, widespread industrial applications are quite likely.

8 Process Monitoring and On-Line Diagnostics

The complexity of many processes and the increasing degree of data acquisition (e.g., thousands of variables with sampling periods of 1-60 seconds) have resulted in an overwhelming amount of current and historical data. It is often said that industrial plants are "data rich, but information poor". Furthermore, there are considerable financial incentives for the development of diagnostic techniques to monitor plant performance and detect abnormal situations. For example, in 1995 an industrial consortium estimated that preventable, abnormal situations cost the U.S. petrochemical industry, $10 billion dollars per year [162]. Improved plant monitoring can play a key role in increasing plant productivity, an important concern in the intensely competitive global economy.

The traditional approach for monitoring processes has been to perform limit checks on measurements and make simple engineering calculations (e.g., furnace efficiencies, mass balance closures). During the 1980s, the renewed emphasis on product quality and the successful experience in Japan, led to the widespread use of statistical quality control (SQC) methods and quality control charts such as Shewhart and CUSUM charts. These methods are typically used to monitor product *quality variables* that are measured infrequently (e.g., once every 4-24 hours). The standard SQC techniques are less suitable for monitoring process variables such as flow rates and temperatures because the commonly used charting techniques rely on the assumption that the measurements are identically and independently distributed (the *IID assumption*). Statistical monitoring based on the integration of SQC methods and standard feedback control has also been proposed [32].

In recent years the development of improved techniques for on-line monitoring and fault detection has rapidly become a major research area, both in process control and in the more general control engineering field. In process monitoring typical objectives are to first to determine whether an abnormal situation has occurred (*fault detection*), and then to locate the source of the problem (*fault diagnosis*). Ideally, this analysis should detect potential problems before they have serious consequences (e.g., resulting in production of off-spec material or a plant shutdown etc). Related problems include *sensor validation* to determine malfunctioning sensors [21], [60], [90] and *data reconciliation* [44] to provide consistent estimates of measured variables.

For purposes of this paper, it is convenient to group process monitoring techniques into four categories:

1. Model-based methods
2. AI-based systems
3. Data analysis techniques
4. Miscellaneous methods

Admittedly, this classification is somewhat arbitrary and various combinations of individual methods can be used, especially for fault diagnosis. More detailed classification schemes are available in survey papers [69], [104], [47], [48], [123], [220] and [169].

If an accurate dynamic model of the process is available, then effective fault detection and diagnosis strategies can be developed based on residual analysis and state & parameter estimation. The dynamic models can be either physical or empirical but most of the theoretical development has been for linear, discrete-time state-space models. Deterministic models have received the most attention but stochastic models can be used to good advantages, especially when correlated or nonstationary disturbances are important. Model matching approaches have been developed for the fortuitous situations where a separate physically-based model can be constructed for each known fault. For an overview of the extensive literature on model-based monitoring strategies, see survey papers [69], [104], [123] and books [95], [170], [22] and [75].

The AI techniques described in Section 7 have been successfully used to monitor industrial processes and can be used in combination with model-based and statistical methods. Recent symposia [49], [51] and survey papers [220], [123] describe specific approaches and applications. Qualitative trend analysis and wavelet analysis has also been successfully used for process monitoring [13] and [236].

8.1 Multivariate Statistical Monitoring Techniques

In situations where the development of a model-based or a monitoring technique would require considerable effort, monitoring techniques based on data analysis provide an attractive alternative. A diverse group of data analysis methods can be used individually or in combinations, as described in comprehensive survey articles [123], [47] and [48]. For example, if historical data are available for a number of known faults, pattern classification techniques [58] or neural networks can then be trained to diagnose faults. Filtering techniques and multiscale analysis can also provide useful information [48] and [138].

In the process control literature, process monitoring based on multivariate statistically based projection techniques, such as *Principal Component Analysis (PCA)* and *Projection to Latent Structures (PLS*), have perhaps received more attention then any other data analysis method. These methods are especially attractive where the available process measurements are highly correlated but only a small number of underlying events produce unusual patterns [145], [120] and [146]. For example, the tray temperatures in a distillation column will respond in a correlated manner after a significant change in feed composition. When the process data are highly correlated, the original process data can be projected onto a smaller number of principal components (or latent variables), thus reducing the dimension of the problem [107]. The PCA and PLS models are linear models and can easily be developed from representative plant data for "normal" (i.e., fault-free) operating conditions. No special plant tests are required but the choice of this reference data set is crucial to the success of the approach. Because multivariate statistical monitoring methods take into account the correlation between process (or product quality) variables, they are able to detect smaller plant upsets than the traditional univariate SQC methods.

PCA and PLS models have been successfully used in a variety of applications which include inferential control [112], [113] and process control [176], [114], [36] as well as process monitoring [145], [120], [146], [119] and [61]. But standard PCA and PLS techniques do have some disadvantages because they are based on linear empirical models and do not explicitly account for process dynamics. Nevertheless, successful applications have been reported for batch processes where both dynamic and nonlinear behaviors are key features [163], [119], [54] and [146]. Also, extensions of the PCA and PLS methods have been proposed to account for nonlinear operation [246], [122] and [53] and dynamic behavior [247], [128] and [185]. Process dynamics can be accommodated by using dynamic models and monitoring techniques based on canonical variables [241] and [161].

The range of multivariate statistical techniques can be enhanced by evaluating model residuals rather than the raw data itself [76] and [238].

8.2 Monitoring Control System Performance

In addition to monitoring the performance of the process, it is very desirable to evaluate how well the control system is performing. This would appear to be an important issue for both standard PID control systems and model-based control strategies such as MPC. As Benson [24] has recently noted, "If as a process industry we do not measure our control performance, maybe it is not surprising that it does not improve". Some key issues in control system performance monitoring are:

1. Is the control system performing up to expectations?
2. If not, can the performance be significantly improved using the present instrumentation and control configuration? (e.g., by sensor or actuator maintenance or by re-tuning the controller.)
3. If the answer to Question 2 is negative, can significant improvement be obtained by application of an "advanced" control strategy?

Despite the importance and centrality of these issues in process control, they have received little attention until the last decade [210], [9], [50] and [64]. Much of the recent research has been based on using statistical techniques, especially assessing how close the control loop performance is to a benchmark, minimum variance control [50], [86], [121], [229], [152] and [184]. This approach was developed for single control loops but multivariable extensions have recently been reported [86], [87], [99] and [100]. A statistically-based monitoring technique for MPC has been reported by Kesavan & Lee [117].

Monitoring techniques have also been developed for specific problems such as the automatic detection of excessively oscillatory control loops [82], [151], [228] and control system evaluation based on a single closed-loop bump test [224].

Conclusions

Process control continues to be a vital, important field with significant unresolved research problems and challenging industrial applications. Nonlinear control strategies, artificial intelligence techniques, and process monitoring are popular research areas with an increasing number of commercial products and industrial applications being reported. The current revolution in information technology and plant automation will continue to have a major impact on process control.

References

1. Agarwal M 1997 Combining Neural and Conventional Paradigms for Modeling Prediction and Control. *Int. J. Systems Science* 28:65-81.

2. Agarwal M 1997 A Systematic Classification of Neural-Network-Based Control. *IEEE Control Systems Magazine,* April 1997, 75-93.

3. Allgöwer F, Doyle III F J 1997 Nonlinear Process Control – Which Way to the Promised Land? *Chemical Process Control-V AIChE Symposium Series* 93(316):24-45.

4. Arbel A, Rinard I H, Shinnar R, Sapre A V 1995 Dynamics and Control of Fluidized Catalytic Crackers. 2. Multiple Steady States and Instabilities. *Ind. Eng. Chem. Res.* 34:3014-3026.

5. Arbel A Rinard I H, Shinnar R 1996 Dynamics and Control of Fluidized Catalytic Crackers. 3. Designing the Control System: Choice of Manipulated and Measured Variables for Partial Control. *Ind. Eng. Chem. Res.* 35: 2215-2233.

6. Åström K J 1989 Toward Intelligent Control. *IEEE Control Systems Magazine,* April, 1989, 60-64.

7. Åström K J, Hägglund T 1995 *PID Controllers: Theory, Design, and Tuning, 2nd Ed.* ISA, Research Triangle Park, NC.

8. Åström K J, Hägglund T, Hang C C, Ho W K 1993 Automatic Tuning and Adaptation for PID Controllers – A Survey. *Control Eng. Practice* 1:699-714.

9. Åström K J, Hang C C, Persson P, Ho W K 1992 Towards Intelligent PID Control. *Automatica* 28:1-9.

10. Åström K J, McAvoy T J 1992 Intelligent Control. *J. Process Control* 2:115-127.

11. Åström K J, Wittenmark B 1994 *Adaptive Control, 2nd ed.* Addison-Wesley, Reading, MA.

12. Atherton R W 1998 Moving Java to the Factory. *IEEE Spectrum* 35(12): 18-23.

13. Bakshi, B R, Stephanopoulos G 1994 Representation of Process Trends: Part IV: Induction of Real-Time Patterns from Operating Data for Diagnosis and Supervisory Control. *Comput. & Chem. Engng.* 18:303-332.

14. Balchen J G 1991 Nonlinear Decoupling in Process Control. *Model Ident. Control* 2:81-94.

15. Balchen J G 1999 How Have We Arrived at the Present State of Knowledge in Process Control? Is There a Lesson to Be Learned? *J. Process* Control 9:101-108.

16. Balchen J G 1999 Private communication (see http://www.cyberlab.org.)

17. Balchen J G, Mummé K 1988 *ProcessControl: Structures and Approximations*, Van Nostrand Reinhold, NY.

18. Banerjee A, Arkun Y 1995 Control Configuration Design Applied to the Tennessee Eastman Plantwide Control Problem. *Comput. Chem. Eng.* 19:453-480.

19. Bartos F J 1997 Artificial Intelligence: Smart Thinking for Complex Control. *Control Eng.*, July 1997, 44-52.

20. Bartusiak R D, Georgakis C, Reilly M J 1989 Nonlinear Feedforward/Feedback Control Structures Designed by Reference System Synthesis. *Chem. Eng. Sci.*, 44: 1837-1851.

21. Basseville M 1998 On-board Component Fault Detection and Isolation Using the Statistical Local Approach. *Automatica* 34:1391-1415.

22. Basseville M, Nikiforov I V 1993 *Detection of Abrupt Changes.* Prentice Hall, Englewood Cliffs, NJ.

23. Bennett S 1996 A Brief History of Automatic Control. *IEEE Control Systems Mag.*, June, 1996, 17-25.

24. Benson R S 1994 The Process Industry Requirements of Advanced Control Techniques: Challenges and Opportunities. *Preprints of the IFAC ADCHEM '94 Symposium* 215-224, Kyoto, Japan.

25. Benson R, Perkins J 1997 The Future of Process Control – A UK Perspective. Chemical Process Control–V AIChE Symposium Series 93(316):192-198.

26. Bequette B W 1991 Nonlinear Control of Chemical Processes – a Review. *Ind. Eng. Chem. Res.* 30: 1391-1413.

27. Bequette B W 1998 Practical Approaches to Nonlinear Control. In *Nonlinear Model-based Process Control,* NATO ASI Series, Berber E R, Kravaris C (Ed.), Kluwer, Dordrecht.

28. Bialkowski W L 1986 Control System Engineering: Where Have We Gone Wrong? *InTech,* February, 1986, 27-32.

29. Biegler L T 1998 Advances in Nonlinear Programming Concepts for Process Control. *J. Process Control* 8:301-311.

30. Blevins T, Wojsznis W 1996 Fieldbus Support for Process Analysis. *ISA Trans.* 35:177-183.

31. Bordons C, Camacho F 1998 A Generalized Predictive Controller for a Wide Class of Industrial Processes. *IEEE Trans. Control System Technol.* 6:372-387.

32. Box G E P, Luceño A 1997 *Statistical Control By Monitoring and Feedback Adjustment*. John Wiley, NY.

33. Buckley P S 1964 *Techniques of Process Control*. John Wiley, NY.

34. Camacho E F, Bordons C 1995 *Model Predictive Control in the Process Industry*. Springer-Verlag, Berlin.

35. Chen G, McAvoy T J 1998 Predictive On-Line Monitoring of Continuous Processes. *J. Process Control.* 8:409-420.

36. Chen G, McAvoy T J, Piovoso M J 1998 A Multivariate Statistical Controller for On-Line Quality Improvement. *J. Process Control* 8:139-149.

37. Chiu S 1997 Developing Commercial Applications of Intelligent Control. *IEEE Control Systems Magazine*, April 1997, 94-97.

38. Chiu S 1998 Using Fuzzy Logic in Control Applications: Beyond Fuzzy PID Control. *IEEE Control Systems Magazine*, October, 1998, 100-105.

39. Clarke D W 1988 Application of Generalized Predictive Control to Industrial Processes. IEEE *Control Systems Magazine*, April, 1988, 49-55.

40. Clarke D W, Mohtadi C, Tuffs P S 1987 Generalized Predictive Control – Part I. The Basic Algorithm. *Automatica* 23:137-148.

41. Cooley B L, Lee J H 1998 Integrated Identification and Robust Control. *J. Process Control* 8:431-440.

42. Costa A C, Alves T L M, Henriques A W S, Maciel Filho R M, Lima E L 1998 An Adaptive Optimal Control Scheme Based on Hybrid Neural Modeling. *Computers Chem. Eng.* 22: Suppl. S859-S862.

43. Cott B J 1991 Progress and Challenges in Nonlinear Model-Based Control. Paper presented at the *41st Annual Can. Chem. Eng. Conf., Vancouver, B.C.*, October 1991.

44. Crowe, C M 1996 Data Reconciliation-Progress and Challenges. *J. Process Control,* 6:89-98.

45. Cutler C R, Ramaker B L 1979 Dynamic Matrix Control – A Computer Control Algorithm. AIChE National Meeting, Houston TX.

46. Dahleh M A, Diaz-Bobillo I J 1995 *Control of Uncertain Systems.* Prentice Hall, Englewood Cliffs, NJ.

47. Davis J F, Bakshi B R, Kosanovich K A, Piovoso M J 1996 Process Monitoring, Data Analysis and Data Interpretation, *AICHE Sympos. Ser.* 92:1-11.

48. *Ibid,* 1999 Process Data Analysis and Interpretation, *Advances in Chem. Eng.*, Academic Press, NY (in press).

49. Davis J F, Stephanopoulos G, Venkatasubramanian V (Ed.) 1996 *Intelligent Systems in Process Engineering.* AIChE Symposium Series 92:312.

50. Desborough L, Harris T 1993 Performance Assessment Measures for Univariate Feedforward/Feedback Control. *Can. J. Chem. Eng* 71:605-616.

51. Dhurjati P S, Cauvin S 1998 (Ed.) *IFAC Workshop on On-Line-Fault Detection and Supervision in the Chemical Process Industries,* Solaize (Lyon), France, June 1998.

52. Dierauer P 1998 Smart Sensors Offer Increased Functionality. *InTech*, May, 1998, 60-63.

53. Dong D, McAvoy T J 1996 Nonlinear Principal Component Analysis – Based on Principal Curves and Neural Networks. *Computers Chem. Eng.* 20:65-78.

54. Dong D, McAvoy T J, Zafiriou E 1996 Batch-to-Batch Optimization Using Neural Network Models. *Ind. Eng. Chem. Res.* 35:2269-2276.

55. Douglas, J 1988 *Conceptual Design of Chemical Processes*, McGraw-Hill, NY.

56. Downs J J and Vogel E F 1993 A Plant-Wide Industrial Process Control Problem. *Comput. Chem. Eng.* 17:245-255.

57. Dubois D 1993 Fuzzy Sets: A Survey of Engineering Applications. *Comput. & Chem. Eng.* 17:S373-380.

58. Duda R O, Hart P E 1973 *Pattern Classification and Scene Analysis.* John Wiley, NY.

59. Dumont G 1986 Application of Advanced Control Methods in the Pulp and Paper Industry – a Survey. *Automatica*: 22:143-154.

60. Dunia R and Qin S J 1998 Joint Diagnosis of Process and Sensor Faults Using Principal Components Analysis. *Control Eng. Practice* 6: 457-469.

61. Dunia R, Qin S J 1998 Subspace Approach to Multidimensional Fault Identification and Reconstruction. *AIChE J.* 44:1813-1831.

62. Edgar T F 1996 Modeling and Control – Back to the Future, Part I. *CAST Communications* 19(1):7-12.

63. Edwards P D, Merkel G 1996 Impact of an Open Architecture Environment on the Design of Software Components for Process Modeling. *AIChE Sympos. Ser.* 92, No. 312, 307-310.

64. Ender D 1993 Process Control Performance: Not as Good as You Think. *Control Eng.*, September, 1993, 180-190.

65. Favoreel W, De Moor B, Van Overschee P 1998 Subspace State Space System Identification for Industrial Processes. *Preprints of the 5th IFAC Sympos. On Dynamics and Control of Process Systems (DYCOPS-5)*, Corfu, Greece 359-364.

66. Fisher D G 1991 Process Control: An Overview and Personal Perspective. *Can. J. Chem. Eng.* 69:5-26.

67. Fisher D G, Seborg D E 1976 *Multivariable Computer Control: A Case Study.* Elsevier, NY.

68. Fisher W R, Doherty M F, Douglas J M 1988 The Interface Between Design and Control. 1, 2 and 3.: 1: Process Controllability, 2: Process Operability 3: Selecting a Set of Controlled Variables. *Ind. Eng. Chem. Res.* 27:597-615.

69. Frank, P M 1990 Fault Diagnosis in Dynamic Systems Using Analytical and Knowledge-Based Redundancy – a Survey and Some New Results. *Automatica* 26:459-474.

70. Frank P M 1994 Application of Fuzzy Logic to Process Supervision and Fault Diagnosis. Proc. *IFAC Sympos. on Fault Detection, Supervision and Safety for Technical Processes*, Espoo, Finland, 1994.

71. Froisy J B 1994 Model Predictive Control: Past, Present and Future. *ISA Trans.* 33:235-243.

72. Garcia C E, Morari M 1982 Internal Model Control I. A Unifying Review and Some New Results *Ind. Eng. Chem. Process Des. & Develop.* 21:308-323.

73. Garcia C E, Prett D M, Morari M 1989 Model Predictive Control: Theory and Practice – A Survey. *Automatica* 25: 335-348.

74. Georgakis C 1986 On the Use of Extensive Variables in Process Dynamics and Control. *Chem. Eng. Sci.* 41:1471-1484.

75. Gertler J 1998 *Fault Detection and Diagnosis in Engineering Systems*. Marcel Dekker Inc., NY.

76. Gertler J, Li W, Huang Y, McAvoy T 1998 Isolation Enhanced Principal Component Analysis. Proc. *IFAC Workshop on On-Line Fault Detection and Supervision in the Chemical Process Industries*, Lyon France, June, 1998.

77. Glanzer D 1998 Take Process Control to a New Level. *Chem. Eng. Progress*, October, 1998, 23-30.

78. Gopal V, Biegler L T 1998 Large Scale Inequality Constrained Optimization and Control. *IEEE Control Systems Mag.*, December, 1998, 59-68.

79. Guay M, McLellan P J, Bacon D W 1997 Measure of Closed-Loop Nonlinearity and Interaction for Nonlinear Chemical Processes. *AIChE J.* 43:2261-2278.

80. Guay M, McLellan P J, Bacon D W 1997 Measurement of Dynamic Process Nonlinearity. *Preprints of the IFAC Symposium ADCHEM '97*, Banff, Canada, 535-540.

81. Haber R, Keviczky L, Unbehauen H 1987 Applications of Adaptive Control on Nonlinear Dynamic Processes – a Survey on Input-Output Approaches. *Proc. 10th IFAC World Congress, Munich.* Pergamon Press, Oxford.

82. Hägglund T 1995 A Control Loop Performance Monitor. *Control Eng. Practice 3: 1543-1551.*

83. Halevi Y, Palmore Z J, Efrati T 1997 Automatic Tuning of Decentralized PID Controllers for MIMO Processes. *J. Process Control* 7:119-128.

84. Hang C C, Lee T H, Ho W K 1993 *Adaptive Control.* Instrument Soc. of America, Research Triangle Park, NC.

85. Harris C J, Moore C G, Brown M 1993 *Intelligent Control – Aspects of Fuzzy Logic and Neural Nets.* World Scientific, London.

86. Harris T J, Boudreau F, MacGregor J F 1996 Performance Assessment of Multivariable Feedback Controllers. *Automatica* 32:1505-1518.

87. Harris T J, Seppala C T, Desborough L D 1999 A Review of Performance Monitoring and Assessment Techniques for Univariate and Multivariate Control Systems. *J. Process Control* 9:1-17.

88. Hashimoto Y, Katoh T, Shiina T, Yoneya A, Togari Y, McGreavy C 1996 Piecewise Linear Controller Improving its Own Reliability. *J. Process Control* 6:129-136.

89. Helbig A, Marquardt W, Allgower, F 1998 Nonlinearity Measures for Chemical Reactors. *Preprints of the 5th IFAC Sympos. on Dynamics and Control of Process Systems (DYCOPS-5)*, Corfu, Greece, 141-150.

90. Henry M P, Clarke D W 1993 The Self-Validating Sensor: Rationale, Definitions and Examples. *Control Eng. Practice*, 1:585-610.

91. Henson M A 1998 Nonlinear Model Predictive Control: Current Status and Future Directions. *Computers and Chem. Eng*,. 23:187-202.

92. Henson M A, Seborg D E 1991 A Critique of Differential Geometric Strategies for Process Control. *J. Process Control* 1:122-139.

93. Henson M A, Seborg D E 1994 Adaptive Nonlinear Control of a pH Neutralization Process. *IEEE Trans. Control Systems Technology* 2:169-182.

94. Henson M A, Seborg D E 1997 (Ed.), *Nonlinear Process Control*, Prentice Hall, Englewood Cliffs, NJ.

95. Himmelblau D M 1978 *Fault Detection and Diagnosis in Chemical and Petrochemical Processes.* Elsevier Scientific, NY.

96. *Loop Scout*™ brochure 1998, Honeywell Hi-Spec Solutions, Thousand Oaks, CA.

97. Hoske M T 1997 Linking Business, Control, and Engineering Systems. *Control Eng.*, October, 1997, 54-59.

98. Hoske M T 1998 Connect to the Benefits of Digital Industrial Networks. *Control Eng.* January, 1998, 48-54.

99. Huang B, Shah S L 1998 Practical Issues in Multivariable Feedback Control Performance Assessment *J. Process Control* 8:421-430.

100. Huang B, Shah S L, Kwok E K 1997 Good, Bad or Optimal? Performance Assessment of Multivariable Processes. *Automatica* 33:1175-1183.

101. Hunt K J, Sbarbaro D, Zbikowski R, Gawthrop P J 1992 Neural Networks for Control Systems – A Survey. *Automatica* 28:1083-1112.

102. Ioannou P A, Sun J 1995 *Robust Adaptive Control.* Prentice-Hall, Englewood Cliffs, NJ.

103. Irwin G W, Warwick K, Hunt K J (Ed.) 1995 *Neural Network Applications in Control.* Inst. of Electrical Engineers, London.

104. Isermann R 1994 Integration of Fault Detection and Diagnosis Methods. *Proc. IFAC Sympos. on Fault Detection, Supervision and Safety for Technical Processes,* Espoo, Finland, 1994.

105. Isermann R, Raab U 1993 Intelligent Actuators – Ways to Autonomous Actuating Systems. *Automatica* 5:1315-1331.

106. Isidori A 1995 *Nonlinear Control Systems. Third Edition.* Springer-Verlag, NY.

107. Jackson J E 1991 *A User's Guide to Principal Components.* Wiley-Interscience, NY.

108. Jamerson R 1998 Enterprise Demands Real-Time Plant-Floor Data. *InTech*, July, 1998, 38-40.

109. Jobling C P, Grant P W, Barker H A, Townsend P Object-Oriented Programming in Control System Design: a Survey. *Automatica* 30:1221-1261.

110. Johansen T A, Foss B A, Sorensen A V 1995 Non-linear Predictive Control Using Local Models – Applied to a Batch Fermentation Process. *Control Eng. Practice* 3:389-396.

111. Juricek B C, Larimore W E, Seborg D E 1998 Reduced Rank ARX and Subspace System Identification for Process Control,. *Preprints of the* 5th *IFAC Sympos. on Dynamics and Control of Process Systems (DYCOPS-5),* 245-250, Corfu, Greece.

112. Kano M, Miyazaki K, Hasebe S, Hashimoto I 1998 Inferential Control System of Distillation Compositions Using Dynamic Partial Least Squares Regression. *Preprints of the 5th IFAC Sympos. on Dynamics and Control of Process Systems (DYCOPS-5),* 377-386, Corfu, Greece.

113. Kano M, Shiren T, Ohshima M, Okamo U, Aoki S, Hashimoto I 1997 On-line Inference System of Tube-Wall Temperature for an Industrial Olefin Pyrolysis Plant. *AIChE Symposium Series No. 316,* 93, 292-296.

114. Kaspar M H, Ray W H 1992 Chemometric Methods for Process Monitoring and High-Performance Controller Design. *AIChE J.* 38:1593-1608.

115. Keeler J D, Martin G, Boe G, Piche S, Mathur U, Johnson D 1996 The Process Perfecter: The Next Step in Multivariable Control and Optimization. *Technical Report 1-35*, Pavilion Technologies, Inc., Austin, TX.

116. Kennedy J P 1996 Building an Industrial Desktop. *Chem. Eng.*, January, 1996, 82-86. (Also, Kennedy J P 1999 Next Generation RIS. *Hydrocarbon Processing*, in press) .

117. Kesavan P, Lee J H 1997 Diagnostic Tools for Multivariable Model-Based Control Systems. *Ind. Eng. Chem. Res.* 36:2725-2738.

118. Koivo H N 1994 Artificial Neural Networks in Fault Diagnosis and Control. *Control Eng. Practice* 2:89-101.

119. Kosanovich K A, Dahl K S, Piovoso M J 1996 Improved Process Understanding Using Multiway Principal Component Analysis. *Ind. Eng. Chem. Res.* 35: 138-146.

120. Kourti T, MacGregor J F 1996 Recent Developments in Multivariate SPC Methods for Monitoring and Diagnosing Process and Product Performance. *J. Qual. Technol.* 28:409-428.

121. Kozub D 1997 Monitoring and Diagnosis of Chemical Processes with Automated Process Control. *Chemical Process Control - AIChE Symposium Series,* 93(316):83-96.

122. Kramer M A 1991 Nonlinear Principal Component Analysis Using Autoassociative Neural Networks. *AIChE J.* 37:233-243.

123. Kramer M A, Mah R S H 1994 Model-Based Monitoring. *Proc. 2nd Int. Conf. on Foundations of Computer Aided Process Operations*, 45-71, CACHE, Austin TX.

124. Kravaris C, Arkun Y 1991 Geometric Nonlinear Control – An Overview. *Chemical Process Control-CPC IV*, Y. Arkun and W.H. Ray (Ed.), AIChE, NY, 477-515.

125 . Kravaris C, Kantor J C 1990 Geometric Methods for Nonlinear Process Control. 1. Background. *Ind. Eng. Chem. Res.*, 29, 2295-2310. 2. Synthesis *Ind. Eng. Chem. Res.*, 29, 2310-2323.

126. Krstic M Kanellakopoulos I, Kokotovic P V 1995 *Nonlinear and Adaptive Control Design*. John Wiley, NY.

127. Kurtz M J, Henson M A 1996 Input-Output Linearizing Control of Constrained Nonlinear Processes. *J. Process Control* 7: 3-17.

128. Lakshminarayanan S, Shah S L, Nandakumar K 1997 Modeling and Control of Multivariable Processes: Dynamic PLS Approach. *AIChE J.* 43:2307-2322.

129. Lange S T 1997 What's All This About Object Technology? *Control Eng.*, August, 1997, 59-64.

130. Larimore W E 1997 Optimal Reduced Rank Modeling, Prediction, Monitoring, and Control Using Canonical Variate Analysis. *Preprints IFAC ADCHEM Sympos,* Banff, Canada, 61-66.

131. Lawrence D A, Rugh W J 1995 Gain Scheduling Dynamic Linear Controllers for a Nonlinear Plant. *Automatica* 31: 381-390.

132. Lee J H, Cooley B 1997 Recent Advances in Model Predictive Control and Other Related Areas. Kantor J C, Garcia C E, Carnahan B (Ed.) *Chemical Process Control - AIChE Symposium Series,* 93(316):201-216.

133. Lee P L (Ed.) 1993 *Nonlinear Process Control: Applications of Generic Model Control.* Springer-Verlag, London.

134. Li S, Lim K Y, Fisher D G 1989 A State Space Formulation for Model Predictive Control. *AIChE J.* 35:241-249.

135. Linkens D A, Nyongesa H O 1996 Learning Systems in Intelligent Control: An Appraisal of Fuzzy, Neural and Genetic Algorithm Control Applications. *IEE Proc.-Control Theory Appl.* 143:367-386.

136. Ljung L 1995 Non-Linear Black Box Models in System Identification. *Preprints IFAC ADCHEM Sympos,* 1-12, Banff, Canada.

137. Ljung L, Ljung E J (1998) *System Identification: Theory for the User, 2nd Edition,* Prentice Hall, Englewood Cliffs, NJ.

138. Luo R, Misra M, Qin S J, Barton R, Himmelblau D M 1998 Sensor Fault Detection via Multiscale Analysis and Nonparametric Statistical Inference. *Ind. Eng. Chem. Res.* 37:1024-1032.

139. Luo R, Qin S J, Chen D 1998 A New Approach to Closed-Loop Autotuning for Proportional-Integral-Derivative Controllers. *Ind. Eng. Chem. Res.* 37:2462-2468.

140. Lundström P, Lee J H, Morari M, Skogestad S 1995 Limitations of Dynamic Matrix Control. *Computers Chem. Eng.* 19:409-421.

141. Luyben W L 1996 Design and Control Degrees of Freedom. *Ind. Eng. Chem. Res.* 35:2204-2214.

142. Luyben W L 1998 Tuning Temperature Controllers on Openloop Unstable Reactors. *Ind. Eng. Chem. Res.* 37:4322-4331.

143. Luyben W L, Tyreus B D, Luyben M L 1997 Plantwide Control Design Procedure. *AIChE J.* 43:3161-3174.

144. Luyben W L, Tyreus B D, Luyben M L 1998 *Plantwide Process Control.* McGraw-Hill, NY.

145. MacGregor J F, Kourti T 1995 Statistical Process Control of Multivariate Processes. *Control Eng. Practice* 3:403-414.

146. Martin E B, Morris A J, Kiparissides C 1998 Multivariate Statistical Process Control and Process Performance Monitoring. *Preprints of the 5th IFAC Sympos. On Dynamics and Control of Process Systems (DYCOPS-5),* Corfu, Greece 349-358.

147. Mayne D Q 1997 Nonlinear Model Predictive Control: An Assessment. *Chemical Process Control - AIChE Symposium Series,* 93(316):217-231.

148. McAvoy T J 1998 A Methodology for Screening Level Control Structures in Plantwide Control Systems. *Computers Chem. Eng.* 22:1543-1552.

149. McLellan P J, Harris T J, Bacon D W 1990 Error Trajectory Descriptions of Nonlinear Controller Designs. *Chem. Eng. Sci.,* 45:3017-3034.

150. Meadows E S, Rawlings J B 1997 Model Predictive Control. In *Nonlinear Process Control.* M. A. Henson and D. E. Seborg (Ed.), Prentice Hall, NJ., 233-310.

151. Miao T, Seborg D E 1998 Automatic Detection of Excessively Oscillatory Feedback Control Loops. *U.S. Patent #5,719,788.*

152. Miller R, Desborough L, Timmons C 1998 CITGO's Experience with Controller Performance Assessment. *NRPA Computer Conf.,* San Antonio, TX.

153. Moore R G 1998 Private Communication.

154. Morari M 1988 Advances in Process Control Theory. *Chem. Eng. Progress,* October 1988, 60-67.

155. Morari M 1994 Current Directions in Process Control Theory. *Proc. of PSE '94,* 849-863, Kyongju, Korea.

156. Morari M 1997 CPC I – CPC V: 20 Years of Process Control Research. *Chemical Process Control - AIChE Symposium Series,* 93(316):179-191.

157. Morari M. Lee J H 1991 Model Predictive Control: The Good, the Bad, and the Ugly. In Y. Arkun and W.H. Ray (Ed.), *Chemical Process Control – CPC IV,* Fourth Int. Conf. on Chemical Process Control, AIChE, NY, 419-444.

158. Morari M, Lee J H 1997 Model Predictive Control: Past, Present and Future. *Proc. of PSE '97*, Trondheim, Norway.

159. Morari M, Zafiriou E 1988 *Robust Process Control.* Prentice Hall, Englewood Cliffs, NJ.

160. Narraway L, Perkins J 1994 Selection of Process Control Structures Based in Economics. *Computers Chem. Engng.* S511-S515.

161. Negiz A, Çinar A 1998 Monitoring of Multivariable Dynamic Processes and Sensor Auditing. *J. Process Control* 8:375-380.

162. Nimmo, I 1995 Adequately Address Abnormal Operations. *Chem. Eng. Progress,* September, 1995, 36-45.

163. Nomikos P, MacGregor J F 1994. Monitoring of Batch Processes using Multi-way Principal Component Analysis. *AIChE J.* 40:1361-1375.

164. Ogunnaike B A, Wright R A 1997 Industrial Applications of Nonlinear Control. In *Chemical Process Control – V, AIChE Symposium Series,* 93(316):46-59.

165. Ohshima M, Ohno H, Hashimoto I 1995 Model Predictive Control – Experiences in the University-Industry Joint Projects and Statistics on MPC Applications in Japan. *Int. Workshop on Predictive and Receding Horizon Control,, Korea,* 91-96.

166. Palm R, Driankov D, Hellendoorn, H 1997 *Model Based Fuzzy Control.* Springer-Verlag, NY.

167. Palmor Z J, Halevi Y, Krasney N 1995 Automatic Tuning of Decentralized PI Controllers for TITO Processes. *Automatica* 31:1001-10.

168. Passino K M, Yurkovich S 1998 *Fuzzy Control.* Addison-Wesley, Reading, MA.

169. Patton R J 1995 Robustness in Model-Based Fault Diagnosis: the 1995 Situation. *Preprints of the IFAC Workshop on On-Line Fault Detection and Supervision in the Chemical Process Industries.* Newcastle, U.K., 55-7.

170. Patton R, Frank P, Clark R (Ed.) 1989 *Fault Diagnosis in Dynamic Systems.* Prentice Hall, NY.

171. Pearson R K 1994 Modern Control: Why Don't We Use It? *InTech,* November, 1984, 47-49.

172. Pearson R K 1999 *Discrete-Time Dynamic Models.* Oxford Univ. Press, Oxford (in press).

173. Pearson R K, Ogunnaike B A 1997 Nonlinear Process Identification. Chapter 4 in *Nonlinear Process Control.* Henson, M.A. and Seborg, D.E. (Ed), Prentice Hall, Englewood Cliffs, NJ.

174. Pedrycz W 1993 *Fuzzy Control and Fuzzy Systems.* 2nd Ed., John Wiley, NY.

175. Perkins J D (Ed.), 1992 *Proc of the Workshop on the Interactions between Process Desgn and Process Control*, Pergamon, Oxford.

176. Piovoso M J, Kosanovich K A 1994 Applications of Multivariate Statistical Methods to Process Monitoring and Controller Design. *Int. J. Control* 59:743-765.

177. Porter J B, Shoda Y 1998 Industrial Plenary Sessions: Industrial Benefits and Unsolved Problems of Process Control. *Preprints of the 5th IFAC Sympos. on Dynamics and Control of Process Systems (DYCOPS-5)*, Corfu, Greece, 223-226.

178. Pottmann M, Seborg D E 1992 Identification of Nonlinear Processes Using Reciprocal Multiquadric Functions. *J. Process Control* 2:189-203.

179. Pottmann M, Seborg D E 1992 Identification of Nonlinear Processes Incorporating *a prori* Physical Knowledge. Paper presented at the 1992 AIChE Annual meeting (Miami Beach, FL).

180. Prett D M, Garcia C E 1988 *Fundamental Process Control*. Butterworths, Boston.

181. Price R M, Georgakis C 1993 Plantwide Regulatory Control Design Procedure Using a Tiered Framework. *Ind. Eng. Chem. Res.* 32:2693-2705.

182. Price R M, Lyman P R, Georgakis C 1994 Throughput Manipulation in Plantwide Control Structures. *Ind. Eng. Chem. Res.* 33:1197-1207.

183. Psichogios D C, Ungar L H 1992 A Hybrid Neural Network – First Principles Approach to Process Modeling. *AIChE J.* 38:1499-1511.

184. Qin S J 1998 Control Performance Monitoring – a Review and Assessment. *Computers and Chemical Engineering* 23:173-186.

185. Qin S J, 1998 Recursive PLS Algorithms for Adaptive Data Modeling. *Computers Chem. Eng.* 22:503-514.

186. Qin S J, Badgwell T A 1997 An Overview of Industrial Model Predictive Control Technology. In: Kantor J C, Garcia C E, Carnahan B (Ed.) *Chemical Process Control - AIChE Symposium Series*, 93(316):232-256.

187. Qin S J, Badgwell T A 1998 An Overview of Nonlinear Model Predictive Control Applications *Proc. IFAC Workshop on Nonlinear Model Predictive Control – Assessment and Future Directions*, Ascona, Switzerland, June 3-5.

188. Rawlings J B, Meadows E S, Muske K 1994 Nonlinear Model Predictive Control: a Tutorial and Survey. *Proc. IFAC ADCHEM*, Kyoto, Japan, 203-214.

189. Reader's Forum, "On Fuzzy Control and Fuzzy Reviewing...," *IEEE Control Systems Magazine*, June 1993, 5-7.

190. Richalet J 1993 Industrial Applications of Model Based Predictive Control. *Automatica* 29: 1251-1274.

191. Richalet J 1993 Model-Based Predictive Control: Who Sells What? *Control Eng.*, May, 1993, 122-129.

192. Richalet J, Rault A, Testud J L, Papon J 1976 Algorithmic Control of Industrial Processes. *Proc. of the 4th IFAC Sympos. on Ident. and System Parameter Estimation*, 1119-1167.

193. Ricker N L 1991 Model Predictive Control: State of the Art. In Y. Arkun and W H Ray (Ed.), *Chemical Process Control – CPC IV*, Fourth Int. Conf. on Chemical Process Control, AIChE, NY, 271-296.

194. Ricker N 1996 Decentralized Control of the Tennessee Eastman Challenge Problem. *J. Process Control*, 6:205-221.

195. Rinard I H, Downs J J 1992 Plant Wide Control – A Review and Critique. Paper presented at the AIChE 1992 Spring National Meeting.

196. Rivera D E, Pollard J F, García C E 1992 Control Relevant Prefiltering: A Systematic Design Approach and Case Study. *IEEE Trans. Automatic Control* 37:964-974.

197. Rowan D A 1992 Beyond FALCON: Industrial Applications of Knowledge-Based Systems. *Proc. IFAC Symposium on On-Line Fault Detection and Supervision in the Chemical Process Industries.* P.S. Dhurjati (Ed.), Newark, DE, USA.

198. Samdani G 1992 Managing Knowledge. *Chem. Eng.*, April, 1992, 5.

199. Samdani G, Fouhy K 1992 Smart Software. *Chem. Eng.*, April, 1992, 30-33.

200. Samdani G, Fouhy K, Moore S 1993 Fuzzy Logic: More Than a Play on Words. *Chem. Eng.*, February, 1993, 30-33.

201. Schaper C D, Larimore W E, Seborg D E, Mellichamp D A 1994 Identification of Chemical Processes Using Canonical Variate Analysis. *Computers & Chem. Eng.* 18:55-69.

202. Schellekens P L 1996 A Vision for Intelligent Instrumentation in Process Control. *Control Eng.*, October, 1996, 89-94.

203. Seborg D E 1994 A Perspective on Advanced Strategies for Process Control. *Model., Ident. and Control,* 15:179-189.

204. Seborg D E, Edgar T F, Mellichamp D A 1989 *Process Dynamics and Control.* John Wiley, NY.

205. Seborg D E, Shah S L, Edgar T F 1986 Adaptive Control Strategies for Process Control: A Survey. *AIChE J.* 32:881-913.

206. Seem J, 1998 A New Pattern Recognition Adaptive Controller with Application to HVAC Systems. *Automatica* 34:969-982.

207. Seider, W D. Seader, J D., Lewin D R. 1998 *Process Design Principles: Synthesis, Analysis, and Evaluation.* John Wiley, NY.

208. Shinnar R 1981 Chemical Reactor Modeling for Purposes of Controller Design. *Cheng. Eng. Commun.* 9:73-99.

209. Shinskey F G 1967 *Process Control Systems, 1st Edition,* McGraw-Hill, NY.

210. Shinskey F G 1990 How Good are Our Controllers in Absolute Performance and Robustness. *Measurement & Control,* 23:114-121.

211. Shinskey F G 1994 *Feedback Controllers for the Process Industries.* McGraw-Hill, Inc., NY.

212. Shinskey F G 1996 Is Fuzzy Logic Appropriate for Process Control Applications? *Chemical Processing* 45, December, 1996.

213. Shoda Y 1998 Abstract for Industry Plenary Session. *Preprints of the 5th IFAC Sympos. On Dynamics and Control of Process Systems (DYCOPS-5),* Corfu, Greece, 223-226.

214. Shunta J P 1995 *Achieving World Class Manufacturing Through Process Control.* Prentice Hall PTR, Englewood Cliffs, NJ.

215. Skogestad S, Larsson T 1998 A Review of Plantwide Control. Internal Report, Department of Chemical Engineering, Norwegian University of Science and Technology, Trondheim, Norway.

216. Skogestad S, Postlethwaite I 1996 *Multivariable Feedback Control: Analysis and Design.* John Wiley, Chichester, UK.

217. Slotine J J E, Li W 1991 *Applied Nonlinear Control*, Prentice Hall, Englewood Cliffs, NJ.

218. Soeterboek R 1991 *Predictive Control – A Unified Approach.* Prentice Hall, Englewood Cliffs, NJ.

219. Stack A J, Doyle F J 1997 Application of a Control-Law Nonlinearity Measure to the Chemical Reactor Analysis. *AIChE J.* 43:425-439.

220. Stephanopoulos G, Han C 1994 Intelligent Systems in Process Engineering: A Review. *Preprints of the 5th Internat. Symposium on Process Systems Eng. (PSE 94)* 1339-1366, Kyongju, Korea.

221. Stephanopoulos G, Ng C 1998 Perspectives on the Synthesis of Plant-Wide Control Structures. *Preprints of the 5th IFAC Sympos. On Dynamics and Control of Process Systems (DYCOPS-5)*, 1-16, Corfu, Greece.

222. Su H T, Bhat N, Minderman P A, McAvoy T J 1992 Integrating Neural Networks with First Principles Models for Dynamic Modeling. *Proc. IFAC Sympos. DYCORD + '92* College Park, MD, 77-82.

223. Su H T, McAvoy T J 1997 Artificial Neural Networks for Nonlinear Process Identification and Control. In *Nonlinear Process Control*. M. A. Henson and D. E. Seborg (Ed.), Prentice Hall, NJ.

224. Swanda, A, Seborg D E 1999 Controller Performance Assessment Based on Setpoint Response Data. *Preprints of the 1999 ACC*, San Diego (in press).

225. Takatsu H, Itoh T, Mituhiko A 1998 Future Needs for the Control Theory in Industries – Report and Topics of the Control Technology Survey in Japanese Industry. *J. Process Control* 8:369-374.

226. Tan S, Hang C-C, Chai J-S 1997 Gain Scheduling: From Conventional to Neuro-fuzzy. *Automatica* 33:411-419.

227. Thompson M L, Kramer M A 1994 Modeling Chemical Processes Using Prior Knowledge and Neural Networks. *AIChE J.* 40:1328-1340.

228. Thornhill N F, Hägglund T 1997 Detection and Diagnosis of Oscillation in Control Loops. *Control Eng. Practice* 5:1343-1354.

229. Thornhill N F, Oettinger M, Fedenczuk P 1999 Refinery-wide Control Loop Performance Assessment. *J. Process Control* 9:109-124.

230. Townsend S, Lightbody G, Brown M D, Irwin G W 1998 Nonlinear Dynamic Matrix Control Using Local Models. *Trans Inst. MC* 20:47-56.

231. Trierweiler J O, Engell S 1997 The Robust Performance Number: A New Tool for Control Structure Design. *Computers Chem. Eng.* 21:S409-S414.

232. Trierweiler J O, Secchi A R 1998 Exploring the Potentiality of Using Multiple Model Approach in Nonlinear Model Predictive Control. *Proc. of COBEQ '98, Porto Alegre, Brazil.*

233. Ungar L H, Hartman E J, Keeler J D, Martin G D 1996 Process Modeling and Control Using Neural Networks. In *AIChE Sympos. Ser.* 92(312) 57-67.

234. Van de Wal M, De Jager, B 1995 Control Structure Design: A Survey. *Proc. Amer. Control Conference*, Seattle, 225-229.

235. Van Overschee P, De Moor B 1996 *Subspace Identification for Linear Systems: Theory, Implementation, Applications.* Kluwer Academic Publishers, Dordrecht.

236. Vedam H, Venkatasubramanian V 1998 A B-Spline Based Method for Data Compression, Process Monitoring and Diagnosis. *Computers Chem. Eng.* 22: Suppl. S827-S830.

237. Verbruggen H B, Krijgsman A J, Bruijn P M 1991 Towards Intelligent Control. *Journal A* 32:35-46.

238. Wachs A, Lewin D R 1998 Process Monitoring Using Model-Based PCA. *Preprints of the 5th IFAC Sympos. on Dynamics and Control of Process Systems (DYCOPS-5)*, Corfu, Greece 86-90.

239. Walsh S, Perkins J 1996 Operability and Control in Process Synthesis and Design. *Advances in Chem Eng.* 23:301-402.

240. Wang Q-G, Zou B, Lee T-H, Bi Q 1997 Auto-tuning of Multivariable PID Controllers from Decentralized Relay Feedback. *Automatica* 33:319-330.

241. Wang Y, Seborg D E, Larimore W E 1997 Process Monitoring Based on Canonical Variate Analysis. Proc. *European Control Conference (ECC '97)*, Paper 775, Brussels, Belgium, July, 1997,

242. White D A, Sofge D A (Ed.) 1992 *Handbook of Intelligent Control: Neural, Fuzzy Adaptive Approaches.* Van Nostrand Reinhold, NY.

243. Widrow B, Rumelhart D E, Lehr M A 1994 Neural Networks: Applications in Industry, Business and Science. *Journal A* 35:17-27.

244. Williams T J 1971 Applied Digital Control – An Adaptive Future. *InTech,* January 1971, 37-42.

245. Williams T J 1977 Two Decades of Change – A Review of the 20-Year History of Computer Control. *Control Eng.:* 71-78, September 1977.

246. Wold S 1989 Nonlinear PLS Modeling. *Chemometrics and Intelligent Laboratory Systems* 7:53-65.

247. Wold S 1994 Exponentially Weighted Moving Principal Components Analysis and Projections to Latent Structures. *Chemometrics and Intelligent Laboratory Systems* 23:149-161.

248. Yamamoto S, Hashimoto I 1991 The View from Japanese Industry. *Chemical Process Control-CPC IV,* Y. Arkun and W.H. Ray (Ed.), AIChE, NY, 1-28.

249. Yen J, Lamagri R, Zadeh L (Ed.) 1995 *Industrial Applications of Fuzzy Logic and Intelligent Systems.* IEEE Press., NY.

250. Yurkovich S 1997 A Control Engineer's Perspective on Fuzzy Control. In *Plenary Lectures and Mini-Courses, ECC '97,* Bastin G and Gevers M (Ed.).

251. Zafiriou, E (Ed.) 1994 *IFAC Workshop on the Integration of Process Design and Control,* (Baltimore, June 1994), Pergamon Press, Oxford. See also special issue of *Computers & Chem. Eng.*, 20:4, 1996.

252. Ziegler J G, Nichols N B 1943 Process Lags in Automatic-Control Circuits. *Trans ASME,* 65:433-444.

5

Advanced Control for Hot Rolling Mills

M. J. Grimble and G. Hearns

Industrial Control Centre, University of Strathclyde, Glasgow, UK

Summary. Hot strip rolling mills are characterised by a process which has many interactions, is distributed with dominant transport delays and has a lack of measurements due to the severe environment. There is ever increasing market demand for better product quality of the steel strip hence the requirement for higher performance controllers. The performance and robustness benefits using H-infinity controllers with a decentralised structure are examined using an advanced simulation of a hot mill.

1 Introduction

A hot strip finishing mill is a complex large scale multivariable process which in general is equipped with single loop regulators. However recently there have been successful demonstrations of multivariable control in, for example, the Japanese industry [1], [6], [7], [11], [12], [13], [14], [15]. Previous applications have focussed on the performance [3], whereas a multivariable controller can also have a robustness advantage, over conventional controllers.

The objective of a multi-stand mill, shown in Fig. 1 is to reduce the thickness (h) of the metal strip while it passes through each stand by controlling the position of the hydraulic capsules (S). Between each stand there is a looper which is a motor driven arm and roll, to keep the strip at a reference tension (σ) which provides stability for the control of massflow and thickness. The angle (θ) of the looper arm provides a measure of the strip massflow and in conventional control the looper angle error is used to control the upstream stand's main drive motor. Due to a lack of space in the mill and the hostile environment, the strip thickness and tension cannot usually be measured. The only measurements available for this purpose are the rolling force (P) and looper angle. These are used in conventional control to indirectly control the thickness and massflow by manipulating the stand capsule and stand motor speed (ω). The strip tension is regulated through the use of an open loop model to predict tension and supply the reference for the looper motor's torque (M). Conventional control uses models to infer the un-measurable outputs, so that the use of a model based approach, such as $\mathcal{H}_\infty$, control is not a great leap in complexity.

Conventional control designs often ignore the interactions between the strip thickness, tension and angle. The angle and the tension are linked through the mechanics of the interstand system whilst the tension is a disturbance on the stand's exit thickness. Any change by the Automatic Gauge Control (AGC) changes the massflow at the entry and exit of the stand, producing disturbances to the tension and angle. In conventional control the thickness performance has to be detuned to enable the regulating loop to remain stable even though there are errors in the estimated mill modulus. It will be shown that an $\mathcal{H}_\infty$ multivariable controller will be robustly stable for an uncertain mill modulus without comprising performance [8]. Previous applications of multivariable control have generally been for the looper/tension system, whereas the work described here includes design of the thickness control to take account of all the interactions.

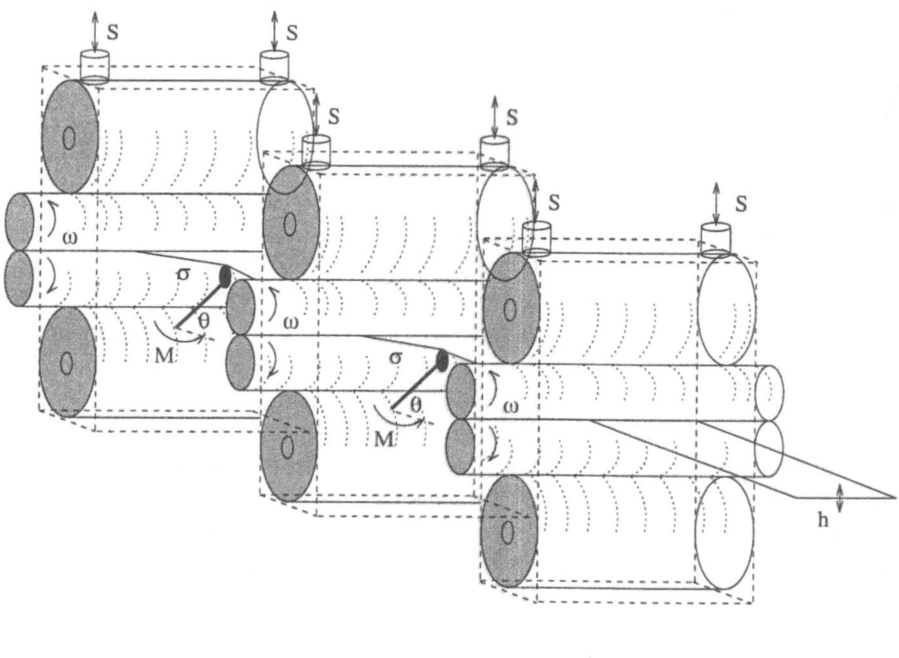

Figure 1: Three Stands of a Tandem Hot Strip Mill

2 Hot Strip Mill Physical Models

2.1 Stand Model

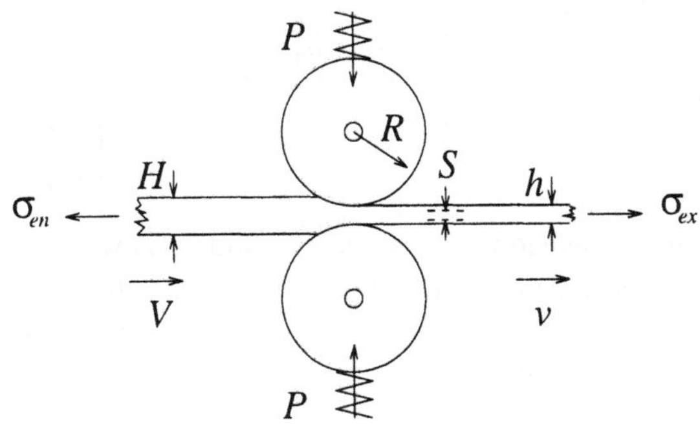

Figure 2: Roll Bite

The physical equations describing the main variables of concern [4] are introduced below. These variables include the rolling force, the strip output thickness and velocity. Much of the modelling of plastic deformation in the roll gap is a mixture of theoretical and empirical methods. Rolling force (P) for a strip width (w) can be approximated as:

$$P=\sqrt{R(H-h)}[\bar{S}wQ_p-m_1\sigma_{en}-m_2\sigma_{ex}] \tag{1}$$

where Q_p is a geometrical factor which is a non-linear function of the work roll radius (R), strip entry temperature (T_{en}) entry thickness (H) and exit thickness (h). Here $\bar{S}$ is the mean constrained yield stress which also depends on the geometry, strip temperature and the work roll velocity. The response of the hydraulic capsule is modelled as a first order lag. The exit thickness (h) depends on the position (S) of the capsule and the mill modulus (M_m):

$$h = S + \frac{P}{M_m} \tag{2}$$

The mill modulus is equivalent to a spring constant for the complete stand structure which relates the rolling force to the stretch of the mill housing and rolls. The mill modulus will change as the rolls wear, after each change of the work rolls, backup rolls and less commonly, after changes to the mill housing.

The strip exit velocity (v) depends on the slip (f) which occurs between the strip and the work roll surface, and is sensitive to the stand entry and exit tensions and thickness, where R is the work roll radius. The massflow through any one stand must also be conserved which results in a change in output thickness producing a change in the strip exit velocity:

$$v = (1+f)R\omega \tag{3}$$

$$VH = vh \tag{4}$$

Linear equations for the force, exit thickness, entry and exit velocities are dependent on the following coefficients:

$$\begin{bmatrix} P^i \\ h^i \\ V^i \\ v^i \end{bmatrix} = \begin{bmatrix} P_1^i & P_2^i & P_3^i & P_4^i & P_5^i & P_6^i \\ h_1^i & h_2^i & h_3^i & h_4^i & h_5^i & h_6^i \\ V_1^i & V_2^i & V_3^i & V_4^i & V_5^i & V_6^i \\ v_1^i & v_2^i & v_3^i & v_4^i & v_5^i & v_6^i \end{bmatrix} \begin{bmatrix} S^i \\ H^i \\ \sigma_{en}^i \\ \sigma_{ex}^i \\ \omega^i \\ T_{en}^i \end{bmatrix} \tag{5}$$

where σ_{en}^i is the entry tension, σ_{ex}^i is exit tension ω^i is work roll angular speed and T_{en}^i is strip stand entry temperature. The main drive motor is modelled as a one mass system with a PI speed controller with the load torque being a function of the rolling force:

$$\dot{\omega}^i = \omega_1^i X_I^i + \omega_2^i \omega^i + \omega_3^i \omega_r^i + \omega_4^i S^i + \omega_5^i H^i + \omega_6^i \sigma_{en}^i + \omega_7^i \sigma_{ex}^i + \omega_8^i T_{en}^i \tag{6}$$

$$\dot{X}_I^i = \omega_r^i - \omega^i \tag{7}$$

2.2 Strip Tension

The strip tension (σ) model is based on longitudinal stress and strain as related by Young's modulus of elasticity (E):

$$\sigma(t) = E\left[\frac{L(\theta(t)) - L_s(t)}{L_s(t)}\right] \tag{8}$$

where,

$$L(\theta(t)) = l_1 + l_2 \tag{9}$$

$$l_1 = [(r - y + l\sin\theta)^2 + (a + l\cos\theta)^2]^{1/2} \tag{10}$$

$$l_2 = [(r - y + l\sin\theta)^2 + (L - a - l\cos\theta)^2]^{1/2} \tag{11}$$

and,

$$L_s(t) = L_o + \int_{t_0}^{t} (v^i(\tau) - V^{i+1}(\tau))d\tau \tag{12}$$

The strain depends on the amount of material in the interstand region which depends on the velocity difference of the strip leaving stand i and entering stand i+1. The extension is the difference between the geometric strip length and the length of strip in the interstand. A linear equation for the strip tension may be expressed as:

$$\dot{\sigma}^i = \sigma_1\dot{\theta} + \sigma_2 v^i + \sigma_3 V^{i+1} \tag{13}$$

2.3 Looper mechanics

If the looper has an inertia J_L with respect to it's pivot, then Newton's second law of motion can be applied to obtain:

$$J_L\ddot{\theta} = M - T_{load} \tag{14}$$

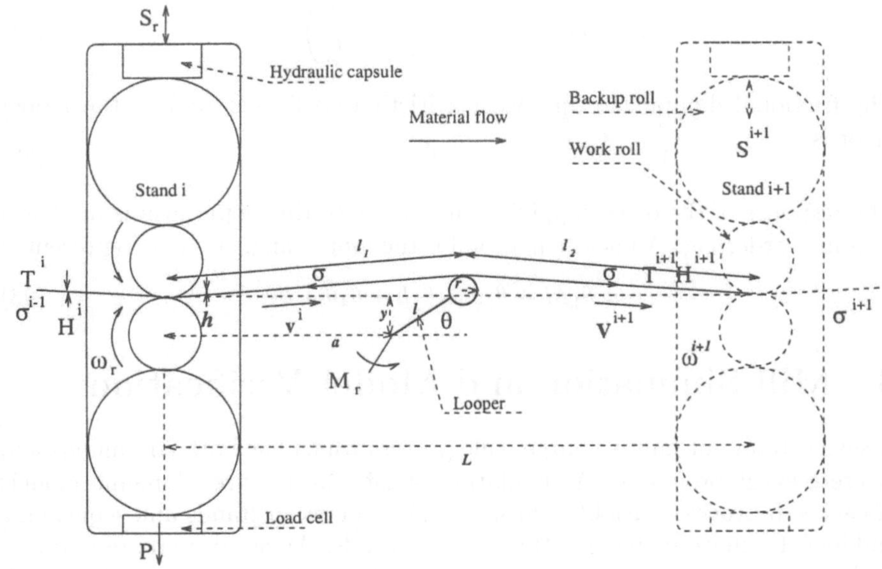

Figure 3: Two Stands and Interstand Looper

where T_{load} is the load torque:

$$T_{load} = T_\sigma + T_s + T_{LW} + T_b + T_d \tag{15}$$

The load torques are functions of the looper angle, strip properties and the interstand geometry as shown in Fig. 3. The load torque due to strip tension is for a strip of width (w):

$$T_\sigma = \sigma w h[l \cos\theta f_1(\theta) + (l \sin\theta + r) f_2(\theta)] \tag{16}$$

where,

$$f_1(\theta) = (r + l\sin\theta - y)\left(\frac{1}{l_1} + \frac{1}{l_2}\right) \tag{17}$$

$$f_2(\theta) = \left(\frac{L - a - l\cos\theta}{l_2} - \frac{a + l\cos\theta}{l_1}\right) \tag{18}$$

The torque to support the strip weight with steel density (ρ) is:

$$T_s = lg\rho w h(l_1 + l_2)\cos\theta \tag{19}$$

The torque required to support the looper arm mass (M_a) and roller mass (M_r) is:

$$T_{LW} = lgM_r \cos\theta + \frac{l}{2} g M_a \cos\theta \tag{20}$$

The torque required to bend the strip with yield stress (σ_{ys}) over the looper roll is:

$$T_b = l\cos\theta\frac{\sigma_{ys}wh^2}{4}\left(\frac{1}{l_1}+\frac{1}{l_2}\right) \tag{21}$$

The frictional damping torque with a friction coefficient (c_f) at the looper pivot is:

$$T_d = c_f\dot{\theta} \tag{22}$$

The response of the drive supplying the torque to the looper arm is modelled by a first order lag. A linear equation for the looper angle may be represented as:

$$\ddot{\theta} = \theta_1 M + \theta_2\sigma + \theta_3 h + \theta_4\theta + \theta_5\dot{\theta} \tag{23}$$

3 Mill Simulation and Model Verification

A seven stand hot mill is a large scale process which has both continuous and discrete event processes. A simulation of the mill requires dynamic models of each sub-process, variable transport delays between stands and sequencing and logic for material passing through the stands. An accurate representation of the conventional control system is required to enable the performance with advanced controllers to be compared. Verification of the simulation is difficult, due to the large number of variables and model parameters, the interaction between models and controllers, and the limited number of measurements available on the mill: force, angle, mill exit thickness and temperature.

Figures 4 and 5 show the simulated and measured strip temperature and the thickness error at the exit of stand 7, respectively. The sinusoidal nature of the exit temperature is a consequence of the bar of metal entering stand 1, having a sinusoidal temperature distribution because of the supports the bar rests on in the reheat furnace and the local cold spots that result. The force and screw position for stand 1 are shown in Fig. 6 and Fig. 7. The high frequency sinusoidal signal shown on the force, is due to the eccentricity of the backup rolls. The rolls can be moved vertically by a screw mechanism or a hydraulic capsule. For this design exercise it was assumed that all stands are equipped with the faster hydraulic capsules, as shown for stand 5 in Fig. 9. The responses of the looper angle and the main drive motor (driving the work roll) are shown in figures 10 to 13 for stands 1 and 5, respectively. The overshoot at the start of the angle response is when the looper is raised to make contact with the strip, after the adjacent stands have been threaded. The increase in the motor speed is to keep the strip mill exit temperature above its reference value and the decrease at the end, is so that the end of the strip can be de-threaded (removed) from the mill at a safe speed.

There are differences between the measured and simulated responses but the objective of having a simulation which replicated the important trends in a mill is achieved. This provides a realistic testbed for the advanced controllers that are developed.

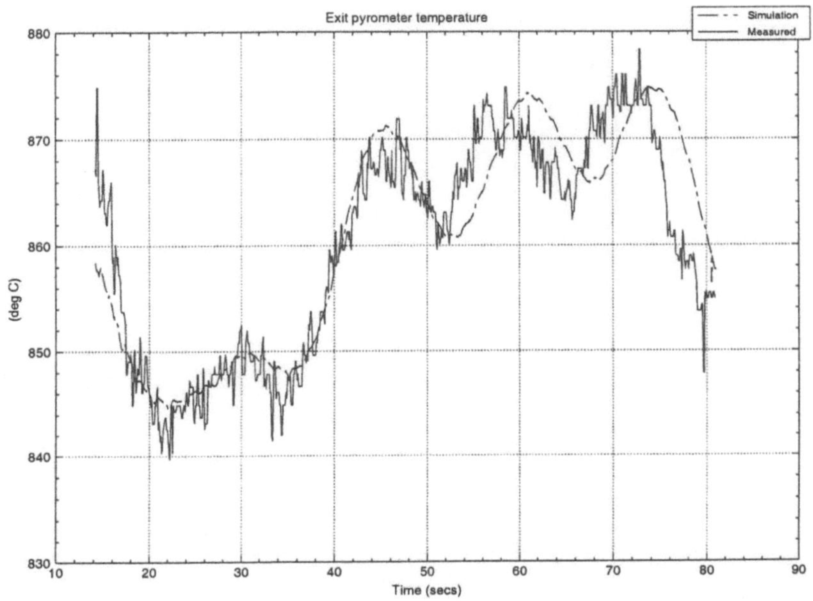

Figure 4: Mill exit strip temperature

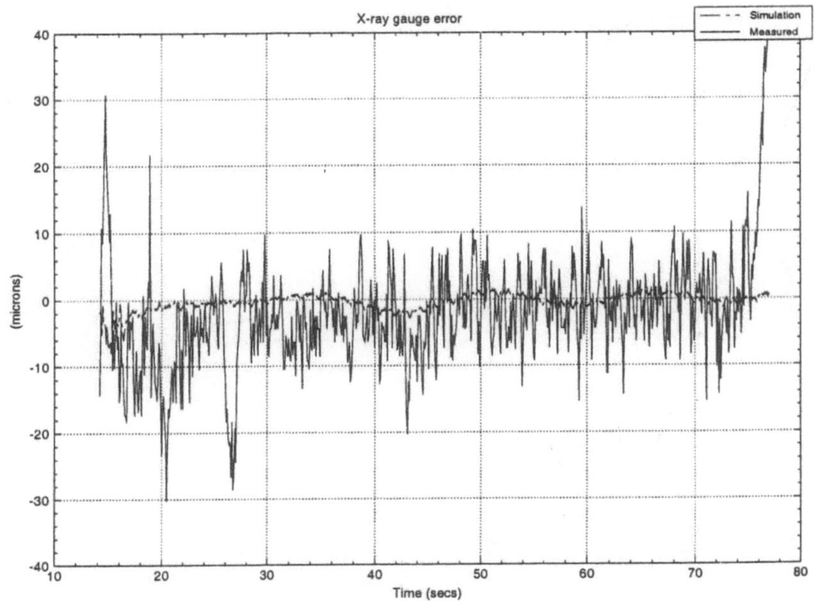

Figure 5: Mill exit strip thickness

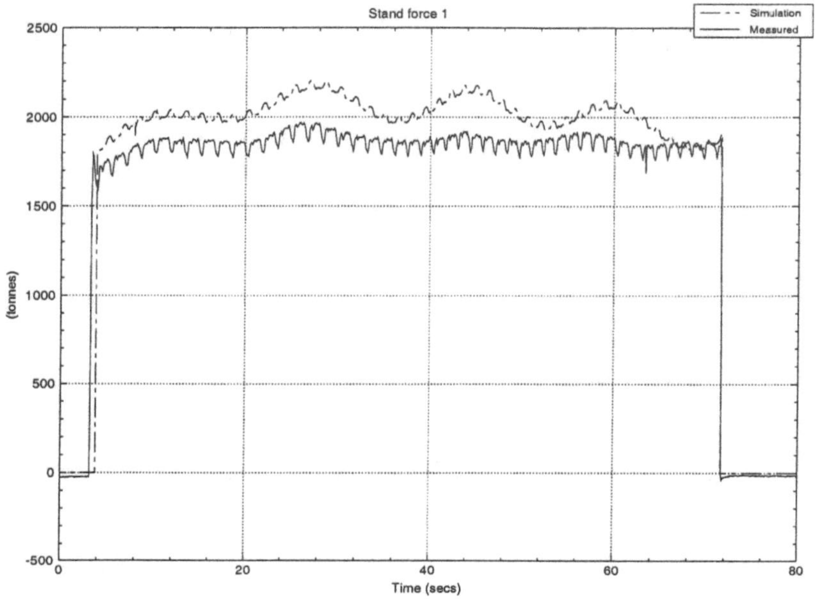

Figure 6: Rolling force stand 1

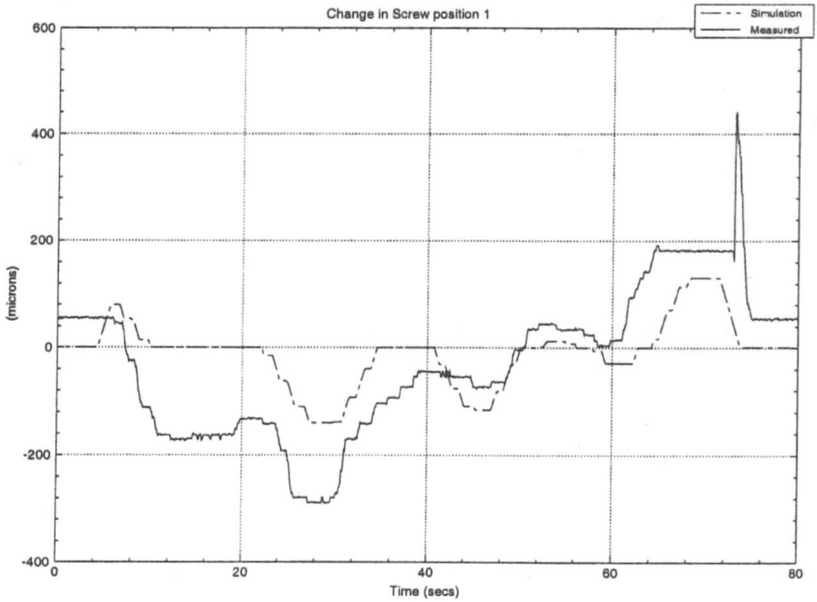

Figure 7: Screw position stand 1

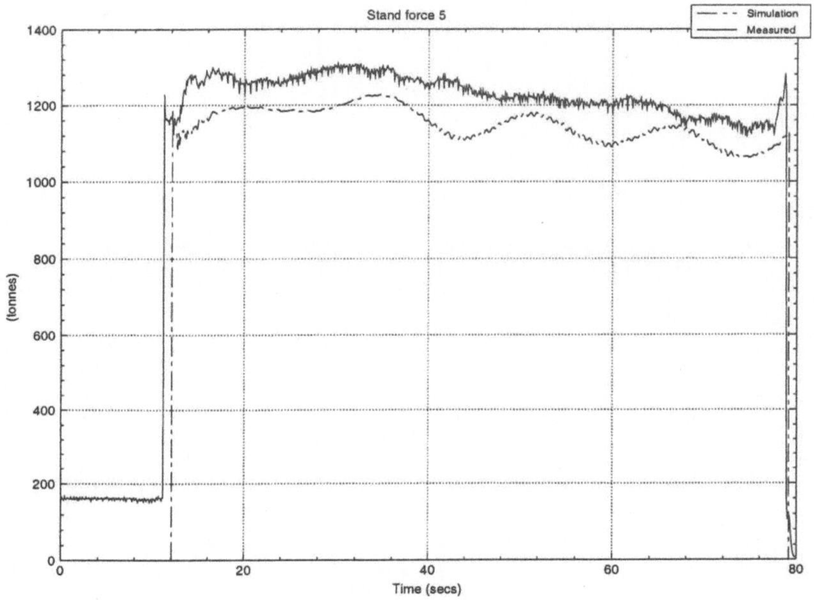

Figure 8: Rolling force stand 5

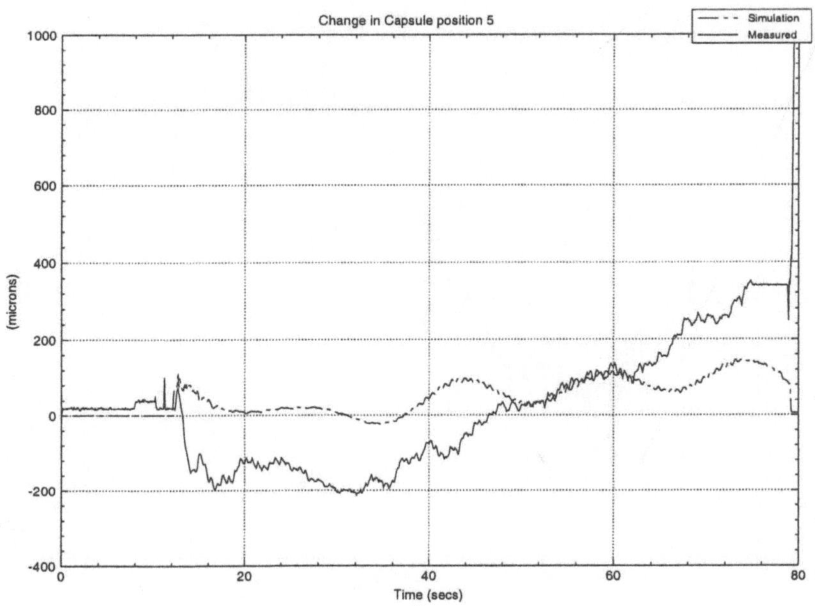

Figure 9: Capsule position stand 5

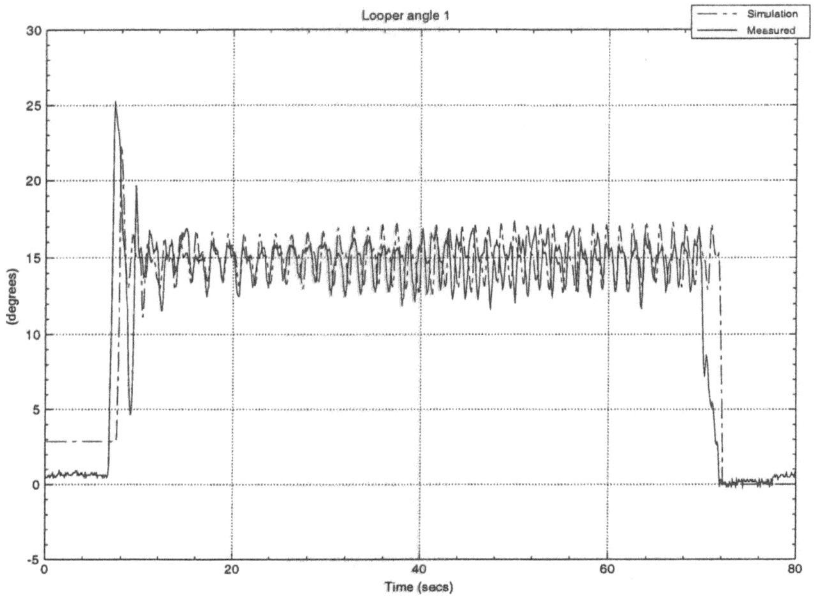

Figure 10: Looper angle 1

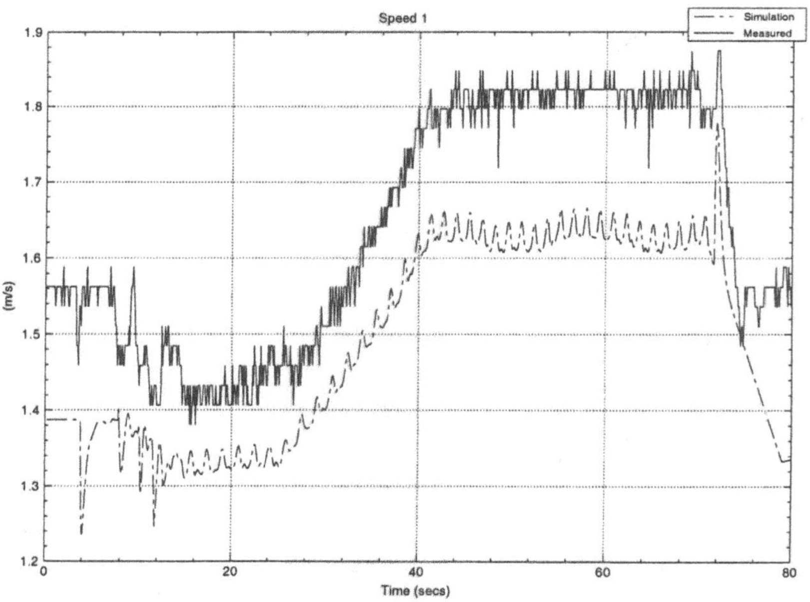

Figure 11: Main drive speed 1

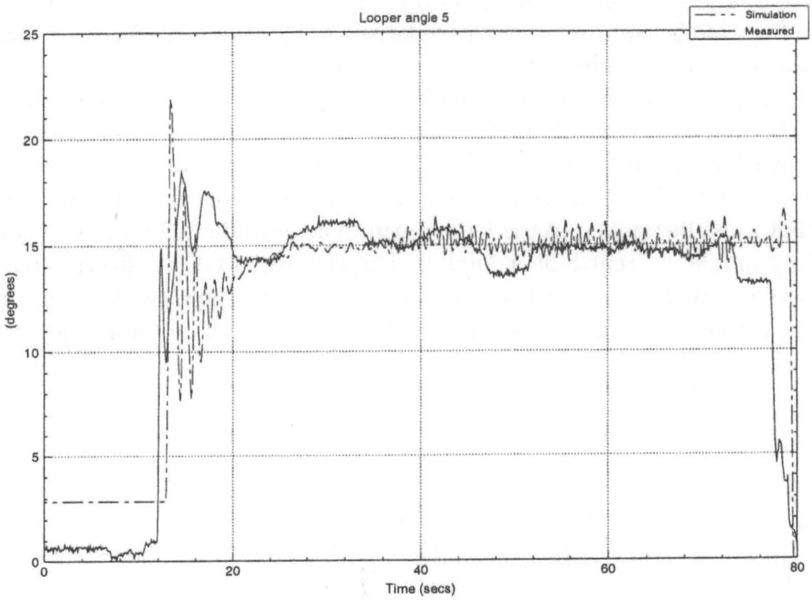

Figure 12: Looper angle 5

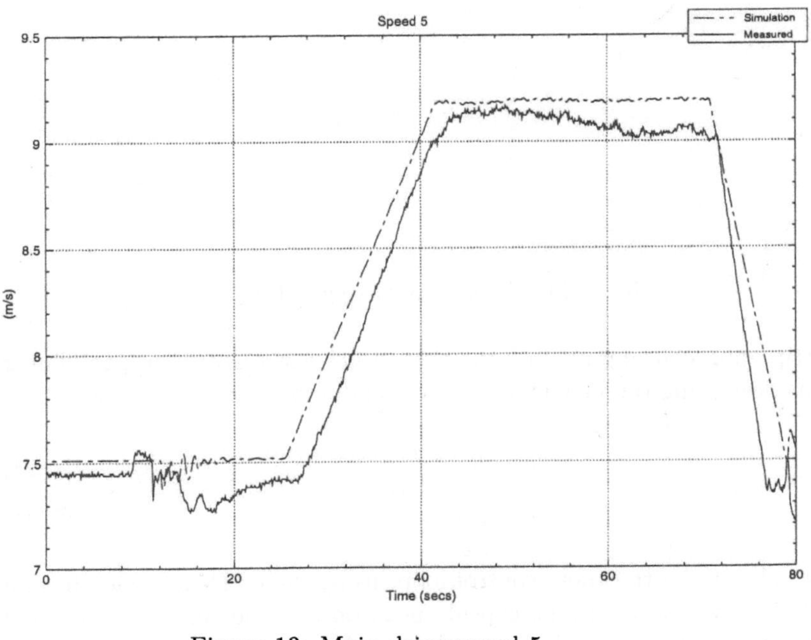

Figure 13: Main drive speed 5

4 Conventional Regulators

The conventional solution for the regulation of thickness, angle and tension is shown in Fig. 14. The automatic gauge control (AGC) uses force feedback via the inverse of the mill modulus to change the capsule position. A PI controller is used to regulate the angle by trimming the speed of the upstream drive motor. The massflow control also includes a speed trim signal from changes of the downstream motor speed. The strip tension is regulated by using a static nonlinear model of the strip geometry and looper load to estimate the torque that must be applied to the looper arm to achieve the reference tension (depending on the current looper angle). The classical AGC is effective but does provide disturbances to the tension loop and more emphasis is given to the angle performance than to the tension. If M_m is the mill modulus and M_s

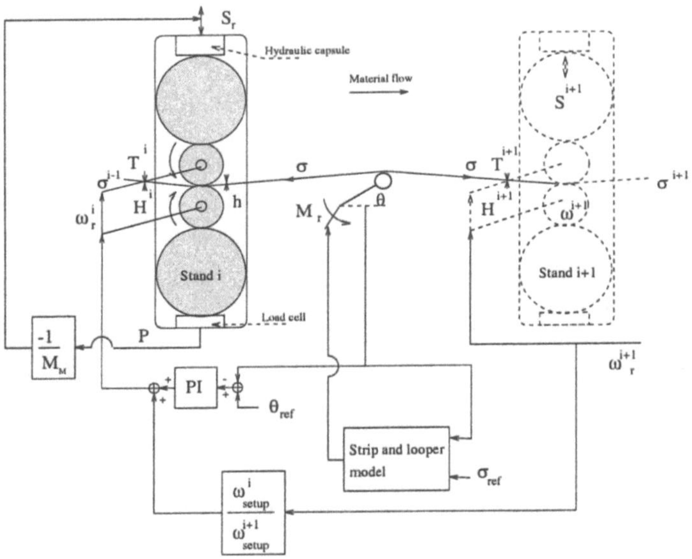

Figure 14: Conventional Regulators

is the strip gain, considering only the entry thickness disturbance, the linear equations describing the behaviour of the stand and strip are given as:

$$h = S + \frac{P}{M_m} \tag{24}$$

$$P = M_s(H - h) \tag{25}$$

Figure 15 shows the thickness control loop, using force (P) feedback from a load cell to drive the hydraulic capsule position S. The AGC is static and the ideal gain is $K = -1/M_m$ since the steady state gain of the thickness sensitivity (S_h) will be zero. The AGC uses the mill modulus to account for the mill stretch, with the ideal case being an infinitely stiff mill (so that the thickness does not change with load). The force feedback loop involves positive feedback at low frequencies, with the loop gain being less than unity. The mill modulus is a variable parameter which lumps many mechanical effects into one measure, and since tests on the stands to estimate the mill modulus are infrequent, then the thickness control loop design must take into account the resulting uncertainty.

For robust stability the closed-loop poles should be analysed for a change of $\pm 20\%\bar{M}$, where $\bar{M}$ is the nominal mill modulus. If $K = -1/\hat{M}_m$ the characteristic equation is given as:

$$s + \frac{1}{\tau_c}\frac{M_s}{M_s + M_m}\left[1 + \frac{M_m}{M_s} - \frac{M_m}{\hat{M}_m}\right] = 0 \tag{26}$$

Since $\tau_c > 0$, $M_m > 0$ and $M_s > 0$, then the closed-loop system is stable, if and only, if:

$$1 + \frac{M_m}{M_s} - \frac{M_m}{\hat{M}_m} > 0 \tag{27}$$

or

$$\hat{M}_m > \frac{M_m M_s}{M_s + M_m} \tag{28}$$

For closed-loop stability, with the worst case conditions:

$$\hat{M}_m > \frac{1.2\bar{M}_m M_s}{M_s + 1.2\bar{M}_m} \tag{29}$$

Now $0 < M_m/Ms << 1$ and robust stability is achieved if:

$$\hat{M}_m \geq 1.2\bar{M}_m \tag{30}$$

The nominal steady-state thickness performance, as determined by the sensitivity $h = S_h H$, with $K = -1/(1.2\bar{M}_m)$, is given as:

$$S_h = \frac{1 + M_m C K}{1 + M_m C K + M_s^{-1} M_m} = \frac{1 - 1.2^{-1}}{1 - 1.2^{-1} + M_s^{-1}\bar{M}_m} \tag{31}$$

Thus, there is a need to trade-off robust stability ($\hat{M}_m$ larger than mill modulus) against nominal performance (determined by S_h).

5 $\mathcal{H}_\infty$ Robust Thickness Control

To analyse the mill modulus robustness problem, the stand dynamics will be considered alone, without the interactions with the tension dynamics. This simplification aids understanding and enables a comparison to be made with the robustness of the conventional AGC. To represent the uncertain mill modulus it is convenient to use an inverse additive uncertainty description. The actual mill modulus M_m is represented by a nominal value $\bar{M}_m$, with feedback through a perturbation Δ, and uncertainty weightings W_I and W_O. The Δ perturbation is bounded: $||\Delta||_\infty \leq 1$, therefore W_I and W_O have to be chosen such that M_m can vary within the desired range:

$$M_m = \frac{\overline{M}_m}{1 - \overline{M}_m W_I \Delta W_O} \tag{32}$$

If $M_m \in [0.8\overline{M}_m, 1.2\overline{M}_m]$ and $||\Delta(j\omega)||_\infty \leq 1 \, \forall\omega$, then letting $W_I = 1$ and $W_O = \frac{0.25}{\overline{M}_m}$ will ensure that as Δ varies from -1 to 1, the mill modulus M_m can cover the specified range. In order to design a robust controller the Δ feedback loop is broken, and the weightings W_I and W_O can be considered as injecting a disturbance signal and weighting an output signal, respectively.

Figure 16 shows the closed loop system, augmented with the weightings for determining performance and representing uncertainty. Here W_s is a low pass filter to weight the exit thickness, W_c is a high pass filter to weight the actuator reference and W_d is a low pass filter to represent the input strip thickness disturbance. If the gain, as seen from w_{mi} to z_{mo}, is less than 1 for the closed-loop system, then the closed-loop system will be stable as the mill modulus changes within its range. The weighting W_O is on the rolling force, whereas W_I can be related to the input thickness model W_d. To reduce the conservativeness W_I can be removed and W_d can be used, both as the uncertainty input weight and the disturbance model, with $W_O = \frac{0.25}{W_d \overline{M}_m}$. If C is the transfer function for the hydraulic capsule, then the three sensitivities to be weighted may be listed as follows. The thickness sensitivity:

$$h = \frac{1 + M_m CK}{1 + M_m CK + M_s^{-1} M_m} H = S_h H \tag{33}$$

The force sensitivity:

$$P = \frac{M_m}{1 + M_m CK + M_s^{-1} M_m} H = S_P H \tag{34}$$

The control sensitivity:

$$S_u = \frac{K M_m}{1 + M_m CK + M_s^{-1} M_m} H = S_u H \tag{35}$$

The $\mathcal{H}_\infty$ sensitivity problem is to find a stabilising controller K such that:

$$\left\| \begin{matrix} W_s S_h W_d \\ W_O S_P W_d \\ W_c S_u W_d \end{matrix} \right\|_\infty \leq 1 \tag{36}$$

If Eqa. 36 is satisfied then the closed-loop will be robustly stable for changes to the mill modulus and will satisfy nominal performance requirements (see [17]).

The uncertainty for M_m was represented within the $\mathcal{H}_\infty$ framework by a norm bounded feedback loop. Since the Δ parameter would be a complex number and the actual uncertainty is real, then this approximation of the uncertainty means the condition for robust stability is only sufficient. The controller will be robustly stable for the specified range of M_m if:

$$||W_O S_P W_I||_\infty \leq 1 \tag{37}$$

The AGC controller $K = -1/(1.2\bar{M}_m)$ will be robustly stable but will not necessarily satisfy (37). The AGC controller achieves the robustness by decreasing the nominal performance. The advantage of the $\mathcal{H}_\infty$ design is that the robust stability and nominal performance objectives are stated as two separate (but connected) objectives in the cost function. Although they are stated separately, to find a controller which will satisfy them, the weights may have to be changed to tradeoff the performance for robustness. This is the case for the SISO case (for just the force feedback loop) but it is the multivariable controller that is of particular interest. If condition (37) is included in the total control design and it can be satisfied, without changing the performance weights, then the multivariable controller will be able to provide robust stability without a corresponding performance reduction.

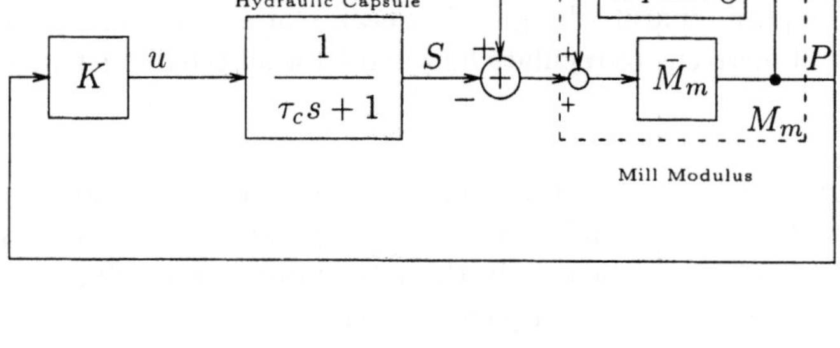

Figure 15: Thickness Control for an Uncertain Mill Modulus

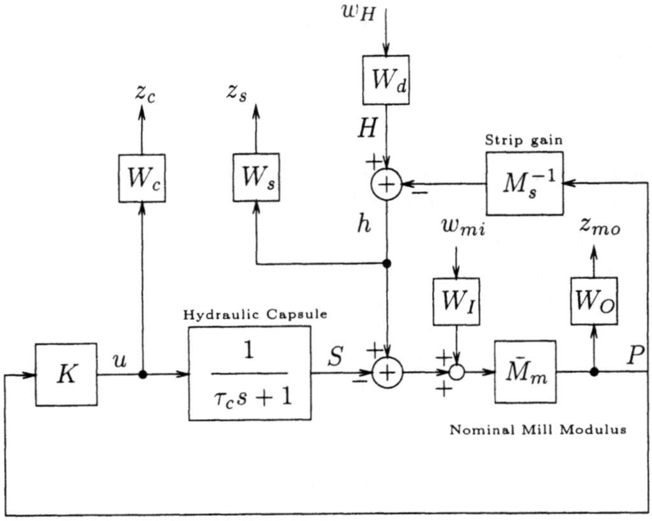

Figure 16: Plant Augmented with Performance and Uncertainty Weights

6 Multivariable Controller Design

6.1 Control Structure

A single multivariable controller for the total seven stand tandem mill is not desirable, since it will not be easy to implement and commission. Moreover if the controller is model based then it will depend on the interstand transport delays which are variable. The proposed multivariable controller is decentralised with respect to the stands which will result in a separate $\mathcal{H}_\infty$ design for each stand/interstand zone. Instead of using a formal result to separate the performance objective into a number of control designs such as [2], the physical structure of the problem will be used to guide the choice of controller structure. Since each zone has the same structure then results for symmetrical systems [10] might be used but the dynamics for each zone are different enough to require separate designs. For each stand/interstand zone shown in Fig. 17 the objective is to regulate h, σ and θ for all the disturbances affecting that zone.

The thickness and temperature entering stand i+1 are the delayed thickness and temperature from stand i, with the temperature also decreasing due to radiation and convection losses, in the interstand region. The tension is disturbed by the velocity of the strip entering the downstream stand V^{i+1}. Using part of the linear model for stand i+1 a linear model can be found for the velocity which depends on S^{i+1}, H^{i+1}, σ^i, σ^{i+1}, T^{i+1} and ω^{i+1}. Since tension cannot be measured, then to achieve good $\mathcal{H}_\infty$ disturbance attenuation, the disturbances should be modelled accurately, in terms of the frequency responses. All disturbances which can be measured should be treated as feed-forward signals in the design. The two disturbances which can be measured are the changes in the capsule position and the work roll speed of the downstream stand. Instead of using the responses of the actuators, if the capsule and motor references are used then this should enable the controller to respond faster to these disturbances. Using the measured force P^{i+1} from stand i+1 for the control of stand i should enable the entry thickness H^{i+1} and entry temperature T^{i+1} to be estimated, and used in the feedforward control for V^{i+1}. All the disturbances affecting that zone have been included, apart from the entry tension and exit tensions to stands i and i+1 respectively. This is due to the tension being a less significant disturbance than the thickness and temperature and neglecting the tension disturbances from up and down stream stands is not significant as long as the tension in each zone is sufficiently regulated. The transport delay for the thickness and temperature is not used explicitly in the design model, since it would require more states to approximate the delay and the delay will change as the mill is accelerated. Assuming that the thickness at stand i and i+1 are not connected by using independent disturbances makes the designs more conservative but of lesser order, and not subject to uncertainty when the mill speed changes. Each of the disturbance signals including the downstream references are passed through 1st order low pass filters to represent the disturbances of different bandwidths. Also included are 1st order high pass filters for measurement noise models to ensure roll off of the controllers at high frequencies.

The basic structure of the mill system is illustrated in Fig. 19 with the controller for each zone separated into the feedback and feedforward components. A feedback loop is formed between zones, due to the transport of the strip (G_{in}) from i to i+1 and the feedback to zone i of the entry velocity of zone i+1 which is represented by G_{11}, G_{12} and G_{13}. The dominant component of the zone feedback is due to G_{13}. Hence this model is used for the feedforward design and G_{in} can be neglected, without a performance loss, and with the benefit that the controllers are speed independent. The regulator structure is shown with three stands of the mill in Fig. 18 with feedforward of the capsule and drive references to the upstream stand, and the use of the stand force measurement for its own stand and upstream stand. The feedforward and feedback components of the controller are calculated in a single integrated $\mathcal{H}_\infty$ optimisation problem which is described in a number of steps next.

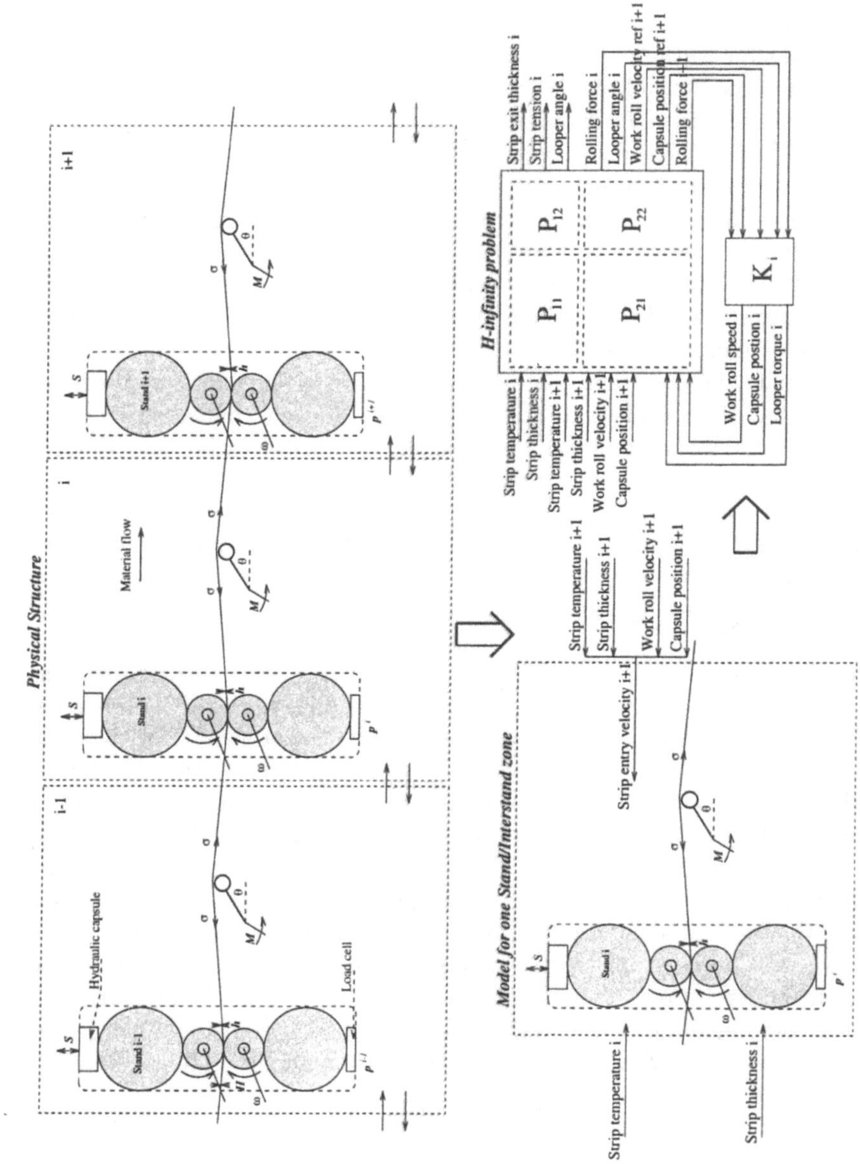

Figure 17: Physical Motivation for Control Structure

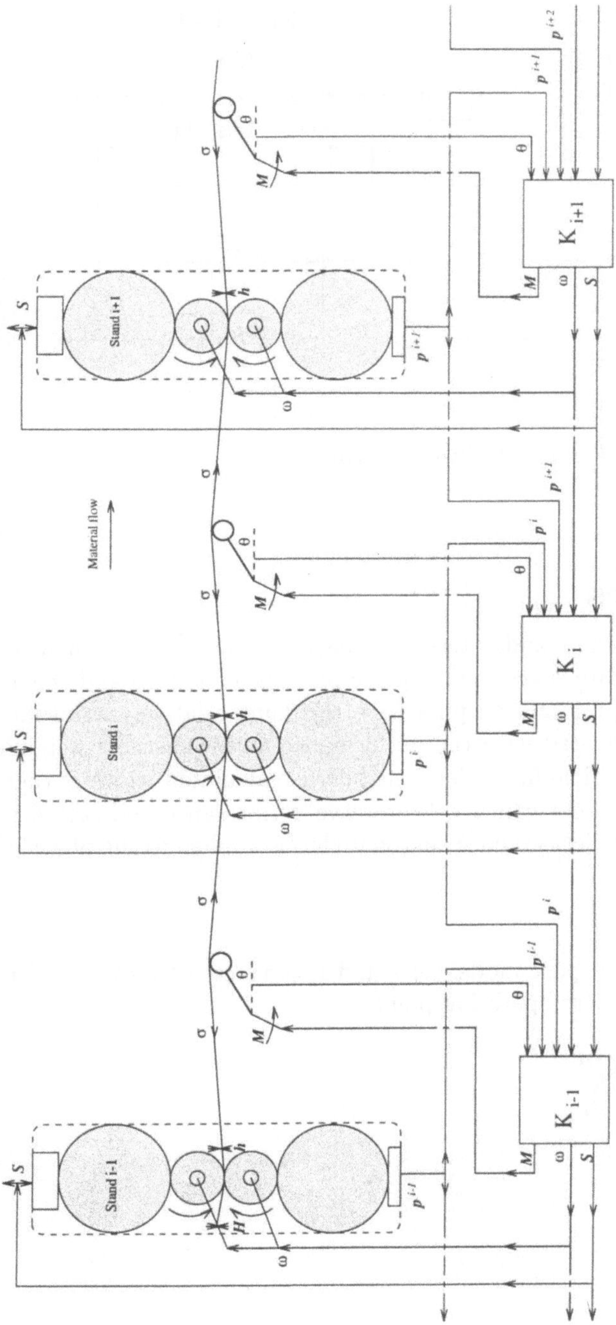

Figure 18: Regulator Control Structure

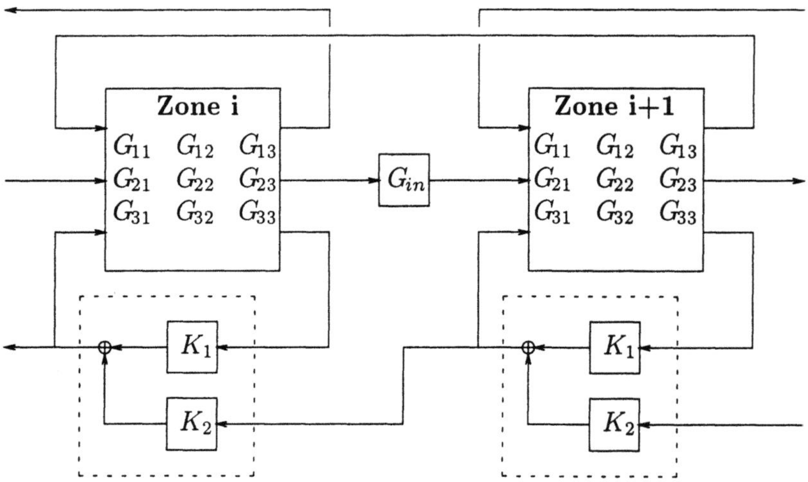

Figure 19: System Structure

6.2 Design Stages

The stages to calculate the stand/interstand controllers will be described sequentially. The linear state-space model of the plant is assembled and this is scaled such that the inputs and outputs have the same relative magnitude. This plant is then augmented with the performance and robustness weights and disturbance models. The full order controller is calculated which is then reduced to the smallest order which will preserve performance and stability. The reduced-order controller is then scaled so it can be applied to the physical system.

(1) Stand/Interstand i

Assemble the linear state-space model for stand i, connected to interstand i, at an estimated temperature operating point.

$x = [\frac{\sigma^i}{E}\ \theta^i\ \dot{\theta}^i\ \omega^i\ M^i\ S^i\ X_I^i]^T$
$w_1 = [H^i\ T_{en}^i \mid \omega_r^i\ M_r^i\ S_r^i]^T$
$w_2 = [V^{i+1}]$
$z_1 = [\sigma^i\ \theta^i\ h^i\ P^i \mid P^i\ \theta^i]^T$
$z_2 = [\sigma^i]$

$$\begin{bmatrix} P_{11} & P_{12} \\ P_{21} & P_{22} \end{bmatrix} := \left[\begin{array}{c|cc} A^p & B_1^p & B_2^p \\ \hline C_1^p & D_{11}^p & D_{12}^p \\ C_2^p & D_{21}^p & D_{22}^p \end{array}\right] =$$

$$\left[\begin{array}{c|c|c}
\begin{bmatrix} \sigma_2 v_4^i & 0 & \frac{\sigma_1}{E} & \frac{\sigma_2}{E} v_5^i & 0 & \frac{\sigma_2}{E} v_1^i & 0 \\ 0 & 0 & 1 & 0 & 0 & 0 & 0 \\ (\theta_2^i + \theta_3^i h_4^i)E & \theta_4^i & \theta_5^i & \theta_3^i h_5^i & \theta_1^i & \theta_3^i h_1^i & 0 \\ \omega_7^i E & 0 & 0 & \omega_2^i & 0 & \omega_4^i & \omega_1^i \\ 0 & 0 & 0 & 0 & M_1^i & 0 & 0 \\ 0 & 0 & 0 & 0 & 0 & S_1^i & 0 \\ 0 & 0 & 0 & -1 & 0 & 0 & 0 \end{bmatrix} &
\begin{bmatrix} \sigma_2 v_2^i & \sigma_2 v_6^i & 0 & 0 & 0 \\ 0 & 0 & 0 & 0 & 0 \\ \theta_3^i h_2^i & \theta_3^i h_6^i & 0 & 0 & 0 \\ \omega_5^i & \omega_8^i & \omega_3^i & 0 & 0 \\ 0 & 0 & 0 & M_2^i & 0 \\ 0 & 0 & 0 & 0 & S_2^i \\ 0 & 0 & 1 & 0 & 0 \end{bmatrix} &
\begin{bmatrix} \sigma_3^i \\ 0 \\ 0 \\ 0 \\ 0 \\ 0 \\ 0 \end{bmatrix} \\ \hline
\begin{bmatrix} E & 0 & 0 & 0 & 0 & 0 & 0 \\ 0 & 1 & 0 & 0 & 0 & 0 & 0 \\ Eh_4^i & 0 & 0 & h_5^i & 0 & h_1^i & 0 \\ EP_4^i & 0 & 0 & P_5^i & 0 & P_1^i & 0 \\ EP_4^i & 0 & 0 & P_5^i & 0 & P_1^i & 0 \\ 0 & 1 & 0 & 0 & 0 & 0 & 0 \end{bmatrix} &
\begin{bmatrix} 0 & 0 & 0 & 0 & 0 \\ 0 & 0 & 0 & 0 & 0 \\ h_2^i & h_6^i & 0 & 0 & 0 \\ P_2^i & P_6^i & 0 & 0 & 0 \\ P_2^i & P_6^i & 0 & 0 & 0 \\ 0 & 0 & 0 & 0 & 0 \end{bmatrix} &
\begin{bmatrix} 0 \\ 0 \\ 0 \\ 0 \\ 0 \\ 0 \\ 0 \end{bmatrix} \\
\begin{bmatrix} E & 0 & 0 & 0 & 0 & 0 & 0 \end{bmatrix} & \begin{bmatrix} 0 & 0 & 0 & 0 & 0 \end{bmatrix} & \begin{bmatrix} 0 \end{bmatrix}
\end{array}\right] \tag{38}$$

(2) Stand $i+1$

Assemble the linear state-space model for stand $i+1$.

$x = [\omega^{i+1} \; X_I^{i+1} \; S^{i+1}]^T$

$\hat{w}_1 = [\sigma^i]$

$\hat{w}_2 = [H^{i+1} \; T_{en}^{i+1} \; \omega_r^{i+1} \; S_r^{i+1}]^T$

$\hat{z}_1 = [V^{i+1}]$

$\hat{z}_2 = [P^{i+1} \; \omega_r^{i+1} \; S_r^{i+1}]^T$

$$\begin{bmatrix} Q_{11} & Q_{12} \\ Q_{21} & Q_{22} \end{bmatrix} := \left[\begin{array}{c|cc} A^q & B_1^q & B_2^q \\ \hline C_1^q & D_{11}^q & D_{12}^q \\ C_2^q & D_{21}^q & D_{22}^q \end{array}\right]$$

$$= \left[\begin{array}{c|c|c}
\begin{bmatrix} \omega_2^{i+1} & \omega_1^{i+1} & \omega_4^{i+1} \\ -1 & 0 & 0 \\ 0 & 0 & S_1^{i+1} \end{bmatrix} &
\begin{bmatrix} \omega_6^{i+1} \\ 0 \\ 0 \end{bmatrix} &
\begin{bmatrix} \omega_5^{i+1} & \omega_8^{i+1} & \omega_3^{i+1} & 0 \\ 0 & 0 & 1 & 0 \\ 0 & 0 & 0 & S_2^{i+1} \end{bmatrix} \\ \hline
\begin{bmatrix} V_5^{i+1} & 0 & V_1^{i+1} \end{bmatrix} & \begin{bmatrix} V_3^{i+1} \end{bmatrix} & \begin{bmatrix} V_2^{i+1} & V_6^{i+1} & 0 & 0 \end{bmatrix} \\
\begin{bmatrix} P_5^{i+1} & 0 & P_1^{i+1} \\ 0 & 0 & 0 \\ 0 & 0 & 0 \end{bmatrix} &
\begin{bmatrix} P_3^{i+1} \\ 0 \\ 0 \end{bmatrix} &
\begin{bmatrix} P_2^{i+1} & P_6^{i+1} & 0 & 0 \\ 0 & 0 & 1 & 0 \\ 0 & 0 & 0 & 1 \end{bmatrix}
\end{array}\right] \tag{39}$$

(3) Stand/Interstand i connected to Stand $i+1$

Since P and Q are LFTs then they can be connected using the star product to make another LFT. The entry thickness and temperature to stand $i+1$ is assumed to be independent of the exit thickness and temperature of stand i. Although this will make the design more conservative, by assuming the disturbances are independent, it will produce a simpler plant model to compute the controllers, since the transport delays do not have to be approximated. Including the delays in the design would also mean the design was strictly only valid for the chosen line speed, so that the controller produced might be less robust than one assuming independent disturbances.

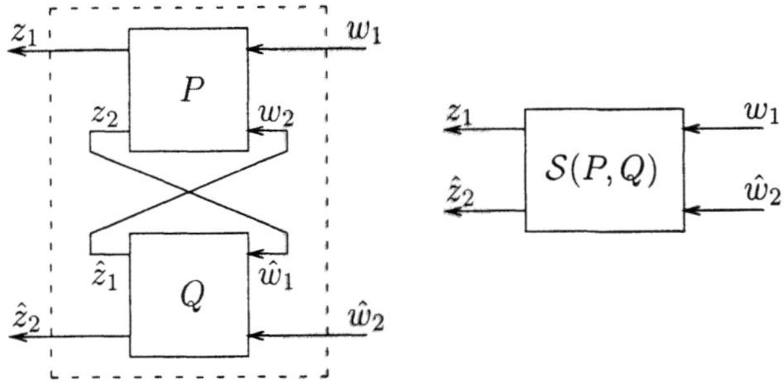

Figure 20: Interconnection of LFTs Using the Star Product

$$N = \mathcal{S}(P,Q) =$$
$$\begin{bmatrix} P_{11} + P_{12}Q_{11}(I - P_{22}Q_{11})^{-1}P_{21} & P_{12}(I - Q_{11}P_{22})^{-1}Q_{12} \\ Q_{21}(I - P_{22}Q_{11})^{-1}P_{21} & Q_{22} + Q_{21}P_{22}(I - Q_{11}P_{22})^{-1}Q_{12} \end{bmatrix} \quad (40)$$

(4) Unscaled Design System

Partition the total system and rearrange to collect the disturbances into one vector.

$z = [\sigma^i \; \theta^i \; h^i \; P^i]^T$

$y = [P^i \; \theta^i \; P^{i+1} \; \omega_r^{i+1} \; S_r^{i+1}]^T \; u = [\omega_r^i \; M_r^i \; S_r^i]^T$

$w = [H^i \; T_{en}^i \; H^{i+1} \; T_{en}^{i+1} \; \omega_r^{i+1} \; S_r^{i+1}]^T$

$$\begin{bmatrix} z \\ y \end{bmatrix} = \begin{bmatrix} \begin{bmatrix} N_{1111} & N_{1211} \end{bmatrix} & \begin{bmatrix} N_{1112} \end{bmatrix} \\ \begin{bmatrix} N_{1121} & N_{1221} \\ N_{2111} & N_{22} \end{bmatrix} & \begin{bmatrix} N_{1122} \\ N_{2112} \end{bmatrix} \end{bmatrix} \begin{bmatrix} w \\ u \end{bmatrix}$$
$$= \begin{bmatrix} M_{11} & M_{12} \\ M_{21} & M_{22} \end{bmatrix} \begin{bmatrix} w \\ u \end{bmatrix} \quad (41)$$

(5) Scaled Design System

Since the system inputs and outputs and are in different units the system has to be normalised to make the error outputs of roughly the same magnitude or importance and to enable a disturbance vector of unity norm to be applied. Scaling the measurement outputs and control inputs will produce a better conditioned system, which should prevent numerical problems. The scaling method involves the use of the expected maximum deviation of the signals to normalise them to unity or norm 1. For this multivariable system the following matrices are diagonal, with the maximum value of each component on the diagonal: $D_z = z_{max}$, $D_y = y_{max}$, $D_u = u_{max}$, $D_w = w_{max}$. The scaled signals are: $\bar{z} = D_z^{-1}z$, $\bar{y} = D_y^{-1}y$, $\bar{u} = D_u^{-1}u$, $\bar{w} = D_w^{-1}w$

If the assumptions about the maximum deviation are correct and a controller is found then $||z||_2 \leq 1$, $||y||_2 \leq 1$, $||u||_2 \leq 1$ and $||w||_2 \leq 1$. The scaled design system is hence:

$$S=\begin{bmatrix} D_z^{-1} & 0 \\ 0 & D_y^{-1} \end{bmatrix}\begin{bmatrix} M_{11} & M_{12} \\ M_{21} & M_{22} \end{bmatrix}\begin{bmatrix} D_w & 0 \\ 0 & D_u \end{bmatrix}=\begin{bmatrix} D_z^{-1}M_{11}D_w & D_z^{-1}M_{12}D_u \\ D_y^{-1}M_{21}D_w & D_y^{-1}M_{22}D_u \end{bmatrix} \tag{42}$$

(6) Weighted Design System

All the weightings are diagonal with first order elements. The feedforward signals have measurement noise added to make the feedforward part of the controller roll off at high frequency, even though the measurable disturbances are known exactly.

$$W_s = \begin{bmatrix} W_\sigma & 0 & 0 & 0 \\ 0 & W_\theta & 0 & 0 \\ 0 & 0 & W_h & 0 \\ 0 & 0 & 0 & W_P \end{bmatrix} \quad W_n = \begin{bmatrix} W_{n1} & 0 & 0 & 0 & 0 \\ 0 & W_{n2} & 0 & 0 & 0 \\ 0 & 0 & W_{n3} & 0 & 0 \\ 0 & 0 & 0 & W_{n4} & 0 \\ 0 & 0 & 0 & 0 & W_{n5} \end{bmatrix}$$

$$W_c = \begin{bmatrix} W_\omega & 0 & 0 \\ 0 & W_S & 0 \\ 0 & 0 & W_M \end{bmatrix} \quad W_d = \begin{bmatrix} W_H & 0 & 0 & 0 & 0 & 0 \\ 0 & W_T & 0 & 0 & 0 & 0 \\ 0 & 0 & W_H & 0 & 0 & 0 \\ 0 & 0 & 0 & W_T & 0 & 0 \\ 0 & 0 & 0 & 0 & W_{\omega_r} & 0 \\ 0 & 0 & 0 & 0 & 0 & W_{S_r} \end{bmatrix}$$

The weighted open-loop design system is:

$$G = \left[\begin{array}{cc|c} W_s S_{11} W_d & 0 & W_s S_{12} \\ 0 & 0 & W_c \\ \hline S_{21} W_d & W_n & S_{22} \end{array}\right] \tag{43}$$

(7) Full Order Controller Calculation

The objective is to find a controller stabilising G, such that:

$$||T_{zw}||_\infty = ||\mathcal{F}_l(G, K)||_\infty = 1 \tag{44}$$

The construction of G and T_{zw} from the individual models and weightings, is shown in Fig. 22. A controller can be found which minimises the ∞ norm but as the controller nears the optimal solution it could be ill-conditioned and have fast dynamics. By solving for the sub-optimal case with a bound of 1 then all the frequency weighted objectives will be satisfied. If improved performance is required, then the weights can be adjusted. For the solution to this problem see, [17].

(8) Reduced Order Controller Calculation

Reducing the dynamic order of a controller is important because it reduces the complexity of the controller to be implemented and removes high frequency poles from the controller which are undesirable. Any systematic method of controller reduction must, for the reduced controller, retain closed-loop stability and if possible preserve the closed-loop performance at the same level as with the full order controller. In particular for $\mathcal{H}_\infty$ controllers [16] and [17] provide a comparison of different techniques and recommend the *Performance Weighted Coprime Factor Reduction* (PWCFR) method which is summarised below.

If K is the full order controller and $\hat{K}$ is the reduced controller, then by making the error $\bar{\sigma}(K - \hat{K})(j\omega)$ small, uniformly over all frequencies, does not necessarily stabilise the closed-loop system and give the performance required. What is required is to make the error small in the critical frequency ranges which affects the stability and performance. This can be done by minimising an error which is weighted by a frequency dependent function. To achieve a design of lower order but with the same level of performance, the parameterisation of all $\mathcal{H}_\infty$ controllers will be used. This means finding a suitable Q (Fig. 21) such that K is of order less than the generalised plant G and the performance is preserved at $\|T_{zw}\|_\infty < \gamma$.

If the central controller is calculated first and M_∞ is partitioned as follows:

$$M_\infty = \left[\begin{array}{c|cc} A & B_1 & B_2 \\ \hline C_1 & D_{11} & D_{12} \\ C_2 & D_{21} & D_{22} \end{array}\right] \tag{45}$$

then the family of all admissible controllers such that $\|T_{zw}\|_\infty < \gamma$ can be written as

$$K(s) = \mathcal{F}_l(M_\infty, Q) = (\Phi_{11}Q + \Phi_{12})(\Phi_{21}Q + \Phi_{22})^{-1} = UV^{-1} \tag{46}$$

where $Q \in \mathcal{RH}_\infty$, $\|Q\|_\infty < \gamma$ and UV^{-1} is a right coprime factorisation over $\mathcal{RH}_\infty$ and

$$\Phi = \left[\begin{array}{c|cc} A - B_1 D_{21}^{-1} C_2 & B_2 - B_1 D_{21}^{-1} D_{22} & B_1 D_{21}^{-1} \\ \hline C_1 - D_{11} D_{21}^{-1} C_2 & D_{12} - D_{11} D_{21}^{-1} D_{22} & D_{11} D_{21}^{-1} \\ -D_{21}^{-1} C_2 & -D_{21}^{-1} D_{22} & D_{21}^{-1} \end{array}\right] \tag{47}$$

The procedure to reduce the order of the controllers is:

(i) Calculate a central ($Q = 0$) $\mathcal{H}_\infty$ controller $K_0 = \Phi_{12}\Phi_{22}^{-1}$ such that $\|T_{zw}\|_\infty < \gamma$.

(ii) Find a reduced order controller $\hat{K} = \hat{U}\hat{V}^{-1}$ such that the following frequency weighted $\mathcal{H}_\infty$ error is satisfied

$$\left\| \begin{bmatrix} \gamma^{-1} I & 0 \\ 0 & I \end{bmatrix} \Phi^{-1} \left(\begin{bmatrix} \Phi_{12} \\ \Phi_{22} \end{bmatrix} - \begin{bmatrix} \hat{U} \\ \hat{V} \end{bmatrix} \right) \right\|_\infty < \frac{1}{\sqrt{2}} \tag{48}$$

(iii) With the reduced order controller $\hat{K}$ the closed loop system is stable and the performance maintained at $||T_{zw}||_\infty = ||\mathcal{F}_l(G, \hat{K})||_\infty < \gamma$. This procedure is based on a sufficient condition and if the bound is not satisfied the reduced order controller may still maintain the closed-loop stability and performance.

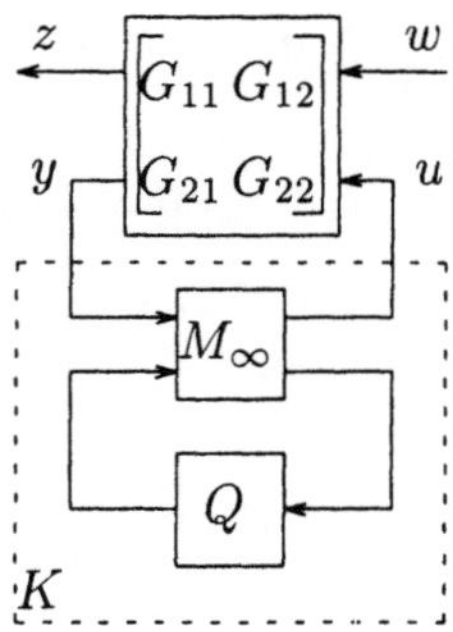

Figure 21: Parameterization of all $\mathcal{H}_\infty$ Controllers

To calculate $\hat{U}$ and $\hat{V}$ frequency weighted balanced model reduction will be used where the objective is to find the reduced order model G_r such that $||W_o(G - G_r)||_\infty < \frac{1}{\sqrt{2}}$ where

$$K(s) = \mathcal{F}_l(M_\infty, Q), \quad Q \in \mathcal{RH}_\infty, \; ||Q||_\infty < 1 \tag{49}$$

Table 1 shows the order, closed-loop performance and the weighted reduction error. The downstream controllers are harder to reduce and need a higher order to provide the desired performance.

K	Order	$\lVert T_{zw}\rVert_\infty$	$\lVert W(G - G_r)\rVert_\infty$
1	18	0.6425	0.1419
2	18	0.4619	0.1726
3	20	0.689	0.1026
4	22	0.9659	0.3265
5	25	0.9892	0.1358
6	25	0.9894	0.1466

Table 1: Controller order reduction

(9) Scaled Controller

To implement the controller on the actual system it must be scaled so that the closed system will correspond with the scaled system used in the design.

$$K_{sc}(s) = D_u K(s) D_y^{-1} \tag{50}$$

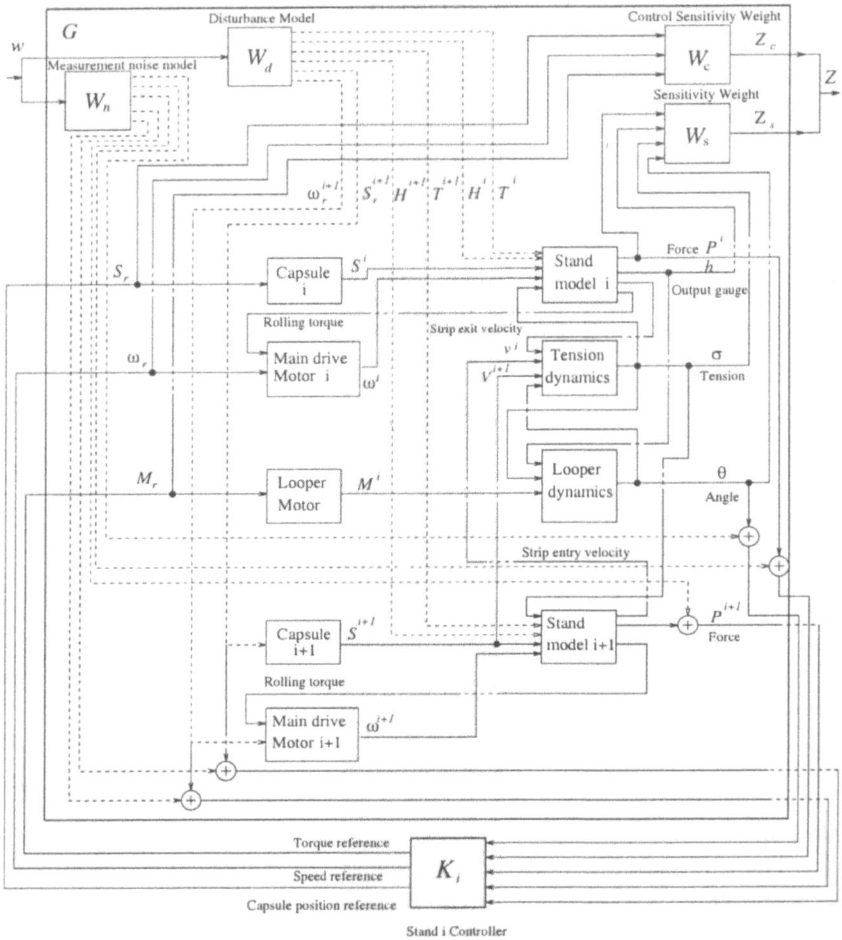

Figure 22: Standard Model Representation for Design

7 Controller Evaluation

The frequency response for the stand 4 controller is shown in figures 23 and 24 for the robust controller and also for the case when the mill modulus robustness objective was not included in the design. The element from force i to the capsule position reference is equivalent to the AGC controller. That channel of the multivariable controller is scaled with $1/\bar{M}$, therefore it has the ideal gain for zero steady-state and the same 180° phase shift as the AGC. The major difference between the robust and "unrobust" multivariable controllers is that for the robust design the gain from the angle to the capsule reference is reduced and there is more phase lag. The channel from the force to the capsule reference does not change if the controller is robustly stable for the mill modulus uncertainty. The robustness of the multivariable controller is due to the angle feedback whilst the thickness performance is provided by force feedback. The closed-loop sensitivities for stand 4 are shown in Fig. 25 with the inverse magnitude response of the weightings which is the design target.

To evaluate the performance of the $\mathcal{H}_\infty$ regulators the full nonlinear simulation of a seven stand finishing mill was used. The simulation includes the threading, zooming and dethreading phases of operation with variable transport delays and has a total of 169 states (excluding controller dynamics). Figure 26 shows the results for stand 4 with the conventional and $\mathcal{H}_\infty$ regulators. The work roll speed shows the acceleration of the mill during zooming to keep the exit temperature of the mill at it's reference level and the deceleration before the strip is de-threaded. The strip temperature shows a corresponding increase due to smaller heat loss at the higher speed. The temperature also shows the sinusoidal disturbance due to the skid chills of the incoming bar. The higher frequency disturbances are due to the eccentricity of the stand rolls. With the $\mathcal{H}_\infty$ design the thickness does show an improvement which mostly corresponds to the reduced tension disturbance affecting the interaction in the stand. The responses for the looper angle and tension reveal a significant improvement when compared to the results for the conventional regulators. The results also demonstrate the robustness of the design to nonlinearities under changing operating points for stand speed and strip temperature, since the regulators were designed at a particular operating point for the expected strip temperature.

Figure 27 shows the results for stand 4 with the conventional and $\mathcal{H}_\infty$ regulators operating for $M_m = \bar{M}$, $M_m = 0.8\bar{M}$ and $M_m = 1.2\bar{M}$. With the nominal mill modulus (M) the multivariable controller keeps the thickness close to its reference, whereas the conventional controller develops a steady-state error, as the temperature increases with mill acceleration. When the actual mill modulus is less than the assumed value $(0.8M)$ then the multivariable controller develops a thickness error with the changing temperature, however the error is less than with the conventional controller. When the actual mill modulus is larger than the assumed value $(1.2M)$, then the conventional controller has the smaller error. This is due to the conventional controller using $\hat{M} = 1.25\bar{M}$ as the inverse gain which is closer to the actual mill modulus (in this case) than the multivariable controller. In all cases the tension and angle response is substantially better with the multivariable controller, since it was designed to take account of the interactions.

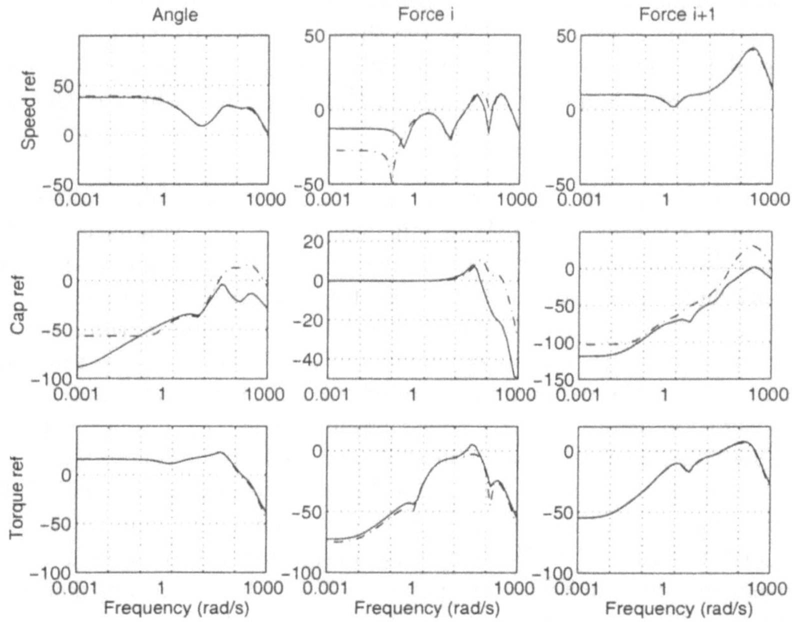

Figure 23: $|K_1^4(j\omega)|$, (dB), - robust, -.- unrobust

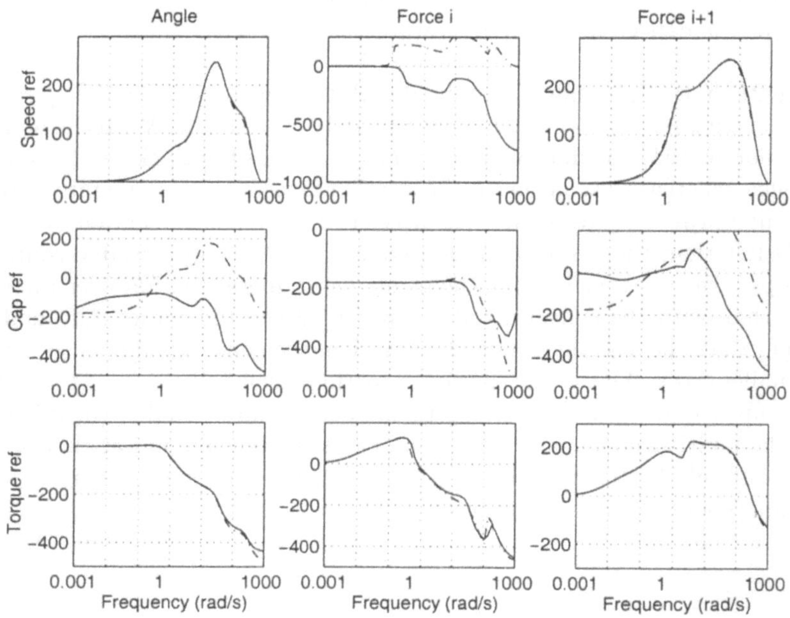

Figure 24: $\angle K_1^4(j\omega)$, (degrees), - robust, -.- unrobust

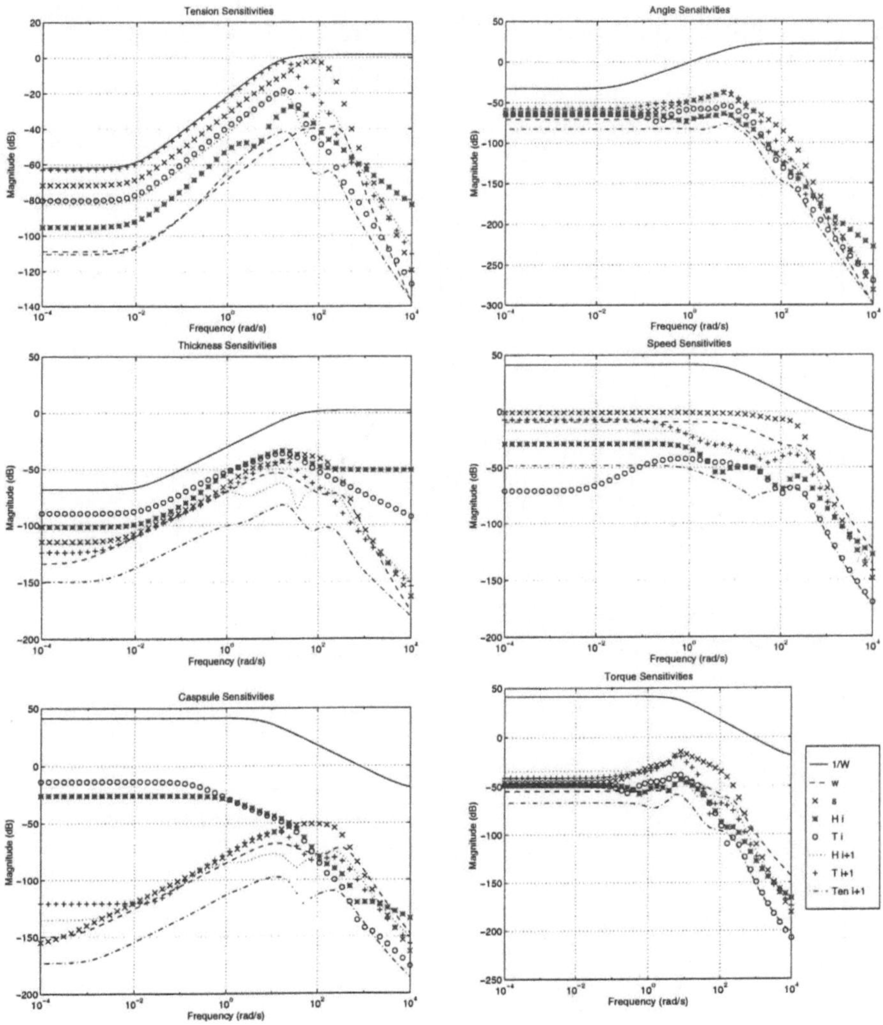

Figure 25: Unit 4 closed-loop sensitivities and inverse weights

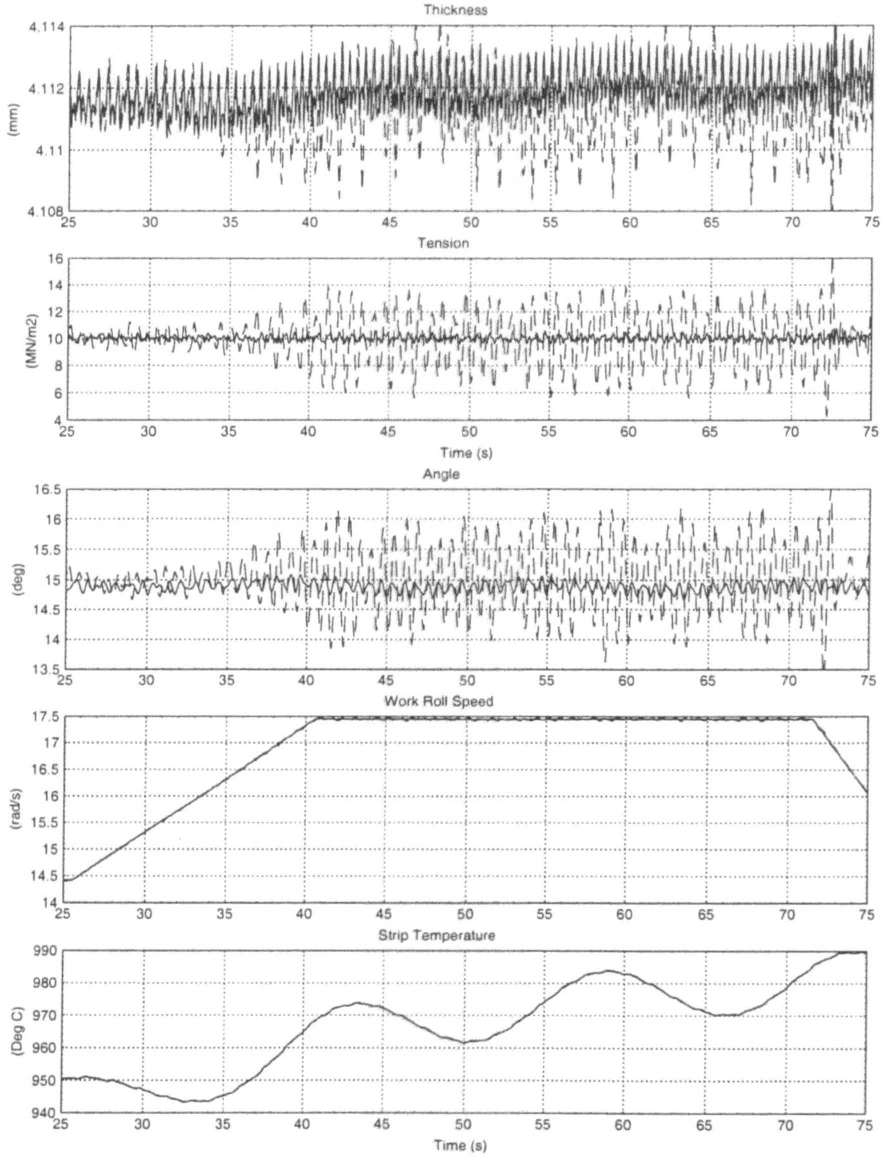

Figure 26: Simulation results, $\mathcal{H}_\infty$ = solid, conventional = dashed

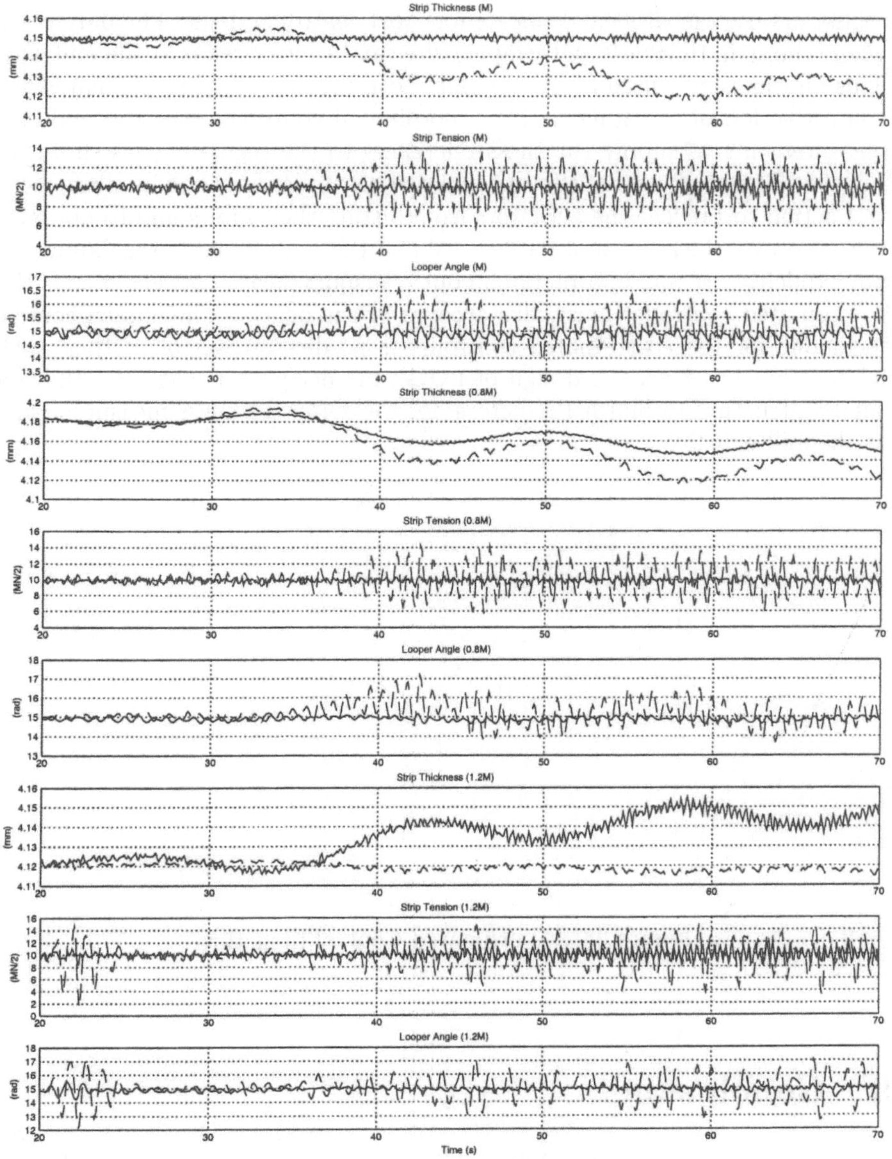

Figure 27: Simulation results for mill modulus uncertainty, $\mathcal{H}_\infty$ controller = solid, conventional controller = dashed

8 Upper Level Co-ordinated Control

For this article attention has concentrated on the regulator level of the mill, where there are many interactions and the dynamics are relatively fast. The regulators used local measurements for local control actions. The only place where the strip thickness and temperature can be measured is at the mill exit. This information is used for slower global control changes to co-ordinated the control action of the stands. The co-ordinated controller uses the mill exit measurements to supply thickness references to the regulators, as shown in Fig. 28. The objectives of the co-ordinated controller are to redistribute control action to prevent the actuators from hitting their limits and to optimise the strip exit thickness and flatness. While the multivariable regulators did not explicitly use any information about transport delays between stands the co-ordinated controller must take account of the delays. The chosen control technique for the co-ordinated controller is predictive control. Other work in this area has been the design of LQG time delay controllers that have a similar structure to Smith Predictors, for the x-ray thickness measurements, at the mill exit [5].

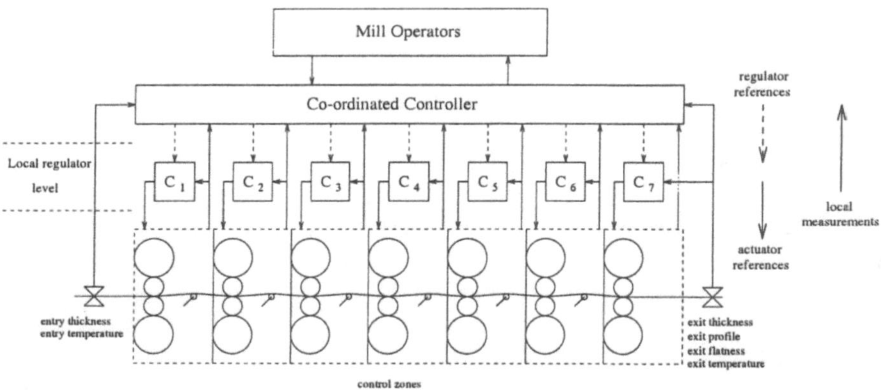

Figure 28: Co-ordinated Control Structure

9 Conclusions

The hot rolling mill control problem involves a very large interacting multivariable system where advanced controllers can offer benefits through improved multivariable control and robustness. However, it is necessary to account for practical problems associated with implementation and commissioning. The overall structure of the solution is therefore important and the strategy for the design process. The approach described here is to split the seven stand mill into a number of zones including a mill stand and the zone between the stands. This enabled each of the individual designs to be completed, relatively independent of the other zones. It also meant that these subsystems could be evaluated and commissioned independently. This does not mean that the link between the stands was not taken into account since feedforward and feedback from other stand zones was included. The resulting structure provided improvements through the use of a true multivariable design approach, and at the same time the resultant solution was sufficiently simple to encourage implementation.

In the solution of the overall problem, several sub-design problems arose which were of independent interest. A very effective mechanism for transport delay compensation was produced involving a Kalman filter type structure. This provided a means of stabilising open loop unstable systems which Smith predictors cannot do. The subject of reliable control [9] was also considered, where the possibility of adding tension measurements was considered but where a possible tension failure could not be allowed to cause a serious malfunction. The reliable control approach utilised was an extension of the $\mathcal{H}_\infty$ design and ensured the system remained stabilising, even in the present of certain sensor failures.

Although the previous material mainly concentrated on thickness and tension control, there is as much interest in so called flatness and profile control. These are very dependent upon the reference levels set for the whole mill and hence the supervisory control system is of substantial importance. It was not possible to go into the developments in supervisory control, but most of the current emphasis is on applying Linear Quadratic Gaussian predictive control methods to the upper level control structure. The ability to handle hard constraints and to optimise the total system performance is valuable. The project results have demonstrated that there are opportunities to improve existing hot mill control schemes, whether it be through design changes to classical systems or through more major structural changes for new systems.

10 Acknowledgements

We are grateful for the support of the EPSRC and the following companies: Alcan International, British Steel, Cegelec Projects, Kvaerner Metals. We should also like to thank Dr Peter Reeve, Mike Clark, Dr Paul Corney and Dr Chris Davenport for their kind advice and help during the project. Our colleagues Prof. Mike Johnson and Dr Reza Katebi have contributed substantially to the project and we are grateful for their help. The help of Phil Reece (British Steel) and Alan Kidd (Cegelec) in setting up the project is also much appreciated.

References

[1] Anbe Y., K. Sekiguchi and H. Imanari, "Tension Control of a Hot Strip Mill Finisher ", *IFAC 13th Triennial World Congress, San Francisco*, pp. 439-444, 1996.

[2] Date R. A., J. H. Chow, "A Parametrization Approach to Optimal $\mathcal{H}_2$ and $\mathcal{H}_\infty$ Decentralized Control Problems", *Automatica*, Vol. 29, No. 2, pp. 457-463, 1993.

[3] Duysters S., J. A. J. Govers and A. J. J. van der Weiden, "Process Interactions in a Hot Strip Mill; Possibilities for multivariable control ?", *3rd Conference on Control Applications, Glasgow*, pp. 1545-1550, 1994.

[4] Ginzburg, V. B. (1989). Steel-Rolling Technology. *Marcel Dekker, New York.*

[5] Grimble M. J. and G. Hearns, "LQG Controllers for State -Space Systems with Pure Transport Delays: Application to Hot Strip Mills", *Automatica*, Vol. 34, No. 10, October 1998, pp. 1169-1184.

[6] Harakawa T. and K. Furuta, "Decentralized Control of Multivariable Looper System in Hot Strip Finisher Mill Control", *4th IEEE Conference on Control Applications, New York*, pp. 856-861, 1995.

[7] Hayashi Y., S. Tanimoto, M. Saito, T. Kataoka, H. Sasao and K. Yabuuchi, "A New Tension Control Method for Hot Strip Finishing Mill", *IFAC 10th Triennial World Congress*, pp. 114-119, 1987.

[8] Hearns G. and M. J. Grimble, "Robust Hot Strip Mill Thickness Control", *9th IFAC Symposium on Automation in Mining, Mineral and Metal Processing, Cologne, Germany*, September 1-3 1998, pp. 115-120.

[9] Hearns G. and M. J. Grimble, "Fault Tolerant Strip Tension Control", *17th American Control Conference, Philadelphia*, June 24th-26th 1998, pp. 2992-2996.

[10] Hovd M., S. Skogestad, "Control of Symmetrically Interconnected Plants", *Automatica*, Vol. 30, No. 6, pp. 957-973, 1994.

[11] Imanari H., Y. Morimatsu, K. Sekiguchi, H. Ezure, R. Matuoka, A. Tokuda and H. Otobe, "Looper H-infinity Control for Hot Strip Mills", *IEEE, IAS, Florida*, pp. 2133-2139, 1995.

[12] Nakagawa S., H. Miura, S. Fukushima and J. Amasaki, "Gauge Control System For Hot Strip Finishing Mill", *29th Conference on Decision and Control, Honolulu*, pp. 1573-1578, 1990.

[13] Okada M., Y. Iwasaki, K. Murayama, A. Urano, A. Kawano and H. Shiomi, "Optimal Control System for Hot Strip Finishing Mill ", *IFAC 13th Triennial World Congress, San Francisco*, 1996.

[14] Seki Y., K. Sekiguchi and Y. Anbe, "Optimal Multivariable Looper Control for Hot Strip Finishing Mill", *IEEE Transactions on Industry Applications*, Vol. 27, No. 1, pp. 124-130, 1991.

[15] Tsuji Y., "High Quality Control in Hot Strip Mill", *IFAC 10th Triennial World Congress*, pp. 120-125, 1987.

[16] Zhou K. , "A Comparative Study of $\mathcal{H}_\infty$ Controller Reduction Methods", *American Control Conference*, Seattle, pp. 4015-4019, 1995.

[17] Zhou K., J. C. Doyle and K. Glover , "Robust and Optimal Control", *Prentice Hall*, 1996.

6

Fault-tolerant Control Systems

M. Blanke

Department of Control Engineering, Aalborg University, Fredrik Bajers Vej 7C, DK 9220, Aalborg, Denmark

Summary. Fault tolerant control offers enhanced availability and reduced risk of safety hazards when component failure and other unexpected events occur in a controlled plant. Fault-tolerant control merges several disciplines into a framework with common goals. The fault-tolerant properties are obtained through on-line fault detection and isolation, automatic condition assessment and calculation of appropriate remedial actions. The final step is activation of the necessary actions through software. The actions to accommodate a fault cover a wide range of possibilities and underlying theory. Appropriate re-tuning can sometimes suffice, estimation of a signal replacing a measurement from a faulty sensor is needed in other events, and some cases require complex re-configuration or on-line redesign. The basis for a remedial action is always detection of an undesired event and the correct assessment of the situation through isolation of the fault. Analysis of the effects of the not-normal conditions, and the possible remedial actions, is a truly complex problem in most cases. The paper gives an overview of recent progress in theory and methods to analyze and develop fault-tolerant control systems. Fault propagation analysis and severity assessment are shown to be the basic means to evaluate safety and dependability. Following this, an analysis of structure will disclose available redundancy and possibilities to recover from faults in the system. These overall tools lead to requirements to fault detection and isolation. Fault detection theory has been the subject of intensive study for two decades. Nevertheless, the requirements from the use in fault-tolerant architectures have caused new challenges and further development. This paper focus on recent results in overall design methods for fault-tolerant control systems. An example shows how the different concepts are used and illustrates the benefits from active fault tolerance as compared to a traditionally designed control architecture.

1 Introduction

Automation systems are vulnerable to faults. Defects in sensors, actuators, in the process itself, or within the controller, can be amplified by the closed loop control systems, and faults can develop into malfunction of the loop. The closed loop may alternatively hide a fault from being observed until eventually developed to a state where loop failure is inevitable. A control loop failure will easily cause production stop or malfunction at a plant level. With economic demand for high plant availability, and an increasing awareness about the risks associated with system malfunction, dependability is becoming an essential concern in industrial automation. A cheap way to obtain increased dependability in automated systems is to introduce fault-tolerant control. This is an emerging area in automatic control where several disciplines and theory areas are combined to obtain a unique functionality. A key issue is that local faults are prevented from developing into failures at a higher level.

Automation for safety-critical applications, where no failure could be tolerated, employ redundant hardware to cope with faults. Fail-operational systems are made insensitive to any single point component failure. Fail-safe systems make controlled shut-down to a safe state when a sensor measurement indicates a critical fault. In contrast, fault-tolerant control (FTC) systems, employ redundancy in the plant and its automation system to make "intelligent" software that monitors behavior of components and function blocks. Faults are isolated, and appropriate remedial actions taken to accommodate critical faults. The objective is to avoid that faults develop into failure. The overall strategy is to keep plant availability and accept reduced performance when critical faults occur.

One way to obtain fault tolerance is to employ fault detection and isolation on-line, get a discrete event signal to a supervisor-agent when a fault is detected, and in turn activate accommodation actions [3]. These can be predetermined for each type of critical fault or obtained from real-time analysis and optimization. A different strategy is to extend robustness properties of a controller to cope with the envelope of possible faults.

Systematic analysis of fault propagation [2], [5] was shown to be a crucial tool for determination of severity of fault effects and for analysis of reconfiguration possibilities. A semantics for services based on generic component models was developed in [33], [15] and a graphic analysis was found to be very useful. Techniques for fault detection and isolation (FDI) are well known, [38], [19], [16], [27], but selected FDI methods are included to show how FDI is integrated in the overall FTC scheme. The properties of fault detection and isolation methods, when used in a fault-tolerant control scheme, were treated in [28] and in [35]. Methods for re-configuration design is a recently started field within the FTC area. A large range of techniques have been suggested and studied in some detail. Some schemes have come into real application. Predetermined design for accommodation, when this is possible, was demonstrated for a small satellite in [6] and [5]. Predictive control using multiple models, rained optimization [26], [18]. Techniques using logic inference on qualitative models are used in [23], [24]. A related area is control of discrete event dynamical systems. DEDS systems have a known structure with predetermined events that occur with unknown instants and sequence. This is a sub-class of FTC, where events are not predetermined and system structure can change when faults occur. DEDS theory is the subject of [39], [37], and [7]. Concerning implementation, a correct and consistent control system analysis should always be followed by equally correct software implementation. This is a particular concern for the supervisory parts of an FTC scheme. Testing of the FTC elements is difficult since real faults can rarely be provoked to occur as they will in practice. Well planned software architecture and implementation are thus crucial issues for FTC implementation. An approach using reflection in object oriented languages was described in [22].

The FTC area is hence very wide and multi-disciplinary. A recent overview [28] emphasized many algorithmic essentials and the role of FDI. Another [3] presented an engineering view on the means to obtain FTC. This paper focus on the methodological issues in analysis for FTC. It treats fault propagation analysis, it links this with analysis of structure, and shows how this can give precise requirements to fault detection and isolation. Recent results on system recoverability from local failures are also introduced. The essential message is to show how these different methods are needed in combination to provide requirements for functionality and performance of an FTC scheme. Finally, cost-benefit assessment for FTC implementation is included to show likely areas of industrial emphasis. An example illustrate salient features of the methods and demonstrates how a fault-tolerant control scheme is used obtain enhanced safety and availability for a marine propulsion plant.

2 Some Definitions

- Fail-operational: the system is able to operate with no change in performance despite any single point failure.
- Fault-tolerant: a system is able to continue operation and a degradation of performance may be accepted.
- Fail-safe: a system fails to a state that is considered safe in the particular context.
- Passive Fault Accommodation: robustness range of a controller is specified to be wide enough to accommodate specified faults within the accepted performance specification.
- Active Fault Accommodation: detection and isolation of a fault leads to a change in the control operation to accommodate a fault. Actions can include but are not limited to re-configuration, controller re-design, change of control path, change of path of measurements.

3 Analysis of Fault Propagation and System Structure

Analysis of failure of parts of a system is a classical discipline and the failure mode and effects analysis (FMEA) is widely used and appreciated in industry.

The traditional FMEA does not support the handling of faults in automated systems, when the goal is fault-tolerance, ie. to continue operation. An extended fault propagation analysis (FPA) was suggested in [2] where an algebraic approach was introduced. The aim of the FPA is to show end effects of faults, and assist in designing for fault tolerance such that end effects with severe consequences are stopped if the system structure makes this possible. If the FPA analysis finds that serious failure can occur due to certain faults, these are included in a list of fault effects to be detected. Whether this is possible is disclosed in a later analysis of structure that shows which redundant information is available in the system [33], [10], [11]. The final step, will then be to find actions, preferably within the control system, that can accommodate the fault and prevent the serious end effects from occurring. Analysis of whether this will be possible is a part of the analysis for recoverability for the system [31], [13].

3.1 Fault propagation analysis

For the reasons given above, fault analysis needs to incorporate analysis throughout a system. In order to do this a component-based method was introduced [2], in which possible component faults are identified at an early stage of design. The method uses an FMEA description of components as a starting point. In this context components are sensors, valves, motors, programmable function blocks etc.. Programmable parts are considered as consisting of separate function blocks that can be treated similarly to physical components in the analysis, bearing in mind that their properties may be changed by software modifications if so desired. All potential faults and their effects are determined.

An FMEA scheme shows how fault effects out of the component relate to faults at inputs, outputs, or parts within the components.

Using f_{ci} for component faults and e_{ci} for the effects, the FMEA scheme can be expressed as:

$$e_{ci} \leftarrow A_i^f \otimes f_{ci} \tag{1}$$

where A_i^f is a Boolean matrix representing the propagation. The index i is a component identifier and $\otimes$ the inner product disjunction operator. The operation carried out by the operator is equivalent to the scalar Boolean disjunction $\vee$ and the inner product to the conjunction $\wedge$, i.e., row no. k of (1) is:

$$e_{cik} \leftarrow (a_{ik1} \wedge f_{ci1}) \vee (a_{ik2} \wedge f_{ci2}) \ldots \vee (a_{ikn} \wedge f_{cin}) \tag{2}$$

When some faults are effects propagated from other components, we get:

$$e_{ci} \leftarrow A_i^f \otimes \begin{bmatrix} f_{ci} \\ e_{c(i-1)} \end{bmatrix} \tag{3}$$

System descriptions are obtained from interconnection of component descriptions. The description of a system with three components and open loop structure is:

$$\begin{aligned} e_{c3} &\leftarrow A_3^f \otimes \begin{bmatrix} f_{c3} \\ e_{c2} \end{bmatrix} \\ e_{c3} &\leftarrow A_3^f \otimes \begin{bmatrix} I & 0 \\ 0 & A_2^f \end{bmatrix} \otimes \begin{bmatrix} f_{c3} \\ f_{c2} \\ e_{c1} \end{bmatrix} \end{aligned} \tag{4}$$

The combined fault-effect description for this example is constructed in three steps:

$$e_{c3} \leftarrow A_3^f \otimes \begin{bmatrix} I & 0 \\ 0 & A_2^f \otimes \begin{bmatrix} I & 0 \\ 0 & A_1^f \end{bmatrix} \end{bmatrix} \otimes \begin{bmatrix} f_{c3} \\ f_{c2} \\ f_{c1} \end{bmatrix} \tag{5}$$

Effects are seen to be propagated to the next level of analysis and act as parts' faults at that level. This is continued until the system level is reached. The schemes give a surjective mapping from faults to effects: there is a unique path from fault to end effect, but several different faults may cause the same end effect.

Reversal is obtainable through finding the generalized transpose A^b of A^f. For the Boolean matrices, the inversion is this is the same the transpose. When there is no feedback involved, the result is the capability of isolation of fault effects at any level. If feedback is involved, loop techniques can be used.

Experience from applications show that in the above expressions, occurrence of one fault and the non-occurrence of another should be considered as an extension to the method [5].

A graphical illustration of the fault propagation analysis is shown in Fig 1.

4 An example: Ship propulsion

Propulsion system availability is crucial for a ship's ability to manoeuver. Today, local safety systems protect the large machinery. They prevent continued operation or start-up if sensors inform that local shut-down conditions are met. While this is fail-safe for each piece of machinery, this local approach is not globally safe for the ship since it looses its manoeuverability if certain sensor faults occur. The result has been many events where consequences vary from irregularity to major economic loss and causalities. Several events could have been prevented if automation systems had been designed to be tolerant to faults, with overall availability in mind. This section introduces the ship propulsion system and relevant mathematical models, as background for further FTC design. The purpose of the modeling is to obtain information to design fault detection and isolation (FDI) modules for essential faults and to give the prerequisites for design of re-configuration. The block diagram in

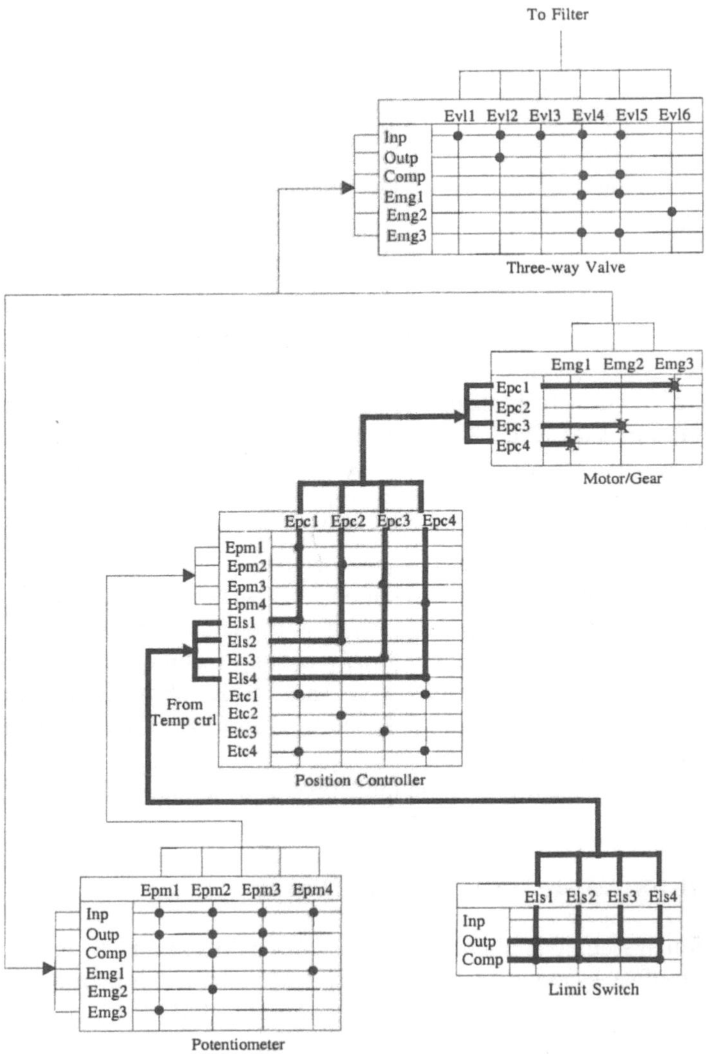

Figure 1: Graphic illustration of fault propagation analysis. Propagation can be stopped along the solid line. FTC is achieved if propagation can be stopped in the programmable components of the system, in this case the controller.

Fig.3 illustrates the structure of the propulsion system. The ship propulsion system was defined as a benchmark [21] for fault-tolerant control in the COSY network [1] where one theme is fault-tolerant control.

4.1 Dynamic models

4.1.1 Propeller thrust and torque

Controllable pitch (CP) propellers have blade angle (pitch) controlled by a hydraulic servo system. Developed thrust and torque are functions of pitch, shaft speed and flow velocity through the propeller

$$\begin{aligned} T_{\text{prop}} &= f_{T\_prop}(\vartheta, n, V_a) \\ Q_{\text{prop}} &= f_{Q\_prop}(\vartheta, n, V_a) \end{aligned} \tag{6}$$

These can be shown to approximately follow quadratic relations, for thrust

$$T_{\text{prop}} = T_{|n|n\vartheta}\vartheta\,|n|\,n + T_{nv}nV_a \tag{7}$$

and for torque

$$Q_{\text{prop}} = Q_0\,|n|\,n + Q_{nn|\vartheta|}\,|\vartheta|\,|n|\,n + Q_{nv\vartheta}\vartheta\,|n|\,V_a \tag{8}$$

These relations give a quite good approximation in the steady state cases whereas they are less applicable during large transients like crash stop. The term $Q_0\,|n|\,n$ accounts for the torque at zero pitch.

4.1.2 Diesel engine prime mover

Elaborate details of the dynamics are not important in this context, but would be for detailed design of FDI for the engine. Here, diesel torque can be considered linearly related to the fuel index, without dynamics involved,

$$Q_{\text{eng}} = K_y Y \tag{9}$$

The dynamics of propeller and shaft is merely that of rotating inertias subjected to torque balance between prime mover torque and load torques,

$$\frac{d}{dt}(\frac{1}{2}I_t n^2) = n(Q_{\text{eng}} - Q_{\text{prop}} - Q_f) \tag{10}$$

The dynamics of the prime mover and its control system is tightly coupled to the speed dynamics of the ship through the propeller, Eq.8. The structure of prime mover control was also shown in Fig.3. The measured shaft speed is compared with a reference speed and the governor (speed controller) regulates the fuel injection to the engine to obtain the desired speed. Limit curves are incorporated for shaft speed dependent torque and air pressure.

[1]The COSY research network is a collaboration between European groups and is supported by the European Science Foundation, ESF.

4.1.3 Hull resistance

Ship's resistance to motion, $R(u)$, through the water can be described to the first order by a resistance curve, which is a third to fifth order polynomial in u. The order of the polynomial is higher the closer the ship operates into the wave making region. The resistance curve is known *a priory* but with some uncertainty. The first order equation

$$m\dot{u} = R(u) + (1-t)T_{\text{prop}} + T_{\text{ext}} \tag{11}$$

is a sufficient approximation in this context.

4.1.4 Actuators for fuel injection and propeller pitch

The actuators can be modeled as first order dynamic systems with limits in rate of change and in output. [21]

4.1.5 Sensors

Sensors for propeller pitch and fuel index are conventional angle transmitters. Shaft speed is usually measured by a set of pulse pickups. A maximum logic selects the higher of the two signals. This protects against drop out of one of the pick ups but not against a "high signal" fault or failure in a common processor/rate counter servicing both channels. The ship speed is measured by magnetic log, Pitot tube or Doppler log.

4.2 Control Hierarchy

The control hierarchy includes controllers for: shaft speed; propeller pitch; diesel overload control; combinator curves from handle position to generate reference values of n and ϑ; efficiency optimization using n and ϑ; constant ship speed control. The signal flow between these function blocks is shown in Fig. 2. Further details can be found in [20].

5 Analysis of System Structure

Structural analysis was defined is [8], [12], [32] as the study of properties which are independent of the actual values of the parameters. Only constraints (relations) between variables and parameters from the operating model are represented. The constraints are independent of the system model and independent of the form under which this model is expressed. The constraints can be analytical or tabular relations, qualitative or quantitative. The links are represented by a graph, or a table, on which the structural analysis is made.

5.1 Description of the model

The system is considered as a set of constraints, $\mathcal{F} = \{f_1^c, f_2^c, , \cdots, f_m^c\}$ that are applied to a set of variables $\mathcal{Z} = X \cup \bar{X}$. X denotes the set of unknown variables

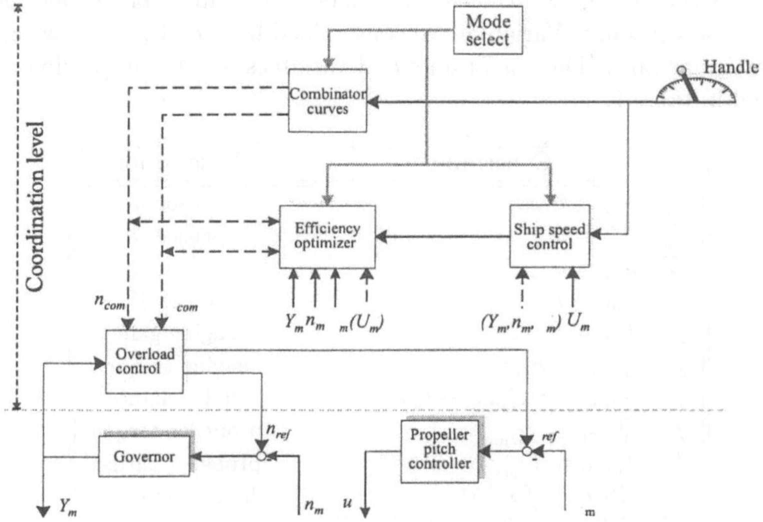

Figure 2: Control of ship propulsion plant with diesel engine and controllable pitch propeller. Several control blocks form a hierarchy of functions used in different operational modes, from manoeuvering to ocean passage.

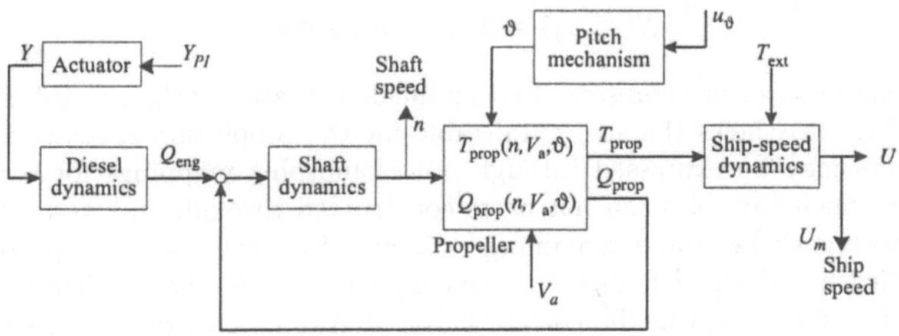

Figure 3: Ship propulsion plant is used for illustration. A diesel engine drives a propeller that gives a force on the ship to obtain the desired ship speed. Failure can be safety critical since plant availability is needed for both manoeuvering and braking of the ship.

while $\bar{X}$ is the set of known variables: sensor measurements, control variables (signals), variables with known values (constants, parameters), and reference variables (signals). The term constraint refers to the imposed relations between values of the variables. Variables are constrained by the physical laws applied to a particular unit. The constraints and variables for the propulsion system are listed in table 1.

Constraint	Description
$f_1^c : n_m = n$	sensor_n
$f_2^c : \vartheta_m = \vartheta$	sensor_ϑ
$f_3^c : u_m = u$	sensor_u
$f_4^c : Y_m = Y$	sensor_Y
$f_5^c : K_y = K_{yc}$	engine gain
$f_6^c : Q_{\text{eng}} = K_y Y$	engine torque
$f_7^c : -Q_{\text{eng}} = Q_{\text{prop}} + Q_f$	shaft balance
$f_8^c : Q_{\text{prop}} = f_{Q_{\text{prop}}}$	propeller torque
$f_9^c : T_{\text{prop}} = f_{T_{\text{prop}}}$	propeller thrust
$f_{10}^c : R(u) = f_{Ru}(u)$	hull resistance.
$f_{11}^c : R(u) = -T_{\text{ext}} - (1-t)\,T_{\text{prop}}$	ship speed

Table 1: Constraints and relations for shaft

5.2 Formal representation

The structure of the system is described by the following binary relation:

$$S : \mathcal{F} \times \mathcal{Z} \rightarrow \{0,1\}$$

$$(f_i^c, z_j) \rightarrow \begin{array}{ll} S(f_i^c, z_j) = 1 & \text{iff } f_i^c \text{ applies to } z_j, \\ S(f_i^c, z_j) = 0 & \text{otherwise.} \end{array}$$

These relations can be represented by an incidence table or the equivalent digraph. Fig.4 a) shows the structural table for the propulsion system. Some constraints may be expressed through non-isomorphic mappings for certain variables. Such variables can not be re-constructed through an inverse mapping from knowledge about remaining variables. Elements with this property are marked by M's (for multiple), replacing the ones in the incidence table and unidirectional arcs in the corresponding digraph. An example of such a constraint is f_8^c: it is always possible to compute the value of Q_{prop} from f_8^c when ϑ, n, and V_a are known. However, knowing the values of Q_{prop}, n, and V_a does not enable calculation of a unique ϑ in all cases. This fact is not apparent from the equations in this paper but is apparent when looking at the underlying propeller characteristics. System structure established through constraints is represented as a graph or structure matrix. Either representation enables analysis of redundancy and recoverability. The information from the structure matrix is binary. The digraph shows which constraints must be applied.

	n_m	ϑ_m	u_m	Y_m	K_{yc}	n	ϑ	u	Y	K_y	Q_{eng}	Q_{prop}	T_{prop}	$R(u)$
f_1^c	1					1								
f_2^c		1					1							
f_3^c			1					1						
f_4^c				1					1					
f_5^c					1					1				
f_6^c									1	1	1			
f_7^c											1	1		
f_8^c						1	M	1				1		
f_9^c						1	M	1					1	
f_{10}^c								1						1
f_{11}^c													1	1

a)

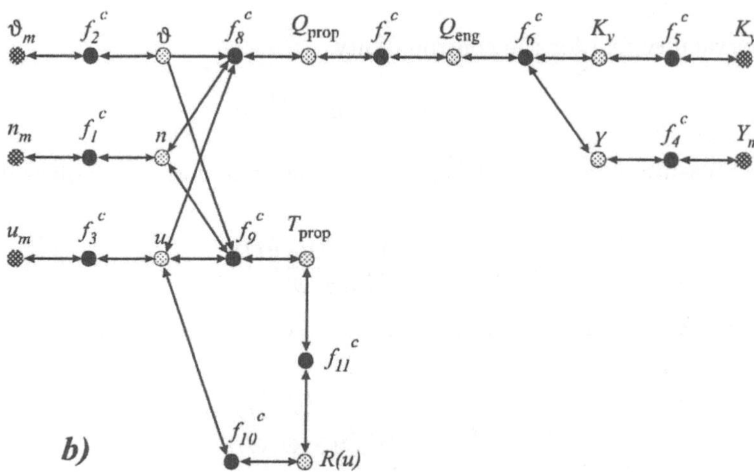

b)

Figure 4: System structure established through constraints represented as a graph or structure matrix. Either representation enables analysis of redundancy and recoverability. The information from the structure matrix is binary. The digraph shows which constraints must be applied.

6 System Recoverability

The structure analysis gives a very effective tool for investigation of the principal properties of both fault detection and accommodation of faults affecting the controllability of the plant, including actuator faults. This is mainly a qualitative analysis. A more detailed look at the degree of controllability led to a recent result for recoverability of systems [13]. This approach gives a quantitative measure of the properties of the faulty system to determine whether the functionality can be recovered after a fault has occurred. The recoverability is treated for linear systems, studying the controllable and observable subspaces of a fault free nominal system $S(\varnothing, \varnothing)$ and the re-configured faulty system $S(q_c, q_f)$ where q_c is the re-configuration and q_f the fault.

A system is defined to be recoverable from a set of faults if the hybrid system (the system where faults can occur as discrete events) is controlable in the regions where the original system was controlable, and the states in these regions, can be reached from any interior point in the hybrid state space.

Weak recoverability is defined similarly but reachability is now defined only from a reduced set of regions.

It is then suggested [13] that the measures to be made could be based on the observability and controlability gramians

$$W_o = \int_0^\infty e^{A^T t} C^T C e^{At} dt \tag{12}$$

for observability and for the controllability,

$$W_c = \int_0^\infty e^{At} B B^T e^{A^T t} dt \tag{13}$$

The measure of how well the system can be excited after a fault is defined by the ratio

$$\rho_c(q_c, q_f) = \frac{|W_c(S(q_c, q_f))|}{|W_c(S(\varnothing, \varnothing))|} \tag{14}$$

A measure of how well the system can be observed after a fault is suggested as

$$\rho_o(q_c, q_f) = \frac{\left|W_o^{-1}(S(q_c, q_f))\right|}{\left|W_o^{-1}(S(\varnothing, \varnothing))\right|} \tag{15}$$

If a measure is zero, the system is not reconfigurable.

6.1 Example - sensor re-configuration

When analytic redundancy is available, as seen by the above methods, accommodation of sensor faults can be obtained by means of reconstruction of a sensor signal using estimation techniques. The structural analysis approach was applied to the ship propulsion problem in [11]. An example for the propulsion system is a critical fault in the shaft speed measurement which can be accommodated by estimating shaft speed from other available measurements.

6.2 Fault in the shaft speed measurement

The constraint f_1^c represents this device in table 1. A fault occurrence means that the constraint f_1^c does not hold, e.g. the values of the variable n_m are not correctly related to the values of the variable n. Figure 4 shows that variable n is involved in 3 relations which are specified by the constraints f_1^c, f_8^c, and f_9^c. Since the constraint f_1^c is not valid, there are two other possible ways of calculating the values of the variable n, namely through constraints f_8^c and f_9^c. The shaft speed can be described as a function of the other known variable as:

$$\hat{n}_q = f_q(\vartheta_m, Y_m, K_{yc}, u_m) \tag{16}$$

$$\hat{n}_t = f_t(\vartheta_m, u_m) \tag{17}$$

The process to apply the sensor fusion based on this approach is the following:

For the interested variable (for instance n) identify the set of related constraints (f_8^c and f_9^c) and

- choose one of the available constraints
- remove all other constraints in the set and their associated arcs
- for all variables connected to the chosen constraint search backward until a known variable for each is reached.

By examining all the constraints, the set of equations/relation by which the variable can be calculated is identified and can be used for re-configuration purposes. For the shaft speed failure, there are two alternatives for estimation of the n-signal when the measurement n_m fails. One is the torque relation, the other uses the relation for thrust of the propeller.

Using the torque relation, n can be estimated from the constraint f_8^c

$$\hat{n}_q = -\frac{Q_{nv\vartheta}(1-w)u_m\vartheta_m}{2Q_{nn\vartheta}\vartheta_m} \left(1 + \sqrt{1 + \frac{4Q_{nn\vartheta}\vartheta_m K_{yc} Y_m}{(Q_{nv\vartheta}(1-w)u_m\vartheta_m)^2}}\right) \tag{18}$$

The estimation of n based on static relations is obviously fairly primitive. The performance of this estimate is shown in Fig. An observer should give better estimation during transients, but an obstacle is the fact that the dynamic equations are non-linear. Before turning to these details, a general architecture for implementation of fault tolerance is needed.

7 An Architecture for Fault Tolerant Control

An architecture for fault-tolerant control systems is illustrated in Fig. 6 It has three layers. A lower with the control loop, the second with detector

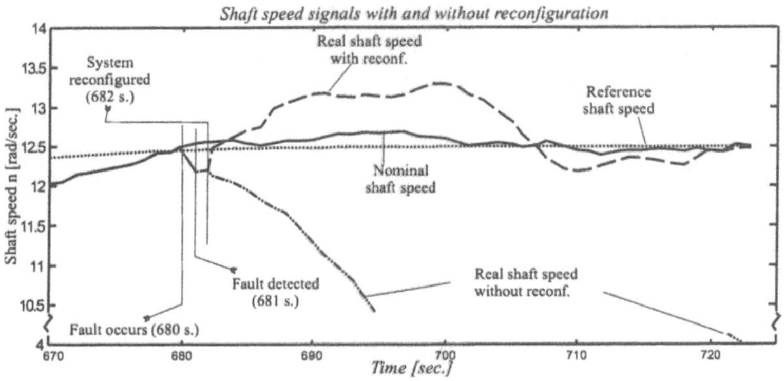

Figure 5: Static constraint used for simple reconfiguration, of shaft speed measurement. Shaft speed overshoot is fairly close to the shut-down limit for the diesel engine, thus not sufficiently robust for a real application.

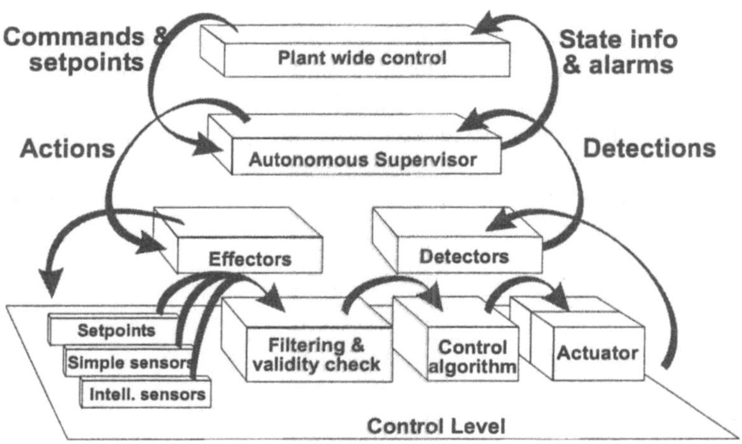

Figure 6: A layered architecture for fault-tolerant control. The lower level comprise usual control and the controlled object. The next two constitute an autonomous supervisor with fault detection, decision logic to determine the required action when triggered by a detected-fault event and a effector module(s) to implement the desired change. The uppermost level is overall plant management/coordination and operator information.

functions and effectors to make reconfiguration, and the third with supervisor functionality. The reasons to separate into three layers are the benefits of clear development structure, independent specification and development of each layer, and last but not least to obtain testability of detector and supervisor functions.-

Lower layer with signal check, control and actuator command

The lower layer comprises the traditional control loop with sensor and actuator interfaces, signal conditioning and filtering, and the controller.

Aiming at obtaining FTC, a key feature for the sensor interfaces should be to support detectability and to include validity check designed to fit the purpose. This includes

- single sensor validity check against limits, maximum slew-rate and statistics:

Second level with detectors and effectors

This level comprises a number of detectors, usually one for each fault effect which shall be detected, and effectors that implement desired re-configuration or other remedy actions initiated from the autonomous supervisor. The functions of the modules are:

- detection based on hardware or analytic redundancy based on FDI methods
- detection of faults in control algorithms and application software
- dedicated effector modules to execute fault handling

Third level with autonomous supervision

The supervisor comprises state-event logic to describe the logical state of the controlled object. Transitions between states are driven by events. The supervisor functionality includes:

- interface to detectors for change detection
- interface to upper level for mode change signals
- demand remedial actions to accommodate a fault: - re-design, re-configuration, configuration, re-tuning, select prespecified actions and software modules
- signal to plant wide coordination or operator about current state

The tasks carried out in the different layers have been subject of separate research efforts. Areas include application of model predictive control [26], qualitative methods [25], [24], and change to predetermined and predesigned controls [6], [5].

The example below shows how a scheme is developed using sensor reconfiguration as a predetermined and predesigned action when a particular fault is isolated [4].

7.1 Example - continued

The propulsion system delivering propeller thrust is the sole source for marine vessel to accelerate and brake. Propellers driven by diesel engines are the most common prime propulsion device for both merchant trade and special purpose vessels. Thrust malfunction will often cause rapid excursion of a vehicle, and in close proximity of fixed structures, collision is a very likely consequence. Failures are often caused by simple faults in local control components, so fault accommodation could be made possible in many cases. Such fault-tolerant design could achieve an increased availability of propulsion thrust, without additional hardware, and this could reduce the likelihood of collisions. The prerequisite for fault accommodation is timely detection and isolation. In the propulsion system, detection of discrepancy between expected and actual performance is easy. It is more difficult to make an isolation. Two faults in the ship benchmark are in particular difficult to isolate: a shaft speed sensor failure and a diesel engine partial malfunction. It is crucial to obtain isolation because the necessary remedial action is quite different in the two cases. This problem is non-linear, one fault is an additive measurement fault, the other gives rise to a parameter change and is thus non-additive.

7.2 FDI for linear and non-linear systems

Fault detection algorithms have been developed for the last two decades and for linear systems, the area has matured, resulting in several text books in the area [1], [16], [27]. Model uncertainty and the associated robustness problems with detectors play an important role since erroneous triggering from a fault detector algorithm will trigger an erroneous action on part of the control system. Robustness issues are well covered in the survey [28] and in later literature [34].

When it comes to non-linear systems, additive faults in states or measurements for non-linear systems were studied in [30]. FDI for bilinear systems were studied by [40] and for general state affine systems by [17]. The problem with combined faults in states, measurements, and system parameters has only been sparsely treated with respect to FDI properties. A general result for combined state estimation and parameter identification was given in [41], for a class of continuous-time non-linear state space systems, and design of an adaptive observer for non-linear systems was derived in [9]. Both of these references treat systems which are linear in the parameters and do not consider the fault detection problem. An overview of FDI results for non-linear systems was given in [14].

Many technical non-linearities are of saturating type and decrease the actuation possibilities at high amplitudes. This means the gain is reduced. Observers for non-linear systems of a form where the gain associated with the non-linearity can be bounded, can be easily dealt with using Lyapunov theory. The Thau observer [36] with a linear measurement update using a model of the exact nonlinearity as time update (predictor) has been used in [29].

A nonlinear system of the form

$$\dot{x} = Ax + Bu + f(x,u)$$
$$y = Cx$$

has an associated observer of the form

$$\tfrac{d}{dt}\hat{x} = A\hat{x} + Bu + f(\hat{x},u) + L(y_m - \hat{y})$$
$$\hat{y} = C\hat{x}$$

The stability conditions require the linear part of the system to be observable and the non-linear function $f(x,u)$ to be bounded (Lipschitz)

$$\|f(x_1,u) - f(x_2,u)\| < \gamma \|x_1 - x_2\|$$

This observer is a state observer, whereas FDI requires an output observation. Further, combined state and parameter change due to faults, require an extension. In the ship benchmark, both the diesel gain and the shaft speed measurement can be subjected to faults.

7.3 Example - continued

The task at hand is to estimate a signal fault in n_m and a parameter change in K_y. The relevant equations for the ship propulsion system are [4]:

$$\begin{aligned} I_t\dot{n} &= Q_{eng} - Q_{prop} - Q_f \\ Q_{eng} &= K_y Y_m \end{aligned} \tag{19}$$

Since we can assume positive shaft speed, $n > 0$ and non-zero fuel index $Y_m > 0$, then

$$Q_{prop} = Q_{\vartheta nn}\vartheta n^2 + Q_{\vartheta nU}\vartheta n U_m$$

and

$$\dot{n} = \frac{1}{I_t}\left[K_y Y_m - Q_{\vartheta nn}\vartheta n^2 - Q_{\vartheta nU}\vartheta n U_m\right] \tag{20}$$

This equation can be written in the form

$$\begin{aligned} \dot{x} &= \Phi(x, u_2, u_3) + \theta u_1 \\ y &= x \end{aligned} \tag{21}$$

using

$$\Phi(n,u) = \tfrac{1}{I_t}\left(-(Q_0 + Q_{\vartheta nn}\,|\vartheta_m|)n^2 - Q_{\vartheta nU}\,\vartheta_m n U_m - Q_f\right) \tag{22}$$

$$x = n,\ u_1 = Y_m,\ u_2 = U_m,\ u_3 = \vartheta_m,\ \theta = \frac{K_y}{I_t} \tag{23}$$

An adaptive observer can be build by using measured inputs: Y_m, U_m, ϑ_m and measured state n_m. The observer is the state estimator

$$\begin{aligned} \tfrac{d}{dt}\hat{n} = \tfrac{-1}{I_t}\left((Q_0 + Q_{\vartheta nn}\left|\vartheta_m\right|)\hat{n}^2 + Q_{\vartheta nU}\,\vartheta_m \hat{n} U_m\right) \\ +\hat{\theta} Y_m + L\left(n_m - \hat{n}\right) \end{aligned} \tag{24}$$

with parameter updating

$$\frac{d}{dt}\hat{\theta} = pY_m\left(n_m - \hat{n}\right) \tag{25}$$

The adaptive observer is semi-globally asymptotically stable with gains L and p specified below. To prove this, a Lipschitz condition can be found for the non-linearity and a Lyapunov argument used. A more general case was treated in [9]. However, the detailed assessment of the Lipschitz conditions, that determine the gains in the adaptive observer, are easily made too conservative to get useful results. A few comments are thus considered appropriate.

The nonlinear torque function Eq. 22 is Lipschitz, $\|\Phi(n,u) - \Phi(\hat{n},u)\| < \gamma\left\|n - \hat{n}\right\|$, since

$$\begin{aligned} \Phi(n,u) - \Phi(\hat{n},u) &= \left((-Q_0 - Q_{\vartheta nn}\left|\vartheta_m\right|)(n+\widehat{n}) - Q_{\vartheta nU}\vartheta_m U_m\right)\frac{1}{I_t}\tilde{n} \\ &= \Big[-\left(\left(Q_0 + Q_{\vartheta nn}\left|\vartheta_m\right|\right)n + Q_{\vartheta nU}\vartheta_m U_m\right) \\ &\quad - \left(Q_0 + Q_{\vartheta nn}\left|\vartheta_m\right|\right)\hat{n}\Big]\frac{1}{I_t}\tilde{n} \end{aligned} \tag{26}$$

where $(n - \widehat{n}) \equiv \tilde{n}$. From practical torque and ship speed constraints it follows that

$$-\left(\left(Q_0 + Q_{\vartheta nn}|\vartheta_m|\right)n + Q_{\vartheta nU}\vartheta_m U_m\right) < \alpha\frac{Q_{eng,\max}}{n_{\min}} \tag{27}$$

and

$$-\left(Q_0 + Q_{\vartheta nn}\left|\vartheta_m\right|\right)\hat{n} < \beta\frac{Q_{eng,\max}}{n_{\max}} \tag{28}$$

Hence

$$\|\Phi(n,u) - \Phi(\hat{n},u)\| < \gamma\left\|n - \hat{n}\right\|$$

and the Lipschitz coefficient is expressed in terms of physical quantities for the engine

$$\gamma = \frac{Q_{eng,\max}}{I_t\, n_{\max}}\left(\alpha\frac{n_{\max}}{n_{\min}} + \beta\right) \tag{29}$$

A Lyapunov function for the observer error

$$\tilde{n} = n - \hat{n}, \tilde{\theta} = \theta - \hat{\theta}$$

is $V = \tilde{n}p\tilde{n}^T + \tilde{\theta}\tilde{\theta}^T$, where $p > 0$ is a scalar. Then, using the notation 23, and details of the proof in [9],

$$\dot{V} < 0 \Leftrightarrow (\Phi(n,u) - \Phi(\hat{n},u))p\tilde{n} + u_1\tilde{\theta}p\tilde{n} - \tilde{n}Lp\tilde{n} + \tilde{\theta}\frac{d\tilde{\theta}}{dt} < 0$$

iff

$$L > \gamma, \quad p > 0$$

and $n \subseteq [0, n_{\max}], \hat{n} \subseteq [0, n_{\max}], Y_m > 0$. The observer is then semi-globally asymptotically stable [4] iff the following holds,

$$\begin{aligned} & f(x,u)\theta - f(\hat{x},u)\hat{\theta} \to 0 \\ & \Rightarrow \theta \to \hat{\theta} \text{ iff } \hat{x} \to x \end{aligned} \tag{30}$$

This requires persistent excitation in $u(t)$. For this reason only one parameter, k_y is the subject of adaptation. The propeller parameters must be identified from batch processing of data during non-faulty conditions.

7.3.1 FDI properties

With the observer equation

$$\frac{d}{dt}\hat{n} = \frac{1}{I_t}\left(-Q_{\vartheta nn}\vartheta_m\hat{n}^2 - Q_{\vartheta nU}\dot{\vartheta}_m\hat{n}U_m\right) + Y_m\widehat{\theta} + L\left(n_{mf} - \widehat{n}\right) \tag{31}$$

and parameter update

$$\dot{\widehat{\theta}} = pY_m\left(n_m - \widehat{n}\right) \tag{32}$$

detection and isolation is analyzed from the properties of the error signals

$$\begin{aligned} e_n &= n - \hat{n} \\ \varepsilon_n &= n_m - \hat{n} = e_n + \Delta n \\ e_\theta &= \theta_0 - \hat{\theta} \end{aligned} \tag{33}$$

Using abbreviations a and b for the system parameters and measured quantities, u for input, the system is

$$\begin{aligned} \dot{n} &= an^2 + bn + \theta u \\ n_m &= n + \Delta n \end{aligned}$$

where $n > 0$ and both a and b are negative. The observer is, with L and p as the gain factors

$$\begin{aligned} \frac{d}{dt}\hat{n} &= a\hat{n}^2 + b\hat{n} + \hat{\theta}u + L(n_m - \hat{n}) \\ \frac{d}{dt}\hat{\theta} &= pu(n_m - \hat{n}) \end{aligned}$$

Simple manipulation, and introduction of the average shaft speed, $\bar{n}$, and average input $\bar{u}$ gives the Laplace transformed operators

$$\begin{aligned} \varepsilon_n(s) &= \frac{(s - (2a\bar{n} + b))\, s}{s^2 - (2a\bar{n} + b + L)s + p\bar{u}^2}\Delta n(s) \\ &\quad + \frac{s\bar{u}}{s^2 - (2a\bar{n} + b + L)s + p\bar{u}^2}\Delta\theta(s) \end{aligned} \tag{34}$$

$$e_\theta(s) = -p\frac{\bar{u}}{s}\varepsilon_n(s) \tag{35}$$

It is clear from these equations that isolation is not possible, unless assumptions are made about the signature of the possible faults.

However, since the diesel gain fault arises due to misfirering or stop of one cylinder, this fault can be assumed to occur as stepwise changes in θ. Eq. 34 and 35 then shows that the $\Delta\theta$ step will give rise to a low-pass filtered step in e_θ. Signature analysis or a GLR test can then help isolate the faults.

7.4 An FTC scheme for the n-fault

The performance of the FTC scheme is illustrated in Fig. 7 where the estimate $\hat{n}$ replaces n_m at the instant the n-fault is detected.

The resulting overshoot in this case is less than 3%, which is well below the critical over-speed shut down limit, and three times better than the static estimator. In the figure, the curves represent the following cases: normal case (solid line), faulty case (dash dotted line), re-configured case (dashed line), and reference signal (dotted line).

This result is very satisfactory, and it has been obtained without additional hardware. Robustness to parameter uncertainty can partly be achieved through on-board batch processing of measurements to identify the propeller parameters. In addition, the overshoot margin is big enough to allow for additional effects of parameter uncertainty.

8 Cost-Benefit Issues

Development of fault-tolerant control schemes must have a pay-back potential to attract industrial interest. The shaft speed fault would eventually need to be accommodated by the governor for the diesel engines. If a failure develops as described above, the result today would be

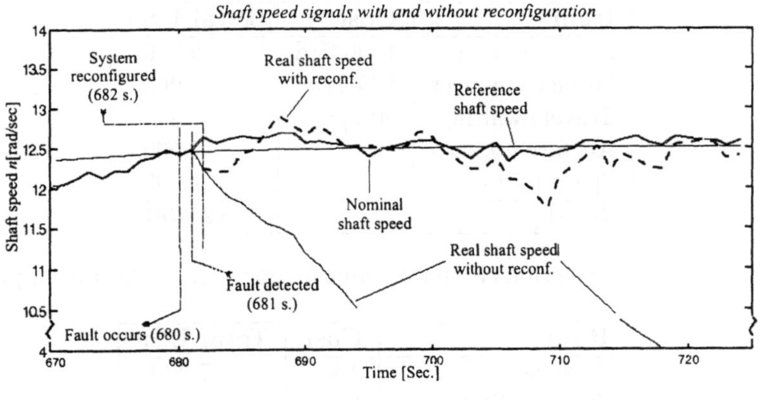

Figure 7: Non-linear observer used to estimate shaft speed signal. The observer gain L is set to zero when the n_m fault is isolated. The control feedback is the estimate provided by the NL observer.

- freezing of the throttle position by engagement of an electronic brake on the actuator shaft
- an over-speed of the diesel engine and an immediate shut-down by an independent safety system on the engine

Either of the two reactions limit the maneuverability of the ship since the propeller driven by the engine is needed to accelerate or stop the ship. Without propeller power, the directional maneuverability is also drastically reduced.

The manual, remedial action is to send a crew member down at the engine and make manual throttle control. This control task requires uninterrupted 24 hour manning of the emergency stand at the engine.

The faults considered in the benchmark could have different severity, depending on external circumstance. If a serious fault appears during manoeuvering close to a quay or to other ships, a major damage could result. If the same fault appeared amid an ocean, the cost is that of 24 hour manning of the emergency stand until the next harbor approach where the faulty component can be replaced.

A cost estimate for this un-dramatic event is listed in Table 2. The total is US$ 11,100.-. If an accident had happened, the cost of damage repair and time where the ship is not in operation can easily sum up to a cost of 500,000.- to 1,000,000.- US$. The likelihood of such event is, however, estimated to only about 0.01 times the likelihood of the fault itself.

A development effort to implement an FTC concept where the fault detection methods from the benchmark were applied, would result in having the actuator continue running when the fault occurs but with reduced performance.

The development effort could be estimated to the figures given in Table 3.

Item	Cost	Total US$
24 h manning	1000 $/d	5,000.-
Service engineer	120 $/h	960.-
Travel+waiting	60 $/h	1,440.-
Air fare etc		4,000.-
Spare parts		1,000.-
Total		**12,400.-**

Table 2: Cost of serious actuator fault at open sea, 5 days from port

Item	Cost	Total US$
Analysis	2 w	7,200.-
Develop FDI	1 m	14,400.-
Implement in SW	2 w	7,200.-
Add testing	2 w	7,200.-
Documentation	2 w	7,200.-
SW over lifetime	2 m	28,800.-
Staff training	2 w	7,200.-
Total - 1000 pcs		**79,200.-**

Table 3: Additional development costs to implement FTC for selected faults on the governor

The development cost is fairly insignificant, so the FTC cost/benefit figures are, a function of the likelihood of events in the fault scenario. With a product lifetime of 20 years and one anticipated event like the one described, every 50 years per ship, FTC would give a net saving over product lifetime of some 6000 US$. This alone might not be significant enough for a ship's owner to request a fault-tolerant design. However, this economy analysis is for one type of failure. Adding others to the range of possible failure modes, lifetime maintenance costs could obviously be significantly reduced and there is a clear intensive for customers to request fault-tolerant design of new automation systems.

9 Conclusions

This paper has treated fault-tolerant control. It has been introduced as a multi-disciplinary area of research where a number of new research areas are emerging. The area has also been demonstrated to be of economic importance for industry and community and offering potential benefits in the form of enhanced safety of automated systems, fault-tolerant control is surely an area that will be firmly established in both research and industry.

References

[1] M. Basseville and I. V. Nikiforov. *Detection of Abrupt Changes: Theory*

and Application. Information and System Science. Prentice Hall, New York, 1993.

[2] M. Blanke. Consistent design of dependable control systems. *Control Engineering Practice*, 4(9):1305–1312, 1996.

[3] M. Blanke, R. Izadi-Zamanabadi, S.A. Bøgh, and C.P. Lunau. Fault-tolerant control systems - a hollostic view. *Control Engineering Practice*, 5(5):693–702, 1997.

[4] M. Blanke, R. Izadi-Zamanabadi, and T.F. Loostma. Fault monitoring and re-configurable control for a ship propulsion plant. *Journal of Adaptive Control and Signal Processing*, pages 671–688, December 1998.

[5] S. A. Bøgh. *Fault Tolerant Control Systems - A Development Method and Real-Life Case Study*. PhD thesis, Aalborg University, Department of Control Engineering, Fredrik Bajers Vej 7. DK 9920 Aalborg Ø . Denmark, November 1997.

[6] S. A. Bøgh, R. Izadi-Zamanabadi, and M. Blanke. Onboard supervisor for the ørsted satellite attitute control system. In *Artificial Intelligence and Knowledge Based Systems for Space, 5th Workshop*, pages 137–152, Noordwijk , Holand, October 1995. The European Space Agency, Automation and Ground Facilities Division.

[7] C.G. Cassandras, S. Lafortune, and G.J. Olsder. Introduction to the modelling, control and optimization of discrete event systems. In A. Isidori, editor, *Trends in Control*, pages 217–291. Springer Verlag, 1995.

[8] J. Ph. Cassar, M. Staroswiecki, and P. Declerck. Structural decomposition of large scale systems for the design of failure detection and isolation procedures. *Systems Science*, 20(1):31–42, 1994.

[9] Y. M. Cho and R. Rajamani. A systematic approach to adaptive observer synthesis for nonlinear systems. *IEEE Transactions on Automatic Control*, 42(4):534–537, April 1997.

[10] V. Cocquempot, J.Ph. Cassar, and M. Staroswiecki. Generation of robust analytical redundancy relations. In *Proceedings of ECC'91*, Grenoble, France, July 1991, pp. 309-314.

[11] V. Cocquempot, R. Izadi-Zamanabadi, M. Staroswiecki, and M. Blanke. Residual generation for the ship benchmark using structural approach. In *IEE Control'98*, Swansea, UK, September 1998.

[12] Ph. Declerck. *Analyse structurale et fonctionnelle des grands systèmes. Application à une centrale PWR 900 MW*. PhD thesis, Université des Science et Technologies de Lille,, Villeneuve D'Ascq, France, December 1991.

[13] C. W. Frei, F. J. Kraus, and M. Blanke. Recoverability viewed as a system property. In *Proc. European Control Conference 1999, ECC'99*, September 1999.

[14] E.A. García and P.M. Frank. Deterministic nonlinear observer-based approaches to fault diagnosis: A survey. *Control Engineering Practice*, 5(5):663–760, 1997.

[15] A. L. Gehin and M. Staroswiecki. A formal approach to reconfigurability analysis - application to the three tank benchmark. In *Proc. European Control Conference 1999, ECC'99*, September 1999.

[16] Janos J. Gertler. *Fault Detection and Diagnosis in Engineering Systems*. Marcel Dekker Inc., Marcel Dekker AG, Hutgasse 4, Postfach 812, CH-4001 Basel, Switzerland, 1 edition, 1998.

[17] H. Hammouri, M. Kinnaert, and E.H. Yaagoubi. Fault detection and isolation for state affine systems. *European Journal of Control*, (4):2–16, 1998.

[18] M. Huzmezan and J.M. Maciejowski. Reconfigurable flight control during actuator failures using predictive control. In *14th IFAC World Congress*, Beijing, P.R. China, July 1999.

[19] R. Isermann. Process fault detection based on modelling and estimation methods: A survey. *Automatica*, 20(4):387–404, 1984.

[20] R. Izadi-Zamanabadi and M. Blanke. A ship propulsion system model for fault-tolerant control. Technical Report R-1998-4262, Dept. of Control Eng., Aalborg University, Denmark, July 1998.

[21] R. Izadi-Zamanabadi and M. Blanke. A ship propulsion system as a benchmark for fault-tolerant control. *Control Engineering Practice*, 7(2):227–239, March 1999.

[22] Charlotte P. Lunau. A reflective architecture for process control applications. In M. Aksit and S. Matsuoka, editors, *ECOP'97 Object Oriented Programming*, pages 170–189. Springer Verlag, 1997. Lecture Notes in Computer Science, Vol. 1241.

[23] Jan Lunze. Qualitative modelling of linear dynamical systems with quantized state measurements. *Automatica*, 30(3):417–431, 1994.

[24] Jan Lunze. Introduction to logic-based fault detection. COSY PhD Course on Fault-Tolerant Control, April 1999. Aalborg University, Denmark.

[25] Jan Lunze and F. Schiller. An example of fault diagnosis by means of probabilistic logic reasoning. In *IFAC Safeprocess'97*, pages 540–545, 1997.

[26] J.M. Maciejowski. *Predictive Control with Constraints.* Addison-Wesley, Wokingham, U.K., 1999.

[27] R.J. Patton, P.M. Frank, and R.N. Clark eds. *Advances in fault diagnosis in dynamic systems.* Springer-Verlag, UK, 1995.

[28] Ron J. Patton. Fault tolerant control: The 1997 situation. In *IFAC Safeprocess'97*, pages 1033–1055, Hull, United Kingdom, August 1997.

[29] G. Schreier, J. Ragot, J. Patton, and P.M. Frank. Observer design for a class of non-linear systems. In *Conference on Safe-Process*, pages 498–503, Hull, United Kingdom, 1997.

[30] R. Seliger and P. M. Frank. Robust component fault detection and isolation in nonlinear dynamic systems using nonlinear unknown input observers. In *Peprints of IFAC/IMACS Symp. SAFEPROCESS'91*, volume 1, pages 313–318, Baden-Baden, Sept 10-13 1991.

[31] M. Staroswiecki, S. Attouche, and M. L. Assas. A graphic approach for reconfigurability analysis. In *Proc. DX'99*, June 1999.

[32] M. Staroswiecki and P. Declerck. Analytical redundancy in non-linear interconnected systems by means of structural analysis. volume II, pages 23–27, Nancy, July 1989. IFAC-AIPAC'89.

[33] Marcel Staroswiecki and Mireille Bayart. Models and languages for the interoperability of smart instruments. *Automatica*, 32(6):859–873, 1996.

[34] Jakob Stoustrup and Niemann H.H. Fault detection for nonlinear systems - a standard problem approach. In *37th IEEE CDC*, pages 96–101, Tampa, Florida, USA, December 1998. Invited paper.

[35] Jakob Stoustrup and Grimble M.J. Integrating control and fault diagnosis: A separation result. In *IFAC Sym. on Fault Detection, Supervision and Safety for Technical Processes*, pages 323–328, Hull, United Kingdom, August 1997.

[36] F.E. Thau. Observing the state of non-linear dynamic systems. *International Journal of Control*, 17:471–479, 1973.

[37] R. J. Veillette, J. V. Medani, and W. R. Perkins. Design of reliable control systems. *Trans. on Automatic Control*, 37(3):290–304, Mar. 1992.

[38] A.S. Willsky. A survey of design methods for failure detection in dynamic systems. *Automatica*, 12(6):601–611, 1976.

[39] W.M. Wonham. A control theory for discrete-event system. In M.J. Denham and A.J. Laub, editors, *Advanced Computing Concepts and Techniques in Control Engineering*, pages 129–169. Springer-Verlag, 1988.

[40] D.L. Yu and D.N. Shields. A bilinear fault detection observer. *Automatica*, 32(11):1597–1602, 1996.

[41] W. W. Zhou and M. Blanke. Identification of a class of nonlinear state space models using rpe techniques. *IEEE Transactions of Automatic Control.*, Vol 34. No.3:312–316, 1989.

7

Are there New Industrial Perspectives in the Control of Mechanical Systems?

J. Lévine

Centre Automatique et Systèmes, École des Mines de Paris,
35 rue Saint-Honoré, 77305 Fontainebleau Cedex, France
E-mail: levine@cas.ensmp.fr

Summary. Three industrial nonlinear control applications in the field of mechanical systems are presented and recently emerged engineering motivated questions addressed: the first one deals with a new approach to aircraft flight control systems, the second one with crane driving assistance and the last one with semi-active and active suspension design for cars. They combine a physical model based approach with singular perturbations and flatness based motion planning and tracking controller design.

1 Introduction

The evolution in electronic industry has lead to powerful and inexpensive microcomputer-based controllers, actuators, and sensors. In addition to hardware improvements, efficient design and simulation software tools are proposed to engineers and researchers, creating a new demand for automation and operating assistance design methods. New needs for more powerful designs are now detected in industrial sectors whose concern in automation was relatively low. Moreover, as hardware costs decrease, control units to assist complex systems operations, that were, until now, too expensive to reach a large audience or whose capacities were too limited, are currently being more and more developed.

Three industrial applications of nonlinear control are presented in three subfields of mechanical systems: aircraft autopilot design, crane control and "intelligent" automotive equipments. The problems and viewpoints here reported have been brought out from questions raised by our industrial partners.

They concern possible simplifications of the nonlinear controller architecture by splitting it into a trajectory generator and a feedback controller, and of the controller itself by using a hierarchical approach. Since, in addition, such controllers are often designed to assist the pilot, driver, or operator, in realizing various tasks, the operator's inputs must be properly taken into account and carefully synchronized to the controller signals.

These control problems have in common a hierarchical structure consisiting in fast and slow subsystems, namely a *singularly perturbed* structure, created or preserved–if originally present– by feedback, and a mathematical model structure, namely *flatness*. Recall that flatness means, roughly speaking, that there exists a particular output, whose dimension is equal to the number of inputs, from which all system variables, including inputs, can be deduced using only time differentiation and without integrating the system differential equations (see [12,13] for a precise presentation of this concept). As a result, it suffices to choose the flat output reference trajectory to deduce the reference trajectories of all other system variables. Moreover, it turns out that the dynamics of the error between the flat output and its reference are linearizable by *endogenous dynamic feedback*, a special type of dynamic feedback (see again [12,13]).

Combining these time scaling and flatness structures significantly simplifies the trajectory generation and the tracking controller design. A cascaded or hierarchical control architecture is naturally obtained. Furthermore, the aforementioned structures faithfully reflect fundamental physical properties of the underlying models which can be exploited in both, motion planning, and feedback design.

Though these examples illustrate some new control perspectives, they are far from drawing an exhaustive picture, even in the field of mechanical systems. Other examples along the same line of thought may be found in [3,4,5,11,17,23,28,29,32,33].

The first example develops a new approach to aircraft autopilot design which is more compact, simpler, and without linearity limitations. It illustrates the improved design needs for lower costs in a traditional control application. A similar situation occurs in the second example concerning crane driving assistance where a controller design set up by a trajectory generator and a PID based output feedback are presented. The third example outlines a general approach to active and semi-active suspension design for cars. Such equipments serve to improve passengers security and comfort but, until now, were too expensive to be installed in commercial cars. The present study is an attempt to adapt the control design of suspensions to these new cost and portability requirements.

These three examples have been partly done in collaboration with industries: the aircraft control one with Sextant Avionique, France, the crane one with the US Office of Naval Research and the University of Maryland, and the suspension one with Valeo Electronique, France. I wish to express my warm thanks to these companies for letting me present this material and to all the engineers and researchers that have been involved in these exciting and fruitful experiences, and particularly my colleagues M. Fliess, Ph. Martin and P. Rouchon. I am also deeply indebted to J. Rudolph and P. Mullhaupt for their most helpful comments on this manuscript.

2 Automatic Flight Control Systems

Automatic Flight Control Systems (AFCS) have been, since the 60's, a major field of application of control methods. The main drawback of most current AFCSs is that they follow the principle of *one function–one controller.* In each case, pitch attitude control, wing leveler, side-slip suppressor, landing, terrain-following, target tracking, etc., a different controller is associated. These controllers may be based on different models, i.e., linear approximation along the trajectory to be followed, linear or nonlinear input-output description, the output of which is chosen such that its regulation corresponds to the control objective (see e.g. [9,26,27,35,36]).

On advanced autopilots, the number of possible combinations of functions and reference trajectories is large. Moreover, strong limitations must be imposed on both the maneuvers and their sequencing rate, which must be kept slow enough to avoid instabilities. Indeed, as far as linear techniques are concerned, one has to restrict to sequences of trajectories for which the tangent approximation is nearly time invariant (pieces of straight lines, circles, or helices with slowly varying altitude, etc.). On the other hand, in most nonlinear approaches, the control objective is translated into an output variable whose reference trajectory, generally constant, is assumed to satisfactorily describe the maneuver. In some cases, however, the stability of the overall closed-loop dynamics causes difficulties. In most cases, designing the switches from one function to another requires a lot of attention. Furthermore, since the highest level loop is the pilot himself, the switching policy may lead to complex tradeoffs in order to take into account the pilot's online queries.

New complex tasks such as collision avoidance, or, more specifically in the case of military or remotely operated aircrafts, terrain following or orientation control are needed to complement the classical ones. Hence, a simpler control architecture is required to both improve reliability and decrease development costs. To this aim, we propose a single controller comprised of a universal reference trajectory generator and a fixed tracking loop. In addition, general tracking tasks, for which linearization about operating points is insufficient, require that the main nonlinearities of the system are properly taken into account. We show that such a controller can be designed thanks to the (almost) flatness property of generic aircraft models. We briefly describe its construction, performances and some perspectives.

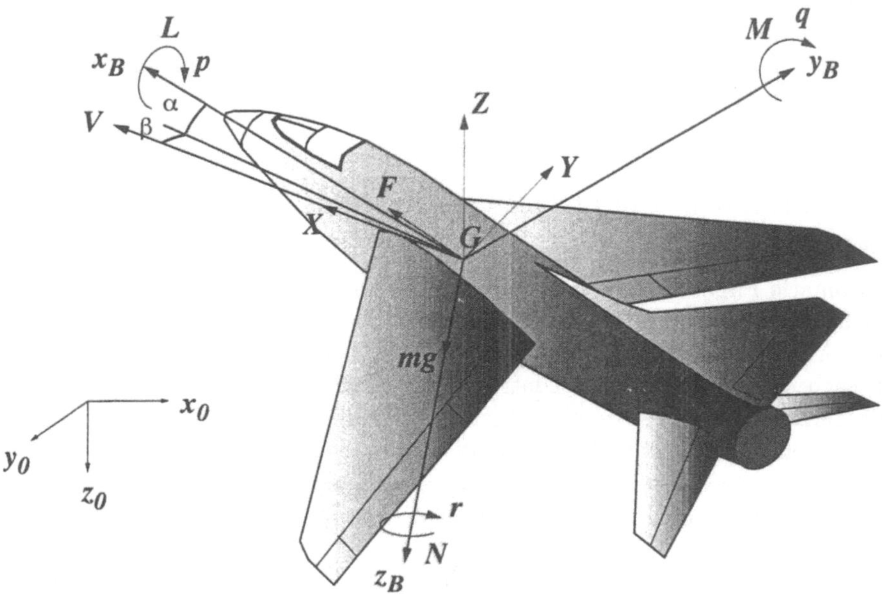

Fig. 1. Aircraft coordinate frames, forces and moments

2.1 Generic Aircraft Model

The aircraft we consider is *generic* in the sense that, as in most civil and military ones, it is actuated through its thrust and deflection surface positions, namely the elevator, ailerons and rudder, in the 3 directions.

A complete discussion of the model can be found in [21,22]. We partly follow [6,35,36] (see also [9,26,27]).

Aircraft dynamics are generally described by a set of 12 variables: the 3 components of the position of the center of mass, 3 variables for the velocity vector, 3 angular positions describing the aircraft attitude, and 3 variables for the corresponding angular velocities.

There are many possible choices, depending on the coordinate frames where positions, velocities, forces and moments are expressed. Here, we are using (see Figure 1):

- x, y, z, components of the center of mass in the Earth axes,
- V, α, β, velocity, angle of attack and side-slip angle in the body axes,
- χ, γ, μ, orientation of the wind axes with respect to the Earth axes,
- p, q, r, components of the angular velocity in the body axes.

The aircraft is conventionally actuated by four independent controls: the thrust F and the positions $(\delta_l, \delta_m, \delta_n)$ of the deflection surfaces. We may also want to include the servo model that produce the required deflection positions, the inputs of which are denoted $(\tilde{\delta}_l, \tilde{\delta}_m, \tilde{\delta}_n)$.

The resultant of the external forces, expressed in the wind axes, is denoted by (X, Y, Z). It is equal to the sum of the aerodynamic, gravitational (weight) and propulsive (thrust) forces:

$$\begin{pmatrix} X \\ Y \\ Z \end{pmatrix} = \underbrace{\frac{1}{2}\rho S V^2 \begin{pmatrix} -C_x \\ C_y \\ -C_z \end{pmatrix}}_{\text{Aerodynamic forces}} + \underbrace{F \begin{pmatrix} \cos(\alpha+\varepsilon)\cos\beta \\ \cos(\alpha+\varepsilon)\sin\beta \\ -\sin(\alpha+\varepsilon) \end{pmatrix}}_{\text{Thrust}} + \underbrace{mg \begin{pmatrix} -\sin\gamma \\ \cos\gamma\sin\mu \\ \cos\gamma\cos\mu \end{pmatrix}}_{\text{Weight}}. \tag{1}$$

The dimensionless so-called *aerodynamic coefficients* C_x, C_y, C_z (which are in fact functions of all the configuration variables, as detailed below) are experimentally determined in a wind tunnel (see [35] for an example). They are naturally expressed in the wind axes, in which the measurements are carried out, which justifies our (quite unusual) choice of coordinate frame. The air density is denoted by ρ. The remaining constants are the reference surface S and the angle ε between the axis of the propulsor and the axis Gx_B (see figure 1) in the symmetry plane of the aircraft.

Similarly, let (L, M, N) denote the components, in the body axes, of the sum of the external torques, generated by the external forces about the center of mass:

$$\begin{pmatrix} L \\ M \\ N \end{pmatrix} = \frac{1}{2}\rho S V^2 \begin{pmatrix} aC_l \\ bC_m \\ cC_n \end{pmatrix} + F \begin{pmatrix} 0 \\ d \\ 0 \end{pmatrix}, \tag{2}$$

where C_l, C_m, C_n are dimensionless *aerodynamic coefficients* (same remark as above) and a, b, c, d are constant (reference lengths).

It is commonly assumed that $C_x, C_y, C_z, C_l, C_m, C_n$ depend on the translational velocity (i.e., V, α, β) and acceleration $(\dot{V}, \dot{\alpha}, \dot{\beta})$, angular velocity (p, q, r), position of the deflection surfaces $(\delta_l, \delta_m, \delta_n)$, and Mach number (i.e., V/c, where c is the velocity of sound in air). They vary a lot with the Mach number during transsonic flight. A crucial remark for the design of our control law will be that, for nearly every aircraft, the dependence of C_x, C_y, C_z on $\dot{V}, \dot{\alpha}, \dot{\beta}$, p, q, r and $\delta_l, \delta_m, \delta_n$ is **weak**. In the same way, though it is not as important, the dependence of C_l, C_m, C_n on $\dot{V}, \dot{\alpha}, \dot{\beta}, p, q, r$ is also **weak**. Of course, on the contrary, C_l, C_m, C_n strongly depend on the deflection surface positions $\delta_l, \delta_m, \delta_n$. Thanks to this dependence, the moments, and hence the aircraft attitude, can be controlled.

Applying Newton's Second Law, we get

$$
\begin{aligned}
\dot{x} &= V\cos\chi\cos\gamma \\
\dot{y} &= V\sin\chi\cos\gamma \\
\dot{z} &= -V\sin\gamma \\
\dot{V} &= \frac{X}{m} \\
\dot{\alpha}\cos\beta &= -p\cos\alpha\sin\beta + q\cos\beta - r\sin\alpha\sin\beta + \frac{Z}{mV} \\
\dot{\beta} &= p\sin\alpha - r\cos\alpha + \frac{Y}{mV} \\
\dot{\gamma} &= -\frac{Y\sin\mu}{mV} - \frac{Z\cos\mu}{mV} \\
\dot{\chi}\cos\gamma &= \frac{Y\cos\mu}{mV} - \frac{Z\sin\mu}{mV} \\
\dot{\mu}\cos\beta &= p\cos\alpha + r\sin\alpha + \frac{Y\cos\mu\tan\gamma\cos\beta}{mV} \\
&\quad - \frac{Z(\sin\mu\tan\gamma\cos\beta + \sin\beta)}{mV} \\
I\begin{pmatrix}\dot{p}\\ \dot{q}\\ \dot{r}\end{pmatrix} &= \begin{pmatrix}L\\ M\\ N\end{pmatrix} - \begin{pmatrix}p\\ q\\ r\end{pmatrix} \wedge I \begin{pmatrix}p\\ q\\ r\end{pmatrix}
\end{aligned}
\tag{3}
$$

where $\wedge$ denotes the wedge product of 3-vectors, with the exterior forces X, Y, Z given by (1) and the torques L, M, N given by (2).

These equations must be completed by the servoactuator dynamics, whose slowest part, though much faster than (3), may be represented by:

$$
\varepsilon^2 \begin{pmatrix}\dot{\delta}_l\\ \dot{\delta}_m\\ \dot{\delta}_n\end{pmatrix} = A\begin{pmatrix}\delta_l\\ \delta_m\\ \delta_n\end{pmatrix} + \begin{pmatrix}\tilde{\delta}_l\\ \tilde{\delta}_m\\ \tilde{\delta}_n\end{pmatrix}
\tag{4}
$$

where ε represents the order of magnitude between the fastest time constants of (3) and these of (4), and where $\tilde{\delta}_l, \tilde{\delta}_m, \tilde{\delta}_n$ are the actual controls. The matrix A is stable. It may be noted, as will be detailed below, that the time constants corresponding to the angular acceleration subsystem of (3) (last equation of (3)) are of order ε compared to these of the remaining part of (3), and that those of (4) are in turn of order ε compared to these of the angular acceleration subsystem. This explains why in (4) the scaling ε^2 is needed.

For the sake of simplicity, we do not consider here the dynamical model relating the thrust F and the throttle.

2.2 Flatness Based Autopilot Design

Assume that the aerodynamic coefficients C_x, C_y, C_z depend only on (z, V, α, β), the altitude, Mach number, angle of attack, and side-slip angle respectively. Assume furthermore that C_l, C_m, C_n depend only on $(z, V, \alpha, \beta, \delta_l, \delta_m, \delta_n)$, namely on the positions of the deflection surfaces in addition to the aforementioned variables. This means in particular that we have neglected the naturally small contribution of the variables $(\dot{V}, \dot{\alpha}, \dot{\beta}, p, q, r, \delta_l, \delta_m, \delta_n)$ in C_x, C_y, C_z, and of of $(\dot{V}, \dot{\alpha}, \dot{\beta}, p, q, r)$ in C_l, C_m, C_n. Following [22], we call these approximated coefficients and the corresponding model (3), (4) *ideal.*

We first sketch the construction of a trajectory generator for the ideal aircraft and then of a unique controller for the real aircraft. This controller will be independent of the trajectories to be tracked.

Trajectory Generation Let us show that all the trajectories of all the variables of the ideal aircraft model are parametrized by (x, y, z, β), and a finite number of their time derivatives. In other words, given a smooth (sufficiently continuously differentiable) arbitrary trajectory for the quadruple (x, y, z, β), one can uniquely deduce the trajectory of each variable of the system, including the control inputs, without integrating the system differential equations. It means that the ideal aircraft is flat with (x, y, z, β) as a flat output (see [12,13,21,22]).

More precisely, let us show that all the system variables $(x, y, z, V, \gamma, \chi, \mu, F, \alpha, \beta, p, q, r, \delta_l, \delta_m, \delta_n, \tilde{\delta}_l, \tilde{\delta}_m, \tilde{\delta}_n)$ can be expressed as functions of $(x, \ldots, x^{(5)}, y, \ldots, y^{(5)}, z, \ldots, z^{(5)}, \beta, \ldots, \beta^{(3)})$, where the superscript $^{(i)}$ stands for the ith order time derivative, $i \in \mathbb{N}$, of the corresponding function.

Inverting the 3 first equations of (3), we get V, γ, χ as functions of $x, \dot{x}, y, \dot{y}, z, \dot{z}$. Differentiating them, and inverting the equations of $\dot{V}, \dot{\chi}, \dot{\gamma}$, we obtain, using formulas (1) and (2) with the ideal coefficients, that F, α, μ are functions of $x, \dot{x}, \ddot{x}, y, \dot{y}, \ddot{y}, z, \dot{z}, \ddot{z}, \beta$. Differentiating again, we find that p, q, r are functions of $x, \dot{x}, \ddot{x}, x^{(3)}, y, \dot{y}, \ddot{y}, y^{(3)}, z, \dot{z}, \ddot{z}, z^{(3)}, \beta, \dot{\beta}$ and that $\delta_l, \delta_m, \delta_n$ are functions of $x, \dot{x}, \ddot{x}, x^{(3)}, x^{(4)}, y, \dot{y}, \ddot{y}, y^{(3)}, y^{(4)}, z, \dot{z}, \ddot{z}, z^{(3)}, z^{(4)}, \beta, \dot{\beta}, \ddot{\beta}$. Finally, by the same technique, it results that $\tilde{\delta}_l, \tilde{\delta}_m, \tilde{\delta}_n$ are functions of $x, \dot{x}, \ddot{x}, x^{(3)}, x^{(4)}, x^{(5)}, y, \dot{y}, \ddot{y}, y^{(3)}, y^{(4)}, y^{(5)}, z, \dot{z}, \ddot{z}, z^{(3)}, z^{(4)}, z^{(5)}, \beta, \dot{\beta}, \ddot{\beta}, \beta^{(3)}$.

This construction is summarized in the following diagram, where the left side column contains the flat output components, the middle column the state variables, and the right side column the input variables:

$$
\begin{array}{ccccc}
(x,y,z) & \xrightarrow{\frac{d}{dt}} & (V,\gamma,\chi) & & \\
 & & \downarrow \frac{d}{dt} & & \\
\beta & \longrightarrow & (\alpha,\mu) & \longrightarrow & F \\
 & & \downarrow \frac{d}{dt} & & \\
 & & (p,q,r) & & \\
 & & \downarrow \frac{d}{dt} & & \\
 & & (\delta_l,\delta_m,\delta_n) & \xrightarrow{\frac{d}{dt}} & (\tilde{\delta}_l,\tilde{\delta}_m,\tilde{\delta}_n).
\end{array}
\tag{5}
$$

Therefore, any trajectory of the ideal aircraft will be easily generated via the construction of a smooth curve $t \mapsto (x(t), y(t), z(t), \beta(t))$: given suitable initial and final conditions (positions, velocities, accelerations, jerks,...) and possibly saturation constraints on the path curvature, velocity, acceleration, etc., it may be constructed by interpolation. Then the state and input variables are recovered according to the above calculations, without need to integrate the system.

For instance, in the case where we want the side-slip angle β to remain equal to 0, the trajectories for x, y, and z may be represented by three 7th degree polynomials with respect to time, with possible simplifications for simple maneuvers as those considered in classical flight control systems. More details are given in [21,22].

Let us sketch the differences between our approach and the classical ones on the simple example of a steady turn. A steady turn consists in setting $\dot{V} = \dot{\alpha} = \dot{\beta} = 0$, $\dot{p} = \dot{q} = 0$ and $\dot{r}$ equal to a constant. The corresponding forces and torques are immediately deduced from these conditions. The reference trajectory is then computed by connecting the corresponding circle to the current aircraft position by drawing a straight line along the current velocity vector (assuming that it is constant or almost constant), that touches tangentially the circle (see Fig. 2). However, at the junction point, the acceleration may jump, thus creating a perturbation to be attenuated by the controller. It can also be filtered to avoid a too shaky behavior.

In our approach, we only need specifying the initial and final state, velocity, acceleration and jerk (see Fig. 3), and translate them into initial and final conditions for the flat output and its derivatives. We generate a flat output reference $(x_{ref}, y_{ref}, z_{ref}, \beta_{ref})$ by polynomial interpolation with respect to time, more precisely, by interpolating this set of initial and final conditions, which involve the derivatives of x, y and z up to the order 3 and those of β up to the order 1. We thus end up with a 7th degree polynomial for x_{ref}, y_{ref} and z_{ref} and a 3rd degree one for β_{ref}. The resulting state and input variable reference trajectories satisfy the ideal system, or, in other words, exactly satisfy the ideal aircraft model differential equations, with an initial error which is 0 by construction, if the initial state is precisely measured. If the ideal system is not too far from reality, which we can reasonably expect,

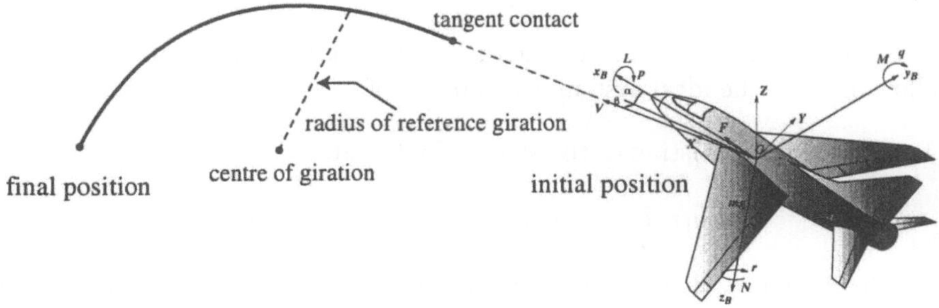

Fig. 2. Steady turn with tangent junction

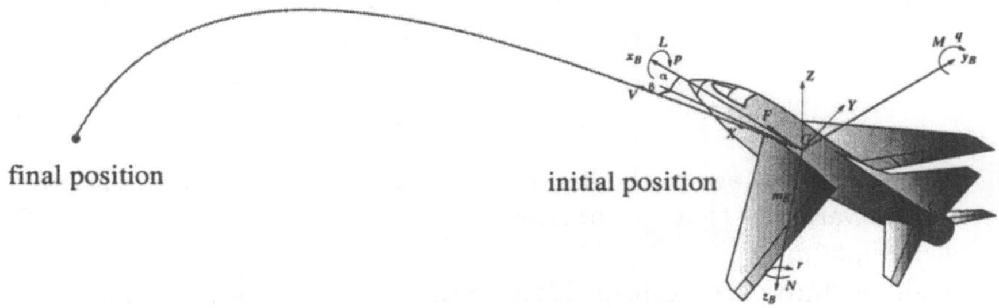

Fig. 3. Flatness based reference trajectory

the perturbations created by the error between the ideal model and reality might be smaller than with the previous method.

Note that this generator is not constrained to sequences of simple geometric paths, corresponding to steady dynamics and, moreover, that constraints on the velocity, acceleration, etc. can be taken into account by tuning the maneuver duration or, if necessary, the clock rate at which the reference trajectory is covered (see e.g. [13,3,31]). Therefore, applications e.g. to terrain following or orientation control are possible in this approach.

Controller Design The flatness property is not only useful for motion planning and trajectory generation, but also for the design of a single universal controller. With this term, we mean a controller able to make the aircraft track any trajectory without reconfiguration of the variables on which we close the loop and without modification of the gains and other parameters involved in the feedback loop. Moreover, it can be synthesized as a three level cascaded

PID, which is particularly surprising compared to the apparent complexity of the model[1].

Recalling the hierarchical structure of (5), it can be interpreted in the following way: the ideal system is composed of

1. a first part consisting of the string of integrators
$$(p,q,r,F) \longrightarrow (\alpha,\beta,\mu) \longrightarrow (V,\gamma,\chi) \longrightarrow (x,y,z) \tag{6}$$
where (F,p,q,r) may be considered as the control variables and where (x,y,z,β) is a flat output; note that the thrust F corresponds to a real control variable, whereas (p,q,r), the angular velocity vector, is a fictitious input, serving as an output for the subsystems described below, and indirectly controlled by their corresponding (lower level) inputs;
2. a second part, corresponding to the angular velocity dynamics, faster than the previous one, consisting of the 3 dimensional subsystem
$$(\delta_l,\delta_m,\delta_n) \longrightarrow (p,q,r) \tag{7}$$
where the positions of the deflection surfaces $(\delta_l,\delta_m,\delta_n)$ play the role of control variables that can be used in order to realize the control requirements for (p,q,r);
3. finally a third part, corresponding to the servoactuator loop, much faster than the two previous ones, constituted by the 3 dimensional local loop
$$(\tilde{\delta}_l,\tilde{\delta}_m,\tilde{\delta}_n) \longrightarrow (\delta_l,\delta_m,\delta_n) \tag{8}$$
where $(\tilde{\delta}_l,\tilde{\delta}_m,\tilde{\delta}_n)$ are the real control variables, that can be used, indeed, to realize the requirements on the fictitious control variables $(\delta_l,\delta_m,\delta_n)$ of the intermediate subsystem (7).

More precisely, denoting by ξ the state vector $(x,y,z,V,\gamma,\chi,\alpha,\beta,\mu,F)$ of the subsystem described by (6) and by ζ the vector $(x,\dot{x},\ddot{x},y,\dot{y},\ddot{y},z,\dot{z},\ddot{z},\beta)$, it results from flatness and (6) that ξ and ζ are diffeomorphic and that the first nine equations of the ideal system, with the dynamic extension $\dot{F}=F_1$, are of the form:
$$\begin{aligned} x^{(3)} &= X_0(\zeta)+X_1(\zeta)w \\ y^{(3)} &= Y_0(\zeta)+Y_1(\zeta)w \\ z^{(3)} &= Z_0(\zeta)+Z_1(\zeta)w \\ \dot{\beta} &= B_0(\zeta)+B_1(\zeta)w \end{aligned} \tag{9}$$

[1] Since flatness implies dynamic feedback linearization, full linearization of the system (3)-(4) can be achieved by dynamic feedback (see [6]). However, the presence of different time scales makes the corresponding feedback ill-conditioned and the above cascaded controller is more adapted to this particular structure.

where $w = (p, q, r, F_1)'$ (the prime denoting transpose) and with the matrix $\Xi_1 = \begin{pmatrix} X_1 \\ Y_1 \\ Z_1 \\ B_1 \end{pmatrix}$ invertible (X_1, Y_1, Z_1 and B_1 are independent line vectors). Note that in $w = (\omega, F_1)'$, the derivative of the thrust F_1, is directly controlled, and $\omega = (p, q, r)'$, the angular velocity, is a fictitious control variable.

One can choose w in order to track a given sufficiently differentiable reference trajectory $(x_{ref}, y_{ref}, z_{ref}, \beta_{ref})$ with stability by feedback linearization:

$$\begin{aligned} X_0(\zeta) + X_1(\zeta)w^* &= x_{ref}^{(3)} - K_{x,0}(x - x_{ref}) - K_{x,1}(\dot{x} - \dot{x}_{ref}) \\ &\quad - K_{x,2}(\ddot{x} - \ddot{x}_{ref}) \\ Y_0(\zeta) + Y_1(\zeta)w^* &= y_{ref}^{(3)} - K_{y,0}(y - y_{ref}) - K_{y,1}(\dot{y} - \dot{y}_{ref}) \\ &\quad - K_{y,2}(\ddot{y} - \ddot{y}_{ref}) \\ Z_0(\zeta) + Z_1(\zeta)w^* &= z_{ref}^{(3)} - K_{z,0}(z - z_{ref}) - K_{z,1}(\dot{z} - \dot{z}_{ref}) \\ &\quad - K_{z,2}(\ddot{z} - \ddot{z}_{ref}) \\ B_0(\zeta) + B_1(\zeta)w^* &= \dot{\beta}_{ref} - K_\beta(\beta - \beta_{ref}) \end{aligned} \tag{10}$$

with the gains $K_{i,j}$, $i = x, y, z$, $j = 0, 1, 2$, such that the polynomials $s^3 + K_{i,2}s^2 + K_{i,1}s + K_{i,0}$ are Hurwitz, for $i = x, y, z$, and with $K_\beta > 0$.

Note that the positions, velocities and accelerations $(x, \dot{x}, \ddot{x}, y, \dot{y}, \ddot{y}, z, \dot{z}, \ddot{z})$ needed to close this loop can be obtained from gyroscopic measurements. Thus, if w^*, satisfying (10), is applied to the system, the closed loop system is

$$\begin{aligned} x^{(3)} - x_{ref}^{(3)} &= -K_{x,0}(x - x_{ref}) - K_{x,1}(\dot{x} - \dot{x}_{ref}) - K_{x,2}(\ddot{x} - \ddot{x}_{ref}) \\ y^{(3)} - y_{ref}^{(3)} &= -K_{y,0}(y - y_{ref}) - K_{y,1}(\dot{y} - \dot{y}_{ref}) - K_{y,2}(\ddot{y} - \ddot{y}_{ref}) \\ z^{(3)} - z_{ref}^{(3)} &= -K_{z,0}(z - z_{ref}) - K_{z,1}(\dot{z} - \dot{z}_{ref}) - K_{z,2}(\ddot{z} - \ddot{z}_{ref}) \\ \dot{\beta} - \dot{\beta}_{ref} &= -K_\beta(\beta - \beta_{ref}) \end{aligned} \tag{11}$$

and is exponentially stable.

In order to realize the desired angular velocity vector ω^* (recall that $w^* = (\omega^*, F_1^*)$), we have to compute $(\delta_l, \delta_m, \delta_n)$, and then $(\tilde{\delta}_l, \tilde{\delta}_m, \tilde{\delta}_n)$, to make ω fastly converge to ω^*.

We can rewrite the subsystem (7) as follows

$$\varepsilon\dot{\omega} = \Omega_0(\zeta, \omega, F) + \Omega_1(\zeta, \omega, F)\delta \tag{12}$$

where ε is the time scaling between the slow part (9) and the faster part corresponding to (7), where $\delta = (\delta_l, \delta_m, \delta_n)'$, and where Ω_1 is an invertible matrix for every (ζ, ω, F) in the flight domain. It suffices then to choose

$$\delta^* = \Omega_1^{-1}(\zeta, \omega, F)(\Lambda(\omega - \omega^*) - \Omega_0(\zeta, \omega, F)) \tag{13}$$

with Λ an arbitrary but otherwise stable matrix, to transform (12) into $\varepsilon\dot{\omega} = \Lambda(\omega - \omega^*)$, ensuring the desired fast convergence of ω to ω^*. Note that (13) is not a high-gain feedback. The fast rate of convergence results only from the time scaling, preserved by this feedback.

Finally, it remains to compute the real servoactuator input vector $\tilde{\delta}^* = (\tilde{\delta}_l^*, \tilde{\delta}_m^*, \tilde{\delta}_n^*)'$ that realizes the required deflection position δ^*. Rewriting (4) as

$$\varepsilon^2\dot{\delta} = A\delta + \tilde{\delta} \tag{14}$$

one can set

$$\tilde{\delta}^* = -A\delta^* \tag{15}$$

which completes our feedback design. This construction is justified in [21], using singular perturbation techniques (see [16,18,34]).

Summing up, we have obtained

$$\tilde{\delta}^* = -A\Omega_1^{-1}(\zeta, \omega, F)\big(\Lambda(\omega - \omega^*) - \Omega_0(\zeta, \omega, F)\big) \tag{16}$$

with ω^* given by (10).

Finally, *the error between the real system*, with non ideal aerodynamic coefficients, *and the ideal one* may be seen as a *perturbation* that the above controller may naturally attenuate, as a consequence of the exponential stability.

Moreover, the control law defined by (10), (16) shows a good robustness versus modelling errors and perturbations if the time scalings are preserved by the choice of (not too high) gains. In particular, the feedback formulas involve several first order partial derivatives of the aerodynamic coefficients, which are not too accurately known, for instance in transsonic flight, but without noticeable consequences on the overall performance.

Let us insist on the fact that this cascaded controller, defined by (10), (13) and (16), and whose structure is displayed in figure 4, is defined for all reference trajectories that remain in the flight domain, and that its gains do not depend on the particular reference trajectory to follow.

The Pilot in the Loop In most control systems, wether civil or military ones, the pilot prescribes the aircraft reference trajectory to follow. The pilot enters, for instance through the instrument panel, the direction towards which the aircraft must be pointing, its velocity, and various angles describing the aircraft's attitude, in particular the angle of attack $\alpha$and the side-slip angle β, as well as several rates describing the desired path. These indications are used to compute the next piece of trajectory to follow during a given interval of time.

The authority allocation between the pilot and the automatic controller remains one of the most difficult questions: *when and in which circumstances can the pilot enforce the autopilot to modify the current reference trajectory?*

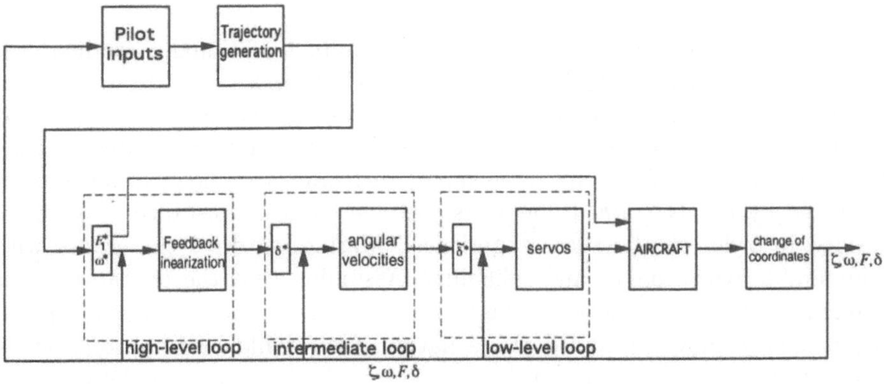

Fig. 4. The cascaded control architecture

Clearly, this external loop might sometimes have destabilizing effects since if the pilot updates the trajectory at a faster rate than the aircraft's response, the servoactuators will have more and more difficulties to follow their reference and, the error increasing, they may soon be saturated, sometimes switching the controller to an emergency mode.

The structure of the above controller is well adapted to design this external loop for at least two reasons:

1. the pilot inputs interact with the internal loops only through the trajectory generator;
2. the time constants of the internal loops being fixed independently of the maneuver, the rate at which the pilot inputs are updated may be chosen slower than the slowest closed-loop time constant, to prevent the controller from injecting too fast dynamics in the slow ones, but otherwise fast enough compared to the pilot's reactions.

The reference trajectory is thus refreshed by computing at each step a new piece of reference trajectory, according to the pilot's indications, starting from the current aircraft flight conditions, and without interfering with the inner loops. For obvious reasons, the same technique may also be used in teleoperated mode.

To conclude this section, let us note that such a controller might also provide important simplifications concerning failure detection aspects.

Comparable approaches may be used for other types of aircrafts such as Vertical Take-Off and Landing aircrafts [13] and helicopters (see [28,29] for example).

3 Crane Control

This class of mechanical systems, though quite traditionally studied (see [25,14]), still needs control assistance to include more operation facilities, security and reliability at low cost. An interesting example is provided by the US Navy cranes [19] (see Figure 5) used to unload heavy materials from large carriers to smaller ships in open sea. Such cranes are nowadays manually operated by highly trained operators. On-line control assistance is therefore needed in order to allow less experienced Marines to drive the cranes with comparable performance and reliability. Assuming that the operator drives the crane by moving a lever in the two horizontal plane directions and pushing pedals to move up and down, a natural idea would be to automatically generate pieces of reference trajectories from the informations delivered by the lever and pedals to tele-operate the crane, with an internal feedback loop taking care of the trajectory tracking. As in the previous section, we are confronted with the design of a universal reference trajectory generator and of a single feedback stabilizing controller, able to track any reference trajectory in a given class.

For simplicity's sake, we assume that the crane's pole does not turn around the vertical axis, which restricts the crane operations to a two dimensional domain. Moreover, the boat, on top of which the crane is set, is assumed to be fixed.

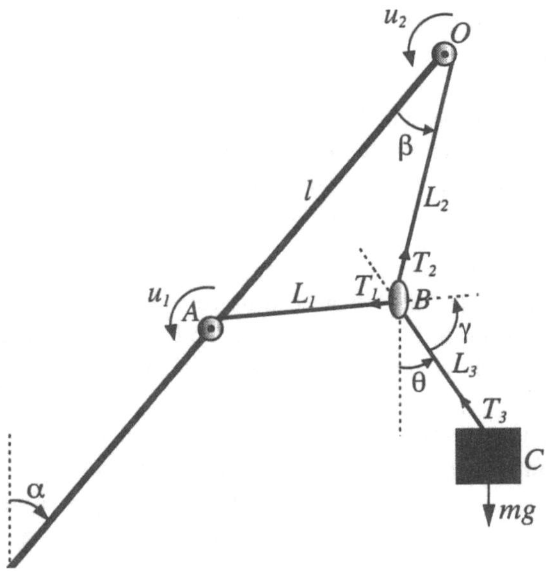

Fig. 5. A U.S. Navy crane

3.1 U.S. Navy Crane Model

The crane displayed in figure 5[2] consists of:

- a pole making a fixed angle α with respect to the vertical, equipped with two winches, one located at the top, denoted by O and chosen as the origin, and the second one located at A, at a fixed distance l from O;
- a first rope of variable length R, starting from O, whose upper part makes an angle β with the vertical, passing through a pulley located at the point B, the lower part of the rope making an angle θ with the vertical. The length of the upper part is denoted by L_2 and the one of the lower part L_3. Since the total length of the rope is R, we have $R = L_2 + L_3$;
- a second rope of variable length L_1 relating the winch A to the pulley B;
- a load of mass m attached to the first rope at the point C, located at a distance L_3 from the pulley B.

On the U.S. Navy cranes there is a third cable which allows the operator to control the distance from B to O; the present model assumes that this cable is used to prevent the pulley from falling down, and to maintain it at its equilibrium position, when the first rope is close to the vertical position, or in other words, when the horizontal component of B approaches 0.

The winch A has radius ρ_1 and inertia J_1 and is submitted to frictions denoted by $\eta_1(L_1, \dot{L}_1)$. The external torque u_1 applied to this winch is the control variable. Similarly, the winch O has radius ρ_2 and inertia J_2, is subject to frictions $\eta_2(R, \dot{R})$ and is controlled by its external torque u_2.

We consider a reference orthonormal frame (O, x, z) with Oz oriented upwards. We denote by g the gravitational acceleration and by (x, z) the coordinates of the load C. The masses of the ropes are neglected and the ropes are assumed unstretchable. The modulus of the tension of the first rope at C, created by the load, is denoted by T_3. The modulus of the tension of this rope at B is denoted by T_2 (note that since the rope passes through a pulley, we must have $T_2 = T_3$). Finally the modulus of the tension of the second rope at B is denoted by T_1.

The model can be divided into two parts, the first one describing the force balance equations for the load, the rope and the winches, and the second one consisting of purely geometric relations (see [19]).

Applying Newton's law to the pendulum BC, we have

$$m\ddot{x} = -T_3 \sin\theta \ , \quad m\ddot{z} = T_3 \cos\theta - mg \ , \tag{17}$$

[2] An experimental small size crane, with comparable characteristics is being developed at the author's lab in the framework of the Nonlinear Control Network, granted by the European Commission's Training and Mobility of Researchers (TMR) Contract # ERBFMRX-CT970137. See the www address: http://www.supelec.fr/lss/NCN/crane.html#PDP for more information.

with the vectorial sum of the tensions equal to 0 and with $T_2 = T_3$:

$$\begin{aligned} & T_2 = T_3 \\ & -T_1 \sin(\gamma+\theta) + T_2 \sin(\alpha-\beta) + T_3 \sin\theta = 0 \\ & T_1 \cos(\gamma+\theta) + T_2 \cos(\alpha-\beta) - T_3 \cos\theta = 0 \ . \end{aligned} \tag{18}$$

The torque balance at the winches, with $R = L_2 + L_3$, reads:

$$\begin{aligned} \frac{J_1}{\rho_1}\ddot{L}_1 &= T_1\rho_1 + \eta_1(L_1, \dot{L}_1) - u_1 \\ \frac{J_2}{\rho_2}\ddot{R} &= T_2\rho_2 + \eta_2(R, \dot{R}) - u_2 \ . \end{aligned} \tag{19}$$

Finally, with (x_B, z_B) the coordinates of B, the geometric constraints between the rope lengths and the angles θ, γ and β read:

$$\begin{aligned} x_B &= -L_2 \sin(\alpha-\beta) \\ z_B &= -L_2 \cos(\alpha-\beta) \\ x - x_B &= L_3 \sin\theta \\ z - z_B &= -L_3 \cos\theta \\ x_B + l\sin\alpha &= L_1 \sin(\theta+\gamma) \\ z_B + l\cos\alpha &= L_1 \cos(\theta+\gamma) \end{aligned} \tag{20}$$

where $\|OB\| = L_2$, $\|AB\| = L_1$, $\|BC\| = L_3$, and $(-l\sin\alpha, -l\cos\alpha)$ are the coordinates of the point A.

The implicit model (17)–(20) comprises 13 equations for the 13 unknown variables $(x, z, L_1, L_2, L_3, \beta, \gamma, \theta, T_1, T_2, T_3, x_B, z_B)$, and the control variables are (u_1, u_2). An explicit model, obtained by eliminating the non differential relations, is not needed here and is omitted.

Remark 1. This analysis may be easily extended to the case where the boat is moving in the swell by adding the corresponding perturbative torques to the torque balance at the winches and to replace the constant angle α by a time-varying one, assumed to be measured.

The above model is characterized by the following invariance property:

Proposition 1. *The point B satisfies:*

(i) *BC is colinear to $(\ddot{x}, \ddot{z}+g)'$, the sum of the load and gravity accelerations;*
(ii) *AB is the bisector of the angle $\widehat{CBO}$.*

Otherwise stated,

$$\frac{x - x_B}{z - z_B} = \frac{\ddot{x}}{\ddot{z} + g} \ , \tag{21}$$

$$\gamma = \frac{1}{2}(\pi + \beta - (\alpha + \theta)) \ , \quad T_1 = 2T_3 \cos\gamma \ . \tag{22}$$

Proof. According to (17), eliminating T_3, we have $-\tan\theta = \frac{\ddot{x}}{\ddot{z}+g}$. Also, from (20), we get $-\tan\theta = \frac{x-x_B}{z-z_B}$ which proves the assertion (i). The second assertion is an easy consequence of the fact that $T_2 = T_3$. Simple trigonometric computations show that (18) is equivalent to (22) (see figure 6). □

3.2 Crane Trajectory Generation

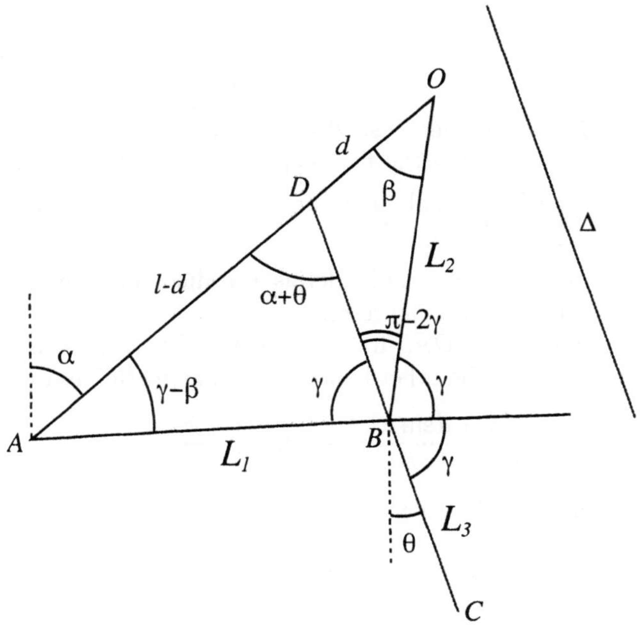

Fig. 6. The geometric characterization of B: knowing A, O and C and the direction Δ, B is such that BC is parallel to Δ and AB is the bisector of $\widehat{OBC}$.

We first show that all the system variables can be expressed as functions of the load coordinates (x, z) and derivatives. This can be done by elementary geometric considerations on triangles.

Proposition 2. *The system variables* $\theta, \beta, \gamma, L_1, L_2, L_3, T_1, T_2, T_3, x_B, z_B$ *and the inputs* u_1, u_2 *can be expressed as functions of* x, z *and their derivatives* *up to the 4th order:*

$$\begin{aligned}
&\tan\theta = -\frac{\ddot{x}}{\ddot{z}+g}\ , \quad \sin\beta = \frac{(\ddot{z}+g)(x+l\sin\alpha) - \ddot{x}(z+l\cos\alpha)}{l\sqrt{\ddot{x}^2+(\ddot{z}+g)^2}}\ ,\\
&\gamma = \frac{1}{2}(\pi+\beta-(\alpha+\theta))\ ,\\
&L_1 = l\frac{\sin\beta}{\sin\gamma}\ , \quad L_2 = l\frac{\sin(\gamma-\beta)}{\sin\gamma}\ , \quad L_3 = \frac{z+L_2\cos(\alpha-\beta)}{\cos\theta}\ ,\\
&x_B = -l\frac{\sin(\gamma-\beta)\sin(\alpha-\beta)}{\sin\gamma}\ , \quad z_B = -l\frac{\sin(\gamma-\beta)\cos(\alpha-\beta)}{\sin\gamma}\ ,\\
&T_2 = T_3 = m\sqrt{\ddot{x}^2+(\ddot{z}+g)^2}\ , \quad T_1 = 2m(\sqrt{\ddot{x}^2+(\ddot{z}+g)^2})\cos\gamma\ ,\\
&u_1 = T_1\rho_1 + \eta_1(L_1,\dot{L}_1) - \frac{J_1}{\rho_1}\ddot{L}_1\ ,\\
&R = L_2+L_3\ , \quad u_2 = T_2\rho_2 + \eta_2(R,\dot{R}) - \frac{J_2}{\rho_2}\ddot{R}\ .
\end{aligned} \tag{23}$$

Proof. From (17), we get that $T_3\sin\theta = -m\ddot{x}$ and $T_3\cos\theta = m(\ddot{z}+g)$. Thus $\tan\theta = -\frac{\ddot{x}}{\ddot{z}+g}$, which proves the first expression of (23). Note that we also have $\sin\theta = -\frac{\ddot{x}}{\sqrt{\ddot{x}^2+(\ddot{z}+g)^2}}$ and $\cos\theta = \frac{\ddot{z}+g}{\sqrt{\ddot{x}^2+(\ddot{z}+g)^2}}$, expressions that will be useful in the sequel.

To compute the angles θ, β and γ, let us introduce the point D intersection of the lines BC and OA (see figure 6).

Since D belongs to OA, its coordinates (x_D, z_D) can be written $(-d\sin\alpha, -d\cos\alpha)$. On the other hand, D belongs to BC and thus satisfies

$$\frac{x+d\sin\alpha}{z+d\cos\alpha} = -\tan\theta = \frac{\ddot{x}}{\ddot{z}+g}\ . \tag{24}$$

Solving with respect to d yields

$$d = \frac{\ddot{x}z - (\ddot{z}+g)x}{(\ddot{z}+g)\sin\alpha - \ddot{x}\cos\alpha}\ . \tag{25}$$

Now, using the relations between the sine of an angle and the length of the opposite segment in a triangle, we have, from the triangle OAB,

$$\frac{\sin\beta}{L_1} = \frac{\sin\gamma}{l} = \frac{\sin(\gamma-\beta)}{L_2}\ . \tag{26}$$

In the triangle ODB, the same relations read:

$$\frac{\sin\beta}{\|DB\|} = \frac{\sin 2\gamma}{d} = \frac{\sin(\alpha+\theta)}{L_2}\ . \tag{27}$$

Finally, in the triangle DAB:

$$\frac{\sin(\alpha+\theta)}{L_1} = \frac{\sin\gamma}{l-d} = \frac{\sin(\gamma-\beta)}{\|DB\|}\ . \tag{28}$$

From (26) and (28), we have $l \sin\beta = L_1 \sin\gamma = (l-d)\sin(\alpha+\theta)$ from which we get

$$\sin\beta = (1 - \frac{d}{l})\sin(\alpha+\theta) \tag{29}$$

which proves the second relation of (23) by replacing d by its expression (25), using $\sin(\alpha+\theta) = \sin\alpha\cos\theta + \cos\alpha\sin\theta$ and the expressions for $\sin\theta$ and $\cos\theta$ established above. It results that θ and β are functions of $(x, z, \ddot{x}, \ddot{z})$ and, according to (22), the same holds true for γ.

Again using (26), we get $L_1 \sin\gamma = l\sin\beta$ and $L_2 \sin\gamma = l\sin(\gamma-\beta)$. The expression of L_3 results from (20). The lengths L_1, L_2, and L_3 are thus functions of $(x, z, \ddot{x}, \ddot{z})$ since β, γ and θ, are.

The expression of T_3 is obtained by using (17) again: from $T_3^2 \sin^2\theta = m^2\ddot{x}^2$ and $T_3^2\cos^2\theta = m^2(\ddot{z}+g)^2$, we get the desired expression. The tensions T_2 and T_3 follow immediately by (18) and (22).

Finally, (19) leads to

$$u_1 = -\frac{J_1}{\rho_1}\ddot{L}_1 + T_1\rho_1 + \eta_1(L_1, \dot{L}_1)\ , \quad u_2 = -\frac{J_2}{\rho_2}\ddot{R} + T_2\rho_2 + \eta_2(R, \dot{R}) \tag{30}$$

and since $\dot{L}_1, \ddot{L}_1, T_1, T_2, \dot{R}, \ddot{R}$ can be expressed as functions of $(x, \dot{x}, \ddot{x}, x^{(3)}, x^{(4)}, z, \dot{z}, \ddot{z}, z^{(3)}, z^{(4)})$ ([3]), the same property holds true for u_1 and u_2. □

Remark 2. We have just proven that the system (17)–(20) is flat with (x, z) as a flat output (see [19]).

Remark 3. Neither the rope lengths L_1, L_2 and L_3 nor the angles θ, β and γ depend on the load mass m. This mass information is only present in the tensions and thus in the torques u_1 and u_2 applied to the winches. Consequently, trajectory generation for the angles and lengths is insensitive to errors on the load mass.

The trajectory generation may be easily done according to proposition 2 since it suffices to design the trajectory of the load in order to deduce all the system variables without integrating the system differential equations. Moreover, the fact that all the trajectories of the system are parametrized by $(x, \dot{x}, \ddot{x}, x^{(3)}, x^{(4)}, z, \dot{z}, \ddot{z}, z^{(3)}, z^{(4)})$ means that, given a pair of points in the space generated by these 10 variables, it is always possible to find a trajectory, satisfying the system equations (17)–(20), starting from the first one and arriving at the second one. Such a trajectory may be simply constructed by polynomial interpolation.

---

[3] Recall that the superscript $^{(i)}$ stands for the ith time derivative, for $i \geq 3$

Assume that we want to start at a given initial time t_i with the load's position, velocity, acceleration, jerk and 4th derivative given by $(x_i, \dot{x}_i, \ddot{x}_i, x_i^{(3)}, x_i^{(4)}, z_i, \dot{z}_i, \ddot{z}_i, z_i^{(3)}, z_i^{(4)})$, and arrive, at time t_f, with the load's configuration $(x_f, \dot{x}_f, \ddot{x}_f, x_f^{(3)}, x_f^{(4)}, z_f, \dot{z}_f, \ddot{z}_f, z_f^{(3)}, z_f^{(4)})$. We can construct 9th degree polynomials

$$
\begin{aligned}
x(t) &= x_i + (x_f - x_i) \sum_{j=1}^{9} a_j \left(\frac{t - t_i}{t_f - t_i} \right)^j \\
z(t) &= z_i + (z_f - z_i) \sum_{j=1}^{9} b_j \left(\frac{t - t_i}{t_f - t_i} \right)^j
\end{aligned}
\tag{31}
$$

where the coefficients a_j, b_j, $j = 1, \ldots, 9$, are computed by solving linear equations, whose entries are combinations of the initial and final conditions.

Assume, in addition, that we must remain in a given subdomain of the (x, z) space and that the curve given by $z = B(x)$, satisfies this constraint. Then x is computed as above by polynomial interpolation to finally obtain $z(t) = B(x(t))$ for all t.

To examplify this situation, let us sketch the trajectory generation for the load starting from a rest point and arriving at a rest point, while avoiding an obstacle by following the parabola

$$z = \lambda x^2 + \mu x + \nu \ . \tag{32}$$

Note that if, at some time t, x is such that $\dot{x}(t) = 0$, then we have $\frac{dz}{dt}(t) = (2\lambda x(t) + \mu)\dot{x}(t) = 0$, which proves that if x satisfies $\dot{x}_i = \ddot{x}_i = x_i^{(3)} = x_i^{(4)} = 0$, then z satisfies the same conditions, and the load is at rest. Therefore, it suffices to compute a trajectory for x by polynomial interpolation:

$$x(t) = x_i + (x_f - x_i) \left(\frac{t - t_i}{t_f - t_i} \right)^4 \sum_{j=0}^{5} a_j \left(\frac{t - t_i}{t_f - t_i} \right)^j \tag{33}$$

and to plug it in (32) to obtain the corresponding trajectory $z(t)$. Note that the factorization in (33) results from the 4 successive derivatives equal to 0 at the starting point. Moreover, since the same holds at the final point, it can be easily checked that the coefficients a_j, $j = 0, \ldots, 5$, do not depend on the initial and final conditions and thus are computed once for all. See [19] for an example of trajectory planning from a rest point to another rest point. Note that, for such a trajectory, the segment BC, though vertical at the initial and final points, does not remain vertical in between (see figure 7), which significantly differs from a quasi-stationnary approach, in which we would require the segment BC to remain vertical all the time. Moreover, the behavior in our case is quite natural since the load, is first pulled by the point B and then catches it up, to finally get ahead B while the latter is slowing

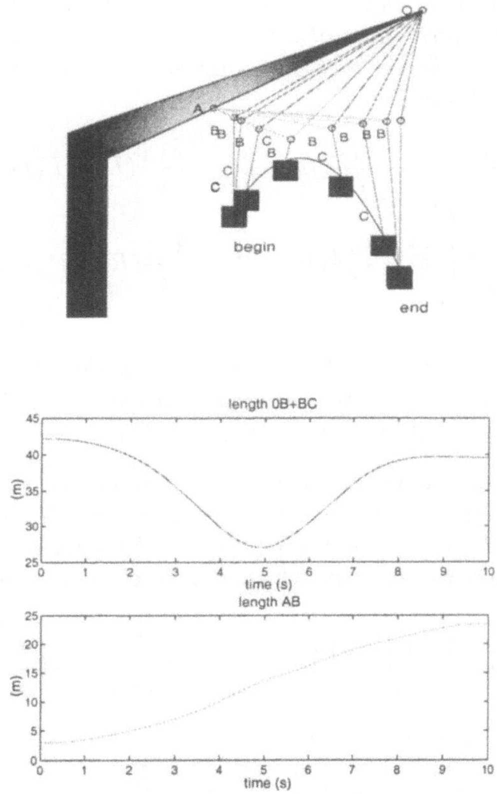

Fig. 7. Moving the load from a rest point to another one while avoiding an obstacle

down to stop at the vertical of the load. It is easily seen, moreover, that, along this trajectory, the mechanical energy remains constant.

Remark, as in the aircraft control case, that velocity and acceleration constraints may be taken into account by tuning the duration $t_f - t_i$, or, if necessary, by controlling the clock as in [3].

To remotely operate the crane, the possible obstacles are managed by the operator. It suffices thus to compute the polynomials (31) online with the initial and final configurations obtained from the lever and pedals actuated by the operator.

3.3 Feedback Controller Design

In this section, we assume that the lengths L_1 and R, and their respective derivatives $\dot{L}_1$ and $\dot{R}$, are measured. These quantities are particularly easy to measure in practice by placing an encoder or a tachymeter on each winch.

Theorem 1. *Assuming that we want to follow an arbitrary reference trajectory (x^*, z^*) the derivatives of which are uniformly bounded up to the 4th order, consider the following high-gain PID controller using only the measurements of L_1, $\dot{L}_1$, R and $\dot{R}$:*

$$\begin{aligned} u_1 &= -\frac{J_1}{\rho_1}\left(\ddot{L}_1^* - \frac{k_{1,0}}{\varepsilon^2}(L_1 - L_1^*) - \frac{k_{1,1}}{\varepsilon}(\dot{L}_1 - \dot{L}_1^*) - k_1\nu_1\right) \\ \dot{\nu}_1 &= L_1 - L_1^* \\ u_2 &= -\frac{J_2}{\rho_2}\left(\ddot{R}^* - \frac{k_{2,0}}{\varepsilon^2}(R - R^*) - \frac{k_{2,1}}{\varepsilon}(\dot{R} - \dot{R}^*) - k_2\nu_2\right) \\ \dot{\nu}_2 &= R - R^* \end{aligned} \tag{34}$$

with $\varepsilon > 0$ a small enough real number and such that the polynomials $s^2 + k_{i,1}s + k_{i,0}$ are Hurwitz and $k_i \geq 0$ small enough, $i = 1, 2$. Then $L_1 - L_1^*$, $L_2 - L_2^*$, $L_3 - L_3^*$, $\beta - \beta^*$, $\gamma - \gamma^*$, $\theta - \theta^*$, $x - x^*$ and $z - z^*$ all converge to 0.

Proof. Using a standard singular perturbation argument, we get that $L_1 - L_1^*$ and $R - R^*$ converge to 0 exponentially and that the integrators ν_1 and ν_2 converge to $\frac{\rho_1}{k_1 J_1}(T_1\rho_1 + \eta_1)$ and $\frac{\rho_2}{k_2 J_2}(T_2\rho_2 + \eta_2)$ respectively.

Using the following relation, which is a direct consequence of (19) and (22):

$$\sin\left(\frac{\beta - (\alpha + \theta)}{2}\right) = -\frac{\rho_2}{2\rho_1}\left(\frac{\frac{J_1}{\rho_1}\ddot{L}_1 + u_1 - \eta_1}{\frac{J_2}{\rho_2}\ddot{R} + u_2 - \eta_2}\right) \tag{35}$$

it results that the limit is given by

$$\sin\left(\frac{\beta^* - (\alpha + \theta^*)}{2}\right) = -\frac{\rho_2}{2\rho_1}\left(\frac{\frac{J_1}{\rho_1}k_1\nu_1 - \eta_1}{\frac{J_2}{\rho_2}k_2\nu_2 - \eta_2}\right) . \tag{36}$$

Then, since γ is given by (23), the limits of β and θ, still using (23), are given by the solution of the system of equations obtained from (36) and

$$L_1^* \cos\left(\frac{\beta^* - (\alpha + \theta^*)}{2}\right) = l \sin\beta^* \tag{37}$$

from which we obtain β^* and then θ^* and γ^*. The limits being locally unique, β, θ and γ converge. Again using (23), we obtain the limits (x_B^*, z_B^*) and thus the lengths L_2^* and $L_3^* = R^* - L_2^*$, which finally provide (x^*, z^*), and the convergence of the respective variables, by continuity, which achieves the proof. □

To conclude, the same kind of approach can be applied to other types of cranes such as the one considered in [10,12] and is easily extendable to 3 dimensional cranes. Notice that, in this example, an explicit state-space representation was not needed for feedback construction. Another interesting and closely related design using the notion of quasi-static feedback may be found in [8,33].

4 Active and Semi-active Suspension Design

Suspension control has recently become a challenging topic for car manufacturers, car equipment industries and academic researchers (see e.g. [1,2,7,15,20,30]). The aim is to improve comfort and security. This corresponds to the current trend consisting in providing more control assistance to classical mechanical devices, such as those already present on usual vehicles, at minimum additional cost.

Here, we aim at improving our mastering of controlled suspensions, studying the possible improvements with respect to passive suspensions, and designing a product adaptable to several classes of vehicles.

By the generic term *controlled suspensions*, we mean:

- Semi-active suspensions; two types are introduced: (i) *switched damper* and (ii) *continuously varying damper*. For the former, characteristics may be modified by a switch with a finite number of positions, from hard to soft. The latter is equipped with a servo-valve able to modify continuously the damping characteristics.
- Semi-active suspensions with an active anti-roll system, which comprises an additional hydraulic circuit relating the front wheels and a pump able to modify the pressure balance between the front wheels in order to compensate the roll angle deviation in curves.
- Fully active suspensions, with hydraulic or electric systems able to deliver prescribed (vertical) forces acting on each wheel for controlling the attitude of the car.

Clearly, pushing on a wheel to compensate the roll angle for instance, amounts to increasing the stiffness of the link between the ground and the car. Therefore, vibrations generated by the ground profile are partially transmitted to the car, and the designer is confronted with a sharp tradeoff between security (keeping the wheels on the road) and comfort (filtering the vibrations created by the road). In fact, the comfort aspect may be divided into two different categories: vibration isolation and keeping the car as horizontal as possible. Note that, in the latter category, security and comfort requirements go in synergy.

A popular strategy for isolating the car from the ground consists in "hooking the car to the sky", which is an idealization of the fact that one wants to design the damping controller such that the closed-loop system is equivalent to a filter with prescribed characteristics. This is a particular case of feedback linearization as will be shown later.

We propose here a unified controller structure, able to work with all kinds of controlled suspensions,. The controller requires measuring five physical quantities (three vertical positions or vertical accelerations and two horizontal accelerations) in order to recover the whole state of the car, used in the feedback loop.

This section is organized as follows: we first present a simplified dynamical model of a quarter-car (single wheel) (see figure 9) and its control. A simplified model of the full three dimensional car (see figure 11) is then introduced and we show how the single wheel controller may be used as a low-level loop. A high-level loop then distributes the necessary forces on the four wheels, to keep the car in the desired attitude.

4.1 Quarter-car Model and its Control

We consider the vertical motion of a single wheel with a spring of stiffness k and a variable damper, supporting a mass m. The tyre dynamics (faster than the rest of the model) are neglected. The road profile (already filtered by the tyre) is denoted by $t \mapsto w(t)$ and is a perturbation, representing the relative height of the road at time t. We denote by z the difference of height between the actual car position and its equilibrium position $\overline{z}$. The control u is supposed to affect the characteristics of the damper: the force created by the damper is a function of the input and the velocity difference in the vertical direction between the car and the road. This force equals $-G(u, \dot{z} - \dot{w})$, where the function G is of the form displayed on the right part of figure 9 for u_{min} (soft) and u_{max} (hard). The curves $\zeta \mapsto G(u, \zeta)$ are non decreasing for every u and $G(u, 0) = 0$, which reflects that, if $\dot{z} - \dot{w} = 0$, the damper is useless for all u. Moreover, the curves of G show that the damper can only oppose itself to vertical motion by creating a positive force, if the car moves up with respect to the road, and a negative force in the opposite case. But it cannot create an arbitrary force, which justifies the name *semi-active suspension*.

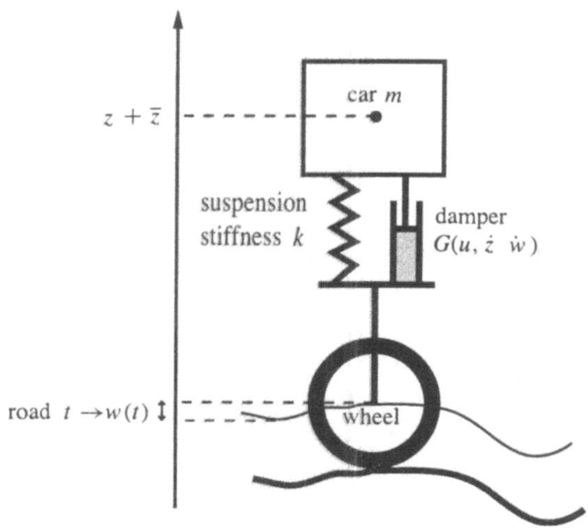

Fig. 8. Quarter-car

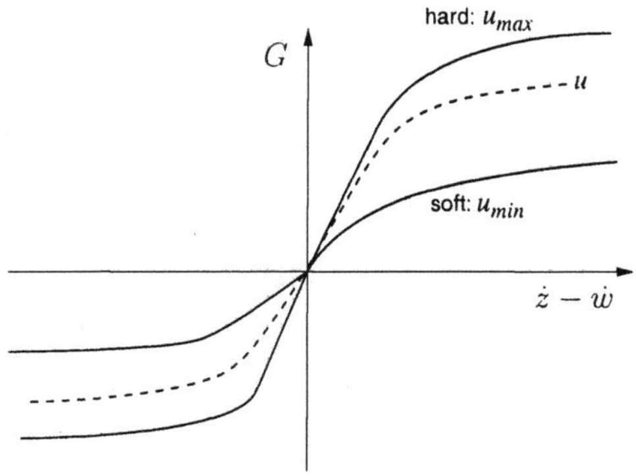

Fig. 9. The damping curves

The model of the quarter-car is thus

$$m\ddot{z} = -k(z - w) - G(u, \dot{z} - \dot{w}) \ . \tag{38}$$

Notice that, as a consequence of the properties of G, *the linear tangent model at an equilibrium point (i.e.* $\dot{z} = \dot{w} = 0, \ddot{z} = 0$*) is uncontrollable.*

The *sky hook controller* is characterized by

$$G(u, \dot{z} - \dot{w}) = \alpha \dot{z} \tag{39}$$

As shown in figure 10, sky hook means that if the damper was attached to the sky (right), the dependence on the perturbative vertical velocity, $\dot{w}$, would be absent and thus, the resulting force on the car would depend on its vertical velocity only. Moreover, this dependence is generally assumed to be linear. The coefficient α of (39) is tuned in order to realize a filter with given frequency response.

If, in addition, we also want to modify the stiffness of the spring by feedback, we may control the damper in such a way that

$$m\ddot{z} = -\alpha_0 z - \alpha_1 \dot{z} \ . \tag{40}$$

This strategy is called *linearizing control.*

If, moreover, a force F produced by an active element is available, then the damping function G may be replaced by $G + F$ in (38).

We immediately obtain the following result:

Theorem 2. *In the semi-active case, with continuously varying dampers, the sky hook strategy may be realized by the control* u^* *given, if* $\dot{z} - \dot{w} > 0$*, by*

$$u^* = \begin{cases} u_{max} & \text{if } \alpha\dot{z} > G(u_{max}, \dot{z} - \dot{w}) \\ U(\dot{z}, \dot{w}) & \text{if } G(u_{min}, \dot{z} - \dot{w}) \leq \alpha\dot{z} \leq G(u_{max}, \dot{z} - \dot{w}) \\ u_{min} & \text{if } \alpha\dot{z} < G(u_{min}, \dot{z} - \dot{w}) \end{cases} \tag{41}$$

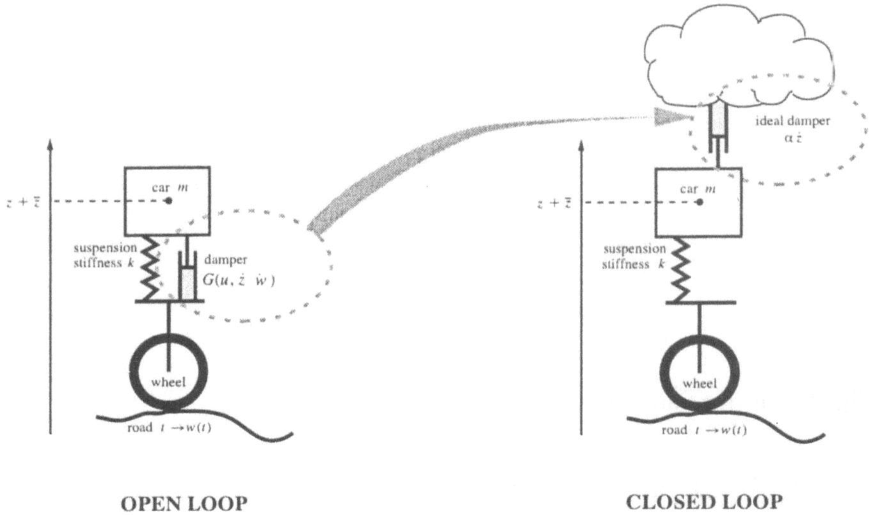

Fig. 10. The sky hook strategy

and, if $\dot{z} - \dot{w} < 0$*, by*

$$u^* = \begin{cases} u_{min} & \text{if } \alpha\dot{z} > G(u_{min}, \dot{z} - \dot{w}) \\ U(\dot{z}, \dot{w}) & \text{if } G(u_{max}, \dot{z} - \dot{w}) \leq \alpha\dot{z} \leq G(u_{min}, \dot{z} - \dot{w}) \\ u_{max} & \text{if } \alpha\dot{z} < G(u_{max}, \dot{z} - \dot{w}) \end{cases} \tag{42}$$

Here, $U(\dot{z}, \dot{w})$ *is the solution, if it exists, of the equation*

$$G(U(\dot{z}, \dot{w}), \dot{z} - \dot{w}) = \alpha\dot{z} \ .$$

Implementing this strategy requires the knowledge of $\dot{z}$ and $\dot{z} - \dot{w}$, which is possible, via an observer, by measuring $\ddot{z}$.

The linearizing controller is given by similar formulas, replacing $\alpha\dot{z}$ by $-k(z - w) + \alpha_0 z + \alpha_1 \dot{z}$.

Finally, in the active control case, the active device must deliver a force that compensates the one missing in the damper: $F^* = \alpha\dot{z} - G(U(\dot{z}, \dot{w}), \dot{z} - \dot{w})$.

4.2 Full Car Suspension Control

We now consider the full car in 3 dimensions and assume that its chassis is rigid, that the 4 wheels are in the same plane, tangent to the road (as before filtered by the wheels and tyres), and that the tyre dynamics and wheel masses are negligible. The profile $w(t)$ is supposed to be equal to the relative height of the wheels.

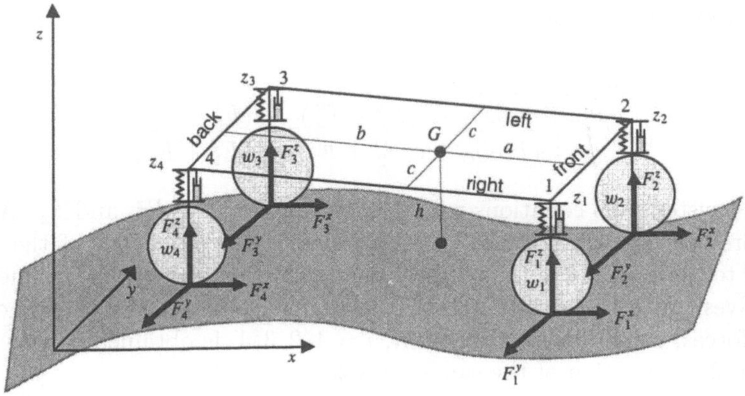

Fig. 11. The full vehicle

The notations are those of figure 11, with θ the pitch angle (rotation around the y axis) and ψ the roll angle (rotation around the x axis). Since we are not interested in the dynamics in the direction of the car's velocity vector, we do not consider the angle of turn (around the z axis). Applying Newton's second law, we get:

$$\begin{aligned} m\ddot{x}_G &= \sum_{i=1}^{4} F_i^x , \quad m\ddot{y}_G = \sum_{i=1}^{4} F_i^y , \quad m\ddot{z}_G = \sum_{i=1}^{4} F_i^z \\ I_y\ddot{\theta} &= -h\left(\sum_{i=1}^{4} F_i^x\right) + b(F_3^z + F_4^z) - a(F_1^z + F_2^z) \\ I_x\ddot{\psi} &= +h\left(\sum_{i=1}^{4} F_i^y\right) + c(F_2^z + F_3^z) - c(F_1^z + F_4^z) \end{aligned} \tag{43}$$

the forces being given by:

$$\begin{aligned} F_1^z &= -k_f(z_1 - w_1) - G_1(u_1, \dot{z}_1 - \dot{w}_1) - c_f(z_1 - z_2 - (w_1 - w_2)) \\ F_2^z &= -k_f(z_2 - w_2) - G_2(u_2, \dot{z}_2 - \dot{w}_2) + c_f(z_1 - z_2 - (w_1 - w_2)) \\ F_3^z &= -k_r(z_3 - w_3) - G_3(u_3, \dot{z}_3 - \dot{w}_3) - c_r(z_3 - z_4 - (w_3 - w_4)) \\ F_4^z &= -k_r(z_4 - w_4) - G_4(u_4, \dot{z}_4 - \dot{w}_4) + c_r(z_3 - z_4 - (w_3 - w_4)) \end{aligned} \tag{44}$$

with c_r and c_f the stiffness of the respective rear and front anti-roll bars.

Assuming that $\ddot{x}_G$ and $\ddot{y}_G$, and the profile w are known functions of time, the system is flat with $(z_G, \theta, \psi, \Gamma)$ as a flat output, where Γ is an arbitrary combination of the damping forces G_1, G_2, G_3 and G_4: the components with respect to x and y of the forces can be eliminated in function of $\ddot{x}_G$ and $\ddot{y}_G$

respectively, and (43) becomes

$$\begin{aligned} m\ddot{z}_G &= \sum_{i=1}^{4} F_i^z \\ I_y\ddot{\theta} &= -hm\ddot{x}_G + b(F_3^z + F_4^z) - a(F_1^z + F_2^z) \\ I_x\ddot{\psi} &= +hm\ddot{y}_G + c(F_2^z + F_3^z) - c(F_1^z + F_4^z) \end{aligned} \tag{45}$$

that is a system of 3 equations with 4 unknowns F_1^z, F_2^z, F_3^z, and F_4^z. Adding the aforementioned function Γ, the four vertical components of the forces applied to the chassis can thus be inverted in function of $(z_G, \theta, \psi, \Gamma)$ and their derivatives up to the second order. Finally, using these vertical components of the forces, each damper control u_i, $i = 1, 2, 3, 4$, is obtained exactly as in theorem 2, in function of the same variables.

Thus, the chassis can be set parallel to a filtered tangent plane that may be seen as the reference trajectory. Note that the additional function Γ can be used to "personalize" the car in the sense that, to a choice of combination of the damping forces, there corresponds a weighting on lateral and longitudinal behaviors of the car.

Moreover, note that, if the accelerations $\ddot{x}_G$, $\ddot{y}_G$ and the vertical accelerations of three distinct points of the chassis are measured, a nonlinear observer may be constructed to compute the road profile and the attitude of the car.

5 Conclusion

We have presented three case studies of control of mechanical systems, with clear engineering and industrial motivations, combining time scaling considerations with flatness, and producing physically well understood controllers. Note that numerous examples in different industrial sectors have successfully been approached with a similar viewpoint (see e.g. [3,5,17,23,24,29,32]).

We may now come back to the question raised in the title. There is an increasing demand for advanced control and driving assistance in various industrial fields such as automotive systems, machine tools, electric drives, magnetic levitation, transportation systems, or medical equipments, to cite but a few of them. Our flatness based approach is relevant to many control systems, and we hope that this presentation will convince the reader of the existence of advanced control methods that are not only applicable, but are also able to propose efficient answers to a variety of frequently asked questions relative to simplicity and compactness of the design, to controller versatility versus system technology evolution, and portability to new systems having comparable characteristics.

References

1. A. Alleyne and J.K. Hedrick. Nonlinear adaptive control of active suspensions. *IEEE Trans. Control Systems Technology*, 3, 1995, pp. 94–101.
2. F. Autran, F. Bessière, J. Lévine and P. Rouchon. La fonction suspension, du passif à l'actif: l'approche Valeo. *Proc. EEA-CNRS workshop "Automatique et Automobile"*, Bordeaux, June 1995.
3. L. Bitauld, M. Fliess and J. Lévine. A flatness based control synthesis of linear systems and application to windshield wipers. *Proc. ECC'97*, Brussels, paper N. 628, 1997.
4. J. Boichot, E. Delaleau, N. V. Diep, J. Lévine and E. Parzy. Modelling and control of a high-precision positionning system actuated by a linear synchronous motor. *Proc. 14th IFAC World Congress*, Beijing, July 1999.
5. J. Boichot, E. Delaleau, N. V. Diep, J. Lévine and E. Parzy. Modelling and control of a 2 d.o.f. high-precision positionning system. *Proc. ECC99*, Karlsruhe, 1999.
6. B. Charlet, J. Lévine and R. Marino. Sufficient conditions for dynamic state feedback linearization. *SIAM J. Control and Optimiz.*, 29, 1, p. 38–57, 1991.
7. A. Constans. L'apport de l'électronique dans la qualité de la suspension, panorama, potentialités, échéances. *Proc. Congrès SIA*, Lyon, 1994.
8. E. Delaleau and J. Rudolph. Control of flat systems by quasi-static feedback of generalized states. *Int. J. Control*, 71, p. 745–765, 1998.
9. B. Etkin. *Dynamics of flight-Stability and control.* John Wiley, 1982. Second edition.
10. M. Fliess, J. Lévine and P. Rouchon. A generalised state variable representation for a simplified crane description. *Int. J. of Control*, 58, p. 277–283, 1993.
11. M. Fliess, J. Lévine, Ph. Martin, F. Ollivier and P. Rouchon. Controlling nonlinear systems by flatness. In *Systems and Control in the Twenty-First Century*, C.I. Byrnes, B.N. Datta, D.S. Gilliam and C.F. Martin, 137–154, Birkhäuser, Boston, 1997.
12. M. Fliess, J. Lévine, Ph. Martin, and P. Rouchon. Flatness and defect of nonlinear systems: introductory theory and applications. *Int. J. of Control*, 61, p. 1327–1361, 1995.
13. M. Fliess, J. Lévine, Ph. Martin, and P. Rouchon. A Lie-Bäcklund approach to equivalence and flatness of nonlinear systems. *IEEE Trans. Automatic Control*, to appear, 1999.
14. T. Gustafsson. On the design and implementation of a rotary crane controller. *European J. Control*, 2, p. 166–175, 1996.
15. D.C. Karnopp. Active damping in road vehicle suspension systems. *Vehicle System Dynamics*, 12, 291-316, 1983.
16. P. Kokotović, H.K. Khalil, and J. O'Reilly. *Singular Perturbations Methods in Control: Analysis and Design.* Academic Press, 1986.
17. J. Lévine, J. Lottin and J.C. Ponsart. A nonlinear approach to the control of magnetic bearings. *IEEE Trans. on Control Systems Technology*, 4, 5, 524–544, 1996.
18. J. Lévine and P. Rouchon. An invariant manifold approach for robust control design and applications. *in "Systems and Networks: Mathematical Theory and Applications", Vol. II, U. Helmke, R. Mennicken, J. Saurer eds.*, Akademie Verlag, p. 309–312, 1994.

19. J. Lévine, P. Rouchon, G. Yuan, C. Grebogi, B.R. Hunt, E. Kostelich, E. Ott and J. Yorke. On the control of US Navy cranes. *Proc. ECC'97*, Brussels, paper N. 717, 1997.
20. B. Lohmann. Application of model order reduction to a hydropneumatic vehicle suspension. *IEEE Trans. Control Systems Technology*, 3, 1995, pp. 102–109.
21. Ph. Martin. *Contribution à l'Étude des Systèmes Différentiellement Plats.* PhD thesis, École des Mines de Paris, France, 1992.
22. Ph. Martin. Aircraft control using flatness. *Proc. CESA Multiconference*, Lille, p. 194–199, 1996
23. P. Martin, R.M. Murray and P. Rouchon. Flat systems. *Plenary Lectures and Minicourses*, G. Bastin and M. Gevers Eds. Proc. ECC 97, Brussels, 1997, pp. 211–264.
24. Ph. Martin and P. Rouchon. Flatness and sampling control of induction motors. *Proc. IFAC World Congress*, San Francisco, 1996.
25. A. Marttinen, J. Virkkunen, and R.T. Salminen. Control Study with a Pilot Crane. *IEEE Trans. Edu.*, 33: 298–305, 1990.
26. D. McLean. *Automatic Flight Control Systems.* Prentice Hall, 1990.
27. D.T. McRuer, I.L. Ashkenas, and D.C. Graham. *Aicraft Dynamics and Automatic Control.* Princeton University Press, 1973.
28. Ph. Mullhaupt, B. Srinivasan, J. Lévine and D. Bonvin. A toy more difficult to control than the real thing. *Proc. ECC'97*, Brussels, paper N. 440, 1997.
29. Ph. Mullhaupt, B. Srinivasan, J. Lévine and D. Bonvin. Cascade control using flatness. *Proc. ECC'99*, Karslruhe, 1999.
30. R. Rajamani and J.K. Hedrick. Adaptive observers for active automotive suspensions; theory and experiments. *IEEE Trans. Control Systems Technology*, 3, 1995, pp. 86–93.
31. R. Rothfuß, J. Rudolph and M. Zeitz. Flatness Based Control of a Nonlinear Chemical Reactor Model. *Automatica*, 32, p. 1433–1439, 1996.
32. R. Rothfuß, J. Rudolph and M. Zeitz. Flachheit: Ein neuer Zugang zur Steuerung und Regelung nichtlinearer Systeme. *Automatisierungstechnik*, 45, p. 517–525, 1997.
33. J. Rudolph and E. Delaleau. Some examples and remarks on quasi-static feedback of generalized states. *Automatica*, Vol. 34, 8, p. 993–999, 1998.
34. A.N. Tikhonov, A.B. Vasil'eva, and A.G. Sveshnikov. *Differential equations.* Springer-Verlag, 1985.
35. M. Vukobratović and R. Stojić. *Modern Aircraft Flight Control.* Springer-Verlag, 1988.
36. J.C. Wanner. *Dynamique du vol et pilotage des avions.* Technical report, École Nationale Supérieure de l'Aéronautique et de l'Espace, Toulouse, France, 1984.

8

Hybrid Systems: Modeling and Supervisory Control

B. H. Krogh and A. Chutinan

Department of Electrical and Computer Engineering, Carnegie Mellon University, Pittsburgh, PA 15213-3890, USA, email: {krogh|ac4c}@cmu.edu

Summary. This paper presents elements of control-oriented models of hybrid systems with input and output signals. Models can be composed through standard block-diagram interconnections. Transition systems with inputs and outputs are introduced to define the semantics for open-loop hybrid systems. Fixed-point iterations can then used to synthesize logical supervisory controllers. This is illustrated for the problem of designing a supervisor to keep the state of the hybrid system out of a specified set of forbidden states. Recently developed tools for computer-aided modeling, simulation, and verification of hybrid systems are reviewed briefly. The paper concludes with a discussion of open research issues and prospects for applications of the theory to problems of practical interest.

1 Introduction

Several factors have stimulated the recent interest in hybrid systems among control theorists. Applications provide one motivation. Some physical systems exhibit distinct discrete modes of behavior, introducing discrete as well as continuous state variables into the models. Often, computer control systems for continuous dynamic systems introduce discrete states in the form of switching logic that selects the control gains based on the current operating conditions. In sequential control applications, the traditional domain of programmable logic controllers (PLCs), continuous process dynamics can influence timing and synchronization of concurrent activities.

But problems of hybrid dynamics and hybrid control have been around for many years, so why the sudden flurry of activity in hybrid systems theory? From the theoretical side, the maturation of control theoretic research in discrete event systems (DES) has contributed significantly to the interest in hybrid systems. Before the early 1980's, control theorists typically did not study discrete-state systems. DES formalisms and theory are now standard in graduate control curricula, conferences and publications. Consequently, familiar tools are at hand to model and investigate the discrete aspects of hybrid systems.

Current tools for computer-aided control system design have provided an additional impetus and support for studying hybrid systems. Effective tools for modeling and simulating dynamic systems with mixed discrete- and continuous-state variables are now standard in most software packages. This makes it much easier to create and study interesting examples, a necessary empirical component of any serious research into the complex behaviors of hybrid dynamic systems.

Coincident to the advent of hybrid systems research among control theorist has been a growing interest in hybrid systems among computer scientists, particularly in the areas of formal methods and computer-aided verification. This reflects the need to extend tools for strictly discrete-state analysis to incorporate the important features of the underlying continuous-dynamics in digital circuits. Computer scientists are also interested in reactive systems, which include real-time control systems. This has stimulated a dialog and collaboration between control theorists and computer scientists that has made hybrid systems a particularly exciting and challenging area of research.

There are several objectives in this paper. First, after briefly reviewing in Sect. 2 the types of phenomena that give rise to hybrid dynamics and the nature of hybrid phenomena, a block diagram approach to modeling hybrid systems is presented in Sect. 3. Second, the concept of transition system semantics that has been used in the verification is extended in Sect. 4 to describe open-loop hybrid systems with discrete inputs and threshold event outputs. Third, a supervisory control problem for hybrid systems is formulated and a formal solution is presented using the transition system semantics in Sect. 5. Fourth, the problem of actually computing the solution to the supervisory control problem is addressed in Sect. 6 using the theory of bisimulations, simulations, and approximations. Finally, computer tools for simulation and analysis are reviewed briefly in Sect. 7, and directions for future research are identified in Sect. 8.

2 Elements of Hybrid System Models

Many physical phenomena are best modeled as hybrid systems. In this section we briefly describe some simple examples that motivate the formal models presented in the remainder of the paper.

Collisions. Mechanical collisions result in instantaneous changes is the velocities of objects. Such jumps in the value of the dynamic state are often modeled as impulsive inputs to the dynamic equations of motion. The interesting aspect of the modeling problem is that the times of the jumps in the state are determined by the state trajectory itself. A bouncing ball illustrates this situation. The horizontal position of the ball determines when the impulse occurs. When the ball hits the ground, the velocity of the ball changes direction instantaneously. Systems with discrete events that cause the continuous state to jump are hybrid systems, particularly when the continuous state trajectory is the cause of the discrete events.

Backlash. Mechanical systems that exhibit backlash operate in different regimes that can be modeled as discrete states. The dynamic equations switch depending on whether or not mechanical parts are touching. When there is contact, and the forces are in a direction to maintain the contact, the parts are constrained to move together. When the parts are not touching, or the forces are in a direction away from the contact, the degrees of freedom increase–the parts can move independently. Hence, the equations of motion must include a discrete state indicating whether there is contact with constrained motion, or no contact with unconstrained motion. Systems with discrete states defining regimes for the continuous state dynamics are hybrid systems. Note that in systems with backlash, a collision occurs when the parts come together, so there will be a jump in a state variable (velocity), in addition to the change in the equations of motion.

Electrical switches. In electrical circuits, switches can introduce jumps in the state of the circuit and changes in the dynamic equations for the circuit. Jumps in the state would occur, for example, if closing a switch creates a short across a capacitor. The discharge would not actually be instantaneous in a physical circuit since the wire in the short would have some resistance. But jumps in the state of a system typically represent the practical idealization of the physical situation where the transients are so much faster than the time scales of interest. A switch might be a discrete input signal to the continuous circuit dynamics. Another situation arises in circuits involving switching devices such as silicon-controlled rectifiers (SCRs). In these cases, the state variables of the circuit (i.e., voltages or currents) determine when the switching, and corresponding jump in the state, will occur.

Gain Scheduling Controllers. The most common source of hybrid dynamics cited in the literature is the application of discrete control logic to control continuous systems. It is often noted that logical operations, to implement such things as exception conditions and safety limits, can take more code than the continuous feedback control computations in a computer program for real-time control. Certain control algorithms are themselves hybrid: discrete logic switches between different continuous feedback laws, a strategy that is often called gain scheduling. The gains are typically selected according to the operating state of the system or the environment, making the value of the discrete state (the gain index) a function of the continuous state variables or continuous input variables.

The examples above illustrate several fundamental features of hybrid systems that need to be captured by a general modeling formalism. These features include:

- discrete and continuous state variables
- discrete and continuous input and output signals
- continuous dynamics that depend on discrete variables
- discrete transitions that depend on continuous variables
- discrete events that cause jumps in the continuous state
- discrete events that depend on the continuous state trajectory

It is also desirable to be able to compose hybrid system models from components. Although simple examples were used above as illustrations, hybrid phenomena actually occur in complex systems that must be modeled as interconnected components. The following section presents a set of input-output blocks that include the features listed above and facilitate creating hybrid system models from simpler subsystems.

3 Block Diagram Models of Hybrid Systems

For control systems engineers, block diagrams have long served as an effective representation of dynamic systems. Models of complex dynamic systems can be constructed with a few fundamental blocks interconnected with input and output signals. This section presents three types of signals and four basic blocks from which models of general hybrid systems can be constructed. Continuous time is assumed throughout this paper, with the independent time variable t being in R. Consequently, the terms *continuous* and *discrete* will always refer to the sets of values (ranges) for the continuous-time signals. Continuous ranges are assumed to be R^n for some $n = 0, 1, \ldots$; discrete ranges are of the form $D_1 \times \cdots \times D_m$, where D_i is a finite set for $i = 1, \ldots, m$. When the $D_i = \{0, 1\}$, the signals will be called *binary signals*.

A *block* is defined by its *input port(s)*, *output port(s)*, *state variables*, *initial states*, and the *input-state-output (ISO) relations*. Each input port and output port is defined by the *signal type* (discrete, continuous, binary, with the associated ranges). The ISO relation for a block defines the relationship between the input signals, internal state trajectories, and output signals through algebraic equations, inequalities, differential equations, and logical expressions.

Switched continuous systems (SCS). An SCS block has one input port for a discrete input signal u, a continuous state x with initial state x_0, and one continuous output port for the state x (i.e., the output is the state vector). The ISO relations are the *continuous state equation*

$$\dot{x}(t) = f(x(t), u(t)),$$

where f is assumed to be Lipschitz with respect to the continuous state, and the *jump equation*

$$x(t) = J(x(t^-), u(t))), \text{ if } u(t) \neq u(t^-).$$

The discrete input to an SCS block selects the continuous dynamics. Given a state x' and discrete input u, $\zeta_{(x',u)}(t)$ denotes the solution to the continuous state equation for $t \geq 0$ with $\zeta_{(x',u)}(0) = x'$. When the discrete input changes (at points where the discontinuity is right continuous), there may be a jump in the continuous state determined by the jump equation.

Finite state machines (FSM). An FSM block has two input ports: one for a discrete input signal d, and one for a (possibly vector-valued) binary signal v; a discrete state r with initial state r_0, and one discrete output port for a discrete signal u. The ISO relations are the *discrete state equation*

$$r(t) = F(r(t^-), d(t^-), v(t), v(t^-)),$$

and the *output equation*

$$u(t) = G(r(t)).$$

It is assumed the discrete state transition function F satisfies

$$F(r, \cdot, v, v) = r,$$

for all r and v. This assumption implies transitions in the discrete state occur only when there is a discontinuity in the binary input signal.

Region blocks. A region block has one input port for a continuous signal x and one output port for a scalar binary signal b. Region blocks are memoryless; they have no state variables. The input-output relation identifies whether the input x is in a region defined by a smooth function $h : R^n \rightarrow R^m$ according to

$$b(t) = \begin{cases} 1 \text{ if } h(x(t)) \leq 0 \\ 0 \text{ otherwise} \end{cases}$$

where $h(x(t)) \leq 0$ means each component of $h(x)$ is less than or equal to zero. A transition from 0 to 1 in the binary output signal indicates the state trajectory has left the region

$$\{x \in R^n \mid h_j(x) > 0 \text{ for some } j = 1, \ldots, m\}.$$

The set of *boundary states* for a region block is defined as

$$\{x \in R^n \mid h(x) \leq 0 \text{ and } h_j(x) = 0 \text{ for some } j = 1, \ldots, m\}.$$

Stacking blocks. A stacking block is a memoryless utility block that can have multiple inputs of a single type of signal (continuous, discrete, binary). It generates a single vector output with subvectors equal to the input vectors in the order in which they are defined as inputs to the block. (See the examples below.)

Figure 1 illustrates a hybrid system defined using the four types of blocks defined above. This figure was generated by a Simulink-based interface for a hybrid system tool in which the region blocks are defined by linear inequalities (see Sect. 7). Since all blocks defined above are deterministic, the model can be simulated to obtain the trajectories of the states and input and output signals in the system. Given a hybrid system H defined by a block diagram, the continuous state space for H, denote X_H, is the product of the state spaces for all the SCS blocks, and the set of boundary states for H, denoted X_H^{∂}, is the union of the boundary states for the region blocks defined in the full state space X_H. The set of possible initial states for the SCS blocks is denoted by $X_0 \subset X_H$.

In this paper we consider a class of supervisory control problems for hybrid systems with the structure shown in Fig. 2. In this system, an SCS is being controlled by a *supervisor* that selects the discrete input u based on observations of the transitions in the signals from the region blocks, concatenated into a single vector v by the stacking block. We refer to the *open-loop hybrid system* with discrete input signal u and the binary vector output signal v in Fig. 2 as H_O in the sequel.

4 Transition System Semantics for Hybrid Systems

Transition systems have been an effective formalism to define semantics for hybrid systems modeled by hybrid automata [14, 15, 12]. In this section we introduce transition systems with inputs and outputs to define the semantics for the open-loop hybrid system H_O in Fig. 2. We are interested in the sequential input-output behavior of H_O, rather than the details of the continuous trajectories between threshold events, since this is all that is observed and affected by the supervisor.

Definition 1. (Transition systems) A *transition system* T is a six-tuple $T = (Q, U, V, \rightarrow, \lambda, Q_0)$, where Q is the (possibly infinite) set of *states*, U and V are the finite sets of *inputs* and *outputs*, respectively, $\rightarrow \subseteq Q \cup Q_0 \times U \times Q$ is the set of *transitions*, $\lambda : Q \rightarrow V$ is the output map, and Q_0 is the (possibly infinite) set of *initial states* with $Q_0 \cap Q = \emptyset$.

Given $q \in Q \cup Q_0$, $q' \in Q$, and $u \in U$, the notation $q \xrightarrow{u} q'$ indicates that $(q, u, q') \in \rightarrow$. Note that we do not require the transition relation $\rightarrow$ to be complete; that is, the transition relation is not necessarily defined at a given state for all inputs. Note also that the set of initial states is an independent set of states to which the system cannot return after the initial state transition.

The following definitions are useful in the subsequent development.

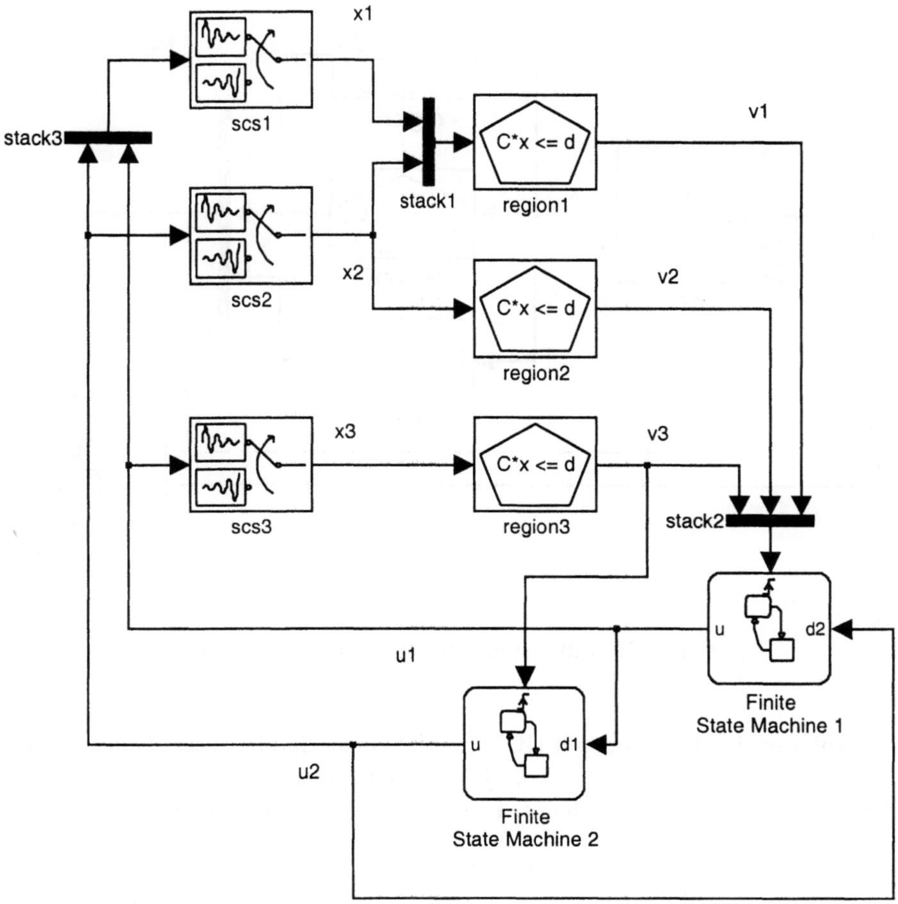

Fig. 1. A block diagram of a hybrid system.

Definition 2. (Pre/Postconditions) Given a transition system $T = (Q, U, V, \rightarrow, \lambda, Q_0)$, $u \in U$ and $P \subseteq Q \cup Q_0$, the *precondition* of P with respect to the input u, denoted $Pre_u(P)$, is defined by

$$Pre_u(P) = \{q \in Q \cup Q_0 \mid \exists p \in P, q \xrightarrow{u} p\}.$$

The *postcondition* of P with respect to u, denoted $Post_u(P)$, is defined by

$$Post_u(P) = \{q \in Q \mid \exists p \in P, p \xrightarrow{u} q\}.$$

Definition 3. (Deterministic transition system) A transition system $T = (Q, U, V, \rightarrow, \lambda, Q_0)$ is *deterministic* if for any $q \in Q \cup Q_0$, $q_1, q_2 \in Q$ and $u \in U$, $q \xrightarrow{u} q_1$ and $q \xrightarrow{u} q_2$ implies $q_1 = q_2$.

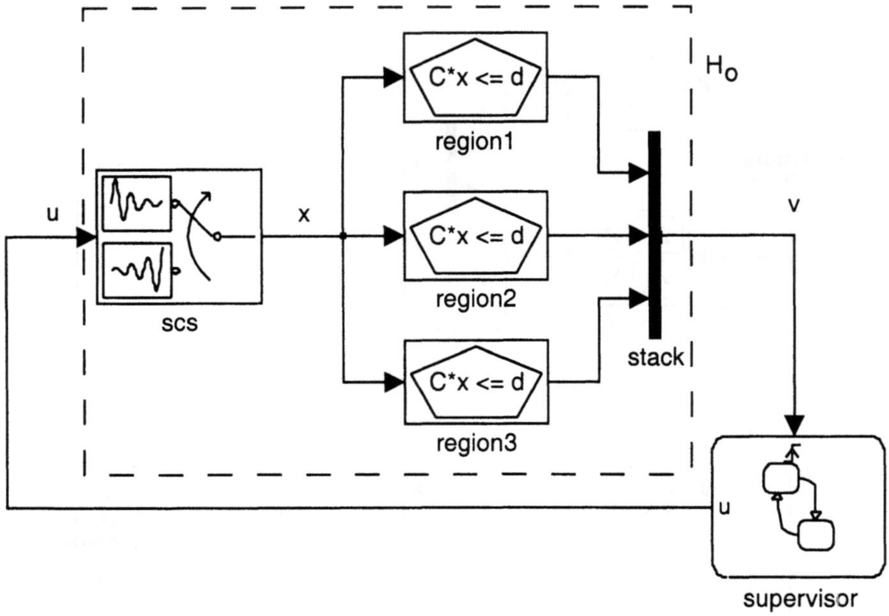

Fig. 2. Hybrid system with supervisory controller.

Definition 4. (Input-output (IO) strings and reachable states) Given a transition system $T = (Q, U, V, \rightarrow, \lambda, Q_0)$, the sets of states that can be reached for IO strings $\sigma \in (UV)^*$, denoted $r_T(\sigma) \subset Q$, are defined recursively by:

i. $r_T(\epsilon) = Q_0$, where ϵ is the empty string;
ii. $r_T(\sigma uv) = \bigcup_{q \in r_T(\sigma)} \{q' \in Q \mid q \overset{u}{\rightarrow} q' \text{ and } \lambda(q) = v\}$, for $\sigma \in (UV)^*$, $u \in U$, $v \in V$.

The set of feasible IO strings for T, denoted $\Sigma_T \subset (UV)^*$, is defined as

$$\Sigma_T = \{\sigma \in (UV)^* \mid r_T(\sigma) \neq \emptyset\}.$$

Henzinger [15] described the complete untimed behavior of a hybrid automaton using the *time-abstract transition system*. Here we are interested in the behaviors of the hybrid system H_O only at the times when discrete transitions occur. To abstract away the continuous dynamics and obtain the projection of the hybrid system behaviors onto the instants of discrete transitions, we define the *discrete-trace transition system* for H_O as follows.

Definition 5. (Discrete-trace transition system) Given an open-loop hybrid system H_O, with continuous state space X_H, initial state set $X_0 \subset X_H$, m region blocks with region functions $h_1, \ldots, h_m$, and boundary states X_H^∂, its *discrete-trace transition system* is given by $T_H = \{Q_H, U, V, \rightarrow_H, \lambda_H, X_0\}$ where $Q_H = X_H^\partial$, and the transition relation $\rightarrow_H$ is defined as follows:

i. if $x \in X_0$ and $x' \in Q_H$, then $x \xrightarrow{u}_H x'$ iff there exists $t_1 > 0$ such that $\zeta_{(x,u)}(t_1) = x'$ and for each $i = 1, \ldots, m$ there exists a $t_i' \in [0, t_1]$,
   (a) $h_i(\zeta_{(x,u)}(t)) > 0 \ \forall t \in (t_i', t_1)$;
   (b) $h_i(\zeta_{(x,u)}(t)) \leq 0 \ \forall t \in (0, t_i']$; and
   (c) for at least one $i' \in \{1, \ldots, m\}$, $h_{i'}(\zeta_{(x,u)}(t_1)) = 1$ and $t_{i'}' < t_1$.
ii. if $\hat{x}, x' \in Q_H$, then $\hat{x} \xrightarrow{u}_H x'$ iff $x = J(\hat{x}, u)$ and the conditions in (i) are satisfied for x, x'.

The discrete-trace transition system gives all possible input-output behaviors for an open-loop hybrid system H_O under all possible values for the discrete input signal u. In the following section we consider the problem of synthesizing a supervisor to control the discrete-trace transition system, and hence H_O, to guarantee a set of forbidden states (a subset of states in X_H^{∂}) is never reached.

5 Supervisory Control

We consider supervisors that observe the IO string for a transition system and select a *set* of admissible next inputs following each output. The particular input that is applied in the implementation could be selected randomly or could be selected to achieve specific objectives beyond the specifications for synthesizing the supervisor. We have the following formal definition for a supervisor.

Definition 6. (Transition system supervisors) A ***supervisor*** S for an transition system $T = (Q, U, V, \rightarrow, \lambda, Q_0)$ is a mapping $S : (UV)^* \rightarrow 2^U$.

A supervisor restricts the set of inputs that can be applied to an transition system following each output. In general, this set of available inputs can depend on the entire input-output history up to that point in time. The initial input value is selected from $S(\epsilon)$. We denote a transition system T under the control of a supervisor S by S/T. The sets of valid IO strings and reachable states for S/T are defined as follows.

Definition 7. (IO sequences and Reachable States for S/T) Given a transition system $T = (Q, U, V, \rightarrow, \lambda, Q_0)$, and a supervisor $S : (UV)^* \rightarrow 2^U$, the sets of states that can be reached for IO strings $\sigma \in (UV)^*$, denoted $r_{S/T}(\sigma) \subset Q$, are defined recursively by:

i. $r_{S/T}(\epsilon) = Q_0$;
ii. $r_{S/T}(\sigma u v) = \bigcup_{q \in r_{S/T}(\sigma)} \{q' \in Q \mid q \xrightarrow{u} q', u \in S(\sigma), \text{ and } \lambda(q) = v\}$, for $\sigma \in (UV)^*$, $u \in U$, and $v \in V$;

The set of feasible IO strings for S/T, denoted $\Sigma_{S/T} \subset (UV)^*$, is defined as

$$\Sigma_{S/T} = \{\sigma \in (UV)^* \mid r_{S/T}(\sigma) \neq \emptyset\}.$$

In words, the behaviors of a supervised transition system can be described as follows. At the initial state, an input is applied to the transition system from the valid set of inputs defined by the supervisor and, if a transition occurs, an output is observed. This IO pair corresponds to the input-output in the transition system transition relation. When the output occurs, another input is applied to the transition system from the set specified by the supervisor and again, if a transition occurs, another output is observed, giving the next input-output pair, and so on.

Cury et al. considered the problem of constructing supervisors for hybrid systems to satisfy specifications for the sequential behavior of the closed-loop system [7, 6]. Here we consider the problem of supervisor synthesis to avoid a set of forbidden states. First, we denote the set of all reachable states for a supervised transition system, S/T, by

$$R_{S/T} = \bigcup_{\sigma \in \Sigma_{S/T}} r_{S/T}(\sigma).$$

We also assume the set of inputs that can be applied to a transition system following a given output can be specified for the supervisor by an input constraint function defined as follows.

Definition 8. (Input constraint function) Given an transition system $T = (Q, U, V, \rightarrow, \lambda, Q_0)$, the supervisor *input constraint function* is a mapping $\mathcal{U} : V \cup \{\epsilon\} \rightarrow 2^U$.

Definition 9. (Forbidden State Problem–FSP) Given a transition system $T = (Q, U, V, \rightarrow, \lambda, Q_0)$, an input constraint function $\mathcal{U}$, and a set of *forbidden states* $Q_F \subset Q$, find a supervisor S such that

i. $S(\epsilon) \subseteq \mathcal{U}(\epsilon)$;
ii. if $\sigma v \in \Sigma_{S/T}$ for some $v \in V$, then $S(\sigma v) \subseteq \mathcal{U}(v)$; and
iii. $R_{S/T} \cap Q_F = \emptyset$.

Furthermore, if S' is any other supervisor satisfying (i)–(ii), then $R_{S'/T} \subseteq R_{S/T}$.

The FSP differs from the standard forbidden state problems in the DES literature in that: (i) the supervisor observes the system input and output signals, rather than the state; (ii) the transition system can be nondeterministic; and (iii) the input that can be applied by the supervisor can be restricted according to the most recent output. Nevertheless, with minor extensions of well known results from the DES literature, it can be shown a solution to the FSP exists. Allowing for nondeterminism in the supervisor guarantees the existence of a maximally permissive supervisor, as required by the last sentence of the FSP definition.

The following is a procedure for computing the solution to FSP. First, we create a *deterministic* transition system using the standard construction from automata theory. We call this system the *transition system state observer*, for reasons that will be clear shortly.

Definition 10. (Transition system state observer) Given a nondeterministic transition system $T = (Q, U, V, \rightarrow, \lambda, Q_0)$, the *transition system state observer* for T is defined as $T^{obs} = (Q^{obs}, U \cdot V, V, \rightarrow_{obs}, \lambda^{obs}, Q_0^{obs})$ where

i. $Q^{obs} = \{P \subseteq Q \mid P = r_T(\sigma) \text{ for some } \sigma \in \Sigma_T, \sigma \neq \epsilon\}$;
ii. $Q_0^{obs} = \{Q_0\}$;
iii. if $\sigma, \sigma uv \in \Sigma_T$, $P = r_T(\sigma)$ and $P' = r_T(\sigma uv)$, then $P \overset{uv}{\rightarrow}_{obs} P'$;
iv. $\lambda^{obs}(P)$ for $P \in Q^{obs}$ is the unique $v \in V$ equal to $\lambda(q)$ for all $q \in P$.

It is evident by construction that T^{obs} is a deterministic transition system and the state of transition system is uniquely determined by an IO sequence in Σ_T. The FSP for transition system T can now be translated into a forbidden state problem for T^{obs} as follows. The observed IO string can be applied to T^{obs} so that the state of T^{obs} is known. This state is the set of possible states for T–and it is the best estimate the supervisor has for the state of T. Given a set of forbidden states Q_F for T, the forbidden states for T^{obs} are given by

$$Q_F^{obs} = \{P \in Q^{obs} \mid P \cap Q_F \neq \emptyset\}.$$

If the supervisor can only observe the IO sequence, it must apply inputs to T so that the state of T^{obs} is not in Q_F^{obs}. To determine the maximal set of inputs that can be applied at any state of T^{obs}, the following definitions are applied and fixed point computations are performed.

Given $P \in Q^{obs} \cup Q_0^{obs}$ and $\mathcal{F} \subseteq Q^{obs}$, define

$$U(P, \mathcal{F}) = \{u \in \mathcal{U}(\lambda^{obs}(P)) \mid Post_u(P) \cap \mathcal{F} = \emptyset\},$$

and

$$G(\mathcal{F}) = \mathcal{F} \cup \{P \in Q^{obs} \mid U(P, \mathcal{F}) = \emptyset\}.$$

Suppose a fixed point $\mathcal{F}$ can be computed for G, i.e., $\mathcal{F} = G(\mathcal{F})$, starting with $\mathcal{F}_0 = Q_F^{obs}$ and applying the iteration

$$\mathcal{F}_{k+1} = G(\mathcal{F}_k).$$

It is evident that for any state inside $\mathcal{F}$ it will be impossible for a supervisor to guarantee the state of T^{obs} will not eventually be in Q_F^{obs}. This implies in turn that, so far as the supervisor can know, the state of T can reach Q_F. On the other hand, if $Q_0^{obs} \notin \mathcal{F}$, the supervisor can guarantee the state of T will not reach Q_F by simply choosing u from the set $U(P, \mathcal{F})$ when the state of T^{obs} is P.

To summarize, Fig. 3 shows the structure for the supervisor that solves the FSP and the construction and operation of the supervisor is given by the following steps:

1. Create the transition system state observer;
2. Find the set of forbidden states for the observer, Q_F^{obs};
3. Perform the iteration described above to find a fixed point $\mathcal{F}$ of the set-valued operator, G;

4. If $Q_0^{obs} \cap \mathcal{F} \neq \emptyset$, there is no solution to the FSP–STOP;
5. If $Q_0^{obs} \cap \mathcal{F} = \emptyset$, select the initial input u_0 from the set

$$S(\epsilon) = U(Q_0, \mathcal{F}).$$

6. Set $k = 0$ and $\sigma = \epsilon$;
7. When output v_k is observed, update $\sigma := \sigma u_k v_k$, and increment k;
8. Compute the observer state $P_k = r_T(\sigma)$ and select the next input u_k from

$$S(\sigma) = U(P_k, \mathcal{F}).$$

9. Return to Step 7.

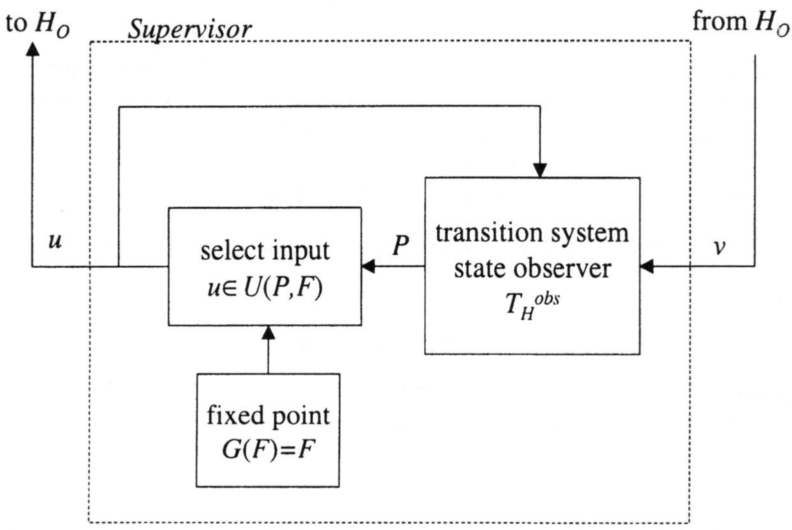

Fig. 3. Structure of the supervisor that solves the FSP.

6 Supervisor Synthesis Using Approximations

To this point we have a formal solution to the supervisory control problem for hybrid systems. It is not a computable solution, however, since T_H is an infinite-state transition system. This makes all of the formal computations in the previous section impossible. In verification, this problem has been addressed using the concept of bisimulations, which may provide a finite-state transition system that is equivalent to an infinite-state transition system. Unfortunately, it may be that there is not a finite bisimulation for T_H. In this case, approximations are needed, which we develop in this section. We begin with the basic theory of simulations and bisimulations for transition systems, extended to include inputs and outputs.

Definition 11. (Simulation) Given transition systems $T_i = (Q_i, U, V, \rightarrow_i, \lambda_i, Q_{0i})$, $i = 1, 2$, a *simulation relation* of T_1 by T_2 is a binary relation $\preceq \subseteq Q_1 \times Q_2$ such that:

i. if $q_1 \preceq q_2$ and $q_1 \xrightarrow{u}_1 q_1'$ for some $u \in U$, then there exists q_2' such that $q_2 \xrightarrow{u}_2 q_2'$, and $q_1' \preceq q_2'$;
ii. if $q_1 \in Q_{01}$ and $q_1 \xrightarrow{u}_1 q_1'$ for some $u \in U$, there exists $q_2 \in Q_{02}$ and $q_2' \in Q_{02}$ such that $q_2 \xrightarrow{u}_2 q_2'$ and $q_1' \preceq q_2'$.
iii. if $q_1 \preceq q_2$, then $\lambda(q_1) = \lambda(q_2)$.

We say T_2 simulates T_1, denoted $T_1 \preceq T_2$, if there exists a simulation relation of T_1 by T_2. T_2 is also called a simulation of T_1. In words, T_2 simulates T_1 when for any trajectory in T_1 there is a trajectory in T_2 with the same input-output sequence and the states along the trajectory in T_2 are related to the states in T_1 by the simulation relation.

Definition 12. (Bisimulation) Given transition systems $T_i = (Q_i, U, V, \rightarrow_i, \lambda_i, Q_{0i})$, $i = 1, 2$, a *bisimulation relation* between T_1 and T_2 is a binary relation $\equiv \subseteq Q_1 \times Q_2$ such that $\equiv$ is a simulation relation of T_1 by T_2 and $\equiv^{-1} \subseteq Q_2 \times Q_1$ is a simulation relation of T_2 by T_1.

We say the transition systems T_1 and T_2 bisimulate each other, denoted $T_1 \equiv T_2$, if there exists a bisimulation relation between T_1 and T_2. An approach to finding a bisimulation of a transition system uses *quotient transition systems* generated by partitions of the transition system state space satisfying the following property.

Definition 13. (Output consistent partition–OCP) Given a transition system $T = (Q, U, V, \rightarrow, \lambda, Q_0)$ a partition $\mathcal{P}$ of $Q \cup Q_0$ is an *output consistent partition* if for any $P \in \mathcal{P}$ either $P \subseteq Q$ or $P \subseteq Q_0$ and λ is constant on any $P \subset Q$.

Definition 14. (Quotient Transition System) [14, 12, 13] Given a transition system $T = (Q, U, V, \rightarrow, \lambda, Q_0)$ and an OCP $\mathcal{P}$ of Q, the *quotient transition system* of T is defined as $T/\mathcal{P} = (\mathcal{P}, U, V, \rightarrow_{\mathcal{P}}, \lambda, Q_0/\mathcal{P})$, where for all $P, P' \in \mathcal{P}$, and $u \in U$, $P \xrightarrow{u}_{\mathcal{P}} P'$ iff there exist $q \in P$ and $q' \in P'$ such that $q \xrightarrow{u} q'$, or equivalently $Post_u(P) \cap P' \neq \emptyset$, and $Q_0/\mathcal{P} = \{P \in \mathcal{P} \mid P \subseteq Q_0\}$.

It is easy to see that the relation $\preceq = \{(q, P) \in Q \times \mathcal{P} \mid q \in P\}$ is a simulation relation of T by $T/\mathcal{P}$. Therefore, $T \preceq T/\mathcal{P}$. The quotient system

$T/\mathcal{P}$ is a bisimulation of T with the relation $\equiv = \{(q, P) \in Q \times \mathcal{P} \mid q \in P\}$ if and only if the partition $\mathcal{P}$ satisfies

$$\forall P, P' \in \mathcal{P} \text{ and } u \in U, \text{ either } P \cap Pre_u(P') = \emptyset \text{ or } P \cap Pre_u(P') = P. \quad (1)$$

In words, (1) states the system behaves uniformly starting from all states in each $P \in \mathcal{P}$; i.e., given another $P' \in \mathcal{P}$, either all states or no state in P can reach some state in P' in one transition. Condition (1) leads to the following general procedure for computing a bisimulation of a transition system using quotient systems.

Bisimulation Procedure (BP) [14, 12, 13]:
set $\mathcal{P} = \mathcal{P}_0$
% check termination condition
while $\exists P, P' \in \mathcal{P}, u \in U$ such that $\emptyset \neq P \cap Pre_u(P') \neq P$ {
  % partition refinement
  split P into $P_1 = P \cap Pre_u(P')$ and $P_2 = P \setminus Pre_u(P')$
  set $\mathcal{P} = (\mathcal{P} \setminus \{P\}) \cup \{P_1, P_2\}$
}

Note that (1) is precisely the termination condition of BP. This guarantees that the quotient system is a bisimulation of T when BP terminates. In each iteration of BP, the partition refinement scheme uses the information obtained from $Pre(P')$ to split P into the part that can reach P' and the part that cannot. Also note that the $Pre_u(\cdot)$ operation in BP is with respect to the transition relation $\rightarrow$ of T, not the transition relation $\rightarrow_{\mathcal{P}}$ of the quotient system $T/\mathcal{P}$.

If BP is applied to the transition system T_H for a hybrid system H_O and it terminates with a bisimulation $T_H/\mathcal{P}$, one can then use $T_H/\mathcal{P}$ to construct a supervisor to solve the FSP following the procedure described in the previous section. This procedure is illustrated in Fig. 4.a. Given a set of forbidden states for T_H, an initial OCP is selected satisfying the following additional condition for the given set of forbidden states in X_H^{∂}.

Definition 15. (F-consistent partition) Given a set of states $F \subset X_H^{\partial}$ for a discrete-trace transition system T_H, a partition P of X_H^{∂} is said to be F-consistent if for all $P \in \mathcal{P}$ $P \cap F = P$ or $\emptyset$.

We assume the set of forbidden states Q_F for T_H admits a Q_F-consistent partition that is finite. For a Q_F-consistent partition, the FSP for T_H can be translated to the quotient system $T_P/\mathcal{P}$ by defining the forbidden states of $T_P/\mathcal{P}$ to be those elements of the partition that are contained in the original set of forbidden states. Starting with the initial partition, BP is applied, and the supervisor synthesis algorithm can then be applied to the resulting quotient system, as indicated in Fig. 4.a.

The problem remains, however, that BP may never terminate, since it is possible that no finite-state bisimulation exists for T_H. If BP never converges, the procedure in Fig. 4.a never reaches the supervisor synthesis step. To circumvent this problem, we propose the alternative procedure illustrated in Fig. 4.b. In this approach, the supervisor synthesis procedure is applied to the intermediate quotient systems obtained on each iteration of BP, rather than waiting until BP terminates. If a solution exists to the FSP for a quotient system that simulates T_H, but is not a bisimulation of T_H, it follows that the resulting supervisor will guarantee the hybrid system H_O will not reach a forbidden state. If no solution exists to the FSP for the quotient system, this does not imply that there is no solution to the original FSP, unless the quotient system is a bisimulation. So, as shown in Fig. 4.b, when there is no solution to the FSP, one can check to see if the current quotient system is a bisimulation, in which case, the fact there is no solution to the FSP for the quotient system implies there is no solution to the FSP for T_H. If, however, the current quotient system is not a bisimulation, the partition can be refined, as in BP, and the procedure can be repeated.

Even if there is a finite bisimulation for T_H, the procedure in Fig. 4.b may find a solution to the FSP much more quickly than the procedure in Fig. 4.a because it does not wait for BP to converge before checking to see if a satisfactory supervisor can be constructed. It may be the case, however, that a supervisor for a quotient system that is not a bisimulation is too conservative due to the coarseness of the partition, so the additional evaluation loop is added to the Fig. 4.b. For example, the supervisor may avoid the forbidden states by choosing an initial input that does not allow any subsequent output. If this is equivalent to shutting the system down, one may want to refine the partition so that a less conservative supervisor can be found. Note that if the quotient system is a bisimulation and a solution exists to the FSP, then the least conservative supervisor possible has been found.

The implementation of the procedure in Fig. 4.b requires the computation of the precondition and postcondition sets. It may not be possible to compute these sets exactly, or the computation method may be too costly to implement. The lack of an effective way to compute these sets naturally leads one to consider replacing them with approximations in the procedure in Fig. 4.b. We now consider the implications of using such approximations.

We call a method to approximate the precondition and the postcondition sets for a transition system T a *reachability approximation method.* We denote the u-precondition and the u-postcondition sets of P computed using a reachability approximation method M by $Pre_u^M(P)$ and $Post_u^M(P)$, respectively.

Definition 16. A reachability approximation method M for a transition system $T = (Q, U, V, \rightarrow, \lambda, Q_0)$ is called *conservative* if for all $P \subseteq Q$ and $u \in U, Pre_u(P) \subseteq Pre_u^M(P)$ and $Post_u(P) \subseteq Post_u^M(P)$.

Without exact precondition and postcondition sets, one can only approximate the quotient system $T/\mathcal{P}$ for a partition $\mathcal{P}$ of Q using a reachability approximation method.

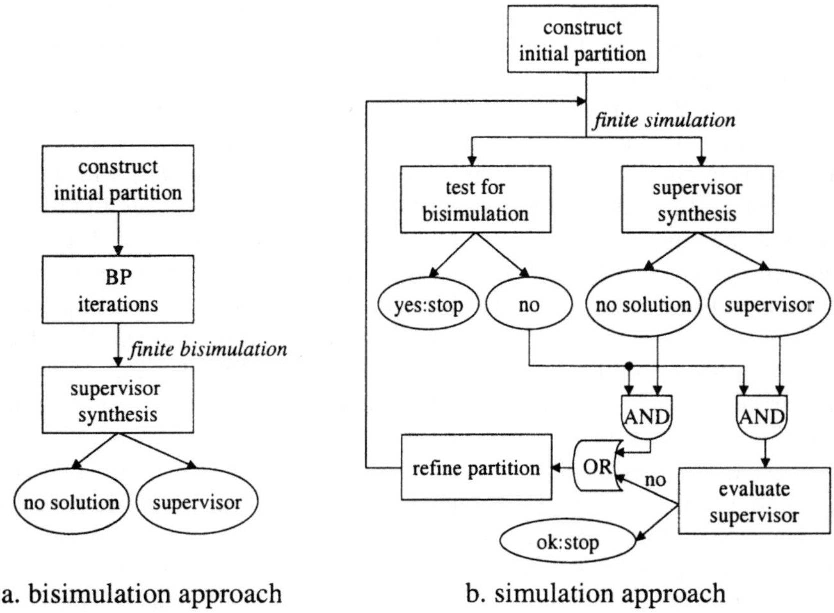

Fig. 4. Two approaches to solving FSP: (a) compute a bisimulation and then apply the supervisor synthesis procedure; (b) apply the supervisor synthesis procedure at each refinement iteration.

Definition 17. Given a reachability approximation method M for a transition system $T = (Q, U, V, \rightarrow, \lambda, Q_0)$, the *$M$-quotient system* for T given a partition $\mathcal{P}$ of Q is defined as

$$T^M/\mathcal{P} = (\mathcal{P}, U, V, \Sigma, \rightarrow^M_{\mathcal{P}}, \lambda, Q_0/\mathcal{P})$$

where $\rightarrow^M_{\mathcal{P}}$ is defined as follows. For all $P, P' \in \mathcal{P}$, $P \overset{u}{\rightarrow}^M_{\mathcal{P}} P'$ iff $Post^M_u(P) \cap P' \neq \emptyset$.

If M is conservative, it follows directly from Defn. 17 that $T/\mathcal{P} \preceq T^M/\mathcal{P}$ with the identity relation on $\mathcal{P} \times \mathcal{P}$.

Since $T^M/\mathcal{P}$ is a conservative approximation of $T/\mathcal{P}$, we can use $T^M/\mathcal{P}$ in the procedure in Fig. 4.b. However, there are some issues that arise from introducing conservative approximation into the verification procedure. First, we may found no solution for FSP for the conservative quotient system when in fact there is a solution for the exact quotient system. Second, if we can solve FSP, we may obtain a supervisor that is more conservative than the one obtained using the exact quotient system. Third, we can no longer use condition (1) to test for bisimulation, i.e. the condition

$$\forall P, P' \in \mathcal{P}, \text{ and } u \in U, \text{ either } P \cap Pre_u^M(P') = \emptyset \text{ or } P \cap Pre_u^M(P') = P \quad (2)$$

does not imply that the M-quotient system $T^M/\mathcal{P}$ is a bisimulation of T. It is worth noting, however, that there are some cases where a test condition for bisimulation can be found by exploiting certain properties of the reachability approximation method.

For the computational cost issue, we observe that any refinement method can be used to refine the quotient system. Different refinement methods merely result in different rates of convergence to a bisimulation. Ideally, we would like to use the refinement method in BP to refine the quotient system since the method always give a strict progress towards bisimulation. That is, if $T/\mathcal{P}_{N+1}$ is obtained by refining $T/\mathcal{P}_N$ using the method in BP, it is guaranteed that $T/\mathcal{P}_{N+1}$ and $T/\mathcal{P}_N$ are not bisimulations of each other, which would mean no progress has been made toward obtaining a tighter simulation of T. However, this method may be too complex and costly to implement as it requires the computation of the precondition sets and the implementation of set subtraction which is complicated for arbitrary sets. Thus, one may prefer using a simpler refinement method in the verification procedure.

7 Computer Tools

Several computer tools have been developed and augmented to the model, simulate, and, in some cases, analyze of certain classes of hybrid systems. Simulation tools include the object-oriented packages OmSim [2] and Shift [9]. The Stateflow toolbox for Simulink from Mathworks provides a block diagram environment with features similar to the general blocks defined in Sect. 3. Verification tools reflect the two basic approaches to verification, namely, model checking (e.g. HyTech [1]), and theorem proving (e.g. STeP [11]). In both cases, given the system description, the objective is to determine if the model satisfies a particular property. Several analysis tools have been developed for timed systems (e.g., KRONOS [8], where the hybrid dynamics are pure integrators.

The principal difficulty in verifying properties of hybrid systems is computing the mappings of sets in the continuous state space. In hybrid system simulators, a single trajectory is computed easily using standard numerical integration. Mapping a set of initial states to a set of boundary states to generate threshold events requires the computation of the entire flow pipe for the continuous dynamics [20]. For so-called linear hybrid automata, the exact flow pipe computation can be carried out using convex polyhedra [1]. Approximations must be used for more general continuous dynamics.

Various procedures have been proposed to approximate the continuous dynamic mappings. One approach is to approximate the continuous state equations with a class of systems for which exact computations can be made. In [16], linear hybrid automata are used to approximate nonlinear dynamic equations so that HyTech can be used for system verification. In [19], timed automata are used to approximate continuous dynamic systems so that KRONOS can be used for verification.

The other approach to approximate the continuous system dynamics is to discretize the continuous system so that tools for purely discrete (finite state) systems can be applied for analysis or synthesis. The most straight forward approach is to create cells in the continuous state space, assign a discrete state to each cell, and create transitions between the cells according to the reachability between cells inferred from the state equation. This approach was proposed at least as early as 1975 [17] and has been analyzed recently by Puri and Variaya [18]. The general shortcoming with the cell approach is that the reachability approximation becomes extremely conservative very quickly with the distance from the set of initial states.

Recently, we proposed a new approach to approximating flow pipes [4, 5] using sequences of convex polytopes. In this approach, a set of direction vectors and a time increment is selected. The smallest convex polytope with faces normal to the direction vectors that contains all flows from the given initial set is then computed using numerical optimization. The advantage of this approach is that the approximation error does not grow with the distance from the initial set. The method is very computationally intensive, however, and the optimization problems are not convex in general, making it difficult to guarantee a global optimum is achieved.

The flow pipe approximation has been incorporated into a verification tool supporting block diagrams of the type described in Sect. 3 using Matlab/Simulink/Stateflow in a custom user interface. This tool generates approximating automata that are the simulations described in Sect. 6. The user can verify the properties of the approximating automata from which properties of the hybrid system can be inferred. The block diagram description language is also being integrated with the tools described above from University of Dortmund, giving access to the timed analysis of KRONOS.

8 Future Directions

This paper has described a block diagram approach to the modeling of hybrid systems, and the use of transition system semantics to formulate and solve a class of supervisory control problems. Computer support for block diagram modeling is available for building simulation models, and more recently, a basic tool has been developed for verification (model checking) of hybrid system block diagrams using autonomous transition systems that simulate the discrete-trace transition system for the hybrid system. Extension of this tool to generate input-output transition systems would allow for supervisor synthesis using tools from DES.

There are several issues to be explored to make this technology useful for practical problems. When approximations are used to solve verification or synthesis problems as proposed in this paper, it is essential that the approximations be conservative. It is not possible to do this using floating point arithmetic without taking measures to assure that the numerical errors are always in the right direction to make the models conservative. Some research is underway on developing such numerical techniques for hybrid systems problems, but a general approach has not been developed to date.

Another issue to be explored is the use of inner approximations to guide the refinement of models, and to solve problems with existential specifications. With outer approximations it is only possible to solve problems with universal specifications (such as the forbidden state problem considered in this paper). To guarantee a trajectory exists with a particular property using approximations, (e.g., there is always some trajectory from the current state that will lead to a safe shutdown state), satisfaction for an inner approximation would guarantee the property is true for the actual system. Some use of inner approximations for timed automata are developed in [10].

Control synthesis involving the continuous as well as the discrete (supervisory) aspects of hybrid systems has received little attention to date, but it is clear that synthesis of practical controllers always includes the continuous variables. In particular, the values of thresholds are often design parameters. One approach to determining thresholds for a class of hybrid control problems is proposed in [3]. In this case, the controller sets thresholds for certain parameters and fixed-point iterations are performed in the continuous space using convex polytopes to find a controller satisfying the specifications. Further research is needed in this direction to develop a comprehensive theory of control synthesis for hybrid systems.

In summary, there are a number of important open problems in hybrid systems of both practical and theoretical significance. Given the complexity of hybrid systems, and the difficulty in solving all but the most trivial problems exactly, approximation techniques could well be the most important direction for future research if real engineering problems are to be solved. It is also crucial that computer tools be developed in conjunction with emerging theory so that the experience with real problems can grow with the field.

Acknowledgments

The first author acknowledges support from the Ford Motor Company and the U.S. National Science Foundation; both authors acknowledge support from the U.S. Defense Advance Research Projects Agency (DARPA).

References

1. R. Alur, T.A. Henzinger, and P.-H. Ho. Automatic symbolic verification of embedded systems. *IEEE Trans. on Software Engineering*, 22(3):181–201, Mar 1996.
2. M. Andersson. *OmSim and Omola Tutorial and User's Manual.* Department of Automatic Control, Lund Institute of Technology, Lund, Sweden, March 1995.

3. E. Asarin, O. Maler, A. Pnueli, and J. Sifakis. Controller synthesis for timed automata. In *IFAC Symposium on System Structure and Control*, pages 469–474, 1998.
4. A. Chutinan and B.H. Krogh. Computing polyhedral approximations to flow pipes for dynamic systems. In *IEEE Conf. On Decision and Control*, Dec. 1998.
5. A. Chutinan and B.H. Krogh. Verification of polyhedral-invariant hybrid automata using polygonal flow pipe approximations. In J. Van Schuppen, editor, *Hybrid Systems: Computation and Control*, Lecture Notes in Computer Science. Springer, March 1999.
6. J.E.R. Cury and B.H. Krogh. Supervisory control of hybrid systems using approximating automata. *Journal of the Society of Instrument and Control Engineers (SICE)*, March 1999.
7. J.E.R. Cury, B.H. Krogh, and T. Niinomi. Synthesis of supervisory controllers for hybrid systems based on approximating automata. *IEEE Trans. on Automatic Control*, 43(4):564–569, April 1998. a preliminary version of this paper also appeared in *Proc. 34th IEEE CDC, pp. 1461-1466.*
8. C. Daws, A. Olivero, S. Tripakis, and S. Yovine. The tool kronos. In *Hybrid Systems III, Verification and Control.* Springer Verlag, 1996.
9. A.R. Deshpande, A. Gollu, and L. Semenzato. *Shift Reference Manual.* University of California and Berkeley, 1996. PATH Report, PRCS projcet SHIC version 2.17.
10. D.L. Dill and H. Wong-Toi. Verification of real-time systems by successive over and under approximation. In *Computer-Aided Verification, 7th International Conf.*, pages 409–422, Liege, Belgium, July 1995.
11. Z. Manna et al. STeP: The Standford temporal prover (educational release), user's manual. Technical report, Computer Science Department, Standford University, 1995.
12. S. Sastry G. Lafferriere, G. J. Pappas. Hybrid systems with finite bisimulations. Technical Report UCB/ERL M98/15, University of California at Berkeley, April 1998.
13. S. Yovine G. Lafferriere, G. J. Pappas. Decidable hybrid systems. Technical Report UCB/ERL M98/39, University of California at Berkeley, June 1998.
14. T.A. Henzinger. Hybrid automata with finite bimulations. In Z. Fülöp and F. Gécseg, editors, *ICALP 95: Automata, Languages, and Programming*, pages 324–335. Springer-Verlag, 1995.
15. T.A. Henzinger. The theory of hybrid automata. In *Proceedings of the 11th Annual Symposium on Logic in Computer Science*, pages 278–292. IEEE Computer Society Press, 1996. Invited tutorial.
16. T.A. Henzinger, P.-H. Ho, and H. Wong-Toi. Algorithmic analysis of nonlinear hybrid systems. *IEEE Transactions on Automatic Control*, 1988. To appear.
17. E.K. Kornoushenko. Finite-automation approximation to the behavior of continuous plants. *Automation and Remote Control*, pages 2068–2074, 1975.
18. A. Puri, P. Varaiya, and V. Borkar. ϵ-approximation of differential inclusions. In R. Alur, T. A. Henzinger, and E. D. Sontag, editors, *Hybrid Systems III: Verification and Control*, pages 362–376. Springer-Verlag, 1996.
19. O. Stursberg, S. Kowalewski, and S. Engell. Generating timed discrete models of continuous systems. In *Proc. 2nd MATHMOD*, pages 203–210, Vienna, Austria, Feb. 1997.
20. Feng Zhao. *Automatic Analysis and Synthesis of Controllers for Dynamical Systems Based on Phase-Space Knowledge.* PhD thesis, MIT Artificial Intelligence Laboratory, 1992.

9

Nonlinear Descriptor Systems

S. L. Campbell[1], *R. Nikoukhah*[2] *and F. Delebecque*[2]

[1]Department of Mathematics, North Carolina State University, Raleigh, NC, USA,. slc@math.ncsu.edu.

[2]INRIA, Rocquencourt BP 105, 78513 Le Chesnay Cedex, France.

Summary. Implicit or descriptor systems, $F(x', x, u, t) = 0$, arise in many applications. Much of the early work on linear time invariant systems was done in the electrical engineering and control literature. Subsequent nonlinear and time varying results have tended to be in other areas. This paper surveys some of the nonlinear theory and then relates it to several control problems. The emphasis is on nonlinear and linear time varying systems.

1 Introduction

Many physical problems are most easily and naturally modeled as implicit systems of differential equations,

$$F(x', x, u, t) = 0 \tag{1}$$

with $F_{x'} = \partial F/\partial x'$ identically singular. This is especially true of the initial form of many computer generated models. Depending on the area of application, these models are variously called differential algebraic equations (DAEs), descriptor systems, behavioral models, singular systems, semi-state equations, implicit models, or differential equations on manifolds.

Usually at control meetings such as the ECC99 when one sees descriptor systems mentioned they are linear time invariant. Many view descriptor or singular systems theory as a part of linear systems theory. After a flurry of research papers a few years ago, there have been relatively few papers on descriptor systems in the control literature. However, in recent years there have been substantial advances made in working with nonlinear and linear time varying implicit systems. Much of this work has either not been done in control terminology or it has been within the context of a specific application community. In particular, nonlinear DAEs are widely discussed in the chemical engineering process control and the mechanical simulation literature. In this talk we will review some of these results and discuss their application.

This talk cannot be comprehensive, that would take another book. Rather it reflects the philosophy that underlies many of the investigations that we have been involved in.

We have been interested in investigating the strengths and limitations of an approach. That is we are sometimes interested in developing control technology rather than solving a specific problem. We are interested in large complex nonlinear systems. Secondly we are interested in approaches where analytic operations that are practical on small systems become replaced by numerical methods. Finally we are especially interested in the rapid development of initial designs based on computer generated implicit models. Such algorithms are potentially of use not only in the early phases of design but also in problems where a system has been altered and must be quickly modeled and a control synthesized in real time.

The nonlinear theory for descriptor systems differs in several key ways from the linear theory. We will begin by briefly reviewing some results from the nonlinear theory and showing how the nonlinear implicit theory differs from the linear time invariant or explicit theories. A brief survey of some of the work on control of nonlinear descriptor systems will be given before we focus on two specific topics and discuss them in more detail.

Most methods for nonlinear control require, at some point, a numerical simulation either as part of an observer or to test the control laws. There has been substantial progress made on the understanding of numerical integrators for descriptor systems [12]. However, many of the methods of first choice only work under restrictive conditions. These conditions mean that the control questions often cannot be answered separately from the numerical integration issues. Those parts of the numerical theory of DAEs that are important for control will be discussed.

As an illustration of how these ideas can interrelate, two specific control problems that we have looked at will be discussed. The first is the use of numerical integrators to generate a nonlinear feedback law for trajectory tracking. The technique which we call DPC will be illustrated with a robotic arm example. The second example will be in the construction of observers using nonlinear descriptor systems. Descriptor observers can occur if the original system is a descriptor system. However, they can also be useful even if the original problem has an explicit model. Specific examples will be given showing how the use of nonlinear descriptor observers gives extra degrees of design freedom that can be exploited.

While results will be carefully stated, throughout this paper the emphasis will be on ideas and relating them to control concepts. A substantial bibliography is included with references for each of the key ideas in the paper. The bibliography concentrates on references that appear after 1994 or are not in [12]. A number of additional references on DAEs and their numerical solution can be found in [12, 39, 40].

2 Needed Descriptor Theory

Before discussing the key issues in DAE theory a few general comments are appropriate.

Ultimately implicit or behavioral models can encompass combinations of differential equations, algebraic equations, delays, PDEs, integral equations, discrete components, and other types of functional equations. There has been some work done on implicit delay systems [2, 13, 14, 86] and mixed PDEs and DAEs [16, 22, 27, 30, 38, 59, 60, 62]. The later are often called PDAEs. Hybrid systems where there are mixtures of discrete and continuous processes are an area of considerable current concern both in simulation and optimization [66]. For conciseness we focus on systems of differential and algebraic equations. We consider only continuous smooth solutions. Discontinuous solutions for nonlinear and time varying DAEs are considered in [68, 69, 70, 71]

More detailed examples will be given later, but one example of a DAE is a constrained mechanical system

$$M(x)v' = G(x,v,t) + c_x^T(x)\lambda + N(x)^T\gamma + B(x,v)u \tag{2a}$$

$$x' = v \tag{2b}$$

$$N(x)v = L(x) \tag{2c}$$

$$c(x) = 0 \tag{2d}$$

$$P(x,t) = 0 \tag{2e}$$

Here (2c) and (2d) are physical velocity and position constraints and hence generate forces in (2a) via the multipliers λ, γ. The constraint (2e) is a desired or program constraint, such as the desire to follow a certain trajectory. It does not generate a force term in (2a). u is a control. The control is not completely free since part or all of u will be determined by the path constraint (2e). The system (2) is a DAE in $(x, v, \lambda, \gamma, u)$.

The mere fact that DAEs occur frequently does not immediately mean that one should develop methods in terms of DAE formulations. There has to be either an advantage or a need to work with DAE formulations. There are several reasons for considering DAEs. Among them are the following. Many systems are easily modeled initially as DAEs. It may not be possible to make the model into an ODE without considerable effort or making additional simplifying assumptions. Reduction can also reduce sparsity in equations and Jacobians. This is why DAE based integrators can sometimes be faster on ODEs than ODE based integrators. In addition, in simulation studies one can work with one implicit model rather than several explicit ones of varying dimension. Finally, there can be extra design flexibility which we will illustrate later.

We turn now to a brief summary of the DAE theory. Additional detail can be found in the references.

2.1 Linear Time Invariant Systems

The most extensively studied descriptor systems in the control literature are the linear time invariant (LTI) systems [12, 57, 58],

$$Ex'(t) + Fx(t) = f(t) \tag{3}$$

If solutions exist for every smooth f, and solutions when they exist are uniquely determined by an initial condition, then (3) is called *solvable*. This happens if and only if $\lambda E + F$ is a regular pencil. That is, $\lambda E + F$ is invertible for at least one value of λ. The theory of the Kronecker canonical form applied to a regular pencil then says that there are invertible matrices V, W so that letting $x = Wy$, and left multiplication by V, give

$$y_1' = Cy_1 + f_1 \tag{4a}$$

$$Ny_2' = y_2 + f_2 \tag{4b}$$

Subsystem (4a) is an ODE. The matrix N is nilpotent of index ν. That is, $N^\nu = 0, N^{\nu-1} \neq 0$. The solution of (4b) is given by

$$y_2 = \sum_{i=0}^{\nu-1} (-N)^i \frac{d^i f}{dt^i} \tag{5}$$

From these formulas we already see two important facts about DAEs. Not all initial conditions are consistent for a given forcing function f and the solution can depend on derivatives of f.

Computation of the full Kronecker form is not always desirable. Computation of the form (4) does not require computing the Kronecker form. An uppertriangular version can be gotten with orthogonal operations and one matrix inversion. See also [79]. A variety of algorithms have been developed which exploit this form without directly computing it. These are some times referred to as shuffle algorithms. We give one that can be done by orthogonal operations.

Suppose that we row compress E to get

$$\begin{bmatrix} E_1 \\ 0 \end{bmatrix} x' + \begin{bmatrix} F_1 \\ F_2 \end{bmatrix} x = \begin{bmatrix} f_1 \\ f_2 \end{bmatrix} \tag{6}$$

Differentiating the bottom equation gives

$$\begin{bmatrix} E_1 \\ F_1 \end{bmatrix} x' + \begin{bmatrix} F_1 \\ 0 \end{bmatrix} x = \begin{bmatrix} f_1 \\ f_2' \end{bmatrix} \tag{7}$$

If $\begin{bmatrix} E_i \\ F_i \end{bmatrix}$ is invertible we stop. Otherwise we repeat the steps (6) and (7).

Theorem 2.1. *Suppose that $\{E, F\}$ is a regular pencil. Let ν be the index of nilpotency of the nilpotent block N in (4). Then the following numbers are equivalent to ν.*

1. *The number of steps of the shuffle algorithm needed to convert the DAE into an ODE.*
2. *1 more than the order of $\lambda = 0$ as a pole of $(E + \lambda F)^{-1}$.*
3. *1 more than the smallest number of derivatives of f that appear in the solution of the DAE.*

The integer ν is called the *index* of the descriptor system. These and other equivalences underlie the numerous papers on LTI control systems.

Various numerical methods have been investigated for LTI DAEs. Details can be found in [12]. The key point is that if the methods do not compute a variant of (4), then they must be implicit. Both backward differentiation (BDF) and Implicit Runge Kutta (IRK) methods have been extensively examined. The index one case is easily solved and poses no special problems for these methods. For systems of higher index, $\nu \geq 2$, BDF methods can have reduced or nonconvergence on the first ν steps. IRK methods can have reduced order of convergence in some or all variables. When actually applied to systems arising from applications one observes that the current codes seem to do fine with $\nu \leq 1$, may work on $\nu = 2$ or 3, but all start having severe numerical problems with $\nu \geq 4$ if not before.

2.2 Linear Time Varying Theory

The LTI theory is very nice and has been exploited extensively. In some sense this success held up the study of linear time varying and nonlinear systems. The linear time varying (LTV) case is important conceptually since it proved possible, after several years of hard technical work, to pretty much fully understand it. This study showed that much of the LTI theory did not directly carry over to this more general case. Since LTV systems can be embedded into the nonlinear systems with the usual $\tau' = 1$ trick, this means that one has to carefully consider these additional problems.

Consider then the linear time varying (LTV) system

$$E(t)x'(t) + F(t)x(t) = f(t) \tag{8}$$

Theorem 2.2. *Suppose that (8) is a solvable DAE. Then the following are equivalent:*

1. *One step of the continuous shuffle algorithm gives an ODE.*
2. *The solution is continuous in f in the sup norm.*
3. *E has constant rank and E, F is an index one pencil for every t.*

In the above situation we say the DAE (8) is index one. Problems involving the index one case are usually straightforward to consider. As soon as one moves beyond the index one case the LTI theory breaks down. We call $\{E, F\}$ a regular pencil if $\{E(t), F(t)\}$ is regular for every t in the interval of interest.

Theorem 2.3. *Suppose that (8) is a solvable DAE and it is not index one. Then*

1. *The pencil $\{E, F\}$ need not be regular.*
2. *E need not have constant rank and it may not be possible to carry out the shuffle algorithm globally.*
3. *If $\{E, F\}$ is regular, then the index of the pencil $\{E, F\}$ need not be independent of t.*
4. *If $\{E, F\}$ is regular and the index of the pencil $\{E, F\}$ is constant, then this number is independent of the number of steps of the shuffle algorithm needed to get an ODE.*

Note that this theorem says that the pencil $\{E, F\}$ which was so fundamental in the LTI theory is not nearly as important for LTV systems. Thus results which try to duplicate the LTI theory will always have limited general applicability although there are some results based on the pencil $\{E, F\}$ which are useful in special circumstances.

It is instructive to consider a specific example

Example 2.1. The DAE

$$\begin{aligned} \beta(t)y' + x &= f(t) \\ \alpha(t)z' + y &= g(t) \\ z &= h(t) \end{aligned}$$

has the solution:

$$\begin{aligned} x &= f(t) - \beta(t)[g'(t) - \alpha'(t)h'(t) - \alpha(t)h''(t)] \\ y &= g(t) - \alpha(t)h'(t) \\ z &= h(t) \end{aligned}$$

Note that the solution not only depends on derivatives of forcing terms like f, g, h, but also on derivatives of the *coefficients* α, β. Also the E matrix need not have constant rank since α, β can be zero for some values of t. That the solution can depend on derivatives of the coefficients means that one cannot do an analysis based on linear time invariant approximations and expect to always capture the correct solution behavior.

In the next subsection we will give a definition of the index of a descriptor system which proves general enough to discuss a wide variety of nonlinear and time varying systems. It turns out that there is one more difficulty. There is a structural theory for linear time varying systems. However, the existence of the time varying coordinate changes restricts its applicability.

Theorem 2.4. (Structure Theorem for LTV) *[15] Suppose that the coefficients of (8) are infinitely differentiable. Suppose that (8) is a solvable system. Then there exist $P(t), Q(t)$ such that $x = Qy$ and premultiplying by P give*

$$\begin{aligned} y_1' + C(t)y_1 &= f_1 && (9a) \\ N(t)y_2' + y_2 &= f_2 && (9b) \end{aligned}$$

where $N\frac{d}{dt} + I$ is an invertible operator on the space of infinitely differentiable functions on $\mathcal{I}$.

Equation (9b) has only one solution for a given f_2. It may not be possible to get N uppertriangular on all of $\mathcal{I}$.

A related theory for linear time varying DAEs is developed by Kunkel and Mehrman in [53, 55]. Some numerical methods exist for computing various structural properties of linear time varying DAEs.

2.3 Differences in Nonlinear Theory

One could try to develop a nonlinear theory by either a nonlinear version of a shuffle algorithm or by generalizing Theorem 2.4. However, our goal is a theory based on computable quantities. Because the computation of nonlinear coordinate changes is often not practical an alternative theory has been developed in terms of quantities that are computationally accessible.

Assume

$$F(x', x, t) = 0 \tag{10}$$

is a sufficiently differentiable *solvable* DAE. That is, there is a well defined manifold of solutions. For notational simplicity we suppress the control variable u during this part of our discussion. Part of the control is absorbed into the x variable because it is constrained and part of it is viewed as a forcing function and reflected by the t dependence. We know that the answer will involve derivatives of the original equations. If we differentiate (10) k times we get the *Derivative Array Equations* (11).

$$G_k(x', w, x, t) = \begin{bmatrix} F(x', x, t) \\ F_t(x', x, t) + F_x(x', x, t)x' + F_{x'}(x', x, t)x'' \\ \vdots \\ \dfrac{d^k}{dt^k}[f(x', x, t)] \end{bmatrix} = 0 \tag{11}$$

where $w = [x^{(2)}, \ldots, x^{(k+1)}]$. We will often write G instead of G_k if the value of k is clear.

The least k for which (11) uniquely determines x' for consistent (t, x) is called the *differentiation index* ν_d. In applications it is often possible to differentiate different equations a different number of times and thus reduce the number of equations and additional derivative variables. This has been extensively studied particularly in the chemical engineering literature [35, 78]. The DAE is *higher index* if $\nu \geq 2$ and *high index* if $\nu > 3$. There are several types of index for nonlinear systems [19] but we will not discuss them here.

In the nonlinear theory there are several key types of structures which appear. One is a semi-explicit descriptor system (12).

$$x' = f(x, y, t) \tag{12a}$$
$$0 = g(x, y, t) \tag{12b}$$

System (12) is index one if and only if g_y is nonsingular.

The most studied higher index systems are the Hessenberg DAEs (13).

$$x' = f(x, y, z, w, t) \tag{13a}$$
$$y' = g(x, y, z, t) \tag{13b}$$
$$z' = h(y, z, t) \tag{13c}$$
$$0 = q(z, t) \tag{13d}$$

System (13) is index 4 if $q_z h_y g_x f_w$ is nonsingular. (13b)–(13d) is index 3 if $q_z h_y g_x$ is nonsingular and (13c)–(13d) is index 2 if $q_z h_y$ is nonsingular.

Hessenberg systems are important for two reasons. First there are positive numerical results for both BDF and some IRK methods. Secondly many control problems are in that form. For example, the mechanical system (2) is often Hessenberg. One simple version of (2) is

$$M(x)x'' + G(x, x') = g_x(x)^T \lambda + B(x)u \tag{14a}$$
$$0 = g(x) \tag{14b}$$

If $v = x'$, and u is considered known, then (14) is index 3 Hessenberg in (x, v, λ).

However, many control problems are not Hessenberg. This is common in chemical process models but can also occur in the control of mechanical systems as the next example shows.

Example 2.2. [76] Consider the planar motion of a crane. The crane consists of a trolley of mass M_1 from which hangs a cable. At the end of the cable is a load of mass M_2. The cable weight is negligible. Controls are the horizontal force u_1 on the trolley, and the torque u_2 on the winch the cable is attached to. Tension in the cable is τ, location of the load is (x, z), and the angle of the cable with the vertical is θ. The trolley location is at d, and the cable length is r. J is the moment of inertia of the winch. The cable is assumed straight so that it acts like a variable length pendulum. The load is to follow a path $(p_1(t), p_2(t))$. The equations are

$$\begin{aligned}
M_2 x'' &= -\tau \sin\theta && \text{(15a)}\\
M_2 z'' &= -\tau \cos\theta + mg && \text{(15b)}\\
M_1 d'' &= -C_1 d' + u_1 + \tau \sin\theta && \text{(15c)}\\
J r'' &= -C_2 r' - C_3 u_2 + C_3^2 \tau && \text{(15d)}\\
0 &= r\sin\theta + d - x && \text{(15e)}\\
0 &= r\cos\theta - z && \text{(15f)}\\
0 &= x - p_1(t) && \text{(15g)}\\
0 &= z - p_2(t) && \text{(15h)}
\end{aligned}$$

System (15) is index 5 in $\{u_1, u_2, x, z, \tau, \theta, x', z', r, d, r', d'\}$. But (15) is not in Hessenberg form since the θ variable appears in the constraint but θ' does not appear in any of the equations.

This example illustrates several key points. It is easy to get DAEs of index 3, 4, 5 and 6. It is also easy to get DAEs which are not in Hessenberg form. Simplification can lower the index, for example by separating the mechanics model from the actuator model [17, 50]. The "real" physical problem may be higher index.

To describe the general theory define the following Jacobians of the derivative array equations (11).

$$\overline{J}_k = [\ G_{x'} \quad G_w\ ], \quad J_k = [\ G_{x'} \quad G_w \quad G_x\ ]$$

Note that $G_k = 0$ is a nonlinear system. The following assumptions form the basis of the general theory. Here $Ax = b$ is 1-full with respect to a projection $\widetilde{x}$ of x if $Ax = b$ uniquely determines $\widetilde{x}$ for consistent b.

(A1) Sufficient smoothness of G_k
(A2) Consistency of $G_k = 0$
(A3) $\overline{J}_k = [\ G_{x'} \quad G_w\ ]$ is 1-full with respect to x' and has constant rank independent of (t, x, x', w).
(A4) $J_k = [\ G_{x'} \quad G_w \quad G_x\ ]$ has full row rank independent of (t, x, x', w)

These Jacobians can be computed by either symbolic or automatic differentiation software [21].

Definition 2.1. *The uniform differentiation index ν_{UD} of the DAE (10) along a solution x is the smallest integer k such that* (A2) *holds and* (A1), (A3), (A4) *hold on an open set containing* $(t, x, x', \ldots, x^{(k+1)})$.

This is the only index that can be reasonably computed for general DAEs. Assumptions (A1)–(A4) are satisfied by many systems that arise in applications. These assumption also turn out to be essentially equivalent to a uniform type of solvability.

Theorem 2.5. (Solvability) *[20] Suppose the derivative arrays G_i of (10) satisfy (A1)–(A4) for $i = k$ and $k + 1$ on the same neighborhood. Then the DAE is geometrically (uniformly) solvable on that neighborhood. In addition there is a characterization of the solution manifold which implies that its dimension is $n - nullity(\overline{J}_k^T)$.*

Linearization is basic to many control methods. The linearization most favored in control is the time invariant linearization which results in a linear time invariant system. There has been considerable work done on linear time invariant DAEs. Unfortunately, this linearization is not always appropriate. In general, the linear time invariant linearization is only correct when linearizing an autonomous system about an equilibrium. Away from an equilibrium point the linear time varying and the linear time invariant linearization can differ in: dimension of solution manifold, amount of differentiation in solution, and dynamical behavior [18]. In these cases a linear time varying linearization must be used.

There are some positive results in the index one case relating stability near an equilibrium (x^*, y^*) to that of the matrix pencil. Krishnan and McClamroch have positive results for constrained mechanical systems in Hessenberg form [51]. The most general result near an equilibrium is the following [72]

Theorem 2.6. (Near an Equilibrium) *Suppose that (DAE) (10) has a constant solution $\overline{x}$ and satisfies (A1)–(A4) on a neighborhood of $(0, 0, \overline{x}, t)$ for t in some finite closed interval $\mathcal{I}$. Then the DAE*

$$A(t)\widehat{x}' + B(t)\widehat{x} = 0$$

where $\widehat{x} = x - \overline{x}$ and $A(t) = f_{x'}(0, \overline{x}, t)$, $B(t) = f_x(0, \overline{x}, t)$, is a solvable DAE with the same dimensional solution manifold near $(0, \overline{x})$ as (10).

There is a similar positive result for linearizing along a trajectory $\overline{x}(t)$.

2.4 Numerical Methods

There has been a considerable amount of work on numerical integrators for DAEs. There are two types of numerical methods. One type is the variants of classical methods applied to the DAE such as a backward differentiation (BDF) or implicit Runge-Kutta (IRK) methods. Other methods are applied to an ODE in some, perhaps implicitly defined, coordinate system. Even for index two linear time varying solvable DAEs it is not true that a BDF method or an IRK is guaranteed to work. There are no known "standard type methods" which work on all solvable index two DAEs even if they are linear time varying.

There are two general purpose numerical approaches based on the derivative array. One, [23] has successfully integrated a number of problems but does not exist as a production code. The other [54] has a time varying code GELDA [56] available.

There are positive results showing that some methods work for certain special classes of DAEs. Thus in a control application one has to worry about whether a DAE has the right structure for the integrator being used. All classical methods are limited to only working with special classes of DAEs.

Discussions of DAE integrators can be found in [12]. We assume that a public domain code is being used. For our purposes it is enough to note that there are general purpose, index one DAE integrators such as DASSL. There are a number of classical codes such as LIMEX (an extrapolation code), various IRK codes such as RADAU5, mechanics codes such as (MEXX), and BVP codes such COLSYS. However, all of these approaches require some of: assumed structure to the equations, different parts of the state variables treated differently for error control, and restrictions on the index. Some standard methods cannot be used on DAEs.

These numerical results have several implications for control. First, if a DAE formulation is to be used, then care must be taken to insure that the equations to be integrated are in the correct form for the chosen integrator. Secondly, some of the methods show an order reduction in some variables. Often this is the control. Also a numerical method that is appropriate for say an index two problem many not work on an index three problem. Finally, most standard methods, other than the general purpose integrators, have error estimates that take the form of

$$O(h^m) + \frac{\epsilon}{h^{\nu-1}} \tag{16}$$

where h is the stepsize of the integrator. Accordingly the common belief that smaller steps are more accurate has to be held with more caution with higher index DAEs.

The MANPAK algorithms [74, 75] while not designed for control could prove useful in the future.

3 Control

We now turn to examining various control issues. We first briefly comment on some areas of research involving nonlinear DAEs and control. Sections 3.3–3.5 discuss in more detail particular control problems to illustrate the interplay between DAE theory, numerical issues, and controller design.

The emphasis in this paper is on the nonlinear case. However, as we have seen above for DAEs the linear time varying theory is much closer in some respects to the nonlinear theory. Accordingly, we will sometimes comment on linear time varying results also.

3.1 Optimization of DAEs

There has been a great deal of interest in the optimization of nonlinear descriptor systems. This is especially true in the chemical process industry and in the control of mechanical systems and in particular those in manufacturing. In both cases the processes are typically first modeled as descriptor systems. This is an important area in which there is much current activity. We mention only [5, 11, 31, 34, 36, 37, 62, 65, 67, 80].

The chemical process problems are typically very high dimension and may arise as discretizations of PDEs by the method of lines. The mechanical systems are usually some what lower dimension but are more frequently high index.

DAEs can arise in optimization problems either because of a DAE modeled process, because the variational equations are a DAE, or because of optimization constraints. We shall illustrate the last point.

Suppose that we have an explicit process

$$x'' = f(x, x', u) \tag{17}$$

and a cost functional $J(x, u)$ which is to be minimized.

In practical problems there are usually a number of additional constraints which may be imposed for performance reasons. For example, there may be manufacturer recommended velocity or acceleration limits. Thus we add to the optimization various constraints

$$g(x', x, t) \leq 0 \tag{18}$$

When (18) is not active we have an ODE. But when a constraint $g_i(x', x, t) = 0$ is active we have a DAE. This has several consequences.

One is that the constraint restricts u so the process is now x and part of u with some, if any, of u left over for optimization. Secondly the accuracy of the control will depend in part on the appropriateness of the discretization scheme being used for the type of DAE that has occurred.

A velocity constraint results in a DAE that is at least index 2 in (x, u). A position constraint gives a DAE that is at least index 3. A position constraint which involves state variables which do not appear in (17) in the same equations that u appears in gives a DAE of index at least 4.

One way that the presence of DAEs is important is that in complex industrial optimization codes the logic is often based on some assumptions about the order of the discretizations. The order reduction due to the presence of DAEs can affect both efficiency and robustness of the software [7].

However, the situation is more complex than just saying that the optimization problem is a DAE. To illustrate, suppose that (17) is $x'' = u$ and (18) is $x \geq q(t)$. When the constraints are active this gives an index three system in the form of (4b). If one does the order estimates then it turns out that the reduction of order is present for two reasons. One is the appearance of derivatives of the forcing terms which in this example is $q(t)$. This means that whether this order reduction occurs can be dependent on the form of the constraints. Simple box constraints would not lead to order reduction in this manner since the higher derivatives of q are zero [7].

Another important factor is the $\epsilon h^{1-\nu}$ term in (16). Mesh refinement strategies are often based on a number of heuristics which often do not take this term into account [4]. Note that ϵ need not be machine tolerance but is often the error in Newton solves, constraint satisfaction, or spline approximation of data which are much larger than machine precision. The order reduction due to this factor is trickier to incorporate in mesh refinement since the order depends on the mesh size. In some cases this form of order reduction is a major factor [7].

3.2 Other Control Questions

Outside of the linear time invariant theory [1, 32] and the optimal control literature, there has been relatively little work done on classical control algorithms for nonlinear higher index DAEs. There has been some work on the topics of feedback linearization [46], estimation [47, 81], and observer design for mechanical systems. We mention here also [29, 51, 52, 61]. Observability and Controllability for linear time varying DAEs is examined in [24, 26, 77].

3.3 Open Loop Trajectory Tracking

Before discussing more elaborate trajectory tracking algorithms we first consider the problem of *exact trajectory tracking* or system inversion. Suppose that we have a plant (1) and an output

$$y = H(x, u, t) \tag{19}$$

Let $r(t)$ be a desired reference trajectory. The problem of exact trajectory tracking is to find a control u so that

$$r(t) = H(x, u, t) \tag{20}$$

Many algorithms instead strive for *asymptotic trajectory tracking*

$$r(t) - H(x, u, t) \to 0 \tag{21}$$

because of increased robustness. However, the exact tracking problem is still fundamental.

In the early phases of design a natural question is whether a given task is even reasonable. That is, given $r(t)$ is there a u that can produce this trajectory? Once u is found the designer can decide whether the control is desirable. It might, for example, require excessive control effort. Thus it is useful to be able to rapidly perform system inversion for complex nonlinear systems.

There are a number of control algorithms for system inversion for the case when the process is explicit. These algorithms typically consist of repeatedly differentiating the output until the control appears. The number of these differentiations is called the *relative degree.* These approaches do not readily extend to the case when the process is a DAE. They also suffer from the fact that the algorithm can be parameter dependent. For a complex system, changing a parameter value may alter the steps of the inversion process and necessitate a recomputation of the control. Finally, some systems may require analytically performed coordinate operations during the inversion process or they will underestimate the relative degree.

From a descriptor system point of view, inversion is straightforward. One treats the process (1) and the desired trajectory (20) as a single system in x, u. A numerical integration of this descriptor system gives an open loop control u. The index of this descriptor system is generally one more than the relative degree. The exception occurs when the zero dynamics are themselves a descriptor system whose index is at least one higher than the relative degree. This can sometimes happen, for example, in a controlled mechanical system where the outputs contain velocity information on a subset of the state variables.

The various numerical algorithms mentioned earlier for determining the index and characterizing consistent initial conditions become algorithms for determining the relative degree, and determining which initial values of x admit exact trajectory tracking. By viewing the problem as a descriptor system in x, u one can quickly get an idea of what kind of control effort will be needed and other important properties which will either lead to immediately considering an alternative design or justify the more time consuming application of traditional approaches if they can be used.

If (20) is a flat output, then the DAE consisting of (1) and (20) has a unique solution for x, u given y. This is sometimes referred to as a totally singular DAE. The earlier results for determining solvability and the dimension of the solution manifold now become algorithms for testing whether a given output is flat [25].

We now turn to two more specific illustrations to serve as examples of what can be involved in trying to use a DAE perspective.

3.4 DAE Integrator Stabilized Trajectory Tracking

Considerable research has been done on the problem of trajectory tracking [73]. Many of the methods for constructing output trajectory tracking controllers, however, run into trouble when applied to large and complex nonlinear plants since they rely on the construction of the inverse dynamics of the plant and, in general, the computation of the inverse dynamics requires differentiating the plant output.

In the previous section we saw that a numerical based descriptor system approach could easily handle many complex high dimensional inversion problems. It would be desirable if the same was true for trajectory tracking.

Jankowski and Van Brussel [43, 44, 45] suggested one such approach where the controller computes the required differentiations numerically, in an implicit manner, rather than symbolically. We shall refer to this approach as Descriptor Predictive Control (DPC). This idea has also been successfully used in the context of nonlinear observer design [33] where approximations of higher order derivatives were obtained using interpolation.

One aspect of this approach is to introduce just enough differentiation so that the index ν is reduced to the point that the DAE can be safely integrated numerically. The flexible-joint problems considered by [43, 44] are index 5 or 6 and are reduced to index 3. Among the potential advantages are: equations are simpler and quicker to integrate than if reduced to index one or zero, possibility of exploiting modern DAE integrators, approach works with much simpler expressions and avoids expression swell for complex systems and analytic computations are replaced by numerical computations. In addition it is easy to adapt to a new or altered system since you don't need to recompute for many parameter or system changes and it avoids need for nonlinear analytic operations. Unfortunately, they only observed stability at long enough sampling rates and lacked a theory to motivate corrections. We shall see that these difficulties can be overcome by modifying DPC to include a specially designed preliminary feedback.

The general idea of DPC would go as follows. We have a process

$$F(x', x, t, u) = 0 \tag{22a}$$

and a desired trajectory

$$H(x, t) = 0 \tag{22b}$$

Physical constraints are included in (22a). If one just solved the DAE (22) as suggested in the last section an open loop control would result. Also the DAE might easily have high index. Merely differentiating (22b) would lower the index but there then could be numerical drift from (22b).

Instead of (22), one could try using

$$F(x', x, t, u) = 0 \tag{23a}$$

and

$$W(\frac{d}{dt})H(x, t) = 0 \tag{23b}$$

where $W(s)$ is a stable matrix polynomial, $\det(W(p)) \neq 0$ for $\mathrm{Re}(p) \geq 0$. Two things are accomplished. One is that (23) will have lower index than (22). Secondly if the control u computed from (23) is used in (22a), then we might hope that $H(x,t) \to 0$ as $t \to \infty$. In order to create feedback, the integration of (23) is frequently reinitialized with estimates of the state x from (22).

There are various ways to implement DPC. However, DPC always includes the three following items:

The *stabilization* of the path constraint. If $W(p) = (p + \alpha)I$, path stabilization is simple Baumgarte stabilization [3]. The path stabilization can be thought of as the partial application of the inversion algorithm that reduces the index and thus improves integration.

The implementation of an *integration method* for the numerical solution of the DAE (23) obtained from the system equations and (stabilized) path constraints.

The numerical solution of the DAE for (x, u) yields a *prediction of the control.* This control is applied for a short period of time called the *prediction horizon* as an open loop control. Then it is recomputed based on a new measurement or estimate of the state.

The fact that the DAE is solved numerically is the key to the potential usefulness of this approach. Current DAE solvers require that the DAE have index no more than three. Thus stabilized differentiation is necessary to reduce the index of the DAE to no more than three. The resulting control is a feedback control since the integration is initialized by the measured state of the plant but open-loop over short periods of time. It is also in a sense predictive because the control is obtained by "looking ahead." However, it is different from the "classical" concept of predictive control [63] in that it does not involve optimization.

The DPC strategy introduces a delay in the control loop equal to the prediction horizon. Numerical computation of the DAE solution imposes a limit on how small this delay can be. The computation of an upper bound on the acceptable delay is an important part of the controller implementation as the rate of convergence of the error turns out to be delay dependent and too large a delay may destabilize the closed loop system.

As noted the early work on DPC was experimental and failed to work for small sampling periods. We now turn to a more careful analysis of DPC. The analysis of DPC has several benefits. For one it helps determine whether the approach can be used. The analysis also helps shed insight on the issues involved in other attempts to replace analytic algorithms with DAE based integrators.

DPC can be used when the process (22a) is a descriptor system. Here we focus on the situation where we have an explicit model.

$$\begin{aligned} x' &= A(x) + B(x)u \\ y &= C(x) + D(x)u \end{aligned}$$

There are two key parameters; h the delay in applying the control and ϵ the length of the time it is applied before updating.

In our implementation we will use backward Euler (BDF-1) as the integrator. The DAE (23) is integrated for k time steps with a stepsize of ϵ/m between samplings. We assume that integration is done faster than real time. The computed control at the end of k integration steps is used as the control during next sampling period. We also assume the sampling period ϵ is constant and the control is kept constant between predictions. Thus $x(t_j)$ at instant t_j yields a prediction $X(\tau)$ of the state $x(\tau)$, and a prediction $U(\tau)$ of the control $u(\tau)$ which is used to construct the piecewise constant control $u(\tau) = U(t_j + h + \rho\epsilon)$, for $t_j + h \leq \tau \leq t_j + h + \epsilon$. This procedure is repeated for each $t_{j+l} = t_j + l\epsilon$, $l = 0, 1, 2, \cdots$. Here $\rho \in [0, 1]$ determines where the prediction of u is computed in the interval on which it is held constant. In [43] $\epsilon = h$ and $\rho = 0.5$ were used.

In turns out that a linear analysis sheds considerable light on what is happening. To assess the stability properties of DPC [82, 84, 85], we initially assume that the plant is the linear system

$$x' = Ax + Bu \tag{24a}$$

$$\widehat{y} = \widehat{C}x + \widehat{D}u \tag{24b}$$

subject to the stabilized path constraint

$$W\left(\frac{d}{dt}\right)\left\{\widehat{y}(t) - \widehat{\xi}(t)\right\} = 0$$

In the remainder of this section we assume that the path constraint has already been stabilized, that is, instead of system (24) we consider

$$(\Sigma)\left\{\begin{array}{rcl} x' &=& Ax + Bu \\ y &=& Cx + Du \end{array}\right. \tag{25}$$

subject to the path constraint $y(t) - \xi(t) = 0$ where $W\left(\frac{d}{dt}\right) \stackrel{\text{def}}{=} \sum_{i=0}^{k} W_i \frac{d}{dt}$ and

$$\begin{array}{llll} D \stackrel{\text{def}}{=} W_0\widehat{D} + \sum_{i=1}^{k} W_i\widehat{C}A^{(i-1)}B\,, & y(t) & \stackrel{\text{def}}{=} & W\left(\frac{d}{dt}\right)\widehat{y}(t) \\ C \stackrel{\text{def}}{=} \sum_{i=0}^{k} W_i\widehat{C}A^i\,, & \xi(t) & \stackrel{\text{def}}{=} & W\left(\frac{d}{dt}\right)\widehat{\xi}(t) \end{array}$$

We assume System Σ is invertible, that is we suppose we have a unique input $u(t)$ for any output $y(t)$. Moreover, we suppose System Σ to be controllable, observable and minimum phase. D is generally singular so that Σ may be higher index in (x, u).

First consider the limiting case of infinitesimal prediction horizon $h \to 0^+$ and infinitesimal sampling period $\epsilon \to 0^+$. The change of coordinates

$$\left[\begin{array}{c} z_1 \\ z_2 \end{array}\right] = Q^{-1}x$$

divides the state into two components: z_1 which corresponds to the Largest Output-Nulling (A, B)-invariant Subspace (LONS) and z_2. System Σ in z-coordinates is

$$\left\{\begin{array}{lcl} \begin{bmatrix} z_1'(t) \\ z_2'(t) \end{bmatrix} & = & \begin{bmatrix} A_1 & A_2 \\ A_3 & A_4 \end{bmatrix}\begin{bmatrix} z_1(t) \\ z_2(t) \end{bmatrix} + \begin{bmatrix} B_1 \\ B_2 \end{bmatrix} u(t) \\ y(t) & = & \begin{bmatrix} C_1 & C_2 \end{bmatrix}\begin{bmatrix} z_1(t) \\ z_2(t) \end{bmatrix} + Du(t) \end{array}\right. \tag{26}$$

If we apply the DPC control as described we get

$$\begin{aligned} \begin{bmatrix} z_1' \\ z_2' \end{bmatrix} &= \begin{bmatrix} \overline{A} & A_2 \\ 0 & A_4 \end{bmatrix}\begin{bmatrix} z_1 \\ z_2 \end{bmatrix} + \begin{bmatrix} B_1 \\ B_2 \end{bmatrix} R(\sigma) y_d \\ y_d &= [0 \ C_2]\begin{bmatrix} z_1 \\ z_2 \end{bmatrix} + DR(\sigma) y_d \end{aligned}$$

where $\overline{A} = A_1 + B_1\overline{E}$ has eigenvalues determined by the zeros of W. However, A_4 may not be stable and the eigenvalues of A_4 cannot be altered by the DPC feedback.

To correct this we design a *preliminary feedback*

$$u = Lx + v \tag{27}$$

such that the application of DPC to the system plus preliminary feedback yields a stable closed loop system. This is easily seen if Σ is transformed in z-coordinates. If we apply first preliminary feedback and then DPC the closed loop system ($h \to 0^+$ and $\xi(t) = 0$) becomes

$$\left\{\begin{array}{lcl} \begin{bmatrix} z_1'(t) \\ z_2'(t) \end{bmatrix} & = & \begin{bmatrix} \overline{A} & A_2 + B_1 L_2 \\ 0 & A_4 + B_2 L_2 \end{bmatrix}\begin{bmatrix} z_1(t) \\ z_2(t) \end{bmatrix} \\ y(t) & = & \begin{bmatrix} 0 & C_2 + DL_2 \end{bmatrix}\begin{bmatrix} z_1(t) \\ z_2(t) \end{bmatrix} \end{array}\right. \tag{28}$$

where $LQ = \begin{bmatrix} 0 & L_2 \end{bmatrix}$. The poles of the closed loop system are given by the eigenvalues of $A_4 + B_2 L_2$ and can arbitrarily be placed by the choice of L_2

This preliminary feedback is based on the assumption that $\epsilon = h = 0$. In the case of infinitesimal sampling periods but nonzero delay ($\epsilon \to 0^+$ and $h > 0$), the control and the closed loop system become the delay differential system

$$x'(t) \;=\; Ax(t) + BK_h x(t-h) + Bw(t) \tag{29}$$

where $K_h, w(t)$ are defined in [84].

The stability analysis of delay differential systems is a delicate problem. This is due to the fact that stability analysis of (29) requires finding the zeros of a characteristic equation of the form

$$\Upsilon(s,h) = \det(sI - A - BK_h e^{-sh}) = \sum_{k=0}^{\rho} \Upsilon_k(s,h) e^{-skh} = 0 \tag{30}$$

that has an infinite number of zeros for $h > 0$. The next lemma assures the existence of a maximal delay below which DPC yields a stable closed loop system.

Lemma 3.1. *Let $\epsilon \to 0^+$. Assume all eigenvalues of A_4 in (26) are stable or have been stabilized by preliminary feedback. Then there exist $h^* > 0$ such that for $h \leq h^*$ the closed loop system is stable.*

An estimate for an upper bound of h^* can be computed, for example, by the iterative method for computing maximal delay intervals given in [28] assuming $K_h \approx K_0$ =const. The exact value for h^* can be computed by multivariable Nyquist plots.

There remains the fundamental question of tracking.

Corollary 3.1. *For $h \to 0^+$ and $\epsilon \to 0^+$, the error equation reduces to*

$$\begin{aligned} r_2'(t) &= A_4 r_2(t) && \text{(31a)} \\ e(t) &= C_2 r_2(t) && \text{(31b)} \end{aligned}$$

for $t \geq 0$ with initial value $r_2(0) = z_2(0) - \overline{\xi}(0)$.

Corollary 3.2. *In the case $\epsilon \to 0^+$ and $h > 0$, DPC achieves $e(t) \to 0$ for $t \to \infty$ if and only if system (29) is stable for $w(t) = 0$ for all t.*

The case $h \to 0^+$ is a singular case since the error dynamics are decoupled from the zero dynamics. That is, the part of the initial conditions in LONS, the z_1-part, has no impact on the error. For $h \to 0^+$ the error is entirely determined by the z_2-part of the initial conditions. However, coupling occurs as soon as $h > 0$.

Theorem 3.1. *If (29) is stable, whether $h \to 0^+$ or not, the tracking error for $\epsilon \to 0^+$ converges to zero exponentially. For $h \to 0^+$ the rate of convergence of the error dynamics can arbitrarily be set by a proper choice of preliminary feedback. For non infinitesimal prediction horizon, the rate is influenced by the zero dynamics of Σ and can no longer be chosen arbitrarily.*

A Nonlinear Example. To illustrate the importance of the preliminary feedback and the influence of delay in the feedback loop we shall consider the planar flexible-joint robotic arm introduced in [43].

The robot is composed of two links. The first link is driven by an actuator in the joint acting on the link through a torsional spring. The angular position of the actuator of the first link is denoted by θ, the absolute angular position of the first link is q_1. The second link is driven by a second actuator which acts directly on the second link with no elasticity. Its relative angular position is denoted by q_2.

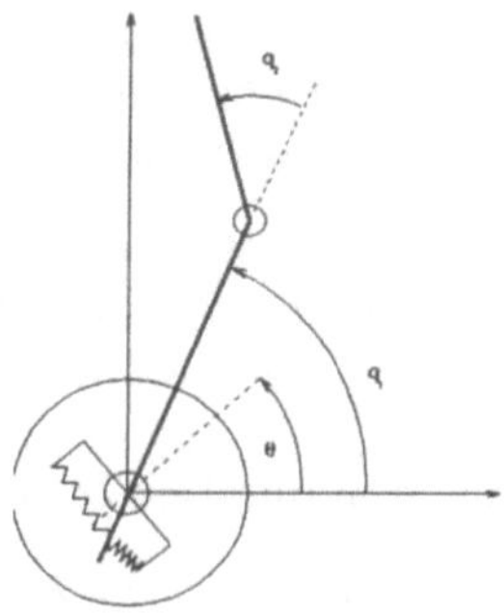

Fig. 1 Two link robotic arm.

As is typical with the path control of flexible joint robots, the index of the original DAE in (x, u) is five. If actuator dynamics were included the index would be six [17]. Two stabilized differentiations will be done to make the DAE index three.

The dynamic model of the system is given by

$$\begin{bmatrix} M_1 + 2M_3\cos q_2 & M_2 + M_3\cos q_2 & 0 \\ M_2 + M_3\cos q_2 & M_2 & 0 \\ 0 & 0 & I_{r_1} \end{bmatrix} \begin{bmatrix} q_1'' \\ q_2'' \\ \theta'' \end{bmatrix} = \begin{bmatrix} M_3 q_2'(2q_1' + q_2')\sin q_2 \\ -B_2 q_2' - x_2 - M_3(q_1')^2 \sin q_2 + T_2 \\ -B_1\theta' - x_1 + T_1 \end{bmatrix} + K_s \begin{bmatrix} \theta - q_1 \\ 0 \\ q_1 - \theta \end{bmatrix} \tag{32a}$$

$$\epsilon_2 x_2' = -|q_2'|x_2 + F_2 q_2' \tag{32b}$$

$$\epsilon_1 x_1' = -|\theta'|x_1 + F_1\theta' \tag{32c}$$

The last two equations model dry friction [10]. Control variables are the torques $\{T_1, T_2\}$. State variables are $\{q_1, q_2, \theta, x_1, x_2\}$. The outputs are q_1, q_2. Values of physical parameters are from [10].

We use DPC to impose the reference trajectories $q_{d_i}(t)$ on the outputs q_i. Thus

$$H(q,t) = \begin{bmatrix} q_1 - q_{d_1} \\ q_2 - q_{d_2} \end{bmatrix}$$

Following [43], we take $W(s) = s^{-1}p(s)I$ where $p(s) = s^3 + 3\alpha s^2 + 3\alpha^2 s + \alpha^3$. The integrator is added to avoid the reconstruction of the third derivative of q_i. Due to the factor $1/s$ we have to add the following integrator equations,

$$w_i' = \alpha_i^3(q_i - q_{d_i}), \quad i = 1, 2 \tag{33}$$

so that the stabilized error equations become

$$0 = (q_i'' - q_{d_i}'') + 3\alpha_i(q_i' - q_{d_i}') + 3\alpha_i^2(q_i - q_{d_i}) + w_i, \quad i = 1, 2 \tag{34}$$

A linear analysis of the DAE is impossible since (32) contains nondifferentiable (dry friction) terms. As proposed in [10], in order to eliminate the dry friction torques x_i we add the nonlinear observer

$$\epsilon_i \widehat{x}_i' = -|v_i'|\widehat{x}_i + F_i v_i' \ , \quad i = 1, 2, \tag{35}$$

where $v_1 = \theta$ and $v_2 = q_2$. Using the observed dry friction torque $\widehat{x}_i$ we can precompensate the effect of dry friction by $T_i = \overline{T}_i + x_{f_i}$, $i = 1, 2$.

Linearizations of the dry friction compensated plant were examined. Numerical evaluation of the eigenvalues of A_4 for linearizations around different nominal points shows that the eigenvalues stay close to $2.63 \pm j29.58$. As a consequence DPC as implemented in [43] yields an unstable closed loop system for small prediction horizon based on our linear analysis. This is consistent with the experimental results in [43].

By applying the preliminary linear feedback

$$u = \begin{bmatrix} \overline{T}_1 \\ \overline{T}_2 \end{bmatrix} = Kx + \begin{bmatrix} \overline{\overline{T}}_1 \\ \overline{\overline{T}}_2 \end{bmatrix} = K \begin{bmatrix} (q_1 - q_{d_1}) \\ (q_2 - q_{d_2}) \\ (\theta_1 - q_{d_1}) \\ (q_1' - q_{d_1}') \\ (q_2' - q_{d_2}') \\ (\theta' - q_{d_1}') \end{bmatrix} + \begin{bmatrix} \overline{\overline{T}}_1 \\ \overline{\overline{T}}_2 \end{bmatrix} \tag{36}$$

where

$$K = \begin{bmatrix} 102.3 & -0.07 & -103.0 & 15.97 & -0.10 & -17 \\ 16.17 & -0.01 & -16.3 & 2.55 & -0.02 & -2.71 \end{bmatrix}$$

(obtained by standard LQ techniques) the eigenvalues of A_4 are moved to $-2.61 \pm j29.58$. $v = (\overline{\overline{T}}_1, \overline{\overline{T}}_2)^T$ is the new control input. The application of DPC to this new system then yields a stable control for all $h < h_{\max}$, $h_{\max} = 0.3$ sec.

Numerical implementation The DAE we use to compute DPC is composed of (32),(33),(34), the inputs T_1 and T_2 are replaced by the controls defined in (36) (friction is considered to be precompensated, that is, the DAE does not include the friction model in $\{x_1, x_2\}$).

In the following simulations we have taken as controller parameters $\alpha_1 = \alpha_2 = 2$, $k = 12$ and $m = 8$. Furthermore, the sizes of the sampling period and the prediction horizon are identical, that is, $h = \epsilon$.

We take the same reference trajectory as in [43]. For sampling periods $\epsilon = h \approx 0.04$, DPC works without preliminary feedback and preliminary feedback yields no major performance improvement. For smaller h, DPC without preliminary feedback destabilizes the controlled system and the use of preliminary feedback is necessary.

DPC without preliminary feedback For given values of ϵ and h, the closed loop system can be determined to be stable for $0.04 \leq h \leq 0.12$. If we assume infinitesimal ϵ, we can determine lower and upper bounds for h, $h_{\text{min}} = 0.053$ and $h_{\text{max}} = 0.158$. The *root locus* in Figure 2 shows the root which determines stability of the the characteristic equation of the delay differential system (29) as a function of h. Note that the root has positive real part on several h intervals of physical interest.

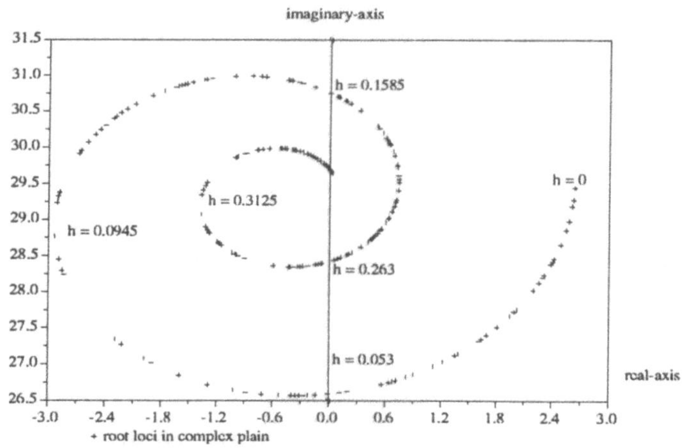

Fig. 2 Critical pole of linearization as h varies: plot of w with $m = 1$.

DPC with preliminary feedback If preliminary feedback is applied we see in Figure 3 one root of the characteristic equation of the delay differential system. The root is stable for all h. However, the closed loop system is unstable for $h > h^*$, since other roots, introduced by the delay, shift to the right hand side of the complex plane. The discrete linear closed loop system is stable for $0 < h < 0.24$.

Preliminary feedback does not improve tracking. However, the error decreases gradually as h and ϵ becomes small.

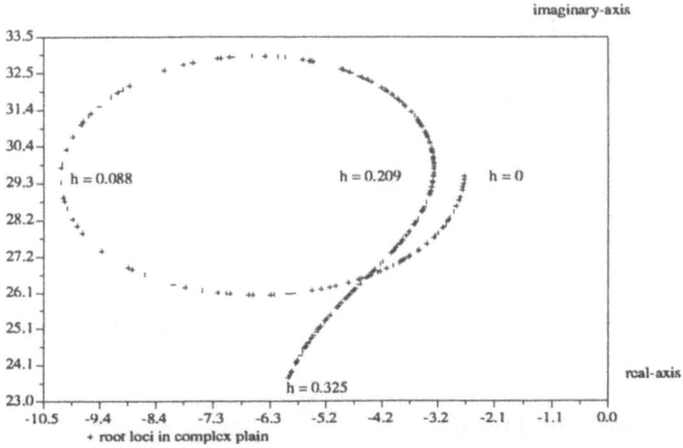

Fig. 3 Critical pole of linearization as h varies: plot of w with $m = 1$.

The application of DPC to the nonlinear system with and without preliminary feedback gives the tracking errors in the first and second links as shown in Figures 4 and 5. The results support the linear analysis.

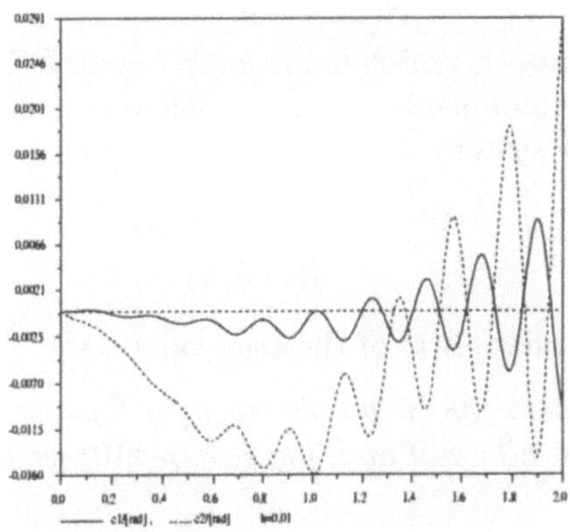

Fig. 4 Nonlinear simulation results for $h = \delta = 0.01$ without preliminary feedback.

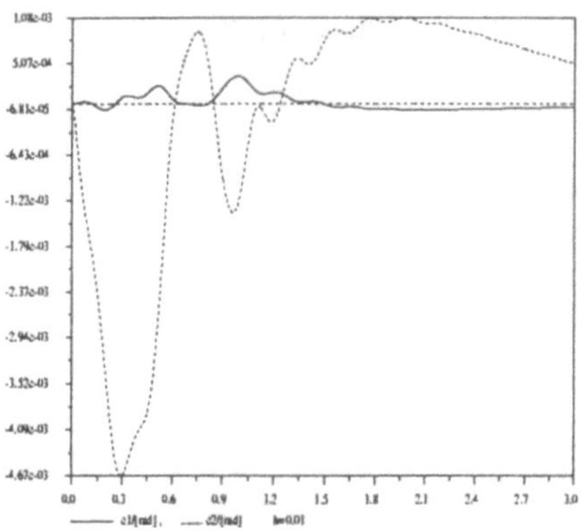

Fig. 5 Nonlinear simulation results for $h = \delta = 0.01$ with preliminary feedback.

3.5 DAE Observer Design

We have looked at observers for linear time varying DAEs [6] and there have been a number of papers on observers for special classes of nonlinear DAEs particularly with respect to mechanical systems [42]. Here we point out how a DAE perspective can be advantageous even for the ODE case. Thus consider the nonlinear system

$$x' = f(x, u) \tag{37a}$$
$$y = h(x) \tag{37b}$$

where f and h are smooth vector fields on $I\!R^n$ and $I\!R^p$, and the p measurements y in (37b) are independent. The problem of observer design consists in finding a nonlinear system

$$0 = \widehat{f}(\omega', \omega, u, y) \tag{38a}$$
$$\widehat{x} = \widehat{g}(\omega, u, y) \tag{38b}$$

that generates an estimate $\widehat{x}(t)$ of the true value $x(t)$.

Definition 3.1. *System (38) is an observer for System (37a), if* $\widehat{x}(0) = x(0)$ *implies* $\widehat{x}(t) = x(t)$ *for all* $t > 0$ *and, for* $\widehat{x}(0) \neq x(0)$ *we have that* $\widehat{x}(t) \to x(t)$ *as* t *goes to infinity.*

The problem of how to design system (38) to be an observer for (37a) has been extensively studied in the past. We show that there can be advantages in formulating the observer (38) as an implicit (descriptor) system which can then be solved using a numerical solver. One advantage is that if (37) has a special form and verifies some algebraic conditions, then we can easily construct an DAE index one observer that has linear time invariant observer error dynamics.

The class of nonlinear systems for which we can get linear error dynamics is even larger if we allow (38) to be an index two Hessenberg DAE. These DAEs can be integrated by fixed step BDF methods or an IRK code like RADAU5. The usefulness of this approach will be shown on a simple example. For a more detailed analysis of DAE index two design and its application to mechanical type problems see [83, 85].

Index one DAE observer. System (37) is a DAE in x (u and y are supposed to be known). This over-determined (n unknowns, $n+p$ equations) DAE describes all the constraints (information) that we have for constructing $\hat{x}$. To make this DAE integrable, we can do relaxation by introducing a p-dimensional vector $\lambda(t)$ into this DAE. One way to introduce λ is in the algebraic part of the DAE which is the observation (37b) to get

$$\widehat{x}' = f(\widehat{x}, u) + g(\lambda) \tag{39a}$$
$$0 = y - h(\widehat{x}) + \lambda \tag{39b}$$

This means that we relax the entire estimate $\widehat{x}$. This is the usual observer design where λ is the residual.

The relationship $y = h(x)$ provides direct information on part of x. Rather than relax this information, we relax the system by introducing λ only in the differential part (37a). This is the way index one DAE observer design is introduced in [64]. Throughout the remainder of this paper we use the following definition.

Definition 3.2. *The DAE of the form*

$$\widehat{x}' + \Phi(\widehat{x})\lambda' = f(\widehat{x}, u) + \Gamma(\widehat{x}, u)\lambda \tag{40a}$$
$$0 = y - h(\widehat{x}) \tag{40b}$$

is called a canonical observer DAE for System (37).

For $\Phi = -h_x(x) = -\partial h(x)/\partial x$, Γ any matrix of appropriate dimensions depending on $\widehat{x}$ and u, we recover the canonical index one DAE observer form of [64].

Note that for $\lambda = 0$ we have the original system equations (37). Under appropriate observability conditions, $\lambda(t) \to 0$ implies that $\widehat{x}(t) \to x(t)$ for $\widehat{x}(0) \neq x(0)$. Thus Γ has to be chosen so that $\lambda(t) \to 0$. To this end we observe that for sufficiently small observation error $\widetilde{x} = \widehat{x} - x$ we have

$$\begin{bmatrix} I & -\Phi(\widehat{x}) \\ 0 & 0 \end{bmatrix} \begin{bmatrix} \widetilde{x}' \\ \lambda' \end{bmatrix} = \begin{bmatrix} f_x(\widehat{x}) & \Gamma(\widehat{x}, u) \\ h_x(\widehat{x}) & 0 \end{bmatrix} \begin{bmatrix} \widetilde{x} \\ \lambda \end{bmatrix} + O(\|\widetilde{x}\|^2) \tag{41}$$

System (41) has an equilibrium point at $(\widetilde{x}, \lambda) = (0, 0)$. For fixed $(\widehat{x}, u)$ the goal is to chose Γ such that this equilibrium point is stable for all fixed $(\widehat{x}, u)$. Since linear systems are generally easier to control it is often desirable to have linear error dynamics if possible. Linear time invariant error dynamics are even better since then there is direct control on how fast the error goes to zero. We shall show that this is possible for certain classes of nonlinear systems using DAE observers.

If system (37) is in the special form

$$\begin{aligned} x_1' &= f_1(x_1,u) + F_1(x_1,u)x_2 \\ x_2' &= f_2(x_1,u) + F_2(x_1,u)x_2 \\ y &= x_1 \end{aligned} \tag{42}$$

we will say it is in *index one DAE observer normal form.* If we chose

$$\Phi = \begin{bmatrix} \Omega_1^{-1}(y) \\ \Omega_1^{-1}(y)\Omega_2(y) \end{bmatrix}, \quad \Gamma = \begin{bmatrix} \Gamma_1(y) \\ \Gamma_2(y) \end{bmatrix}$$

then (40) is an index one DAE and (41) becomes the linear time varying system

$$\begin{bmatrix} \lambda' \\ \widetilde{x}_2' \end{bmatrix} = \begin{bmatrix} \Omega_1\Gamma_1 & \Omega_1 F_1 \\ \Gamma_2 - \Omega_2\Gamma_1 & F_2 - \Omega_2 F_1 \end{bmatrix} \begin{bmatrix} \lambda \\ \widetilde{x}_2 \end{bmatrix} \tag{43}$$

There exist a number of methodologies to stabilize (43) by use of Ω and Γ such as extended linearization and Lyapunov design. Here we consider the case where (43) can be made time invariant. For that we need that the matrices F_1 and F_2 have a particular structure.

Theorem 3.2. *Let $F_2(\widehat{x}_1,u) = \overline{F}_2 + \widetilde{F}_2(\widehat{x}_1,u)$ where $\overline{F}_2$ is constant. If*

1. *there exists an invertible matrix Ω_1 such that $\Omega_1 F_1$ is constant,*
2. *$\overline{F}_2$ can be chosen such that the matrix pair $\{F_1, \overline{F}_2\}$ is observable and there exists a matrix F_2^* such that $\widetilde{F}_2 = F_2^* F_1$*

then the error equation (43) can be made time invariant. Its modes can be arbitrarily placed by proper choices of Γ_1 and Γ_2 for $\Omega_2 = F_2^*$.

Index two DAE observer. The idea of using DAEs as observers for (37) can be extended to the case where (40) is a system that can be put in Hessenberg semi explicit index two form. This generalization of DAE index one observer design is, for example, of interest in the case when the original system (37) can be put in the form

$$\begin{aligned} x_1' &= x_2 \\ x_2' &= f_2(x_1,x_2,u) + F_2(x_1,u)x_3 \\ x_3' &= f_3(x_1,x_2,u) + F_3(x_1,u)x_3 \\ y &= x_1 \end{aligned} \tag{44}$$

Note that (42) required the unobserved variable x_2 to appear linearly while (44) allows the unobserved variables x_2 to appear nonlinearly. This particular form is frequently encountered in mechanical systems. For example, flexible joint robots can be in this form [85]. We will be refer to (44) as the *index two DAE observer normal form.* We take the observer to be the canonical observer DAE (40) with

$$\Phi = \begin{bmatrix} 0 \\ \Omega_1^{-1}(y) \\ \Omega_2(y)\Omega_1^{-1}(y) \end{bmatrix}, \qquad \Gamma = \begin{bmatrix} 0 \\ \Gamma_1(y,\widehat{x}_2,u) \\ \Gamma_2(y,\widehat{x}_2,u) \end{bmatrix}$$

which we will refer to as *index two canonical DAE observer form.* $\Omega_1, \Omega_2, \Gamma$ remain to be chosen. As noted earlier not all index two DAEs can be safely integrated. To insure integrability, we will show that the index two canonical DAE observer form can be transformed into Hessenberg form.

The observer equations are the following DAE in x_1, x_2, x_3, λ.

$$\begin{aligned} \widehat{x}_1' &= \widehat{x}_2 && \text{(45a)} \\ \widehat{x}_2' &= f_2(\widehat{x}_1,\widehat{x}_2,u) + F_2(\widehat{x}_1,u)\widehat{x}_3 - \Omega_1^{-1}(y)\lambda' + \Gamma_1(y,\widehat{x}_2,u) && \text{(45b)} \\ \widehat{x}_3' &= f_3(\widehat{x}_1,\widehat{x}_2,u) + F_3(\widehat{x}_1,u)\widehat{x}_3 - \Omega_2(y)\Omega_1^{-1}(y)\lambda'_\Gamma(y,x_2,u) && \text{(45c)} \\ y &= \widehat{x}_1 && \text{(45d)} \end{aligned}$$

Lemma 3.2. *The change of coordinates*

$$\begin{aligned} \omega_1 &= \widehat{x}_1 \\ \omega_2 &= \widehat{x}_2 \\ \omega_3 &= \widehat{x}_3 - \Omega_2(\widehat{x}_1)\widehat{x}_2 \\ \omega_4 &= \lambda + \Omega_1(\widehat{x}_1)\widehat{x}_2 \end{aligned} \tag{46}$$

transforms (45) into the Hessenberg semi explicit index two form

$$\begin{aligned} \omega_1' &= \omega_2 \\ \omega_3' &= \widehat{f_2}(\omega_1,\omega_2,\omega_4,u) + \widehat{F_2}(\omega_1,\omega_2,u)\omega_3 \\ \omega_4' &= \widehat{f_3}(\omega_1,\omega_2,\omega_4,u) + \widehat{F_3}(\omega_1,\omega_2,u)\omega_3 \\ 0 &= y - \omega_1 \end{aligned} \tag{47}$$

To see that (47) is Hessenberg according to the criteria of (13) we take $z = (\omega_1, \omega_3, \omega_4)$ and $y = \omega_2$. Due to the assumption that (37a) is in index two DAE observer normal form (44), the error equation is the linear time varying system.

$$\begin{bmatrix} \lambda' \\ \widetilde{x}_3' \end{bmatrix} = \begin{bmatrix} \Omega_1\Gamma_1 & \Omega_1 F_2 \\ \Gamma_2 - \Omega_2\Gamma_1 & F_3 - \Omega_2 F_2 \end{bmatrix} \begin{bmatrix} \lambda \\ \widetilde{x}_3 \end{bmatrix} \tag{48}$$

To get a time invariant linear error equation we need that F_2 and F_3 have a particular structure.

Theorem 3.3. *Let $F_3(\widehat{x}_1, \widehat{x}_2, u) = \overline{F}_3 + \widetilde{F}_3(\widehat{x}_1, \widehat{x}_2, u)$ where $\overline{F}_3$ is a constant matrix. Then, if*

1. *there exists an invertible matrix Ω_1 such that $\Omega_1 F_2$ is constant*
2. *$\overline{F}_3$ can be chosen such that the matrix pair $\{F_2, \overline{F}_3\}$ is observable and there exists a matrix F_3^* such that $\widetilde{F}_3 = F_3^* F_2$,*

then the error equation (48) can be made time invariant and its modes can arbitrarily be placed by proper choice of Γ_1 and Γ_2 for $\Omega_2 = F_3^*$.

A Nonlinear Example. We use as example the model of a three-phase current motor [9, 64]. We show on this model how to apply index one DAE observer design if (37) is in index one DAE normal form. Furthermore, we show that DAE index two observer design may yield linear time invariant error dynamics even if DAE index one observer design does not. This shows that the extension of DAE observer design to index two DAEs allows us to enlarge the class of nonlinear systems with linearizable error dynamics. The model is

$$x_1' = x_2 \tag{49a}$$

$$x_2' = B - A_1 x_2 - A_2 x_3 \sin(x_1) + \frac{1}{2}\sin(2x_1) \tag{49b}$$

$$x_3' = u - D_1 x_3 + D_2 \cos(x_1) \tag{49c}$$

where $x = (x_1, x_2, x_3)^T$ is the state, u a control input and B, A_1, A_2 , D_1 and D_2 are constants.

For the index one DAE observer design we use the output

$$y = [x_1, x_2]^T \tag{50}$$

System (49) is in index one DAE normal form (42) with

$$f_1(y, u) = \begin{bmatrix} x_2 \\ B - A_1 x_2 + \frac{1}{2}\sin(2x_1) \end{bmatrix}, \quad F_1(y, u) = \begin{bmatrix} 0 \\ -A_2 \sin(x_1) \end{bmatrix},$$

$$f_2(y, u) = u + D_2 \cos(x_1), \quad F_2(y, u) = -D_1$$

To see that (49) has for the output (50) a linear time invariant error equation we need the F_1 and F_2 satisfy the assumptions of Theorem 3.2 with $F_2(y) = \overline{F}_2$ and

$$\Omega_1 = \begin{bmatrix} 1 & 0 \\ 0 & -\frac{k_3}{A_2 \sin(x_1)} \end{bmatrix} \tag{51}$$

k_3 is a constant scalar. The poles of the error equation can be placed at arbitrary locations since the matrix pair $\{F_1, \overline{F}_2\}$ is observable.

For

$$\Gamma_1 = \begin{bmatrix} k_1 & 0 \\ 0 & -\frac{k_2}{k_3} A_2 \sin(x_1) \end{bmatrix}, \; \Gamma_2 = [k_4, k_5], \; \Omega_2 = [0, 0]$$

and (51) we obtain the DAE index one observer

$$\begin{aligned}\widehat{x}_1' &= \widehat{x}_2 - \lambda_1' + k_1\lambda_1\\ \widehat{x}_2' &= B - A_1\widehat{x}_2 - A_2\widehat{x}_3\sin(\widehat{x}_1) + \frac{1}{2}\sin(2\widehat{x}_1) + A_2\sin(\widehat{x}_1)\frac{1}{k_3}(\lambda_2' - k_2\lambda_2)\\ \widehat{x}_3' &= u - D_1\widehat{x}_3 + D_2\cos(\widehat{x}_1) + k_4\lambda_1 + k_5\lambda_2\\ y_1 &= \widehat{x}_1\\ y_2 &= \widehat{x}_2\end{aligned}$$

which has the linear time invariant error equation

$$\begin{bmatrix}\lambda_1'\\ \lambda_2'\\ \widetilde{x}_3'\end{bmatrix} = \begin{bmatrix}k_1 & 0 & 0\\ 0 & k_2 & k_3\\ k_4 & k_5 & -D_3\end{bmatrix}\begin{bmatrix}\lambda_1\\ \lambda_2\\ \widetilde{x}_3\end{bmatrix}$$

This result is consistent with [9] where it is shown that System (49) admits a linear time invariant error equation for the output (50).

Now, suppose that instead of two observations, we have

$$y = x_1 \tag{52}$$

System (49) is no longer in index one DAE observer normal form (42). More importantly, the conditions of nonlinear observer design of [9] or [49] (or [48] for the single output case) show that System (49) cannot be transformed into nonlinear observer form and, consequently, that it has no linear error equation for the output (52).

System (49) is in index two DAE observer normal form (46), where

$$\begin{aligned}f_2(y,y',u) &= B - A_1x_2 + \frac{1}{2}\sin(2x_1)\\ F_2(y,y',u) &= -A_2\sin(x_1)\\ f_3(y,y',u) &= D_2\sin(x_1) + u\\ F_3(y,y',u) &= -D_3\end{aligned}$$

Furthermore, F_3 and F_2 satisfy Theorem 3.3 since we have $F_3 = \overline{F}_3$. The choice

$$\Omega_1 = -\kappa_2/(A_2\sin(x_1)) \tag{53}$$

yields

$$\begin{aligned}\widehat{x}_1' &= \widehat{x}_2\\ \widehat{x}_2' &= B - A_1\widehat{x}_2 - A_2\widehat{x}_3\sin(\widehat{x}_1) + \frac{1}{2}\sin(2\widehat{x}_1) + A_2\sin(x_2)\frac{1}{\kappa_2}(\lambda' - \kappa_1\lambda)\\ \widehat{x}_3' &= u - D_1\widehat{x}_3 + D_2\cos(\widehat{x}_1) + \kappa_3\lambda\\ y &= \widehat{x}_1\end{aligned}$$

which is an index two DAE provided $\widehat{x}_1 \neq \pi k,\ k = 0, \pm 1, \pm 2, \cdots$. The error equation is

$$\begin{bmatrix} \lambda' \\ \tilde{x}_3' \end{bmatrix} = \begin{bmatrix} \kappa_1 & \kappa_2 \\ \kappa_3 & -D_1 \end{bmatrix} \begin{bmatrix} \lambda \\ \tilde{x}_3 \end{bmatrix}$$

To insure integrability we need to transform the index two DAE into Hessenberg form, that is we need that the observer (38) be in the special form (47). We have for the observer (38)

$$0 = \begin{bmatrix} \omega_1' - \omega_2 \\ \omega_3' - \widehat{f}_2(\omega_1, \omega_2, \omega_4, u) - \widehat{F}_2(\omega_1, \omega_2, u)\omega_3 \\ \omega_4' - \widehat{f}_3(\omega_1, \omega_2, \omega_4, u) - \widehat{F}_3(\omega_1, \omega_2, u)\omega_3 \\ y - \omega_1 \end{bmatrix}$$
$$(\widehat{x}_1, \widehat{x}_2, \widehat{x}_3) = (\omega_1, \omega_2, \omega_3)$$

where

$$\begin{aligned}
\widehat{f}_2 &= D_2 \sin(\omega_1) + u + \kappa_3 \lambda \\
\widehat{f}_3 &= \frac{-\kappa_2}{A_2 \sin(\omega_1)} \left(B - A_1 \omega_2 + \frac{1}{2} \sin(2\omega_1) + \frac{\kappa \omega_2^2 \cos \omega_1}{A_2 \sin^2(\omega_1)} \right) + \kappa_1 \lambda \\
\widehat{F}_2 &= -A_2 \sin(\omega_1) \\
\widehat{F}_3 &= \kappa_2 \\
\lambda &= \omega_4 + \kappa_2 \omega_1 / (A_2 \sin(\omega_1)
\end{aligned}$$

3.6 Future Problems

There are several areas where DAE formulations have proved useful in the linear time invariant case but work needs to be done in the nonlinear or linear time varying case. One of these is in infinite dimensional descriptor systems including, for example, PDEs. Another consideration will be more complex "mixed" systems with a large number of constraints which will be going active and inactive during system evolution.

4 Conclusion

The preceding discussion has shown that there are many potential advantages from using descriptor theory. Sometimes the insight or design freedom gained is helpful even for explicit models. However, there are several differences between the theory for general descriptor systems and the theory of either explicit systems or linear time invariant descriptor systems. These differences concern both analytical and numerical issues and must be taken into account. Considerable progress has been made in the application of descriptor systems but many questions remain.

Acknowledgement. Many students and colleagues have contributed to the ideas in this paper but we would like to especially acknowledge Dirk von Wissel who was intimately involved with some of the research discussed here. Research of the first author was supported in part by the National Science Foundation under DMS-9714811, DMS-9802259, and INT-9605114.

References

1. J.D. Aplevich. *Implicit Linear Systems.* Lecture Notes in Control and Information Sciences, Springer-Verlag, 1991.
2. U. Ascher and L. Petzold. The numerical solution of delay-differential-algebraic equations of retarded and neutral type. *SIAM J. Numer. Anal.*, 32:1635-1657, 1995.
3. J. Baumgarte. Stabilization of constraints and integrals of motion in dynamical systems. *Comp. Math. Appl. Mech. Engr.*, 1:1-16, 1972.
4. J.T. Betts and W.P. Huffman. Mesh Refinement in Direct Transcription Methods for Optimal Control. *Optimal Control Applications & Methods*, 19:1-21, 1998.
5. L.T. Biegler. Optimization strategies for complex process models. *Advances in Chemical Engineering*, 18:197-256, 1992.
6. N. Biehn, S.L. Campbell, F. Delebecque, and R. Nikoukhah. Observer Design for Linear Time Varying Descriptor Systems: Numerical Algorithms. In *Proc. 37th IEEE-CDC*, Tampa, 1998, 3801-3806 .
7. N. Biehn, S.L. Campbell, L. Jay, and T. Westbrook. Some comments on DAE theory for IRK methods and trajectory optimization. preprint, 1999.
8. J. Birk and M. Zeitz. Extended Luenberger observer for non-linear multivariable systems. *Int. J. Control*, 47:1823-1836, 1988.
9. P. A Bliman and M. Sorine. A system-theoretic approach of systems with hysteresis. Application to friction modelling and compensation. In *Proc. ECC93*, 1993, 1844-1849.
10. H.G. Bock, H.X. Phu, and J.P. Schlöder. Extremal solutions of some constrained control problems. *Optimization*, 35:345-355, 1995.
11. K.E. Brenan, S.L. Campbell, and L.R. Petzold. *Numerical Solution of Initial-Value Problems in Differential-Algebraic Equations.* SIAM, 1996.
12. S.L. Campbell. Singular linear systems of differential equations with delays. *Applicable Analysis*, 11:129-136, 1980.
13. S.L. Campbell. 2-D (differential-delay) implicit systems. In *Proc. IMACS World Congress on Scientific Computation* Dublin, 1991, 1828-1829.
14. S.L. Campbell. A general form for solvable linear time varying singular systems of differential equations. *SIAM J. Math. Anal.*, 18:1101-1115, 1987.
15. S.L. Campbell. DAE approximations of PDE modeled control problems. In *Proc. IEEE Mediterranean Symposium on New Directions in Control and Automation*, Crete, 1994, 407-414.
16. S.L. Campbell. High index differential algebraic equations. *J. Mech. Struct. & Machines*, 23:199-222, 1995.
17. S.L. Campbell. Linearization of DAEs along trajectories. *Z. angew Math. Phys. (ZAMP)*, 46:70-84, 1995.
18. S.L. Campbell and C.W. Gear. The index of general nonlinear DAEs. *Numerische Mathematik*, 72:173-196, 1995.
19. S.L. Campbell and E. Griepentrog. Solvability of general differential algebraic equations. *SIAM J. Scientific Computation*, 16:257-270, 1995.

20. S.L. Campbell and R. Hollenbeck. Automatic differentiation and implicit differential equations. In *Computational Differentiation: Techniques, Applications, and Tools*, SIAM, Philadelphia, 1996, 215-227.
21. S.L. Campbell and W. Marszalek. The index of an infinite dimensional implicit system. *Mathematical Modelling of Systems.* to appear.
22. S.L. Campbell and E. Moore. Constraint preserving integrators for general nonlinear higher index daes. *Numerische Mathematik*, 69:383-399, 1995.
23. S.L. Campbell and W.J. Terrell. Observability of linear time varying descriptor systems. *SIAM J. Matrix Analysis*, 12:484-496, 1991.
24. S.L. Campbell and with W.J. Terrell. Determining flatness for complex nonlinear systems. In *Proc. IEEE Southeastcon '95*, Raleigh, NC, 1995, 118-122.
25. S.L. Campbell, N.K. Nichols, and W.J. Terrell. Duality, observability, and controllability for linear time varying descriptor systems. *Circuits, Systems, Signal Process*, 10:455-470, 1991.
26. M. Caracotsios and W.E. Stewart. Sensitivity analysis of initial-boundary-value problems with mixed PDEs and algebraic equations. Applications to chemical and biochemical systems. preprint, 1994.
27. J. Chen. On Computing the Maximal Delay Intervals for Stability of Linear Delay Systems. *IEEE Trans. on Automatic Control*, 40(6):1087-1093, 1995.
28. C.C. Cheah and D. Wang. Learning control for a class of nonlinear differential-algebraic systems with application to constrained robots. In *Proc. 94ACC*, Baltimore, June, 1994, 1737-1742.
29. P.D. Christofides and P. Daoutidis. Finite dimensional control of parabolic PDE systems using approximate inertial manifolds. preprint, 1997.
30. J.E. Cuthrell and L.T. Biegler. On the optimization of differential-algebraic process systems. *AIChE Journal*, 33:1257-1270, 1987.
31. L. Dai. *Singular Control Systems.* Lecture Notes in Control and Information Sciences, Springer-Verlag, 1989.
32. S. Diop, J.W. Grizzle, P.E. Moraal, and A. Stefanopoulou. Interpolation and numerical differentiation for observer design. In *Proc 94 ACC*, 1994, 1329-1333.
33. W.F. Feehery and P.I. Barton. Dynamic simulation and optimization with inequality path constraints. *Comp. and Chem Engng.*, 20:S707-S712, 1996.
34. W.F. Feehery and P.I. Barton. A differentiation-based approach to dynamic simulation and optimization with high-index differential-algebraic equations. preprint, 1996.
35. W.F. Feehery, J.R. Banga, and P.I. Barton. A novel approach to dynamic optimization of ODE and DAE systems as high-index problems. *AICHE annual meeting*, Miami, 1995.
36. V. Gopal and L.T. Biegler. A Successive linear programming approach for initialization and reinitialization after discontinuities of differential algebraic equations. preprint, 1996.
37. M. Günther and U. Feldmann. The DAE-index in electric circuit simulation. *Mathematics and Computers in Simulation,* 39:573-582, 1995.
38. E. Hairer, C. Lubich, and M. Roche. *The Numerical Solution of Differential-algebraic Systems by Runge-Kutta Methods.* Springer Lecture Notes in Mathematics No. 1409, 1989.
39. E. Hairer and G. Wanner. *Solving Ordinary Differential Equations II, Stiff and Differential Algebraic Problems*, Springer-Verlag, 1991.

40. R.M. Hirschorn and J.H. Davis. Global output tracking for nonlinear systems. *SIAM J. Contr. Opt.*, 26:1321-1330, 1988.
41. M. Hou, Th. Schmidt, R. Schüpphaus, and P.C. Müller. Normal form and Luenberger Observer for linear mechanical descriptor systems. *J. of Dynamic Systems, Measurement, and Control*, 115:611-620, 1993.
42. K.P. Jankowski and H. Van Brussel. An approach to discrete inverse dynamics control of flexible-joint robots. *IEEE Transactions on Robotics & Automation*, 8:651-658, 1992.
43. K.P. Jankowski and H. Van Brussel. Inverse dynamics task control of flexible joint robots-part I: continuous-time approach. *Mechanism and Machine Theory*, 28:741-750, 1993.
44. K.P. Jankowski and H. Van Brussel. Inverse dynamics task control of flexible joint robots-part II: discrete-time approach. *Mechanism and Machine Theory*, 28:751-763, 1993.
45. S. Kawaji and E.Z. Taha. Feedback linearization of a class of nonlinear descriptor systems. In *Proc. 33 IEEE-CDC*, 1994, 4035-4042.
46. J.Y. Keller, S. Nowakowski and M. Darouach, State estimation and failure detection in singular systems. *Control-Theory and Advanced Technology*, 8(4):755-762, 1992.
47. A.J. Krener and A. Isidori. Linearization by output injection and nonlinear observers. *Sys. Contr. Letters*, 3:47-52, 1983.
48. A.J. Krener and W. Respondek. Nonlinear observers with linearizable error dynamics. *SIAM J. Cont. Opt.*, 23(2):197-216, 1985.
49. H. Krishnan, *Control of Nonlinear Systems with Applications to Constrained Robots and Spacecraft Attitude Stabilization*, Ph.D. Thesis, Department of Aerospace Engineering, University of Michigan, 1992.
50. H. Krishnan and N.H. McClamroch. Computation of state realizations for control systems described by a class of differential-algebraic equations. *Int. J. Control*, 55:1424-1441, 1992.
51. A. Kumar and P. Daoutidis. Control of nonlinear differential-algebraic process systems. In *Proc. 94 ACC*, 1994, 330-334.
52. P. Kunkel and V. Mehrmann. Canonical forms for linear differential-algebraic equations with variable coefficients. *J. Comp. Appl. Math.*, 56:225-251, 1994.
53. P. Kunkel and V. Mehrmann, Local and global invariants of linear differential-algebraic equations and their relation, *Electron. Trans. Numer. Anal.*, 4:138-157, 1996.
54. P. Kunkel and V. Mehrmann, A new class of discretization methods for the solution of linear differential-algebraic equations with variable coefficients. *SIAM J. Numer. Anal.*, 33:1941-1961, 1996.
55. P. Kunkel, V. Mehrmann, W. Rath and J. Weickert. A new software package for linear differential-algebraic equations. *SIAM J. Sci. Comp.*, 18:115-138, 1997.
56. F.L. Lewis. A survey of linear singular systems. *Circuits Systems & Signal Processing*, 5:3-36, 1986.
57. F.L. Lewis. Geometric design techniques for observers in singular systems. *Automatica*, 26:411-415, 1990.
58. W. Lucht, K. Strehem, and C. Eichler-Liebenow. Linear partial differential algebraic equations Part I: Indexes, consistent boundary/initial conditions. Report No. 17, Universität Halle, 1997.

59. W. Lucht, K. Streheml, and C. Eichler-Liebenow. Linear partial differential algebraic equations Part II: Numerical solution. Report No. 18, Universität Halle, 1997.
60. N.H. McClamroch. Feedback stabilization of control systems described by a class of nonlinear differential-algebraic equations. *Systems & Control Letters*, 15:53-60, 1990.
61. W. Martinson and P. Barton. A differential index for partial differential equations. preprint, 1998.
62. H. Michalska and D. Mayne. Receding horizon control of nonlinear systems without differentiability of the optimal value function. *Systems & Control Letters*, 16:123-130, 1991.
63. R. Nikoukhah. A new methodology for observer design and implementation. *IEEE Trans. Automatic Control*, 43:229-234, 1998.
64. C.C. Pantelides, R.W.H. Sargent, and V.S. Vassiliadis. Optimal Control of Multistage Systems Described by High-Index Differential-Algebraic Equations. *Computational Optimal Control* , International Series of Numerical Mathematics, Birkhauser Publishers, Basel, 115:177-191, 1994.
65. T. Park and P.I. Barton. State event location in differential-algebraic models. *ACM Transactions on Modeling and Computer Simulation*, 6:137-165, 1996.
66. R. Pytlalk, M.J. Mohideen, and E.N. Pistikopoulos. Numerical procedure for optimal control of differential-algebraic equations. In *Proc. Computational Engineering in Systems Applications (CESA96)*, Lille, France, 1996, 398-402.
67. P.J. Rabier and W.C. Rheinboldt. Time-dependent linear DAEs with discontinuous inputs. *Linear Algebra Appl.*, 247:1-29, 1996.
68. P.J. Rabier and W.C. Rheinboldt. Discontinuous solutions of semilinear differential-algebraic equations. II. *P*-consistency. *Nonlinear Anal.*, 27:1257-1280, 1996.
69. P.J. Rabier and W.C. Rheinboldt. Discontinuous solutions of semilinear differential-algebraic equations I. Distribution solutions. *Nonlinear Anal.*, 27:1241-1256, 1996.
70. P.J. Rabier and W.C. Rheinboldt. Classical and generalized solutions of time-dependent linear differential-algebraic equations. *Linear Algebra Appl.*, 245:259-293, 1996.
71. S. Reich. On the local qualitative behavior of differential-algebraic equations. *Circuits Systems & Signal Processing*, 14:427-443, 1995.
72. W. Respondek and H. Nijmeijer. On local right-invertibility of nonlinear control systems, *Control-Theory and Advanced Technology*, 4(3):325-348, 1988.
73. W.C. Rheinboldt. MANPAK: a set of algorithms for computations on implicitly defined manifolds. *Comput. Math. Appl.*, 32:15-28, 1996.
74. W.C. Rheinboldt. Solving algebraically explicit DAEs with the MANPAK-manifold-algorithms. *Comput. Math. Appl.*, 33:31-43, 1997.
75. P. Rouchon, M. Fliess, J. Lévine, and P. Martin. Flatness, motion planning and trailer systems. In *Proc. 32 IEEE-CDC*, San Antonio, 1993, 2700-2705.
76. W.J. Terrell. Output-nulling space, projected dynamics, and system decomposition for linear time varying systems. *SIAM J. Control*, 32:876-889, 1994.
77. J. Unger, A. Kröner and W. Marquardt. Structural analysis of Differential-algebraic equation systems - Theory and Applications. *Comp. Chem. Eng.*, 19:867-882, 1995.

78. P. Van Dooren. The computation of Kronecker's Canonical Form of a singular pencil. *Linear Algebra and its Applications*, 27:103-140, 1979.
79. V.S. Vassiliadis, R.W.H. Sargent, and C.C. Pantelides. Solution of a class of multistage dynamic optimization problems 2. problems with path constraints. *Ind. Eng. Chem. Res.*, 33:2123-2133, 1994.
80. R. von Schwerin, M. Winckler, and V. Schulz. Parameter estimation in discontinuous descriptor systems. preprint.
81. D. von Wissel, R. Nikoukhah, F. Delebecque, and S.L. Campbell. Numerically generated path stabilizing controllers: use of preliminary feedback. *Kybernetika*, 6:657-666, 1995.
82. D. von Wissel, R. Nikoukhah, S.L. Campbell, and F. Delebecque. Nonlinear observer design using implicit system descriptions. In *Proc. Computational Engineering in Systems Applications (CESA96)*, Lille, France, 1996, 404-409.
83. D. von Wissel, R. Nikoukhah, F. Delebecque, and S.L. Campbell. The effects of computational delay in descriptor based trajectory tracking control. *Int. J. Control*, 67:251-273, 1997.
84. D. von Wissel, *DAE Control of Dynamical Systems: Example of a Riderless Bicycle*, Docteur Thèse, L'École Nationale Supérieure Des Mines De Paris, 1996.
85. W. Zhu and L. Petzold, Asymptotic stability of linear delay differential algebraic equations and numerical methods. *Applied Numerical Mathematics*, to appear.

10

Two-Dimensional Linear Systems

T. Kaczorek

Warsaw University of Technology, Faculty of Electrcial Engineering,
Institute of Control and Industrial Electronics,
00-662 Warszawa, Koszykowa 7, Poland
e-mail: kaczorek@nov.isep.pw.edu.pl

Preface. In the last two decades considerable effort and progress have been made in the two-dimensional (2D) and multidimensional systems theory and its industrial applications. This part of the book summarizes some recent developments in 2D linear systems theory and its applications.

The consequences of imposing various symmetry conditions on the bounds of wavenumber response of interval finite impulse response of 2D digital filters are investigated in the paper „Robust wavenumber response of multidimensional FIR filters" by Bose and Yang. Some results of the interval finite response filter frequency analysis presented in 1991 by Bose and Kim are extended for the multidimensional case.

2D approach to analysis and design of resource to task allocation processes is proposed in the paper „2D-Analysis of hybrid systems" by Franke. A notion of the set stability is introduced and an 2D approach for testing the stability is proposed. An attempt to establish a general theory of discrete systems by successive specializations of basic structures is presented in the paper „Towards a theory of general discrete systems" by Gregor. The presented approach is based on some ideas developed by Mesarovic and Takahara.

Positive 2D discrete and continuous-discrete linear systems are considered in the paper „Positive 2D discrete and continuous-discrete linear systems" by Kaczorek. Necessary and sufficient conditions for reachability and controllability of the positive 2D Roesser type model and positive 2D continuous-discrete systems are established and the minimum energy control problem is solved.

A survey of recent results concerning the controllability of 2D linear systems is presented in the paper „Controllability of 2D linear systems" by Klamka. Several controllability conditions for different types of linear 2D systems with constant coefficients are established and relationship between different concepts of controllability of 2D linear systems are discussed. A survey of recent results concerning the repetitive processes is presented in the paper „Control theory and applications for repetitive processes" by Owens, Rogers and Galkowski. Repetitive processes are a special class of 2D continuous-discrete linear systems of both theoretical and practical interest. They have their origins in the mining and metal processing industries (the dynamics of the long-wall coal cutting process).

2D models for web processes and industrial web control systems are considered in the paper „Two-dimensional systems concepts for industrial web processes" by Wellstead, Zarrop and Duncan. The web processes form a general class of engineering sysetms which includes for example the paper making and plastic extrusion. The analysis of web processes can be unified via the 2D approach to gives a fresh perspective to the modelling control and sensing of class of industrial processes.

I hope that the reader will find this part of the book highly informative, thought provoking and intellectually stimulating for further research.

10.1

Robust Wavenumber Response of Multidimensional FIR Filters*

N. K. Bose and C. Yang

The Spatial and Temporal Signal Processing Center,
Department of Electrical Engineering, The Pennsylvania State University,
University Park, PA 16802, U. S. A.

Summary. The consequences of imposing various symmetry conditions on the bounds of wavenumber response of interval finite impulse response (FIR) multidimensional digital filters are investigated. For the sake of brevity, discussions are centered around the two-dimensional (2-D) case. However, some of the results arrived at generalize naturally to the multidimensional situation.

1 Introduction

The parameters characterizing a linear shift-invariant system are usually not known precisely. Uncertainties, parametric or non-parametric, are modeled for analysis in various ways. Robustness analysis of systems with parametric interval uncertainty, where each coefficient in the transfer function is allowed to vary within a finite interval, independent of other coefficients, has been a popular topic of research and an extensive body of literature exists with respect to the robust interval stability property of both continuous-time and discrete-time systems [1], [2]. Generalizations of one-dimensional (1-D) results documented in [1], [2] to the multidimensional case has been partially documented [3], [4] and this topic is also a focus for current research [5]. With respect to frequency/wavenumber assessment of interval systems, the results available are more limited [6], [7] and one objective of this paper is to generalize the interval finite impulse response (FIR) filter frequency analysis results in [6] to the multidimensional case. Such generalizations have only been reported for very special cases [8].

---

* This research was conducted under the sponsorship of National Science Foundation grant ECS-9711590

2 General 2-D FIR Filter

The wavenumber response $H(\omega_1,\omega_2)$ of an FIR filter is given by the Fourier transform of the unit impulse response sequence $\{h[n_1,n_2]\}$. If N_1, N_2 are arbitrary but fixed positive integers, then

$$H(\omega_1,\omega_2) = \sum_{k_1=-N_1}^{N_1} \sum_{k_2=-N_2}^{N_2} h[k_1,k_2]e^{-j\omega_1 k_1}e^{-j\omega_2 k_2}. \tag{1}$$

In image processing, we often want to maintain the property of zero phase,

$$H(\omega_1,\omega_2) = H^*(\omega_1,\omega_2), \quad h[n_1,n_2] = h^*[-n_1,-n_2]. \tag{2}$$

where the star superscript denotes complex conjugation. Let R be the region of support of $\{ \{h[n_1,n_2]\} \}$. Consider R to be the union of three non-overlapping index sets: (0,0), R_1 and R_2, where the elements of R_2 are negatives of those of R_1. Then the wavenumber response is

$$H(\omega_1,\omega_2) = h[0,0] + \sum_{(n_1,n_2)\in R_1}\sum h[n_1,n_2]e^{-j(\omega_1 n_1+\omega_2 n_2)} + h[-n_1,-n_2]e^{j(\omega_1 n_1+\omega_2 n_2)}.$$

Let the real and imaginary parts of $h[n_1,n_2]$ be $h_R[n_1,n_2]$ and $h_I[n_1,n_2]$ and suppose that their respective Fourier transforms are $H_R(\omega_1,\omega_2)$ and $H_I(\omega_1,\omega_2)$. Due to the zero-phase condition in Eq. (2),

$$h_R[n_1,n_2] = h_R[-n_1,-n_2], \quad h_I[n_1,n_2] = -h_I[-n_1,-n_2],$$

where

$$h[n_1,n_2] = h_R[n_1,n_2] + jh_I[n_1,n_2].$$

Then,

$$H_R(\omega_1,\omega_2) = h_R[0,0] + \sum_{(n_1,n_2)\in R_1}\sum h_R[n_1,n_2]\cos(\omega_1 n_1+\omega_2 n_2),$$

and

$$H_I(\omega_1,\omega_2) = -2j \sum_{(n_1,n_2)\in R_1}\sum h_I[n_1,n_2]\sin(\omega_1 n_1+\omega_2 n_2).$$

The wavenumber response of the complex coefficient 2-D FIR filter is

$$H(\omega_1,\omega_2) = H_R(\omega_1,\omega_2) + jH_I(\omega_1,\omega_2). \tag{3}$$

In this paper, we focus on certain types of FIR filters, i.e., filters with twofold (half-plane), fourfold (quadrantal), and eightfold (octagonal) symmetries. If the filter is twofold symmetric, then

$$h[n_1,n_2] = h[-n_1,-n_2]. \tag{4}$$

Since the 2-D filter must also satisfy the zero phase constraint in Eq.(2), therefore

$$h[n_1,n_2] = h^*[n_1,n_2]. \tag{5}$$

This means that the FIR filter with twofold symmetry (and, therefore, also fourfold and eightfold symmetric filters) must have real coefficients in the transfer function. In the following section, we will limit our discussion to FIR filters with real coefficients.

3 Wavenumber Response with Fourfold and Eightfold Symmetries of 2-D Filter

More constraints on the coefficients of the filter, than in Eqs. (2) and (5) are imposed to get fourfold and eightfold symmetries. The fourfold symmetry conditions are

$$h[n_1,n_2] = h[-n_1,n_2] = h[n_1,-n_2]. \tag{6}$$

The equivalent constraints in the wavenumber domain are

$$H(\omega_1,\omega_2) = H(-\omega_1,\omega_2) = H(\omega_1,-\omega_2). \tag{7}$$

A fourfold symmetric FIR filter whose $\{h[n_1,n_2]\}$ has a rectangular region of support of size $(2N_1+1)\times(2N_2+1)$ has a wavenumber response of the form

$$\begin{aligned} H(\omega_1,\omega_2) &= h[0,0] + \sum_{n_1=1}^{N_1} 2h[n_1,0]\cos(\omega_1 n_1) + \sum_{n_2=1}^{N_2} 2h[0,n_2]\cos(\omega_2 n_2) \\ &+ \sum_{n_1=1}^{N_1}\sum_{n_2=1}^{N_2} 4h[n_1,n_2]\cos(\omega_1 n_1)\cos(\omega_2 n_2). \end{aligned} \tag{8}$$

The wavenumber response can be specified by an appropriate first-quadrant support sequence $\{a[n_1,n_2]\}$ whose relation to $\{h[n_1,n_2]\}$ is obvious from the preceding and following equations,

$$H(\omega_1, \omega_2) = \sum_{k_1=0}^{N_1} \sum_{k_2=0}^{N_2} a[k_1, k_2] \cos(k_1\omega_1) \cos(k_2\omega_2). \tag{9}$$

Letting $a[k_1, k_2] \in [\underline{a}[k_1, k_2], \overline{a}[k_1, k_2]]$, the interval wavenumber response is

$$P(\omega_1, \omega_2) = \sum_{k_1=0}^{N_1} \sum_{k_2=0}^{N_2} [\underline{a}[k_1, k_2], \overline{a}[k_1, k_2]] \cos(k_1\omega_1) \cos(k_2\omega_2). \tag{10}$$

The product separability of the kernel $\cos(k_1\omega_1)\cos(k_2\omega_2)$ suggests that like in the 1-D case [6], the number of distinct zero-crossing points along the ith axis in the first quadrant, $0 \le \omega_i < \pi$, $i = 1,2$, of the base period is

$$N_{C_i} = \sum_{k=1}^{N_i} \phi(2k),$$

where $\phi(n)$ is the Euler totient function. So, the number of non-overlapping rectangular regions (cells, bands) that partition the base region defined by $0 \le \omega_1 < \pi$ and $0 \le \omega_2 < \pi$ is

$$N_B = \left(\sum_{k_1=1}^{N_1} \phi(2k_1) + 1 \right) \left(\sum_{k_2=1}^{N_2} \phi(2k_2 + 1 \right). \tag{11}$$

In each band, we only need to test the wavenumber response associated with two sets of extreme coefficients of the interval response $P(\omega_1, \omega_2)$ in Eq. (10). Consequently, $2N_B$ extreme functions of the generic form in Eq.(8) are needed to bound the wavenumber response for the 2-D FIR interval filter in Eq.(10). Coefficients of these extreme functions are set from appropriate choices of the upper or lower extreme values of the coefficients $h[n_1, n_2]$ (or $a[n_1, n_2]$). In the eightfold symmetry case, where $N_1 = N_2 = N$,

$$h[n_1, n_2] = h[-n_1, n_2] = h[n_1, -n_2] = h[n_2, n_1], \tag{12}$$

or, equivalently,

$$H(\omega_1, \omega_2) = H(-\omega_1, \omega_2) = H(\omega_1, -\omega_2) = H(\omega_2, \omega_1), \tag{13}$$

The number of distinct zero-crossing points must be the same along the two axes. Because of the symmetry about the diagonal joining points *(0,0)* and (π, π) in the wavenumber domain, we need only consider the responses of the bands below the diagonal. The total number of bands required is, therefore,

$$N_B = \frac{1}{2}\left(\sum_{k=1}^{N}\phi(2k)+1\right)\left(\sum_{k=1}^{N}\phi(2k)+2\right) \tag{14}$$

As in the case of fourfold symmetry, one needs to evaluate $2N_B$ extreme functions in order to bound the wavenumber response for the 2-D FIR interval filter.

4 Wavenumber Response with Twofold Symmetry of 2-D Real Coefficient FIR Filter

Since FIR filters with real coefficients, brought about by the imposition of two-fold symmetry, are the focus of discussion, therefore the general representation of the wavenumber response for a complex coefficient 2-D FIR filter in Eq.(3) will specialize to

$$H(\omega_1,\omega_2) = H_R(\omega_1,\omega_2) = h[0,0] + 2\sum_{(n_1,n_2)\in R_1}\sum h_R[n_1,n_2]\cos(\omega_1 n_1 + \omega_2 n_2). \tag{15}$$

For simplicity, the choice of R_1 is uniquely set by imposing the index constraint $n_1 \geq 0, n_2$ arbitrary, and $[n_1,n_2] \neq [0,0]$ .The generic form of the interval wavenumber response for this case is, therefore (where we focus on the case when $N_1 = N_2 = N$),

$$P(\omega_1,\omega_2) = \sum_{n_1=0}^{N_1}\sum_{k_2=0}^{N_2}[\underline{a}[n_1,n_2],\bar{a}[n_1,n_2]]\cos(n_1\omega_1 + n_2\omega_2]. \tag{16}$$

The generic function $\cos(n_1\omega_1 + n_2\omega_2)$ in Eq.(16) is zero when

$$n_1\omega_1 + n_2\omega_2 = \frac{\pi(2k-1)}{2}, \quad k \in N_+, \tag{17}$$

Kernel term	$\cos(\omega_1)$	$\cos(\omega_2)$	$\cos(\omega_1+\omega_2)$	$\cos(\omega_1-\omega_2)$
Cells with positive valued kernel term	1,2,5,6, 9,10,13,14	1,2,3,4, 9,10,11,12	1,8,9,10, 11,14,15,16	1,2,3,6, 7,8,9,16

Table 1 : Cells with positive valued kernel term in Figure 1

where N_+ denotes the set of positive integers. The quadrant in the base domain over which the wavenumber response is calculated has for its four corners the points *(0,0)*, *(0,π)*, *(π,π)* and *(π,0)*. The results in this section which are the counterparts of the restricted cases of fourfold and eightfold symmetries in the previous sections are developed next by proceeding from lower order to higher order FIR filters.

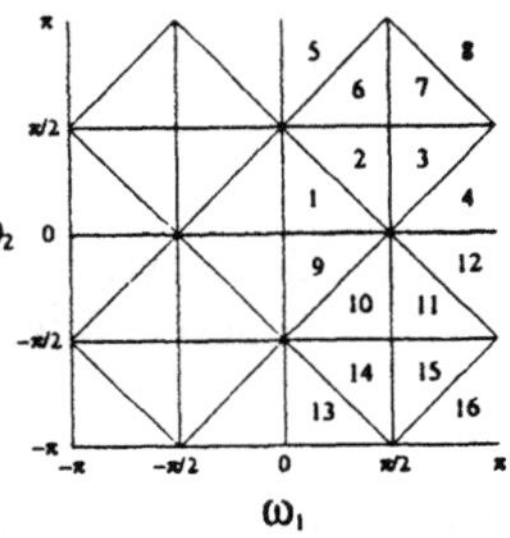

Figure 1

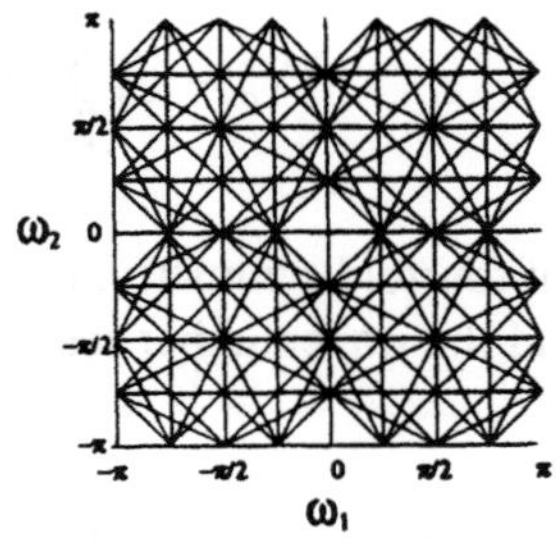

Figure 2

Example 1: Consider the (3 x 3)-D FIR filter with twofold symmetry for which $N_1 = N_2 = N = 1$. The wavenumber response in this case, where $R_1 = \{[1,0],[0,1],[1,1],[1,-1]\}$, is

$$P(\omega_1,\omega_2] = a[0,0] + a[1,0]\cos(\omega_1) + a[0,1]\cos(\omega_2)$$
$$+a[1,1]\cos(\omega_1+\omega_2) + a[1,-1]\cos(\omega_1-\omega_2).$$

The tessellation of the quadrant defined by $0 \le \omega_1 < \pi$ and $0 \le \omega_2 < \pi$ into cells, so that the sign of each kernel term (each trigonometric function occurring in $P(\omega_1,\omega_2)$) in each cell is invariant is shown in Figure 1.
Table 1 identifies the cells where each of the kernel terms are positive valued. The cells not identified in a particular column are those where the relevant trigonometric function (kernel term) is negative valued. To obtain the least upper (greatest lower) bound of the wavenumber response in a certain cell, e.g. cell 3 in Figure 1, it is necessary to pick $h[n_1,n_2]$ or $a[n_1,n_2]$ ($\underline{h}[n_1,n_2]$ or $\underline{a}[n_1,n_2]$) when the kernel $\cos(n_1\omega_1 + n_2\omega_2)$ is positive (negative) in the cell. So, the least upper bound of the interval filter wavenumber response for cell 3 is

$$\overline{H_3}(\omega_1,\omega_2) = \overline{a}[0,0] + \underline{a}[1,0]\cos(\omega_1) + \overline{a}[0,1]\cos(\omega_2)$$
$$\underline{a}[1,1]\cos(\omega_1+\omega_2) + \overline{a}[1,-1]\cos(\omega_1-\omega_2)$$

and the greatest lower bound for the same cell is

$$\underline{H_3}(\omega_1,\omega_2) = \underline{a}[0,0] + \overline{a}[1,0]\cos(\omega_1) + \underline{a}[0,1]\cos(\omega_2)$$
$$\overline{a}[1,1]\cos(\omega_1+\omega_2) + \underline{a}[1,-1]\cos(\omega_1-\omega_2).$$

Similarly, the greatest lower bound of the interval filter wavenumber response in another cell, say cell 5 in Figure 1, can be realized by choosing the lower extreme values of the coefficients associated with the kernel terms that are positive valued.

$$\underline{H}_5(\omega_1,\omega_2) = \underline{a}[0,0]+\underline{a}[1,0]\cos(\omega_1)+\overline{a}[0,1]\cos(\omega_2)$$
$$\overline{a}[1,1]\cos(\omega_1+\omega_2)+\overline{a}[1,-1]\cos(\omega_1-\omega_2).$$

Consequently, the tightest bounds of the wavenumber responses can be described by the procedure used in this example. The total number of cells in this case is 16 and at most 32 extreme functions are sufficient, but not necessary, to obtain the least upper and greatest lower bounds of the interval filter wavenumber responses. Once the extreme functions describing the wavenumber responses in all cells are listed with the help of Table 1, repetitions of extreme functions in different cells may be observed. For example, cells 1 and 9 required identical extreme functions to bound the wavenumber response. The same applies to cells 5 and 13, cells 4 and 12, and cells 8 and 16. Therefore, only 24 distinct extreme functions are necessary as well as sufficient for bounding tightly the wavenumber responses in this example. The point $[n_1,n_2]$ in the region of support R_1 will generate a line with slope $-n_1/n_2$ in the (ω_1,ω_2), as Eq.(17) is satisfied. The number of lines with non-negative slope is constrained by the number of integers $k\in N_+$ that satisfy

$$0<\frac{\pi(2k-1)}{2}<(n_1+n_2)\pi \quad \text{or} \quad 0<k<n_1+n_2+\frac{1}{2}. \tag{18}$$

The number of lines with finite negative slopes can be similarly determined and, in fact, equals the number of lines with finite positive slopes. It is easy to verify that the number of lines generated in the (ω_1,ω_2)- plane by the generic trigonometric functions: $\cos\omega_1, \cos\omega_2, \cos(\omega_1+\omega_2), \cos(\omega_1-\omega_2)$ in Figure 1 with associated $[n_1,n_2]$ 2-tuples, [1,0], [0,1], [1,1], and [1,-1] are 1(vertical), 1(horizontal), 2(each of negative slope) and 2(each of positive slope), respectively.

<u>Example 2:</u> Consider the (5 x 5)2-D FIR filter with twofold symmetry for which $N_1=N_2=N=2$. Here R_1 = {[1,0], [0,1], [2,0], [0,2], [1,1], [1,-1], [1,2], [2,1], [1,-2], [2,-1], [2,2], [2,-2]} and the generic wavenumber response is

$$\begin{aligned}P(\omega_1,\omega_2) &= a[0,0]+a[1,0]\cos(\omega_1)+a[0,1]\cos(\omega_2)+a[2,0]\cos(2\omega_1)\\ &+a[0,2]\cos(2\omega_2)+a[1,1]\cos(\omega_1+\omega_2)+a[1,-1]\cos(\omega_1-\omega_2]\\ &+a[2,1]\cos(2\omega_1+\omega_2)+a[2,-1]\cos(2\omega_1-\omega_2)+a[1,2]\cos(\omega_1+2\omega_2)\\ &+a[1,-2]\cos(\omega_1-2\omega_2)+h[2,2]\cos(2\omega_1-2\omega_2)+h[2,-2]\cos(2\omega_1-2\omega_2).\end{aligned}$$

Figure 2 shows the wavenumber domain tessellation of cells each of which is sign-invariant for each kernel term. The parallel lines with slope -1/2 are generated by the kernel term $\cos(\omega_1+2\omega_2)$. In the first quadrant the number of lines with slope 1/2 is 3 as derivable from the constraint in Eq.(18). This result matches the derivation stated previously.

5 Conclusions

The generic expression for the wavenumber response of complex coefficient 2-D FIR filters has a multidimensional r-D counterpart which, after the imposition of the zero-phase and symmetry constraints,

$$h[n_1,...,n_r] = h^*[n_1,...,n_r], \quad h[n_1,...,n_r] = h[-n_1,...,-n_r]$$

on the unit impulse response $h[n_1, n_2,...,n_r]$, is of the form

$$P(\omega_1,...,\omega_r) = h[0,...,0] + 2\sum ... \sum h_R[n_1,...,n_r] \cos(\sum_{i=1}^{r} \omega_i n_i).$$

The r-D counterpart of the 2-D case of fourfold symmetry is a natural generalization of Eqs. (9), (10), (11). Some progress has been reported for bounding the wavenumber response of 2-D zero-phase FIR interval filters having twofold symmetry. The procedure indicated by proceeding from low to high order filters can, in principle, be implemented for bounding the wavenumber response but the task becomes increasingly computation-intensive and impractical for filters of even moderate order. Research is needed either to reduce the computational chore or develop good sub-optimal strategies.

References

1. Bhattacharyya, S. P., Chapellat, H., Keel, L. H., 1995, *Robust Control -The Parametric Approach,* Prentice Hall, Englewood Cliffs, New Jersey.
2. Barmish, B. R., 1994, *New Tools for Robustness of Linear Systems*, Maxwell MacMillan International, New York.
3. Bose, N. K., 1988, *Robust multivarite scattering Hurwitz interval polynomials*, Linear Algebra and Its Applications 98:123-136.
4. Basu, S., 1989, *On the multidimensional generalization of robustness of scattering Hurwitz property of complex polynomials*, IEEE Trans. on Circuits and System 36:1159-1167.
5. Kharitonov, V. L., Torres MuÔoz, J. A., 1998, *Robust stability of complex multivariate polynomials*, IFAC Conference on System Structure and Control, Nantes, France, pp. 19 – 24.
6. Bose, N. K., Kim, K. D., 1991, *Boundary implications for frequency response of interval FIR and IIR filters*, IEEE Trans. on Signal Processing 39:2167-2173.
7. Levkovich, A., Zeheb, E., Cohen, N., 1995, *Frequency response envelopes of a family of uncertain continuous-time systems*, IEEE Trans. on Circuits and Systems I 42:156-165.
8. Bose, N. K., 1991, *Invariance of interval system properties*, in: Bhattacharyya, S. P., Keel, L. H. (eds), 1991, Control of Uncertain Dynamic Systems, CRC Press, Inc., Boca Raton, Florida, pp. 417 – 421.

10.2

2D-Analysis of Hybrid Systems

D. Franke

University of the Federal Armed Forces, Hamburg, Germany

Summary. This paper addresses resource to task allocation processes as a special class of hybrid dynamical systems. An adequate notion of set stability is used, and a 2D-approach for testing stability is proposed.

1 Introduction

Resource to task allocation problems in manufacturing control are quite often hybrid in nature since the tasks (or subtasks) are continuous-time processes whereas the resource allocation is a discrete-event scheduling process. Typically the number of resources is small as compared to the number of tasks to be processed.

Resource allocation poses a scheduling problem. The scheduler is a Boolean control device (Figure 1.1) which communicates with the continuous-time subsystems via Boolean variables only. Boolean controls **u** are activated by the occurance of plant events **v** oriented at thresholds for continuous variables. The overall system takes the structure of a hybrid feedback loop. This is a variable structure system as can be seen from the above examples. However, different from classical variable structure control [1], [2] one has switchings between different plants (tasks) rather than between different controllers.

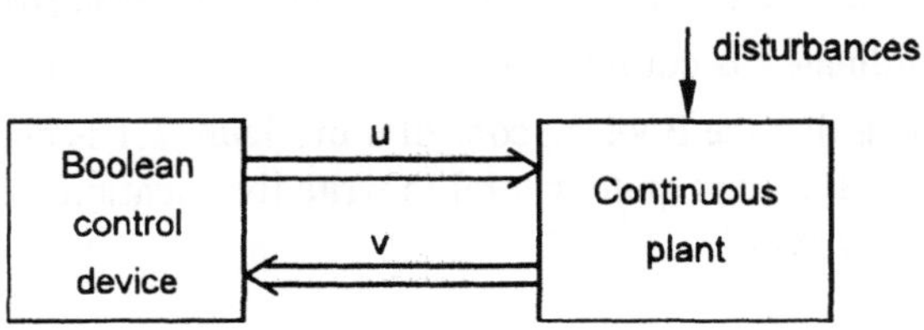

Figure 1.1. Structure of the hybrid system

The switch strategy should be designed such as to avoid undesirable chattering between tasks [3]. To this end the Boolean rule-basis for switching should contain a rule which assures that any subtask once having started will be completed.

2 Set Stability

As far as stability of hybrid systems is concerned most authors adopt the classical notion of asymptotic Lyapunov stability [4], [5], [6]. However, as can be seen from the above examples, resource allocation problems with competing subtasks usually do not attain an equilibrium state. Inherent disturbances make the overall system a non-autonomous one. Therefore an adequate notion of set stability is proposed.

Let $\mathbf{x}(t) \in \mathcal{R}^n$ be the state of the (linear or nonlinear) continuous-time subsystems in Figure 1.1, with initial state $\mathbf{x}(0) \in \mathcal{B}_0 \subset \mathcal{R}^n$. Let these subsystems be subject to external disturbances $\mathbf{z}(t) \in \mathcal{Z} \subset \mathcal{R}^m$ and to parametric uncertainties $\mathbf{q} \in Q \subset \mathcal{R}^r$. Let the Boolean control device be a finite Boolean automaton. Then the following definition of set stability is proposed.

Definition:
The hybrid system in Figure 1.1 is said to be $\mathcal{B}$-stable w.r.t. $\{\mathcal{B}_0, Q, \mathcal{Z}\}$ *if* $\mathbf{x}(t) \in \mathcal{B}\ \forall\, t > 0$ *provided that* $\mathbf{x}(0) \in \mathcal{B}_0 \subset \mathcal{B}, \mathbf{q} \in Q$ *and* $\mathbf{z}(t) \in \mathcal{Z}\ \forall\ t > 0$.

Based on this definition the following types of problems may be stated:

I) Given $\{Q, \mathcal{Z}\}$ and prescribed $\mathcal{B}$, which is the largest allowable domain $\mathcal{B}_0$?

II) Given $\{Q, \mathcal{Z}\}$ and $\mathcal{B}_0$, which is the smallest domain $\mathcal{B}$ that can be guaranteed ?

III) Given $\mathcal{B}_0$ and prescribed $\mathcal{B}$, which is the largest allowable set $\{Q, \mathcal{Z}\}$?

3 Stability Analysis via a 2D-Approach

Originally the system shown in Figure 1.1 is a 1D-system where time t is the independent variable. Let switchings of the Boolean vector **v** occur at times T_k, $k = 0, 1, 2, \ldots$, due to thresholds which provide a mapping from **x** to **v**. Therefore the time coordinate t is cut into intervals of individual length $t_k = T_{k+1} - T_k$. Taking into account that the Boolean control **u** in Figure 1.1 is constant on each of these intervals, **u** = **u**(k), it is quite natural to write the equations of the mixed overall system as a 2D-system:

$$\begin{aligned}\dot{\mathbf{x}}(t,k) &= \mathbf{A}(\mathbf{q}, \mathbf{u}(k))\mathbf{x}(t,k) + \mathbf{B}(\mathbf{q}, \mathbf{u}(k))\mathbf{u}(k) + \\ &\quad + \mathbf{B}_z(\mathbf{q}, \mathbf{u}(k))\mathbf{z}(t,k),\end{aligned} \tag{1}$$

$$\mathbf{x}(0,0) = \mathbf{x}_0 \in \mathcal{B}_0,$$
$$k = 0, 1, 2, \ldots;\ 0 \le t < t_k,\ \text{ where } t_k = T_{k+1} - T_k.$$

Events: $\mathbf{x}(t_k, k) \in \partial\mathcal{B}_0$. (2)

Boolean coding of the state: $\mathbf{v}(k) = \mathbf{g}(\mathbf{x}(t,k)),\ 0 \le t < t_k$. (3)

Feedback control: $\mathbf{u}(k+1) = \mathbf{f}(\mathbf{v}(k+1), \mathbf{v}(k), \mathbf{u}(k))$. (4)

Boundary condition: $\mathbf{x}(0,k+1) = \mathbf{x}(t_k, k)$. (5)

Remarks:
In Eq. (1), linearity of each continuous-time subsystem has been assumed for simplicity. The switching of resources makes **u**(k) enter into the system matrices, hence Eq. (1) is piecewise linear. In Eq. (2), $\partial \mathcal{B}_0$ denotes the boundary of domain $\mathcal{B}_0$. Whenever the trajectory $\mathbf{x}$ crosses $\partial \mathcal{B}_0$, this is considered an event. Eq. (3) provides a relay type mapping **g**: $\mathcal{R}^n \to \mathcal{V} \subseteq \mathcal{B}^q$. Eq. (4) describes the discrete dynamics of a finite Boolean automaton. Eq. (5) has a simple meaning. At each occurance of an event, time t is reset to zero, and k is replaced by k + 1. In this way the state **x**(t, k) evolves row by row in the t, k-plane (Figure 3.1).

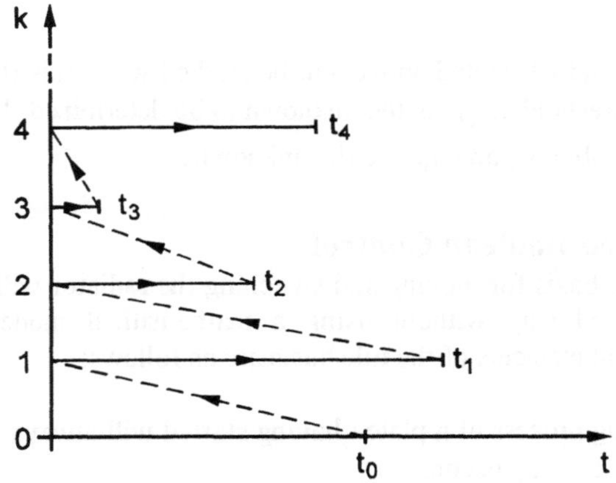

Figure 3.1. Event-driven evolution in the t, k-plane

4 Example of a Switched Radiator Heating Process

4.1 Problem Formulation

The above methodology will be applied to the following thermal process:

Three plates should be heated by turns by one moving radiator such that the temperature $x_i(t)$ of each plate $(i = 1,2,3)$ is guaranteed to remain in the prescribed interval $[x_a, x_b]$. Therefore

$$\mathcal{B}=\left\{\mathbf{x}(t) \mid x_a \le x_i(t) \le x_b \,, i=1,2,3\right\}, \tag{6}$$

$$\mathcal{B}_0=\left\{\mathbf{x}(0) \mid x_{min} \le x_i(0) \le x_b \,, i=1,2,3\right\}, \tag{7}$$

respectively, where $x_{min}\left(> x_a\right)$ and x_b are the thresholds for Boolean coding of the temperatures $x_i(t), i=1,2,3$. In addition to the moving control action the radiator can be switched on and off. The process is subject to disturbance

$$\mathcal{Z}=\left\{x_u(t) \mid x_{um} \le x_u(t) \le x_{uM}\right\}, \tag{8}$$

where x_u is the surrounding temperature. Also parameter uncertainty

$$Q=\left\{a_u \mid a_{um} \le a_u \le a_{uM}\right\} \tag{9}$$

is involved, where a_u is the heat transfer coefficient from the plates to the surrounding air.

The problems I) and III) stated above will be studied w.r.t. this thermal process. In problem I) the threshold x_{min} is the unknown to be determined. In problem III) the worst-case admissible x_u and a_u are the unknowns.

4.2 Rule- Based Boolean Control

The Boolean rule basis for moving and switching the radiator will be established in a knowledge-based way without using a mathematical model of the thermal process. Important elements of the rule basis are as follows:

i) A reheating process of a plate i having started will always be continued until the event $x_i = x_b$ occurs.

ii) A reheating process will be started if at least one $x_i < x_{min}$ and no other plate is just being reheated.

iii) No reheating process will be started as long as $x_i \ge x_{min}$, $i=1,2,3$.

These rules guarantee an as calm as possible switching behaviour and exclude chattering actions. The rule basis can be written as a Boolean switching table. For details see the related example reported in [7]. The rule basis turns out to be of the type of a finite Boolean automaton acc. to Eq. (4).

4.3 Model-Based 2D - Analysis

The problems of stability stated above cannot be treated by a merely rule-based design. A model-based analysis has to be added in order to determine the crucial threshold x_{min} in problem I) or the admissible values of x_u and a_u in problem III). The following mathematical model is used for plate i being heated:

$$\dot{x}_i(t) = a_s \cdot [x_s - x_i(t)] + a_u \cdot [x_u - x_i(t)], \tag{10}$$

and for plate i not being heated:

$$\dot{x}_i(t) = 2a_u \cdot [x_u - x_i(t)], \tag{11}$$

where x_s is the constant temperature of the radiator when switched on, and a_s is the corresponding heat transfer coefficient. The following numerical values have been assumed:

$$x_s = 2, \ a_s = 1, \ x_a = 0.7, \ x_b = 1.$$

The values of x_u and a_u have not been fixed in advance.

Since there are three possible positions of the radiator and since the radiator can be switched on and off, there results a total of four different state space models of the plant acc. to Eqs. (10) and (11). The Boolean rule basis provides switchings between these state space models resulting in a complex behaviour of the over-all hybrid dynamical system. The 2D-approach introduced above will be used for a worst-case analysis. This requires to solve at each occurance of an event the continuous state space model subject to the Boolean rule basis, starting with the worst-case initial state $x_i(t=0,\ k=0) = x_{min}$, $i = 1,2,3$. At the end $(t = t_k)$ of each step k the temperatures $x_i(t_k, k)$, $i = 1,2,3$, have to be determined. From the requirement that $x_i(t_k, k) \geq x_a$, $i = 1,2,3$, admissible lower bounds for the threshold x_{min} are obtained numerically.

4.4 Numerical Results

For the fixed value $x_u = 0.1$ Figure 4.1 outlines the lower bounds on x_{min} corresponding to the discrete steps k = 1 to k = 5, as a function of parameter a_u. It can be seen that k = 2 is the crucial step for small a_u whereas for larger a_u the situation becomes more and more crucial with growing k. Therefore the long-term behaviour $(k \to \infty)$ has to be included in the approach. The thermal energy condition

$$a_s \cdot (x_s - x_b) > 5a_u \cdot (x_b - x_u)$$

provides

$$a_u < \frac{a_s \cdot (x_s - x_b)}{5 \cdot (x_b - x_u)} \tag{12}$$

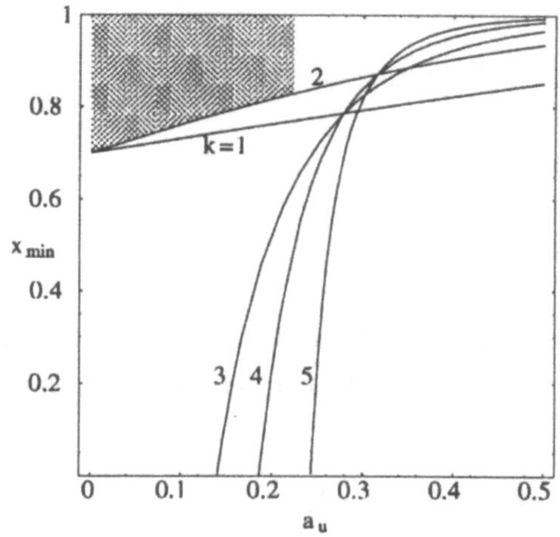

Figure 4.1. Lower bounds on x_{min} for discrete steps $k = 1$ to $k = 5$

as a sufficient condition for proper long-term behaviour. In the above numerical example $a_u < 0.222$, and therefore the hatched domain in Figure 4.1 is an admissible one.

By repeating the calculations for different values of x_u and by transfering the numerical information into the plane spanned by a_u and x_u, Figure 4.2 is obtained. The fat line indicates the upper bound of a_u acc. to Eq. (12). The other curves indicate upper bounds of a_u for different fixed values of x_{min}. This graphical result solves both problem I) and problem III) stated above. An example of parameter uncertainty is illustrated in Figure 4.2 (hatched domain) which requires $x_{min} \geq 0.82$ for a proper operation of the process.

5 Conclusions

An approach to analysis and design of resource to task allocation processes has been proposed which is hybrid in a twofold sense. On the one hand hybrid mathematical models are considered to be adequate. They are composed of continuous subsystems (subtasks) and a discrete-event scheduler for the switching of resources. On the other hand the proposed design method is hybrid in the sense that it combines a knowledge-based (rule-based) design and a model-based stability analysis. An adequate notion of stability has been introduced and the mixed approach has been illustrated at a hybrid thermal process. A 2D-interpretation of the hybrid overall system has been proposed in order to facilitate the dynamical

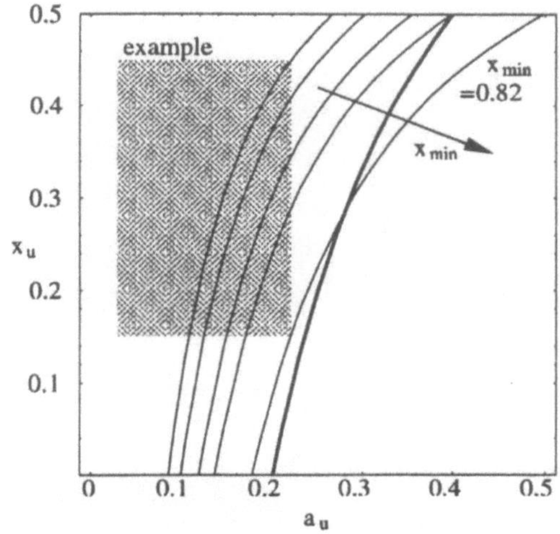

Figure 4.2. Numerical result in the plane of uncertain parameters

analysis and to obtain some insight into both the continuous-time and the discrete-event behaviour of the process.

References

1. Emeljanov S V 1969 *Automatische Regelsysteme mit veränderlicher Struktur.* R. Oldenbourg Verlag, München
2. Utkin V I 1977 Variable Structure Systems with Sliding Modes. *IEEE Transactions on Automatic Control* 22: 212-222
3. Petterssen S, Lennartson B 1996 Time Optimal Control and Disturbance Compensation for a Class of Hybrid Systems. In: Gertler J J et al. (eds) 1996 *Preprints of the 13th World Congress of IFAC, Vol. J*, San Francisco, pp 281-286
4. Branicky M S 1994 Stability of Switched and Hybrid Systems. In: 1994 *Proc. IEEE CDC*, pp 3498-3503
5. Fierro R, Lewis F L, Abdallah C T 1996 Common, Multiple and Parametric Lyapunov Functions for a Class of Hybrid Dynamical Systems. In: 1996 *Proc. 4th IEEE Mediterranean Symp.*, pp 77-82
6. Peleties P, Decarlo R A 1991 Asymptotic Stability of m-Switched Systems Based on Lyapunov-like Functions. In: 1991 *Proc. of the American Control Conference*, pp 1679-1683
7. Franke D 1997 Switched Flow Systems Redesign. In: Hamza M H (ed) 1997 *Proc. IASTED Conf. CONTROL'97*, Cancun, Mexico, pp. 23-26

10.3

Towards a Theory of General Discrete Systems

J. Gregor
Dept. of Mathematics, Faculty of Electrical Eng., Czech University,
Technická 2, 166 27 Praha 6 Czech Republic,
email: gregorj@math.feld.cvut.cz

Summary. The main goal of the paper is an attempt to establish a general theory of discrete systems by successive specializations of basic structures. This way the most general conditions of validity of achieved results can be revealed, the scope of applicability can be widened and program of further research activities can be outlined

1 Introduction

Real world systems, as relatively isolated entities, can often be characterized by measuring various quantities connected with their behaviour. Systems theory wants to establish their mathematical models so as to predict the behaviour of such systems or to design systems with desired behaviour.

We proceed as follows: The measured quantities are supposed to be functionals (or functions) defined on certain sets. These sets ought to be decomposed onto suitable (and 'homogeneous') parts and considered as independent variables of the model. In some cases, the number of parts seems to be predetermined (e.g. time, spatial coordinates etc.), in others it can be artificial. A system is then characterized by various types of functional equations. We shall call a mathematical model of a real world system a *discrete system*, if it can be fully characterized by functions defined on a countable set Σ.

The enormous variability of real world systems forces us to avoid unnecessary specializations of models as far as possible. This approach, adopted by Mesarovic and Takahara [14], [15], enables to include nonstandard systems and new applications.

A theory of discrete systems is being developed for several decades on the basis of a complete theory of difference equations (see e.g. [1], [5] and many others). Most of the efforts have been devoted to linear systems. Convolution and z-transform techniques lead to basic results in synthesis of digital filters and other important applications. These are now the classics of discrete systems theory. The importance of nonlinear systems grew in accordance with their applications in many unexpected areas.

The attempts to develop a 2-D discrete systems theory, first of all for the set $N_+ \times N_+$, as a generalization of the corresponding 1-D theory were widely successful. But soon it has been realized that such generalizations cannot characterize and describe all specific properties of these systems, even more, many of these generalizations are responsible for the errors and mistakes, which had to be later corrected. Among others, the lack of a relatively complete theory of difference equations in several independent variables can be taken responsible here. The space of solutions of linear difference equations in several independent variables is, in general, a linear space of infinite dimension. This is the source of most of the difficulties encountered so far.

Remarkable general results for n-D linear systems have been obtained using Willems behavioral approach (see e.g. [19], [6], [20]) or using Gröbner bases (see [16]). Also the polynomial approach, based on formal power series lead to many important results ([13]). Shift-invariant 2-D and n-D linear systems are described in several monographs (see [3], [4], [11], [12]).

For a sufficiently general approach to n-D discrete systems it seems indispensable to investigate where are the ideas common to all of the above mentioned successful theories and what kind of restrictions are enforced by their basic ideas. Most of them are restricted to linear systems and many of them to systems with constant parameters. But in the so called 'one-sidedness' of some classical systems, in applications of polynomial techniques and even in Gröbner bases approach the basic idea of a linear order of the basic set Σ seems to be the principal ingredient. As a result, this linear order and correspondingly, the idea of recurrence could be used as the starting point. In [2], [9] basic existence and uniqueness results (EU theorems) have been obtained by this recursive approach also for some nonlinear systems, in [18] by a similar approach the growth rate of any solution of linear partial difference equation has been established. Prerequisites for the use of convolution, of the concept of transfer functions and henceforth limitations of z-transform techniques have been found [7], [8]. Many other significant results have been proved and can be proved when considering the idea of linear order. We shall follow here this idea and present a sufficiently general framework of discrete systems following basically the reasoning of [15] and [10].

2 Basic Concepts

In what follows, Σ will denote a countable set. Capitals will denote subsets of Σ, greek letters $\alpha, \beta, \ldots$ stand for elements of Σ. The letters a, b, x, y will be mappings as in $a: A \to \boldsymbol{R}$ or $b: \Sigma \to \mathbf{C}^n$. We shall call these mappings sequences. The set of all $x: A \to \boldsymbol{R}$ will be denoted by X_A. When Σ is a semigroup containing a 'zero element', we shall use the symbol Σ_+. For $A, B \subset \Sigma_+$ the set

$A+B=\{\gamma=\alpha+\beta,\alpha\in A,\beta\in B\}$. We shall also consider Σ_+ as a direct sum of several semigroups with the structure of Σ accordingly defined. Their number will be denoted by n and henceforth $\alpha\in\Sigma_+,\alpha=(\alpha_1,\alpha_2,...,\alpha_n)$.

Definition 1 *Let X,Y be sets of mappings into linear spaces L_X,L_Y, respectively, over the field $\mathcal{T}$. Then $S\subset X\times Y$ is called a system. If*

$$X=\{x:A\to L_X\},Y=\{y:B\to L_Y\}$$

with $A,B\subset\Sigma$ then S is called discrete system.

In special cases the system S is a function $S:X\to Y$. Such systems could be called function type systems. Important examples of such systems are the so called FIR filters, convolutional type systems $y=h*x$ with $x,y\in l^p$ forming a Banach algebra, and many others. We will give only marginal significance to such systems here, although the basic theorems remain valid.

Theorem 1 *For any system there exists a set G and a mapping $\rho:G\times X\to Y$ such that*

$$(x,y)\in S \;\; if \;\; \exists g\in G \;\; such\ that \;\; y=\rho(g,x) \tag{1}$$

For the proof see [14]. The function ρ and the element $g\in G$ in this theorem are called response function and state object of the system *S*, respectively. Note that the nontriviality of this result is hidden in its 'only if' part: there is no such element $g\in G$ for which $(x,\rho(g,x))\notin S$. The elements of G can be viewed as a parametrization of the system *S*.

Definition 2 *The system S is called linear, if*

$$\forall(x_i,y_i)\in S,\ i=1,2\ \&\ \forall c\in\mathcal{T}\Rightarrow\ (cx_1,cy_1)\in S,\quad (x_1+x_2,y_1+y_2)\in S$$

For linear systems the following basic result has been proven in [14]:

Theorem 2 *Let X,Y be linear spaces over the same field $\mathcal{T}$. Then $S\subset X\times Y$ is a linear system if and only if there exists a linear space $G=\{g:A\to\mathcal{T}\}$ and a function $R:G\times X\to Y$ such that:*

1. *$(x,R(g,x))\in S$ for every $x\in X$,*
2. *there exists a pair of linear mappings $R_1:G\to Y$ and $R_2:X\to Y$ such that for all $(g,x)\in G\times X$ there is $R(g,x)=R_1(g)+R_2(x)$.*

Note that the concept of state objects, linearity and its basic property as given in Theorem 2 needs no specific structure of the set Σ.

Definition 3 *A discrete system S will be called*

- *(semigroup) structured if Σ is a semigroup,*
- *n-dimensional if the semigroup Σ_+ has n generators,*
- *dynamic if in the semigroup Σ_+ a linear order $\le$ is defined.*

Dynamic systems are also called time systems and the corresponding order is called the abstract time of the system. For given $\alpha, \beta \in \Sigma_+$ the set $I(\alpha,\beta) = \{\gamma : \alpha \le \gamma \le \beta\}$ is called a time interval. A time interval may consist of an infinite number of elements.

Dynamic systems may exhibit some invariance properties. For the structure of a dynamic system the most important one is the invariance with respect to time shifts.

Definition 4 Let $A \subset \Sigma_+$ The operator r^σ defined for all $x: A \to L_x$ and for all $\sigma \in A$ by $r^\sigma x = \hat{x}$,

$$\hat{x}(t) = \begin{cases} x(t-\sigma) & \text{for } t \in (A+\sigma) \cap A \\ 0 & \text{for } t \in A \setminus (A+\sigma) \end{cases}$$

is called the translation operator. Hence, $\hat{x}: A \to L$.

A modified translation operator r^σ is called completed with ξ when a mapping $\xi: (t \in A \setminus (A+\sigma)) \to L_X$ is given and $r^\sigma x = \xi$ on this set.

Definition 5 *A discrete (free or structured) system S is called translation-invariant if*

$$\forall \sigma \in A \forall (x,y) \in S, y = \rho(g,x) \text{ there is } \rho(g, r^\sigma x) = r^\sigma \rho(g,x).$$

When excluding functional discrete systems, the most important way to model a system is by a functional equation (or a system of such equations). These equations include the values of the input X_A at some points of the set $A \subset \Sigma$ and the values of the output Y at some points of the set Σ, or Σ_+.

We shall discuss equations of the form

$$F(x_1, x_2, \ldots, x_{|E|}, y_1, y_2, \ldots, y_{|M|},) = 0, \tag{2}$$

where $E = \{\varepsilon_i, i = 1,2,\ldots,|E|\}, M = \{\mu_i, i = 1,2,\ldots,|M|\}$ and

$$\bar{x}_i = x(\alpha + \varepsilon_i), \bar{y}_i = y(\alpha + \mu_i), \alpha \in A \subset \Sigma_+, E, M \subset \Sigma_+,$$

and $x: A + E \to$ *C*, $y: A + M \to$ *C*. ($|E|$ stands for cardinality of the set E, here it means the number of elements of the set E.) We shall also refer to the equation

(2) in a shorter form $F(\underline{x},\underline{y}) = 0$. Since e.g. x_i and other arguments of the function F can be expressed as linear combinations of backward and forward partial differences of the mappings x, y, (recall the similar situation in ordinary difference equations) we may call equation (2) for historical reasons a partial difference equation. Evidently, any partial difference equation defines a mathematical model of a discrete system S. Often the mappings x, y are called the input and output, respectively, and the finite sets E, M are the input and output masks. Systems defined by (2) will be called basic discrete systems and, if Σ_+ has n generators, basic n-D systems.

3 Basic Results

Theorem 3 *The basic n-D discrete system (2) is linear if there exists a linear mapping G from* $\mathbf{C}^q$, $q = |M| + |E|$ *into* $\mathbf{C}$ *such that*

$$F(\underline{x},\underline{y}) = 0 \Leftrightarrow G(\underline{x},\underline{y}) = 0 .$$

To specialize basic n-D discrete systems as dynamic systems a linear ordering should be defined, if possible. According to a general result of set theory any set can be well-ordered, but here we face a more specific problem: the order and the semigroup structure must be compatible. We have to demand that for any $\alpha, \beta, \gamma \in \Sigma_+$ the implication $\alpha \le \beta \Rightarrow \alpha + \gamma \le \beta + \gamma$ holds true. A following relevant special result has been proved in [2], [9].

Theorem 4 *Let* A, M *be subsets of* $\mathbf{Z}^n$, *M finite with at least two elements. Then there exists a mapping* $\mu: A \to M$ *and a linear ordering* $\le$ *of A such that A is well-ordered with respect to* $\le$ *and*

$$\alpha' + \mu(\alpha') \in \alpha + M \Rightarrow \alpha' \le \alpha \text{ for all } \alpha, \alpha' \in A . \tag{3}$$

In this theorem the mapping μ choses a ‘leading element’ of the output mask such that exactly one value of the output can be considered as unknown provided all the previous ones are known. This is a central result showing that the equation (2) can be considered as a recurrence relation provided $F(\underline{x},\underline{y}) = 0$ has a unique solution for $y(\alpha + \mu(\alpha))$. Calling such equations quasilinear we may prove the following (see [2]).

Theorem 5 *For a quasilinear system S let the following set D of points be defined*

$$D = (A + M) \setminus \bigcup_{\alpha \in A} \{\alpha + \mu(\alpha)\} , \tag{4}$$

and let a set of mappings $G = \{g: D \to \mathbf{C}\}$ *be considered.*

Then there exists a mapping $\rho: G \times X \to Y$ *such that* $\forall x \in X$, $\forall g \in G$ *there is* $(x, \rho(g,x)) \in S$.

Although the last few results have been proved in a special case of $\Sigma_+ = Z^n$, they show that the abstract time concept and the corresponding recursive evaluation of the output works also in more general setting, perhaps with some modifications. In fact, Theorems 4, 5 are rather general existence and unicity theorems for partial difference equations. A method of evaluation of the output is the basic step in analysis and synthesis of general discrete systems. The more specific systems we want to investigate, the more details of their behaviour can be revealed. In the limited space of this paper we can present only some examples.

For linear systems defined by

$$\sum_{\mu \in M} a_\mu(\alpha) y(\alpha + \mu) = \sum_{\varepsilon \in E} d_\varepsilon(\alpha) x(\alpha + \varepsilon), \quad \alpha \in A \subset \mathbf{Z}^n. \tag{5}$$

the EU theorem follows from Theorem 4 with the assumption $a_{\mu(\alpha)} \neq 0$. It has been proved [18] that, under very mild restrictions, the output is sub-exponential, i.e. there exist constants $M > 0, g \in Z^n$ such that the output satisfies the inequality $|y(\alpha)| \leq Mg^\alpha$ for all $\alpha \in A$ provided the input x and the initial state object g satisfies similar inequalities.

Further specialization is to shift-invariant systems, i.e. those of (5) with constant coefficients. For such systems S there exists a sequence h such that $y = h * x \forall x$ if S is translation-invariant [7]. Contrary to the well known 1-D case, the conditions herewith imposed on the mask M are rather restrictive. However, their fulfilment is a prerequisite for introducing the concept of transfer (or weight) function and also for application of multidimensional z-transform [8].

The initial state object $g \in G$ (see Theorem 5) guarantees that some values of the output can be obtained by a single step of recursion. Denoting the domain of this set by D_I, we may proceed similarly from any time instant α. Denote D_α the resulting domain. We obtain an ordering $D_\alpha \subset D_{\alpha'}$ for any $\alpha < \alpha'$. A mapping $g_\alpha : D_\alpha \to C$ could be consistent with the system S in the sense that it is identical with y up to the time instant α. With this construction, g_α characterizes the state of the system at the point α, the partial order in the set A, defined by $D_\alpha \subset D_{\alpha'}$, can be called state time and an operator Φ as in $g_{\alpha'} = \Phi g_\alpha$ is the state transition rule. E.g. in LSI systems the state transition rule is the first of the two equations commonly called the state space description of the system. With this state-transition rule the concepts of controllability, observability etc. can be introduced and analysed.

At least three aspects of systems theory have been omitted here for the sake of brevity. The first one (demanding new results) means to consider sequences as elements of a normed space and to define a topology in the space of inputs and outputs. This way stability investigations may and have to be included. The second one is to consider vector functions as in $f: A \to C^n$ and correspondingly, systems of equations etc. Most of the above results remain valid in this situation and their reformulation is straightforward. The third one is optimal control, or, more generally speaking, goal-seeking systems as in [15], where general results are rather scarse so far.

The overview of gradually introduced specializations of the general mathematical model of discrete systems given here aimed at enlarging the scope of possible applications. It also tries to give a (possibly non complete) summary of results available now and show the gaps, which remain to be filled up by further research.

References

1. Agarval, R. P., 1992 *Difference Equations and Inequalities*, N.Y., Hong-Kong.
2. Bos, K. M., Gregor, J., 1987, *On generalized difference equations*, Apl. Mat., (Prague), Vol.32, No. 3, pp. 224 - 239.
3. Bose, N. K., 1982, *Applied Multidimensional Systems Theory*, Van Nostrand, N.Y.
4. Dudgeon, D. E., Merserau, R. M., 1984, *Multidimensional Digital Signal Processing*, Prentice Hall Inc., Eng. Cliffs, NJ.
5. Elaydi, S. N., 1996, *An Introduction to Difference Equations*, Springer, N.Y.
6. Fornasini, E., Rocha, P., Zampieri, S., 1993, *State Space Realization of 2-D Finite-Dimensional Behaviours*, SIAM Control and Opt., Vol. 31., No. 6., pp. 1502 - 1517.
7. Gregor, J., 1991, *Convolutional Solutions of Partial Difference Equations*, Math.of Control, Signals and Systems, No. 4, pp. 205 - 215.
8. Gregor, J., 1988, *The multidimensional z-transform and its use in solution of partial difference equations*, Kybernetika (Prague), Suppl., No. 1, 2, pp. 1 - 40.
9. Gregor, J., 1998, *The Cauchy Problem for Partial Difference Equations*, Acta Appl. Mathematicae (to appear).
10. Gregor, J., 1997, *A Formal Approach to discrete systems Theory*, Appl. Math. and Comp.sci., Vol. 7. No. 4, pp.775-795.
11. Huang, T. S. (Editor), 1981, *Two-Dimensional Digital Signal Processing I*, Springer, Berlin.
12. Kaczorek, T., 1985, *Two-Dimensional Linear Systems*, Springer Verlag, Englewood Cliffs.
13. Kucera, V., 1991, *Analysis and Design of Discrete Linear Control Systems*, Prentice Hall, London.
14. Mesarovic, M. D., Takahara, Y., 1975, *General Systems Theory: Mathematical Foundations*, Academic Press, New York.
15. Mesarovic, M. D., Takahara, Y., 1989, *Abstract Systems Theory*, Springer Verlag, Berlin.
16. Oberst, U., 1990, *Multidimensional Constant Linear Systems*, Acta Appl.Math., 20: pp. 1-175.
17. Veit, J., 1995, *Fundamental Solution of a Multidimensional Difference Equation with Periodical and Matrix Coefficients*, Aequ. Math., 49: pp. 47 - 56.
18. Veit, J., 1997, *Sub-exponential solutions of multidimensional difference equations*, Multidim.Systems and Signal Proc., 8: pp. 369 - 385.
19. Willems, J. C., 1991, *Paradigms and Puzzles in the Theory of Dynamical Systems*, IEEE Trans. Aut.Control, Vol. 36, 3: pp. 259 - 294.
20. Zampieri, S., 1994, *A Solution of the Cauchy Problem for Multidimensional Discrete Linear Shift-Invariant Systems*, Lin.Alg. and its Appl., 202: pp. 143 - 162.

10.4

Positive 2D Discrete and Continuous-Discrete Linear Sytems

T. Kaczorek

Warsaw University of Technology, Faculty of Electrical Engineering, Institute of Control and Industrial Electronics, 00-662 Warszawa, Koszykowa 75, Poland, email: kaczorek@nov.isep.pw.edu.pl

Summary. New classes of positive 2D discrete and continuous-discrete linear systems are introduced. Necessary and sufficient conditions for reachability and controllability of the positive 2D Roesser type model and positive 2D continuous-discrete linear systems are established. The minimum energy control problem for the positive 2D linear systems are formulated and solved.

1 Introduction

The most popular models of two-dimensional (2D) linear systems are the models introduced by Roesser [26], Fornasini and Marchesini [3], [4] and Kurek [23]. The reachability, controllability and minimum energy control of 2D linear systems have been considered in many papers and books [21], [6], [7], [20]. The minimum energy control problem for the classical 2D Roesser model was formulated and solved by Klamka [21] and next the method was extended for 2D linear systems with variable coefficients [19] and other type of 2D models [22]. The reachability, controllability and observability for linear positive 1D systems have been investigated in [25], [1], [2]. The realisation problem for 2D positive systems has been investigated in [5]. The regularization of singular 2D systems by state and output feedback has been considered in [10], [12], [13] and the stabilisation problem of singular 2D continuous-discrete linear systems by state and output feedbacks in [9], [11].In this paper an overview of some recent author's contributions and some new results on positive 2D discrete and continuous-discrete linear systems will be presented. New classes of positive 2D discrete and continuous-discrete linear systems will be introduced. Necessary and sufficient conditions for reachability and controllability of the positive 2D Roesser type model and positive 2D continuous-discrete linear systems will be established. The minimum energy control problem will be solved.

2 Discrete Systems

Let $R_+ := [0,+\infty)$ be the set of nonnegative numbers and let $Z_+ := \{0,1,2,...\}$ be the set of nonnegative integers. The set of $n \times m$ nonnegative matrices will be denoted by $R_+^{n\times m}$ and $R_+^n := R^{n\times 1}$.

Consider the 2D Roesser model

$$x_{ij}^{(1)} = Ax_{ij} + Bu_{ij}, \quad y_{ij} = Cx_{ij} + Du_{ij}, \quad x_{ij}^{(1)} = \begin{bmatrix} x_{i+1,j}^h \\ x_{i,j+1}^v \end{bmatrix}, \quad x_{ij} = \begin{bmatrix} x_{ij}^h \\ x_{ij}^v \end{bmatrix}$$

$$i,j \in Z_+ \tag{1}$$

where $x_{ij}^h \in R^{n_1}$ is the horizontal state vector, $x_{ij}^v \in R^{n_2}$ is the vertical state vector, $u_{ij} \in R^m$ is the input vector, $y_{ij} \in R^p$ is the output vector,

$$A = \begin{bmatrix} A_1 & A_2 \\ A_3 & A_4 \end{bmatrix}, B = \begin{bmatrix} B_1 \\ B_2 \end{bmatrix}, C = [C_1 \ C_2], \qquad \begin{matrix} A_1 \in R^{n_1\times n_1}, B_1 \in R^{n_1\times m}, C_1 \in R^{p\times n_1} \\ A_4 \in R^{n_2\times n_2}, B_2 \in R^{n_2\times m}, C_2 \in R^{p\times n_2} \end{matrix}$$

Definition 1. The 2D Roesser model (1) is called positive 2D Roesser type model if for all boundary conditions

$$x_{0j}^h \in R_+^{n_1}, j \in Z_+, \ x_{i0}^v \in R_+^{n_2}, i \in Z_+ \tag{2}$$

and for all $u_{ij} \in R_+^m$, $i,j \in Z_+$ we have $x_{ij} \in R_+^n, n = n_1 + n_2$ and $y_{ij} \in R_+^p$ for $i,j \in Z_+$.

The 2D Roesser type model (1) is positive if and only if [8]

$$A \in R_+^{n\times n}, B \in R_+^{n\times m}, \qquad C \in R_+^{p\times n}, D \in R_+^{p\times m} \tag{3}$$

The transition matrix T_{ij} for (1) is defined as follows [26], [21], [7]

$$T_{ij} = \begin{cases} I_n \ (\textit{the } n\times n \ \textit{identity matrix}) & \textit{for} \ \ i=j=0 \\ T_{10}T_{i-1,j} + T_{01}T_{i,j-1} & \textit{for} \ \ i,j \ge 0 \ (i+j \ne 0) \\ T_{ij} = 0 \ (\textit{the zero matrix}) & \textit{for } i<0 \textit{ or / and } j<0 \end{cases} \tag{4}$$

where

$$T_{10} := \begin{bmatrix} A_1 & A_2 \\ 0 & 0 \end{bmatrix}, T_{01} := \begin{bmatrix} 0 & 0 \\ A_3 & A_4 \end{bmatrix}$$

The transition matrix T_{ij} for the positive 2D Roesser type model (1) is a nonnegative matrix, $T_{ij} \in R_+^{n\times n}$ *for all* $i,j \in Z_+$ [8].
The solution to (1) with boundary conditions (2) is given by [26], [21], [7]

$$x_{ij} = x_{bc}(i,j) + \sum_{(p,q)\in D_{ij}} M_{i-p,j-q} u_{pq} \tag{5}$$

where

$$x_{bc}(i,j) := \sum_{p=0}^{i} T_{i-p,j} \begin{bmatrix} 0 \\ x_{p0}^{v} \end{bmatrix} + \sum_{q=0}^{j} T_{i,j-q} \begin{bmatrix} x_{0q}^{h} \\ 0 \end{bmatrix}, \quad M_{i-p,j-q} := T_{i-p-1,j-q} \begin{bmatrix} B_1 \\ 0 \end{bmatrix} + T_{i-p,j-q-1} \begin{bmatrix} 0 \\ B_2 \end{bmatrix} \tag{6}$$

$$D_{ij} := \{(p,q) \in Z_+ \times Z_+,\ 0 \le p \le i,\ 0 \le q \le j \ \ and \ \ p+q \ne i+j\} \tag{7}$$

Definition 2. The positive 2D Roesser type model (1) is called reachable for zero boundary conditions (ZBC) at the point $(h,k), h,k \in Z_+$, if for zero boundary conditions (2) and every $x_f \in R_+^n$ there exists a sequence of inputs $u_{ij} \in R_+^m$ for $(i,j) \in D_{hk}$ such that $x_{hk} = x_f$.

Definition 3. A set of all nonnegative linear combinations of columns of the matrix $A \in R^{n\times m}$ belonging to R_+^n is called the positive image of A (denoted by $\mathrm{Im}_+ A$)

$$\mathrm{Im}_+ A := \{y \in R_+^n : y = Ax \ \ for \ all \ \ x \in R_+^m\} \tag{8}$$

A matrix $R_n \in R^{n\times n}$ is called the generalised positive permutation matrix (or nomomial) if and only if R_n has only one positive entry in each row and column and the remaining entries are equal zero.

Theorem 1. [15] The positive 2D Roesser type model (1) is reachable for ZBC at the point (h,k) if and only if

i) $$\mathrm{Im}_+ R_{hk} = R_+^n \tag{9}$$

or equivalently

ii) there exists a generalised positive permutation matrix R_n consisting of n columns of R_{hk} where

$$R_{hk} := \left[M_{hk}, M_{h-1,k}, M_{h,k-1}, \ldots, M_{10}, M_{01}\right] \tag{10}$$

Theorem 2. [15] The positive 2D Roesser type model (1) is reachable for ZBC at the point (h,k) if $rank\, R_{hk} = n$ and there exists a R_{hk}^r such that:

$$R_{hk}^r \in R_+^{[(h+1)(k+1)-1]m \times n} \tag{11}$$

where R_{hk}^r is a right inverse of $R_{hk}\left(R_{hk} R_{hk}^r = I_n\right)$.

The matrix A of (1) is called nilpotent if $\det\begin{bmatrix} I_{n_1} z_1 - A_1, -A_2 \\ -A_3, I_{n_2} z_2 - A_4 \end{bmatrix} = z_1^{n_1} z_2^{n_2}$

Definition 4. The positive 2D Roesser type model (1) is called controllable (to zero) if and only if there exists a point $(h,k), h,k \in Z_+$, and a sequence of inputs $u_{ij} \in R_+^m$ for $(i,j) \in D_{hk}$ for any nonzero boundary conditions (NBC) such that $x_{hk} = 0$.

Theorem 3. The positive 2D Roesser type model (1) is controllable (to zero) if and only if the system matrix A of (1) is nilpotent matrix.
Consider the positive 2D Roesser type model (1) and the performance index

$$I(u) := \sum_{(p,q) \in D_{hk}} u_{pq}^T Q u_{pq} \tag{12}$$

where Q is the $m \times m$ symmetric positive definite weighting matrix such that

$$Q^{-1} \in R_+^{m \times m} \tag{13}$$

The minimum energy control problem for the positive 2D Roesser type model with zero boundary conditions (2) can be stated as follows. Given the matrices A, B of (1), the weighting matrix Q and the point (h,k), find a sequence $u_{ij} \in R_+^m$ for $(i,j) \in D_{hk}$ which transfers the model from zero boundary conditions to the desired local state $x_f = x_{hk}$ and minimises the performance index (12).
To solve the problem we define the matrix

$$W_Q(h,k) := \sum_{(p,q)\in D_{hk}} M_{h-p,k-q} Q^{-1} M^T_{h-p,k-q} = R_{hk} Q_d R^T_{hk} \tag{14}$$

where $M_{h-p,k-q}$ and R_{hk} are defined by (6) and (10), respectively,

$$Q_d := diag\left[Q^{-1},\dots,Q^{-1}\right] \in R_+^{[(h+1)(k+1)-1]m\times[(h+1)(k+1)-1]m}$$

The matrix $W_Q(h,k) \in R_+^{n\times n}$ is nonsingular if and only if the matrix R_{hk} has full row rank.

Define the sequence of inputs

$$\hat{u}_{ij} := Q^{-1} M^T_{h-i,k-j} W_Q^{-1}(h,k) x_f \quad \text{for} \quad (i,j)\in D_{hk} \tag{15}$$

Note that $\hat{u}_{ij} \in R_+^m$ for any $x_f \in R_+^n$ if and only if

$$W_Q^{-1}(h,k) \in R_+^{n\times n} \tag{16}$$

Theorem 4. Let us assume that

i) the positive 2D Roesser type model is reachable for ZBC at the point (h,k),
ii) (13) and (16) hold,
iii) $\overline{u}_{ij}, (i,j)\in D_{hk}$ is any sequence of inputs which transfers the model from zero boundary conditions to the desired local state $x_f = x_{hk}$.

Then the sequence of inputs (15) accomplishes the same task and

$$I(\hat{u}) \le I(\overline{u})$$

Moreover, the minimum value of (12) is given by

$$I(\hat{u}) = x_f^T W_Q^{-1}(h,k) x_f$$

3 Continuous-discrete systems

Consider the 2D continuous-discrete linear system

$$\dot{x}(t,k+1) = A_0 x(t,k) + A_1 \dot{x}(t,k) + A_2 x(t,k+1) + Bu(t,k),\ t\in R, k\in Z_+ \tag{17a}$$

$$y(t,k) = Cx(t,k) + Du(t,k) \tag{17b}$$

where $\dot{x}(t,k)=\dfrac{\partial x(t,k)}{\partial t}, x(t,k)\in R_+^n$ is state vector, $u(t,k)\in R^m$ is the input vector, $y(t,k)\in R^p$ is the output vector and $A_i\in R^{n\times n}$, $i=0,1,2$, $B\in R^{n\times m}, C\in R^{p\times n}, D\in R^{p\times m}$

Definition 5. The system (12) is called positive if for any boundary conditions

$$x(t,0)\in R_+^n, \dot{x}(t,0)\in R_+^n, t\geq 0 \quad \text{and} \quad x(0,k)\in R_+^n, k\geq 1 \tag{18}$$

and all input vectors $u(t,k)\in R_+^m, t\geq 0, k\in Z_+$ the state and output vectors satisfy the condition $x(t,k)\in R_+^n, y(t,k)\in R_+^p$ for $t\geq 0, k\in Z_+$.

A matrix $A\in R^{n\times n}$ is called a Metzler matrix if all its off-diagonal entries are nonnegative. It is easy to show that $e^{At}\in R_+^{n\times n}$ for $t\geq 0$ if and only if $A\in R^{n\times n}$ is a Metzler matrix.

Theorem 5. The system (17) is positive if
$A_0\in R_+^{n\times n}, A_1\in R_+^{n\times n}, A:=A_0+A_1A_2\in R_+^{n\times n}, B\in R_+^{n\times m}, C\in R_+^{p\times n}, D\in R_+^{p\times m}$
and A_2 is a Metzler matrix.
The system (17) is positive only if A_2 is a Metzler matrix

Theorem 6. The solution $x(t,k)$ of the equation (17a) with boundary conditions (18) has the form

$$x(t,k)=e^{A_2t}x(0,k)+\int_0^t e^{A_2(t-\tau)}F(\tau,k-1)d\tau, \quad t\geq 0, k\in Z_+ \tag{19}$$

where $F(t,k)$ is given by

$$F(t,k)=P_t^k\left[A_0x(t,0)+A_1\dot{x}(t,0)+Bu(t,0)\right]+\sum_{i=0}^{k-1}P_t^{k-i-1}\left[Ae^{A_2t}x(0,i+1)+Bu(t,i+1)\right] \tag{20}$$

and P_t is an operator defined by

$$P_tF(t):=A\int_0^t e^{A_2(t-\tau)}F(\tau)d\tau+A_1F(t) \tag{21}$$

Theorem 7. The positive system (17) is locally reachable in the rectangle D_{hr} if and only if

$$\mathrm{Im}_+ R_{hr}=R_+^n \tag{22}$$

where

$$R_{hr} = \left[R_0, R_1, \ldots, R_{r-1}\right],\ R_i := \int_0^h e^{A_2(h-\tau)} P_t^{r-i-1} B d\tau \qquad i = 0,1,\ldots,r-1 \tag{23}$$

and the operator P_t is defined by (21).

Theorem 8. If

$$P_h := \int_0^h e^{A_2\tau} BB^T e^{A_2\tau} d\tau \in R_+^{n\times n} \tag{24}$$

is a generalised permutation matrix then the positive model (17) is reachable in the rectangle $D_{h1}(t = h, k = 1)$ and the desired input vector is given by

$$u(t,0) := B^T e^{A_2^T(h-t)} P_h^{-1} x_f \quad \text{for} \quad 0 \le t \le h \tag{25}$$

Consider the positive system (17) and the performance index

$$I(u) := \sum_{k=0}^{r-1} u_k^T Q u_k \tag{26}$$

where Q is the $m \times m$ symmetric positive definite weighting matrix such that

$$Q^{-1} \in R_+^{m\times m} \tag{27}$$

and $u_k = u(t,k)$ for $0 \le t \le h,\ 0 \le k < r$.

Given the matrices $A_k, k = 0,1,2$ and B of (17), the weighting matrix Q and the point (h,r), find a sequence $u_i \in R_+^m$ for $i = 0,1,\ldots,r-1$ which transfers the system (17) from ZBC to the desired local state $x_f = x_{hr}$ and minimizes the performance index (26).
To solve the problem we define the matrix

$$W_Q(h,r) := R_{hr} Q_d R_{hr}^T \tag{28}$$

where R_{hr} is defined by (23) and

$$Q_d := diag\left[Q^{-1}, \ldots, Q^{-1}\right] \in R_+^{rm\times rm}$$

Using (28) it is easy to show that for the positive system (17) the matrix (28) is nonsingular if and only if the matrix R_{hr} has full row rank.

Define the sequence of inputs

$$\hat{u}_{0r} := Q_d R_{hr}^T W_Q^{-1}(h,r) x_f \tag{29}$$

Note that $\hat{u}_{0r} \in R_+^{rm}$ for any $x_f \in R_+^n$ if and only if

$$W_Q^{-1}(h,r) \in R_+^{n \times n} \tag{30}$$

Theorem 9. Let us assume that

i) the positive system (17) is reachable for ZBC at the point (h,r),

ii) (27) and (30) hold,

iii) $\bar{u}_{0r}$ is any sequence of inputs which transfer the system (17) from zero boundary conditions to the desired local state $x_f = x_{hr}$.

Then the sequence of inputs (29) accomplishes the same task and

$$I(\hat{u}_{0r}) \le I(\bar{u}_{0r})$$

Moreover, the minimum value of (26) is given by

$$I(\hat{u}_{0r}) = x_f^T W_Q^{-1}(h,k) x_f$$

4 Conclusions

New classes of positive 2D discrete and continuous-discrete linear systems have been introduced. Necessary and sufficient conditions for reachability (theorem 1,2) and controllability (theorem 3) of the positive 2D Roesser type model and positive 2D continuous-discrete linear systems (theorem 7) have been established. Sufficient conditions for positiveness) theorem 6) of the 2D continuous-discrete linear system (17) have been given. The minimum energy control problem for the positive 2D continuous-discrete linear system (17) have been solved. With slight modifications the considerations can be extended for n-D (n>2) linear systems. An extension of these considerations for singular 2-D and n-D linear systems is an open not trivial problem.

References

1. Fanti M.P., Maione B., Turchiano B. 1989, *Controllability of linear single-input positive discrete-time systems,* Int. J. Control, vol. 50, No 6, pp. 2523-2542.
2. Fanti M.P., Maione B. and Turchiano B. 1990, *Controllability of multi-input positive discrete-time systems,* Int. J. Control, vol. 51, No 6, pp. 1295-1308.
3. Fornasini E., Marchesini G. 1976, *State space realization of two-dimensional filters,* IEEE Trans. Autom.Control, **AC-21,** pp. 484-491.
4. Fornasini E. and Marchesini G. 1978, *Doubly indexed dynamical systems: State space models and structural properties, Math. Syst.Theory* **12.**
5. Fornasini E. and Valcher M.E. 1997, *Recent developments in 2D positive system theory,* Applied Mathematics and Computer Science, vol. 7, No 4, pp. 713-736.
6. Kaczorek T. 1985, *Two-Dimensional Linear Systems*, Springer-Verlag, Berlin, New York, Tokyo.
7. Kaczorek T. 1993, *Linear Control Systems,* vol. 2, Research Studies Press and J. Wiley, New York.
8. Kaczorek T. 1996, *Reachability and controllability of non-negative 2-D Roesser type models*, Bull. Pol. Acad. Techn. Sci., vol. 44. No. 4, pp. 405-410.
9. Kaczorek T. 1996, *Stabilization of singular 2-D continuous-discrete systems by state-feedbacks controllers,* IEEE Trans. Autom. Contr. AC-41, No 7, pp. 1007-1009.
10. Kaczorek T. and Nguyen Bang Giang 1997, *Regularization of a singular 2-D Roesser model by output-feedback*, Bull. Pol. Acad. Techn. Sci., vol. 45, No 3, pp. 417-426.
11. Kaczorek T. 1997, *Stabilization of singular 2-D continuous-discrete systems by output-feedback controllers*, SAMS,vol. 28, pp. 21-30.
12. Kaczorek T. and Nguyen Bang Giang 1997,*Regularisation of a singular 2-D Fornasini-Marchesini model by output-feedback,* Applied Mathematics and Computer Science, vol. 7, No 4, pp. 817-834.
13. Kaczorek T. 1997, *Regularisation of a singular 2-D linear models by output-feedback*, Applied Mathematics and Computer Science, vol. 7, No 4, pp. 795-816.
14. Kaczorek T. 1997, *Reachability and minimum energy control of positive 2D Roesser type models*, Proceedings of the 15th IMACS World Congress on Scientific Computation, Modelling and Applied Mathematics, Berlin, August 24-29, vol. 1, pp. 279-284.
15. Kaczorek T. 1998, *Reachability and controllability of positive 2D Roesser type models*, Proc. 3rd Intern. Conf. Automation of Mixed Processes, 19-20 March, Reims-France, pp.164-168
16. Kaczorek T. 1998, *Reachability and minimum energy control of positive 2D continuous-discrete systems*, Bull. Pol. Acad. Techn. Sci., vol. 46, No. 1, pp. 85-104.
17. Kaczorek T. 1998, *Positive 2D continuous-discrete linear systems*, Proc. IFAC Conf. System Structure and Control, Nantes, 8-10 July , vol 2, pp. 327-331.
18. Kaczorek T. 1998, *Reachability and minimum energy control of positive 2D continuous-discrete systems*, Proc. CESA'98 IMACS Multiconference in Tunisia, April 1-4, pp. 606-610
19. Kaczorek T. and Klamka J. 1986, *Minimum energy control of 2-D linear systems with variable coefficients,* Int. J. Control, vol. 44, No. 3, pp. 645-650.
20. Kaczorek T. and Klamka J. 1988, *Minimum energy control for general model of 2-D linear systems,* Int. J. Control, vol. 47, No. 5, pp. 1555-1562.
21. Klamka J. 1991, *Controllability of Dynamical Systems,* Kluwer Academic Publ., Dordrecht.

22. Klamka J. 1993, *Minimum energy control problem for general linear 2-D systems in Hilbert spaces,* Proc. IEEE Symp. Crete.
23. Kurek J. 1985, *The general state-space model for a two-dimensional linear digital systems,* IEEE Trans. Autom. Contr., vol. AC-30, No. 6, pp. 600-601.
24. Lewis L.E., *Techniques in 2-D implicit systems,* Control and Dynamic Systems, vol. 69, pp. 89-131.
25. Ohta Y., Madea H. and Kodama S. 1984, *Reachability, observability and realizability of continuous-time positive systems,* SIAM J. Control and Optimization, vol. 22, No 2, pp. 171-180.
26. Roesser P.R. 1975, *A discrete state-space model for linear image processing,* IEEE Trans. Autom. Contr., vol. AC-20, No. 1, pp. 1-10.
27. Valcher M.E. and Fornasini E. 1995, *State Models and Asymptotic Behaviour of 2D Positive Systems,* IMA Journal of Mathematical Control & Information, No 12, pp. 17-36.

10.5

Controllability of 2-D Linear Systems

J. Klamka

Institute of Automation, Technical University, street Akademicka 16,
44-100 Gliwice, Poland, email: jmoscinski@lemon.ia.polsl.gliwice.pl

Summary.. In this paper a survey of the recent results concerning controllability of 2-D systems is presented. Different types of linear 2-D systems with constant coefficients are discussed. Several controllability conditions for various kinds of 2-D systems are formulated without proofs. Moreover, many supplementary remarks and bibliographical comments are given. The relationships between different concepts of controllability are also explained.

1 Introduction

Controllability is one of the fundamental concept in modern mathematical control theory. Many dynamical systems are such that the control does not affect the complete state of the dynamical system but only a part of it. Therefore it is important to determine whether or not complete system control is possible. Roughly speaking, controllability generally means, that it is possible to steer dynamical system from an arbitrary initial state to an arbitrary final state using the set of admissible controls. In the literature there are many different definitions of controllability which depend on the type of dynamical system. The list of publications concerning various controllability problems found in the monograph [1]. Moreover, a survey of recent results and the current state of the controllability theory for different types of dynamical systems can be found in the paper [2].

A growing interest has been developed over the past few years in problems involving signals and systems that depend on more the one independent variable. The motivations for studying 2-D systems have been well justified in monographs [1], [3] and in several papers. Most of the major results concerning the multidimensional signals and systems are developed for two-dimensional cases. Discrete dynamical systems with two independent variables so called 2-D systems are important in image processing, multivariable network realizability and in multidimensional digital filters [3]. During last two decades controllability of 2-D systems has been considered in many papers and books. The main purpose of this paper is to present a compact review over the existing controllability results for linear 2-D systems with constant coefficients.

The paper is organized as follows. Section 2 contains systems descriptions and fundamental results concerning unconstrained controllability for the most popular linear 2-D models with constant coefficients. In section 3 unconstrained controllability of linear singular 2-D systems with constant coefficients is discussed. The section 4 is devoted to a study of constrained controllability of linear 2-D systems. Special attention is paid for the so called positive controllability. Section 5 presents results on positive controllability for linear positive 2-D systems. In section 6 controllability of the so called continuous-discrete linear systems is investigated. Finally, in section 7 concluding remarks and comments concerning possible extensions are presented. Since the paper should be limited to a reasonable size, it is impossible to give a full survey on the subject. In consequence, only selected fundamental results without proofs will be presented. Sections consist of a few major results, and some additional bibliographical comments and supplementary remarks.

2 Unconstrained Controllability

In the theory of 2-D systems several different models are considered. The most popular and the most frequently used are Fornasini-Marchesini model and Roesser model.

First, let us consider Fornasini-Marchesini model of linear 2-D system with constant coefficients given by the following difference equation

$$x(i+1,j+1) = A_0x(i,j) + A_1x(i+1,j) + A_2x(i,j+1) + Bu(i,j) \tag{2.1}$$

where $i,j \in Z_+ = \{0,1,2,3,...\}$-a set of nonnegative integers, $x(i,j) \in R^n$ is a local state vector, $u(i,j) \in U \subset R^m$ is an input vector, U is a given set, A_0, A_1, A_2, and B are real matrices of appropriate dimensions. Boundary conditions for equation (2.1) are given by the following equalities

$$x(i,0) = x_{i0} \in R^n \quad for\ i \in Z_+ \quad x(0,j) = x_{0j} \in R^n \quad for\ j \in Z_+ \tag{2.2}$$

In order to present the general response formula for equation (2.1) in a convenient compact form, it is necessary to introduce (n×n)-dimensional so called state transition matrix $A^{i,j}$ defined as follows.

(i) $A^{0,0} = I$ - (n×n) identity matrix,
(ii) $A^{-i,j} = A^{i,-j} = A^{-i,-j} = 0 \quad for\ i,j > 0$,
(iii) $A^{i,j} = A_0A^{i-1,j-1} + A_1A^{i,j-1} + A_2A^{i-1,j} = A^{i-1,j-1}A_0 + A^{i,j-1}A_1 + A^{i-1,j}A_2 \quad for\ i,j>0$

Therefore, general response formula for equation (2.1) with boundary conditions (2.2) and given admissible controls sequence has the following form.

$$x(i,j)=A^{i-1,j-1}A_0x_{00}+\sum_{p=1}^{p=i}(A^{i-p,j-1}A_1+A^{i-p-1,j-1}A_0)x_{p0}+$$

$$+\sum_{q=1}^{q=j}(A^{i-1,j-q}A_2+A^{i-1,j-q-1}A_0)x_{0q}+\sum_{p=0}^{p=i-1}\sum_{q=0}^{q=j-1}A^{i-p-1,j-q-1}Bu(p,q) \tag{2.3}$$

It is well known, that for 2-D systems it is possible to introduce several different notions of controllability. For example, we may consider global controllability of 2-D systems or the so called straight line controllability of 2-D systems (Kaczorek, 1987b; 1987c; 1987d).

Now, let us recall the most popular and frequently used fundamental definition of unconstrained controllability in a given rectangle $[(0,0),(r,s)]$ for linear 2-D systems with constant coefficients (Fornasisni and Marchesini, 1982; Kaczorek, 1985; 1992a; 1994b; 1994e; 1996e; Klamka, 1984a; 1991a).

Definition 1. System (2.1) with boundary conditions (2.2) is said to be controllable in a given rectangle $[(0,0),(r,s)]$ if for every boundary conditions (2.2) and every vector $x_{rs}\in R^n$, there exists a sequence of controls $u(i,j)\in R^m$, $(0,0)\leq(i,j)<(r,s)$ such that $x(r,s)=x_{rs}$.

From the general formula (2.3) immediately follows, that for zero boundary conditions $x(i,0)=x_{i0}=x(0,j)=x_{0j}=0$, for $i,j\in Z_+$, the solution $x(i,j)$ of the equation (2.1) is simply given by the following equality

$$x(i,j)=W_{ij}u_{ij}$$

where $W_{ij}=[A^{i-1,j-1}B\,,A^{i-2,j-1}B\,,\ldots,A^{0,j-1}B\,,A^{i-1,j-2}B\,,\ldots,A^{1,0}B\,,B]$

W_{ij} is $(n\times ij)$ -dimensional matrix with constant coefficients and the sequence of admissible controls u_{ij} is given by

$$u_{ij}=[u^T(0,0)\,,u^T(1,0)\,,\ldots,u^T(i-1,0)\,,u^T(0,1)\,,\ldots,u^T(i-2,j-1)\,,u^T(i-1,j-1)]^T$$

Theorem 1. [3] System (2.1) is controllable in a given rectangle $[(0,0),(r,s)]$ with unconstrained controls if and only if
$rank\ W_{rs}=n$

Corollary 1. [3] System (2.1) is controllable in a given rectangle $[(0,0),(r,s)]$ if and only if the n×n-dimensional symmetric matrix $W_{rs}W^T_{rs}$ is nonsingular.

Now, let us consider the Roesser model of 2-D linear system given by the set of two difference equation

$$\begin{bmatrix} x^h(i+1,j) \\ x^v(i,j+1) \end{bmatrix} = \begin{bmatrix} A_{11} & A_{12} \\ A_{21} & A_{22} \end{bmatrix} \begin{bmatrix} x^h(i,j) \\ x^v(i,j) \end{bmatrix} + \begin{bmatrix} B_1 \\ B_2 \end{bmatrix} u(i,j) \tag{2.4}$$

where $x^h(i,j) \in R^{n_1}$ is the horizontal state vector,
$x^v(i,j) \in R^{n_2}$ is the vertical state vector,
$A_{11}, A_{12}, A_{21}, A_{22}, B_1, B_2$ are real matrices of appropriate dimensions.
Boundary conditions for (2.4) are given by

$$x^h(0,j) = x^h_{0j} \in R^{n_1} \quad for\ j \in Z_+ \qquad x^v(i,0) = x^v_{i0} \in R^{n_2} \quad for\ i \in Z_+ \tag{2.5}$$

Let $n=n_1+n_2$ and let us introduce n-dimensional vector $x' \in R^n$, n×n-dimensional matrix A' and n×m-dimensional matrix B' defined as follows

$$x' = \begin{bmatrix} x^h(i,j) \\ x^v(i,j) \end{bmatrix} \qquad A = \begin{bmatrix} A_{11} & A_{12} \\ A_{21} & A_{22} \end{bmatrix} \qquad B' = \begin{bmatrix} B_1 \\ B_2 \end{bmatrix}$$

The state transition matrix $A'^{i,j}$ for Roesser 2-D system is defined as follows

(i) $A'^{0,0} = I$, n×n-dimensional identity matrix

(ii) $A'^{1,0} = \begin{bmatrix} A_{11} & A_{12} \\ 0 & 0 \end{bmatrix}$, $A'^{0,1} = \begin{bmatrix} 0 & 0 \\ A_{21} & A_{22} \end{bmatrix}$

(iii) $A'^{i,j} = A'^{1,0}A'^{i-1,j} + A'^{0,1}A'^{i,j-1}$ for $i,j = 1,2,3,\ldots$ $i,j > 0$
(iv) $A'^{i,j} = 0$ for $i<0$ or/and $j<0$.

Using the state transition matrix $A'^{i,j}$ it is possible to express the solution of the Roesser model in a convenient, compact form quite similar to those given by the formula (2.3) for Fornasini-Marchesini model .

Definition 2. Roesser model of 2-D system (2.4) is said to be controllable in a given rectangle $[(0,0),(r,s)]$ if for every boundary conditions (2.5) and every vector $x'_{rs} \in R^n$, there exists a sequence of controls $u(i,j) \in R^m$, $(0,0) \leq (i,j) < (r,s)$ such that $x'(r,s) = x'_{rs}$.

Using the state transition matrix $A'^{i,j}$ and the formula for the solution it is possible to formulate the following necessary and sufficient condition for controllability of Roesser model in a given rectangle $[(0,0),(r,s)]$.

Theorem 2. [3] Roesser model (2.4) is controllable in a given rectangle $[(0,0),(r,s)]$ if and only if

$$rank W'_{rs} = n$$

where matrix $W'_{rs} = [M'(0,1), M'(1,0),..., M'(i,j),..., M'(r,s)]$
and n×m-dimensional matrices $M'(i,j)$ are defined as follows

$$M'(i,j) = A'^{i-1,j}B'^{1,0} + A'^{i,j-1}B'^{0,1}$$

Corollary 2. [3] Roesser model (2.4) is controllable in a given rectangle $[(0,0),(r,s)]$ if and only if the n×n-dimensional symmetric matrix $W'_{rs}W'^{T}_{rs}$ is nonsingular.

It should be pointed out, that more general concept of controllability namely, the so called straight line controllability has been considered in the papers. Moreover, using the concept of the attainable set, it is possible to introduce the notion of reachability for 2-D systems [4]. The relationships between the concepts of controllability and reachability are explained in the papers [4], [5].

3 Singular Systems

Recently the classical regular 2-D systems have been extended for singular 2-D systems also known as implicit, descriptor or generalized 2-D systems [4], [5], [6], [7], [8] and [9]. The motivations for studying singular 2-D systems have been well justified in several papers where the reader may find many examples of possible practical applications of singular 2-D systems in network theory, chemical engineering, robotics or economics. Two-dimensional singular systems, unlike the familiar state-space 2-D systems, do not require any notion of causality or recursibility. Instead of, they require the milder notion of regularity which is required for existence and uniqueness of solutions. Thus, the singular 2-D systems are more natural and suited for the description of naturally occurring 2-D systems as for applications in image processing. It is also possible to consider linear singular 2-D systems with delays and infinite-dimensional singular 2-D systems. A survey of recent results and the current state of the theory of 2-D singular systems can be found e.g. in [4].

Let us consider a linear singular 2-D system with constant coefficients described by the following difference equation

$$Ex(i+1,j+1) = A_0x(i,j) + A_1x(i+1,j) + A_2x(i,j+1) + Bu(i,j) \tag{3.1}$$

where A_0, A_1, A_2 are the same matrices as in the equation (2.1) and E is an n×n-dimensional possibly singular matrix.

It is known that singular 2-D system has unique solution only for admissible boundary conditions. The linear subspace of admissible boundary conditions depends on matrices E, A_0, A_1, A_2 and input sequence. However, the following lemma holds.

Lemma 1. [7] Any boundary conditions are admissible in any rectangle and any input sequence if and only if

$$rank[z_1z_2E - A_0 - z_1A_1 - z_2A_2] = rank[z_1z_2E, A_0, z_1A_1, z_2A_2, B], \text{ for some } z_1,z_2\in C$$

The solution for admissible boundary conditions and for given admissible control sequence has the following form

$$Ex(i,j) = A^{i-1,j-1}A_0x(0,0) + \sum_{p=0}^{p=i}(A^{i-p,j-1}A_1 + A^{i-p-1,j-1}A_0)x(p,0) +$$

$$+\sum_{q=0}^{q=j}(A^{i-1,j-q}A_2 + A^{i-1,j-q-1}A_0)x(0,q) + \sum_{p=0}^{p=i-1}\sum_{q=0}^{q=j-1}A^{i-p-1,j-q-1}Bu(p,q) \qquad (3.2)$$

Theorem 3. [6] Singular system (3.1) is locally controllable in the rectangle $[(0,0),(r,s)]$ if and only if

$$rank\, M_{rs} = n$$

where
$M_{rs}=[M(0,0),M(1,0),...,M(r,0),M(0,1),...,M(0,s),M(1,1),...,M(r,1),...,M(r,s-1)]$
and n×m-dimensional constant matrices $M(i,j)$ are defined as follows
$M(i,j) = A^{r-i-1,s-j-1}B \qquad for \qquad i=1,2,3,...,r \qquad j=1,2,3,...,s$

Corollary 3. [6] System (3.1) is controllable in a given rectangle $[(0,0),(r,s)]$ if and only if the n×n-dimensional symmetric matrix $M_{rs}M^T{}_{rs}$ is nonsingular.

4 Conclusions

In recent years controllability problems for different kinds of 2-D systems have been considered in many publications. Most papers so far has been concerned however, with the so called unconstrained controllability problems. Only a few papers deal with the so called constrained controllability problems, i.e. with the case when the control functions are restricted to take their values in a prescribed admissible set [10], [11], [12]. Moreover, it should be also stressed, that up to now constrained controllability problems for abstract retarded dynamical systems defined in infinite-dimensional Hilbert spaces has not been considered in the literature. It should be stressed that, controllability of positive 2-D systems has been discussed in [13] and 2-D continuous-discrete systems have been considered in [14], [15], [16] and [17].

The paper contains fundamental theorems concerning unconstrained and constrained controllability problems both for linear and nonlinear 2-D systems with constant coefficients. In the literature there are many other controllability results, derived for more general 2-D systems. For example, controllability of the multi-dimensional discrete systems has been recently considered in the paper [18].

Up to the present time the problem of controllability in continuous and discrete time linear dynamical systems has been extensively investigated in many monographs and papers. However, this is not true for the nonlinear dynamical systems specially with constrained controls. Only a few papers concern constrained controllability problems for 2-D nonlinear systems [19], [20]. The paper [20] contains results concerning local controllability of nonlinear 2-D systems without differentiability assumptions. Finally, it should be mentioned, that formula for solution and controllability results concerning 2-D bilinear systems can be found in the paper [19].

References

1. Kaczorek, T., 1985, *Two-dimensional Linear Systems*, Springer-Verlag, Berlin.
2. Kaczorek, T., 1989, *Decomposition of Roesser model and new conditions for local controllability and local observability*, International Journal of Control, V. 49, No. 1, pp. 65 - 72.
3. Kaczorek, T., 1992, *Local controllability, reachability, and reconstructibility of the general singular model of 2-D systems*, IEEE Transactions on Automatic Control, V. AC-37, No. 10, pp. 1527 - 1530.
4. Kaczorek, T., 1994, *When the local controllability of the general model of 2-D linear systems implies its local reachability*, Systems and Control Letters, V. 23, No. 3, pp. 442 - 452.
5. Kaczorek, T., 1994, *Reachability and controllability of 2-D continuous-discrete linear systems*, Proceedings of First Symposium on Mathematical Models in Automation and Robotics, Miedzyzdroje, pp. 24 - 28.
6. Kaczorek, T., 1995, *U-reachability and U-controllability of 2-D Roesser model*, Bulletin Polish Academy of Sciences, Technical Sciences, Electronics and Electrotechnics, V. 43, No. 1, pp. 31 - 37.
7. Kaczorek, T., 1991, *Some recent results in singular 2-D systems theory*, Kybernetika, V. 27, No. 3, pp. 253 - 262.
8. Kaczorek, T., 1995, *General response formula, local reachability and local controllability of a class of 2-D bilinear systems*, Bulletin of the Polish Academy of Sciences.Technical Sciences,Electronics and Electrotechnics, V. 43, No. 2, pp. 191 - 201.
9. Kaczorek, T., 1995, *Controllability and minimum energy control of 2-D continuous-discrete linear systems*, Applied Mathematics and Computer Science, V. 5, No. 1, pp. 5- 21.
10. Kaczorek, T., 1996, *Reachability and controllability of nonnegative 2-D Roeser type models*, Bulletin of the Polish Academy of Sciences, Technical Sciences, Electronics and Electrotechnics, V. 44, No. 4, pp. 405 - 410.
11. Klamka, J., 1988, *Constrained controllability of 2-D linear systems*, Proceedings of 12 World IMACS Congress, Paris, V. 2, pp. 166 - 169.

12. Klamka, J., 1989, *Controllability of singular 2-D systems*, Proceedings of IFAC Workshop on System Structure and Control, Praga, pp. 245 - 248.
13. Klamka, J., 1991, *Controllability of Dynamical Systems*, Kluwer Academic Publishers, Dordrecht.
14. Klamka, J., 1991, *Complete controllability of singular 2-D system*, Proceedings of. 13 World IMACS Congress, Dublin, pp. 1839 - 1840.
15. Klamka, J., 1993, *Controllability of dynamical systems - a survey*, Archives of Control Sciences, V. 2, No. 3 - 4, pp. 281 - 307.
16. Klamka, J., 1994, *Controllability of 2-D continuous-discrete linear systems*, Proceedings of the Third Conference on Automation, Robotics and Computer Vision, Singapur, V. 2, pp. 1268 - 1272.
17. Klamka, J., 1995, Controllability of 2-D continuous-discrete abstract linear systems. *Proceedings of European Control Conference*, Roma, V. 3, pp. 2208 - 2213.
18. Klamka, J., 1996, *Constrained controllability of 2-D systems*, Proceedings of IMACS International Symposium on Modeling, Analysis and Simulation, Lille, pp. 376 - 379.
19. Klamka, J., 1996, *Controllability of 2-D nonlinear systems*, Proceedings of Second World Congress of Nonlinear Analysis, Athens, pp. 1339 - 1348.
20. Klamka, J., 1997, *Controllability of M-D linear systems*, System Analysis, Modelling, Simulation, V. 25, No. 1, pp. 119 - 128.

10.6

Control Theory and Applications for Repetitive Processes[1]

D. H. Owens, ✠ *E. Rogers* † *and K. Galkowski* ✳

✠ School of Engineering and Computer Science, University of Exeter, Exeter, EX4 4QF, UK.
† Department of Electronics and Computer Science, University of Southampton, Southampton, SO17 1BJ, UK.
✳Department of Robotics and Software Engineering, Technical University of Zielona Góra, ul. Podgórna 50, 55-246 Zielona Góra, Poland.

Summary. Repetitive processes are a class of two-dimensional (2D) systems of both theoretical and practical interest. These processes cannot be controlled by (direct) application of standard (termed 1D here) theory since such an approach neglects their unique dynamic characteristics. In this paper we present an introduction to repetitive processes and survey the work to-date on the development of a control theory for these processes and its application to, in particular, classes of iterative learning control schemes.

1 Introduction

The essential unique characteristic of a repetitive process (also known as a multipass process in the early literature) can be illustrated by considering machining operations where the material or workpiece involved is processed by a series of sweeps, termed passes, of the processing tool. On each pass an output, termed the pass profile, is produced which acts as a forcing function on, and hence contributes to, the next pass profile. Clearly these processes have a 2D systems structure, i.e. information propagation along a given pass, which is finite by definition, and from pass to pass.

[1]This work has been sponsored by EPSRC and The British Council-KBN Collaborative Research Programme

Repetitive processes have their origins in the mining and metal processing industries [1], and, in particular, the dynamics of the long-wall coal cutting process which is the most commonly encountered method of extracting coal from deep cast mines. The essential unique control problem for these processes arises from the explicit interaction of successive pass profiles with the result that the sequence of pass profiles produced can contain oscillations that increase in amplitude in the pass to pass direction.

Repetitive processes do share certain structural similarities with 1D systems and, in the case when linear dynamic models are the basis of controller design, this initially led to attempts to control these processes (and, in particular, the long-wall coal cutting example) by direct application of classical 1D linear systems techniques. This general approach fails, except in a few very restrictive special cases, precisely because it effectively ignores the effects of the finite pass length and the resetting of the initial conditions before the start of each new pass, i.e. the two most distinguishing features of these processes!

To remove this basic difficulty, a rigorous stability theory for these processes has been developed based on an abstract model of their dynamics in a Banach space setting which includes all examples with linear dynamics and a constant pass length as special cases [1]. This theory coupled with the continual emergence of new applications areas, such as classes of iterative learning control schemes (see, for example, [2]), has established repetitive processes as a distinct branch of 2D control systems theory and applications. In this paper, we provide an overview of progress to-date together with 'pointers' to stimulate further research/development. We will concentrate on the methodology and major results and omit proofs which can be found in the cited references.

2 The Abstract Model and Stability Theory

Clearly any stability/control theory for repetitive processes must be based on a representation of their dynamics which explicitly includes their unique features. The following model structure meets this objective for the class of processes with linear dynamics and a constant pass length [1].

Definition 1 *A linear repetitive process $S(E_\alpha, W_\alpha, L_\alpha)$ of constant pass length $\alpha > 0$ consists of a Banach space E_α, a linear subspace W_α of E_α, and a bounded linear operator L_α mapping E_α into itself. The system dynamics are described by linear recursion relations of the form*

$$y_{k+1} = L_\alpha y_k + b_{k+1}, \; k \geq 0 \tag{1}$$

where $y_k \in E_\alpha$ is the pass profile on pass k and $b_{k+1} \in W_\alpha$. Here the term $L_\alpha y_k$ represents the contribution from pass k to pass $k+1$ and b_{k+1} represents initial conditions, disturbances and control input effects on pass $k+1$.

The stability theory has been defined in terms of this model and necessary and sufficient conditions for the various properties expressed in terms of conditions on the bounded linear operator L_α. The task then for a particular example or sub-class is to translate these abstract conditions into conditions on the corresponding model. We now introduce the stability theory and apply it to the important sub-class of so-called differential linear repetitive processes.

The dynamics of a differential linear repetitive process are described by the following state space model defined over $0 \leq t \leq \alpha$, $k \geq 0$

$$\dot{x}_{k+1}(t) = A\, x_{k+1}(t) + B\, u_{k+1}(t) + B_0\, y_k(t),\; y_{k+1}(t) = C\, x_{k+1}(t) \qquad (2)$$

Here on pass k, $x_k(t)$ is the $n \times 1$ state vector, $y_k(t)$ is the $m \times 1$ vector pass profile and $u_k(t)$ is the $l \times 1$ vector of control inputs. To complete the process description it is necessary to specify the initial, or boundary, conditions, i.e. the state initial conditions on each pass and the initial pass profile respectively. The most general form for these over $0 \leq t \leq \alpha$, $k \geq 0$, is

$$x_{k+1}(0) = d_{k+1} + \sum_{j=1}^{M} K_j\, y_k(t_j),\; y_0(t) = y(t) \qquad (3)$$

where $0 \leq t_1 < t_2 < \cdots < t_M \leq \alpha$ are M sample points along the previous pass profile, K_j, $1 \leq j \leq M$, is an $n \times m$ matrix with constant entries, d_{k+1} is an $n \times 1$ vector with constant entries, and $y(t)$ is an $m \times 1$ vector whose entries are known functions of t.

Consider now the choice of $E_\alpha = L_2^m[0,\alpha] \cap L_\infty[0,\alpha]$. Then for a process described by (2) and (3), it is easy to show that over $0 \leq t \leq \alpha$

$$\begin{aligned} (L_\alpha y)(t) &= Ce^{At}\Big(\sum_{j=1}^{M} K_j y(t_j)\Big) + C\int_0^t e^{A(t-\tau)} B_0 y(\tau)\, d\tau \\ b_{k+1} &= Ce^{At} d_{k+1} + C\int_0^t e^{A(t-\tau)} B u_{k+1}(\tau)\, d\tau \end{aligned} \qquad (4)$$

The stability theory for $S(E_\alpha, W_\alpha, L_\alpha)$ consists of two distinct concepts, termed asymptotic stability and stability along the pass respectively. Of these, stability along the pass is the stronger and, in effect, demands that bounded (in the sense of the norm on the underlying function space) sequences of inputs produce bounded sequences of pass profiles independent of the pass length. The following is one of several equivalent characterizations of this property (where $||.||$ denotes both the norm on E_α and the induced operator norm).

Definition 2 *$S(E_\alpha, W_\alpha, L_\alpha)$ is said to be stable along the pass if, and only if, $\exists$ finite real scalars $M_\infty > 0$ and $\lambda_\infty \in (0,1)$ which are independent of α and satisfy*

$$||L_\alpha^k|| \leq M_\infty \lambda_\infty^k,\; \forall\, \alpha > 0,\; \forall\, k \geq 0 \qquad (5)$$

Theorem 1 *$S(E_\alpha, W_\alpha, L_\alpha)$ is stable along the pass if, and only if,*
(a)

$$r_\infty := \sup_{\alpha>0} r(L_\alpha) < 1 \tag{6}$$

where $r(.)$ denotes the spectral radius of its argument, and
(b)

$$M_0 := \sup_{\alpha>0} \sup_{|z|\geq\lambda} ||(zI - L_\alpha)^{-1}|| < +\infty \tag{7}$$

for some real number $\lambda \in (r_\infty, 1)$.

In the case of processes described by (2) and (3) we assume that M and t_j, $1 \leq j \leq M$, are fixed at the outset.

Theorem 2 *Suppose that $\{A, B_0\}$ is controllable and $\{C, A\}$ is observable. Then $S(E_\alpha, W_\alpha, L_\alpha)$ generated by (2) and (3) is stable along the pass if, and only if,*
(a) *all solutions of*

$$|zI_n - \sum_{j=1}^{M} K_j C e^{\hat{A}(z)t_j}| = 0 \tag{8}$$

where $\hat{A}(z) = A + z^{-1}B_0C$, $z \neq 0$, have modulus strictly less than unity.
(b) *all eigenvalues of the matrix A have strictly negative real parts, and*
(c)

$$\sup_{\omega\geq 0} r(G(\imath\omega)) < 1 \tag{9}$$

where

$$G(s) := C(sI_n - A)^{-1}B_0 \tag{10}$$

Other characterizations of stability along the pass are known - some of which also provide computable information on expected system performance. For example, the so-called 1D Lyapunov equation (the defining matrices of this equation are functions of a complex variable) produces such information [3].

3 Systems Theoretic Properties

Significant progress has been made on the development of systems theoretic concepts for differential and discrete linear repetitive processes. Topics which have seen the most progress in the differential case include the fundamentals of a 2D polynomial matrix theory [4]. In general, however, the most progress to-date on the structure and design of control schemes for differential and discrete linear repetitive processes has been made in the context of iterative learning control schemes - see section 4.

The remainder of this section gives an overview of the main results to-date on reachability/controllability of discrete linear repetitive processes. This general area is much more complicated than the 1D linear systems case, a feature which is shared by other classes of 2D discrete linear systems.

The state space model of a discrete linear repetitive process has the following form over $0 \leq p \leq \alpha$, $k \geq 0$

$$\begin{aligned} x_{k+1}(p+1) &= A\,x_{k+1}(p) + B\,u_{k+1}(p) + B_0 y_k(p) \\ y_{k+1}(p) &= C\,x_{k+1}(p) + D\,u_{k+1}(p) + D_0\,y_k(p) \end{aligned} \tag{11}$$

It is also easy to see that the stability theory based on the abstract model also applies to this sub-class. Hence the details are omitted here and can found in [1].

This state space model has 'strong' structural links with 2D discrete linear systems recursive over the positive quadrant. In particular, define the so-called augmented state vector for (11) as

$$z(k,p) = \left[x_k^T(p), y_k^T(p)\right]^T \tag{12}$$

Then the dynamics of this process can be written, see [6], as

$$Ez(k+1,p+1) = A_1 z(k+1,p) + A_2 z(k,p) + B_1 u(k+1,p) \tag{13}$$

where $u(k+1,p) \equiv u_{k+1}(p)$ and the matrix E is singular.

This description is a special case of the singular version of the Fornasini Marchesini model [5], but it should also be noted that singularity is not an intrinsic feature (in a well defined sense) of discrete linear repetitive processes. In particular, a 'transformation theory' exists [6] for constructing nonsingular (also termed either standard or regular) Fornasini Marchesini state space models from their singular counterparts. (This also holds for alternative state space models to the Fornasini Marchesini structure.)

Using (12)-(13) as a starting point, a transition matrix (or fundamental matrix sequence) has been developed and used to establish an analytic expression for the trajectories of (11) in response to given control input sequences and boundary conditions. This has then led to matrix rank based characterizations of so-called local reachability and controllability of these processes. Essentially, these properties concern the existence or otherwise of an admissible control input sequence which will drive the vector $z(k,p)$ to a specified vector - reachability - or the zero vector - controllability - at a specified point on a specified pass. As noted above, these properties are, in general, distinct concepts but well known results can be applied (after suitable modification) to study the relationships between them.

The problem of constructing an equivalent 1D discrete linear systems representation of the dynamics of 2D linear systems has been the subject of research effort, see, for example, [7]. Such representations have been shown to exist but have the highly undesirable feature that the dimensions of the matrices and vectors involved increase as the underlying process evolves, and this fact alone means that they are of little real use in problem solving. In the case of discrete linear repetitive processes, however, it is possible to find a 1D equivalent representation expressed in terms of matrices and vectors with constant dimensions [6].

This 1D model is defined in terms of the so-called global state, input and pass profile vectors. Using this 1D representation, it has been possible to solve the fundamental pass controllability problem for (11) whose formal statement is as follows.

Definition 3 *The discrete linear repetitive process (11) is said to be pass controllable if $\exists$ an admissible control input sequence which will drive the process to an arbitrarily specified pass profile on an arbitrarily specified pass number.*

As with local reachability/controllability, the solution is in the form of matrix rank based conditions. The solution of this problem has opened up new lines of research in terms of basic structural theory and the structure and design of control schemes for these processes which are the subject of on-going research.

4 Conclusions and Further Work

At this stage, key elements of a systems/control theory for linear repetitive processes are well established, but further development work is still required in most cases. To-date, however, relatively little is available by way of a theoretical basis for the design of control schemes for these processes.

The importance of this last point is strengthened by the continual emergence of new applications areas and, in particular, iterative learning control (ILC) schemes. These schemes have their origins in the analysis and control of systems which are required to continually perform the same task. Motivated by human learning, the basic idea of ILC is to use information from previous trials to improve performance from trial to trial. The ultimate aim (in theory) is to learn the precise input needed to achieve the desired output over the trial length which is, by definition, finite. In practice, the objective will be to reach the reference signal within a given accuracy or tolerance (suitably defined).

ILC has been the subject of intensive research activity since the mid 1980's and many different approaches to both the linear and nonlinear cases have been studied and actually applied to real processes. A very convenient reference point for all of this work is the web-site [8]. It is clear that ILC schemes have a natural repetitive process structure and this fact has been the sole basis for the development and application of linear ILC schemes which use a combination of feedforward action from previous trials and feedback action on the current trial, where one immediate benefit from the presence of the latter is that all the advantages of such action (i.e. improved robustness) are potentially available.

The papers which describe this and subsequent developments on the application of linear repetitive process concepts to ILC schemes (in addition to [2]) are also listed on [8]. Currently work is proceeding on the extension of this and other approaches from the particular special cases of differential and discrete linear repetitive processes used in ILC schemes to more general cases. This is a very large problem area, even in the linear case, and much profitable research/development work remains to be done.

References

1. E. Rogers and D. H. Owens. *Stability Analysis for Linear Repetitive Processes*, Springer-Verlag, Lecture Notes in Control and Information Sciences Series, Vol. 175, 1992.

2. N. Amann, D. H. Owens and E. Rogers. Predictive optimal iterative learning control. *International Journal of Control*, 69(2): 203-226, 1998.

3. D. H. Owens and E. Rogers. Frequency domain Lyapunov equations and performance bounds for differential linear repetitive processes. *Systems and Control Letters*, 26: 65-68, 1995.

4. D. S. Johnson, A. C. Pugh, E. Rogers, G. E. Hayton, and D. H. Owens. A polynomial matrix theory for a class of 2D linear systems, *Linear Algebra and its Applications*, 241-243: 669-703, 1996.

5. E. Fornasini and G. Marchesini. Doubly-indexed dynamical systems: state space models and structural properties, *Mathematical Systems Theory*, (12): 59-72, 1978.

6. K. Galkowski, E. Rogers, and D. H. Owens. Matrix rank based conditions for reachability/controllability of discrete linear repetitive processes, *Linear Algebra and its Applications*, 275-276: 201-224, 1998.

7. J. L. Aravena, M. Shafiee, and W. A. Porter. State models and stability for 2-D filters, *IEEE Transactions on Circuits and Systems*, CAS-37(12): 1509-1519, 1990.

8. Dedicated Web Server for Iterative Learning Control, URL http://ilc.ee.nus.edu.sg/

10.7

Two Dimensional Systems Concepts for Industrial Web Processes

P. E. Wellstead, † *M. B. Zarrop,* † *and S. R. Duncan* ✳

†Control Systems Centre, UMIST, Manchester, M60 1QD. UK.
✳Engineering Science, Oxford University, Oxford, OX1 1AA, UK.

Summary. Two dimensional (2D) system and signal process ideas are used as a paradigm for the analysis of industrial web processes. This applications driven approach to 2D systems gives a special perspective which directs the development of corresponding 2D identification and control algorithms. Web processes form a general class of engineering systems, which includes paper making, plastic extrusion and many treatment and conversion processes for sheet products. The analysis of web processes can be unified via the 2D paradigm to gives a fresh perspective to the modelling, control and sensing of an important class of manufacturing processes.

1 Introduction

The industrial processes associated with the manufacture of continuous webs (or sheets) of material constitute an area of significant growth in which improved control and signal processing can yield large economical and technical advantages. Specifically, customers and retailers require continuous improvements in product quality, flexibility and price from the supplier of paper, plastics film, sheet steel. This places great competitive pressure upon manufacturers, who in turn require continuous improvements in performance from the control system suppliers. As remarked above, well known examples of industrial web forming processes are paper making, steel sheet forming, polymer film extrusion and related coating and converting processes. However examples of web forming processes also occur in the emerging technologies, such as biotechnology, surface treatment and composite materials. In fact anywhere that organic or inorganic material is manufactured or processed in sheet form. Further areas where 2D methodologies can fruitfully be developed are where 2D sensing is used for control and analysis. This area of 'vision in the loop' (e.g. [15]) will grow rapidly in the next decade as 2D sensing techniques are further developed.

The applications of 2D methods in industry is potentially vast, and yet the area is underdeveloped in certain key technological respects. It is thus timely and important for control engineers to study web forming processes as a generic class of two dimensional systems - particularly in view of the competitive pressures and the scientific opportunities mentioned above. At the Control Systems Centre, we began in the early 1980's to look at what 2D methods are appropriate to scientific and industrial applications, and to develop modelling, identification and control/prediction methods which worked on the applications which were then being studied by the Centre. Specifically, our work in 2D began with extending one dimensional self tuning minimum variance prediction to two dimensions for the prediction of periodic processes [25] this led to extensions to 2D smoothing and applications in image restoration for clinical studies of retina damage and deblurring of finger print images [26]. Good results with 2D image prediction and interpolation led us to develop the dual algorithms for control and regulation, using input-output methods [27], [9], [10], [11]. More general state space methods with their natural linkage to predictive control tools were then developed [34]. Sensing of web processes is a major research topic, since it is costly and difficult to measure the entire 2D characteristics and compromises are made by only sensing a fraction of the web surface. This led to the development of sparse data estimators and control methods [7] for the sensing systems found in web forming processes. Current Control Systems Centre research on 2D aspects of industrial web processes concerns the problems of sensing and the technical implications of actuator/sensor systems for control theory of this class of practical 2D processes [5].

The aim of this presentation is to review the results of the research process outlined above. The theme is the use of a 2D paradigm as the basis for the analysis and control of one class of industrial processes (e.g. industrial web forming systems) and to show the benefits and difficulties of this approach. After a long missionary effort [28], [30] the treatment of web processes as a two dimensional problem is now gaining support in the web forming industries. Interestingly this is being stimulated not only by 2D control research developments but also by the development of 2D sensor systems [18], [1], [13], [15], [19], [21]. Not withstanding this growth of interest in a 2D view of web forming systems the practical control methods used are not expected to change for some time. All existing web manufacturing systems are controlled using multivariable control systems in which interaction is characterised by a so-called interaction matrix [33], [8], [4], [31]. This situation is expected to continue for some time, but the pressure to improve quality and provide control over ever finer features of web products, is such that it is inevitable that a 2D form of control/analysis will in the long run become established in the industry. This change will be led by 2D data analysis, since 2D sensing is already coming, and followed by 2D control as actuation methods evolve.

2 Industrial Web Processes

An archetypal industrial web forming process is one in which a delivery mechanism supplies material through a narrow slit to form a continuous web of material. The narrow slit is controlled by an array of actuators which control the properties of the web, such that a movement in one of the actuators in the array influences the product over an area in the region of the spatial location and time of the actuation. The spatial direction across the process is called the cross direction (CD) and the direction of material web movement is the machine direction (MD). The properties of the product are measured by a downstream sensor which is usually a scanning device which measures the web properties along a zig zag path. This basic description of an industrial web forming process is elaborated in [22], [2], [16] for the paper, steel and plastics industries respectively. The converting industries which perform additional processes (such as applying adhesive, paint or metallic coatings) to these basic sheet materials is a further important web processing area. Many others exist in new manufacturing and processing methods and are growing in number and importance.

In continuous time and space, the output of a web process is denoted by $y(x,t)$, where x is the CD position of the measurement and the time t indicates the location of the measurement in the MD direction. Likewise the input actuation is denoted $u(x,t)$. The output variable $y(x,t)$ depends upon the specific web process, see [29] for a general review. However, a primary 'output' in any web forming system is product thickness, although this is usually accompanied by other important process properties (e.g. moisture content in paper making). The input actuation signal $u(x,t)$ is usually applied using a spatial array of actuators which displace either a continuous beam (which forms the gap through which the material web is formed), and/or a heater set which continuously varies the thermal properties of the material web. The actuation process can be described by a partial differential equation (PDE) derived from direct analysis of the actuation and web forming process, [32]. However, this is a complex modelling process and even when the model has been obtained, the final equation set cannot be reliably parametrized. For this reason discrete 2D difference equations (or state space forms) are preferred as models since they can be parametrized using 2D-ARMAX identification techniques with data measured from the particular web forming process to be analysed.

3 2D Process Models For Web Processes

As noted above it is important to base practical control studies upon model structures that can be directly used in estimation, analysis and design. The forms of model which we use are the 2D discrete time equations reviewed below. Although our aim is to directly estimate the parameters of these models, the models can also be related to the underlying PDE description via the discretisation procedures used in finite element analysis [17].

3.1 2D-ARMAX Model

Because of the linkage between the PDE's which define the web forming dynamics and two dimensional difference equations, it is assumed that a two-dimensional web forming process can be represented by a 2D-ARMAX model:

$$\begin{aligned} &A(w^{-1},z^{-1})y(m,n) = z^{-\upsilon}B(w^{-1},z^{-1})u(m,n) \\ &+C(w^{-1},z^{-1})e(m,n) \end{aligned} \qquad 1$$

Where *y(m,n)* is the output, *u(m,n)* is the control input and *e(m,n)* is white noise. The two dimensional polynomial $A(w^{-1},z^{-1})$ is given by:

$$A = \sum_{i=0}^{M}\sum_{j=0}^{N} a_{i,j}w^{-i}z^{-j} + \sum_{i=1}^{M}\sum_{j=1}^{N} a_{-i,j}w^{i}z^{-j} \qquad 2$$

And similarly for B and C. The co-ordinates (m,n) represent position in the plane with respect to some arbitrary origin. The shift operators z^{-1} and w^{-1} operate in the cross (CD) and machine (MD) directions respectively. The integers M and N define the structure or support [12] of the 2D-ARMAX model. The support type and the corresponding causality depend upon the application. The A,B,C polynomials are frequently chosen to have non-symmetric half plane (NSHP) support, which has the advantage of being causal for raster scan date collection and recursively computable. The web manufacturing process is non-causal in the horizontal (CD) direction and a symmetric half plane (SHP) model is appropriate for this form of system causality. This can be block recursively computed for finite width processes [3], moreover, this block recursive feature can be used in recursive estimation of 2D-ARMAX model parameters, (see section 4). In web forming systems the material web will usually be made in a finite width, (the blown film process is a particular exception [16], here the web is formed in a continuous tube). For finite width webs the 2-D model is only approximate because the support for the 2D-ARMAX model, (1), must be truncated for data near the edges [9].

3.2 2D State Space Models

The 2D-ARMAX model can used for estimation (section 4) and control (section 5) based upon input-output models. However to take full advantage of predictive control methods a state space innovations form is preferred. Popular forms of 2D state models are Roesser's [20] and Fornasini - Marchesini's model [6] and elsewhere in this collection of papers). 2D-ARMAX models are easily transformed into state space forms. Details of the state space innovations form are given in [23] and the use of 2D innovations models in output prediction, data reconstruction and control are covered in [34].

4 2D-ARMAX Estimation

The PDE which defines a web forming process provides a comprehensive but complex systems model and the corresponding 2D difference equation obtained by finite element analysis or transform methods can only rarely be parametrized using direct measurements of system parameters. Thus, as noted above, it is necessary to use identification and parameter estimation techniques to determine the coefficients of (1). If the data is available in a true 2D form then direct estimation is used. If however the data is collected by a scanning gauge then only partial data is available and a sparse data estimation method is used [7].

The 2D-ARMAX model (1) of a data field is a useful form for analysis and control since it is possible to recursively compute and estimate parameters of the model. Recursive computing is important since this allows a 2D forgetting factor to be incorporated into the estimator, such that adaptation to changes in the underlying dynamics can be tracked. With minor house keeping in the estimator, a conventional recursive maximum likelihood algorithm [24] can be applied to estimate the 2D-ARMAX model parameters. If a 2D forgetting factor is required then a modified numerical and algorithmic implementation is required which allows data to be discounted according to their radial distance form the current data point. This requires that the recursive estimator uses an information matrix update method, rather than the Kalman procedure, [9].

Estimation of a system model generates a residual process $\hat{e}(m,n)$ which can be used as an estimate of the process noise, $e(m,n)$ as required for the synthesis of the minimum variance controller. A consequence of this is that the estimation and controller synthesis process can be easily combined into a self-adaptive algorithm whereby the estimation and control are performed at each point by scanning raster-style across the data field. If a SHP model is used then the generation of the residual process requires a little more work, but is straightforward.

If the true two-dimensional memory mentioned previously is required then a modified numerical and algorithmic implementation is required. Specifically, an information matrix update is used which allows data to be discounted according to their distance from the current data point. Specifically the cost function which is minimised at each time step is:

$$J(m,n) = \sum_{i=0}^{\infty}\sum_{j=0}^{\infty} \lambda^i \mu^j \hat{e}(m-i,n-j)^2 + \sum_{i=0}^{\infty}\sum_{j=0}^{\infty} \lambda^i \mu^j \hat{e}(m+i,n-j)^2 \qquad 3$$

This cost function includes two-dimensional forgetting, where λ, μ are the forgetting factors which weight the data according to the distance from the current data point. The structure of the model (M,N) is usually unknown and is determined using information theoretic criteria.

Standard scanning gauges usually lead to sparse data sets. This can be considered as a missing data problem. For example, a single sensor generating a zigzag path across the web can be viewed as performing a standard raster scan in which only one datum line is accessible. Again, this situation is most easily dealt with within the state space framework as in [7].

5 2D Controller Synthesis

Both input-output and state space methods have been developed for 2D control of web processes. In the input-output approach the optimal (minimum variance) controller for the process described by equation (1) uses the fact that it is possible to construct a minimum variance predictor for data k_1 columns and k_2 rows ahead of the current data point, [9]. For equation (1) with delay υ the corresponding minimum variance controller involves the use of a predictor with $k_1 = 0$ and $k_2 = \upsilon$ [10], [11]. The state space version algorithms for 2D predictive control a gives a more versatile and algorithmically simpler solution than the input-output approach. Full details of the two dimensional predictive control method are given in [23], [34].

6 Developments In Industrial Web Control Systems

Thus far we have discusses only the algorithm development for the application of 2D systems to industrial processes, and although many algorithmic problems remain associated with issues such as edge effects and sensitivity properties of the controller/estimation methods, of greater relevant to the medium term are the process developments in 2D sensing for web forming processes. Specifically if the aim is an increased ability to control product at greater production speeds, then bandwidth of the process loop must be increased. The bandwidth issue is crucial since it is when the controllable bandwidth increases that the separability assumption [14] which underpins the interaction matrix approach will begin to breakdown. The achievable closed loop bandwidth is determined by (a) the time delay in the process, (b) the actuator speed and (c) the sensor bandwidth.

In paper machines for example the actuation takes place at the wet-end while sensing occurs after a large time delay at the dry-end [22]. The transport delay between wet-end and dry-end severely limits the achievable closed loop bandwidth that can be achieved by any controller. To a degree, this can be overcome using additional sensing based upon dry-line measurement at the paper machine wet-end [15], [18]. The actuator response time is also being reduced using (for example) piezo-electric and magneto-strictive technologies as well as innovations in process technology.

The third and most interesting limiting factor is the sensor bandwidth. The bandwidth (spatial and temporal) associated with traditional scanning sensors is completely inadequate for 2D control, since the speed of scanning is very much slower than that of the product passing under the scanner. This shortcoming is being addressed by new gauging technologies which will allow a much greater density. For example, dry line sensing methods mentioned above give true two dimensional information which can be used for control and diagnostic work, this in addition to reducing the transport delay in a web forming system.

In a further developments stationary arrays of sensors have been developed by a number of instrumentation companies in the USA and Europe, [1], [19], [21]. These array sensing devices allow the full (CD) width properties of the web to be measured and at a data density appropriate for 2D-ARMAX estimation. Also important is the scanning array approach in which of a set of gauges uniformly spaced on a single scanning frame. The frame moves a limited amount from side to side such that the sensors describe a chevron scanning pattern over the web [13]. This method has the advantage that the density of the data is significantly increased over the single gauge, but without the expense of a full sensor array. The data is not fully two dimensional, but by selecting the gauge number for each application the system can be customised to give the measurement density required.

7 Acknowledgements

The work described here was supported by EPSRC (grants GR/K04378, GR/M21692, GR/L33283, GJR/L18600), NATO, the Royal Society, the Royal Academy of Engineering, 3M Corporation and NDC/Infrared Engineering Ltd.

References

1. Anon. Gauging System Gives Small Converter a Big Edge, *Converting Magazine*, Sept. 1993.
2. G. F. Bryant. *Automation of Tandem Mills*. Iron and Steel Institute, London, UK. 1973.
3. M. M. Daniel and A.Willsky. *Efficient Implementation of 2D Non causal IIR Filters*, MIT research report. 1995.
4. S. R. Duncan and G. F. Bryant. Spatial Controllability of Cross Directional Control Systems for Web Processes, *Automatica*, 33, 2, pp 139-153. 1996.
5. S. R. Duncan, P. E. Wellstead and M. B. Zarrop. *Actuation, Sensing and Two Dimensional Control Algorithms: Fundamental Limitations on Achievable Bandwidths in MD/CD Control*, Proceedings 1998 TAPPI/ISA- PUPID Process Control, Electrical and Information Conference, Vancouver. 1998.
6. E. Fornasini and Marchesini. Doubly Indexed Dynamical Systems: State Space Models and Structural Properties, *Mathematical Systems Theory*, 12, pp 59 - 75. 1978.
7. D. M. Gacon, M. B. Zarrop and R. Kulhavy. *EM Estimation and Control of Two Dimensional Systems*, Proceedings of Control '96, IEE Conference, Exeter. 1996.

8. A. Halouskova, M. Karny and I. Nagy. Adaptive Cross Directional Control of Paper Basis Weight, *Automatica*, 29, pp 125-429. 1993.

9. W. P. Heath. *Self-Tuning Control of Two - Dimensional Systems*, Research Studies Press and Wiley. 1994.

10. W. P. Heath and P. E. Wellstead. Self-Tuning Prediction and Control for Two Dimensional Processes Part 1: Fixed Parameter Algorithms, *Int. J. Control*, 62, 1, pp 65-107. 1995a.

11. W.P. Heath and P. E. Wellstead. Self-Tuning Prediction and Control for Two Dimensional Processes Part 1: Self-Tuning Algorithms", *Int. J. Control*, 62, 1, pp 65-107. 1995b.

12. A. Jain. *Fundamentals of Image Processing,* Information and Systems Science Series, Prentice Hall. 1989.

13. P. H. Hindle and C. R. R. Smith. Ultra Fast Web Thickness and Coating Measurement with True Two Dimensional Control, *Proc. NIR '94, International Conference on Spectroscopy*, Sydney, April. 1994.

14. A. P. Kjaer. Modelling, Sensing and Identification of Web Forming Processes, PhD Thesis, Control Systems Centre, UMIST, Manchester, UK. 1994.

15. A. P. Kjaer, P. E. Wellstead and W. P. Heath. On-Line Sensing Of Paper Machine Wet-End Properties Dry-Line Detection, IEEE Trans. on Control Systems Technology, 35, 8, pp 232-247. 1997.

16. S. Levy and J. F. Carley. Plastics Extrusion Technology Handbook, Industrial Press Inc. New York, NY 2nd Edition. 1989.

17. A. R. Mitchell and D. F. Griffiths. The Finite Difference Method in Partial Differential Equations, Wiley. 1980.

18. A. Niemi and C. Backstrom. Automatic Observation of Dry Line on Wire for Wet End Control of the Paper Machine, *Proc. Control Systems '92, CPPA* Whistler, Canada, pp 261 - 265. 1992.

19. R. J. Pfeifer, P. Williams and R. Holmes. Systems Considerations for a Full Sheet Imaging System, *Proc Joint TAPPI Process Control and Information Conf./38th ISA, PUPID Annual Symposium*, Vancouver, BC, Canada, pp 131-144. 1998.

20. R. P. Roesser. A Discrete State Space Model for Linear Image Processing, *IEEE Trans. Aut. Control,* AC-2, pp 1-10. 1975.

21. S. I. Shapiro, R. N. Vyse and C. Hagart-Alexander. New Wet End Weight Measurement System Opens Papering Optimisation Opportunities, *Proc Joint TAPPI Process Control and Information Conf./38th ISA, PUPID Annual Symposium*, Vancouver, BC, Canada, pp 207-214. 1998.

22. G. A. Smook. *Handbook for Pulp and Paper Technologists*, Tappi Press. 1982.

23. X. Troyas. *Control and Condition Monitoring of Two Dimensional Dynamical Processes,* PhD Thesis, Control Systems Centre, UMIST, Manchester, UK. 1992.

24. G. Wagner. *Self-Tuning Algorithms for Two-Dimensional Signal Processing*, Ph.D. Thesis, Control Systems Centre, UMIST, Manchester. U.K. 1987.

25. P. E. Wellstead and J. R. Caldas Pinto. Self-Tuning Filters and Predictors for Two-Dimensional Systems, Part 1: Algorithms, Int. J. Control, 42, 2, pp 457-473. 1985.

26. P. E. Wellstead and G. R. Wagner. Extended Self-Tuning Predictors, Smoothers and Filters for Two-Dimensional Data Fields, Int. Journal of Adaptive Control and Signal Processing, Vol. 4, pp 299-329. 1990.

27. P. E. Wellstead and M. B. Zarrop. *Self Tuning Systems: Control and Signal Processing*, John Wiley. 1991.

28. P. E. Wellstead and W. P. Heath. *Two-Dimensional Control Systems : Application to the CD and MD Control Problem*, Proceedings of "Control '92", Biennial Control Conference of the Pulp and Paper Industry, British Columbia Canada. 1992.

29. P. E. Wellstead and M. H. Waller. Control and Signal Processing for Two Dimensional Systems in Web Forming Industries, *Proc. IEE International Conference, Control '96*. pp 1272-1277, Exeter. 1996.

30. P. E. Wellstead, M. B. Zarrop, W. P. Heath , A. P. Kjaer and. X. Troyas. Two Dimensional Methods for Machine Direction and Cross Machine Direction Estimation and Control: Recent Progress, *Pulp and Paper Canada*, 98, 9, T331, pp 72-76. 1997.

31. P. E. Wellstead, A. P. Kjaer and W. P. Heath. Identification and Control of web processes: polymer film extrusion, *IFAC Journ. of Control Eng. Practice*, 6,3, pp 21-25. 1998.

32. W. Westermeyer. *Modellbildung und Simulation for Papiermaschine zur Reglung des Flächenbezognenen Massquerprofils der Papierbahn*, PhD Thesis der Technischen Fakultät der Universität Erlangen, Germany. 1987.

33. R. G. Wilhelm and M. Fjeld. Control Algorithms for Cross-Directional Control: the State of the Art, *Proc 5th Conference on Instrumentation and Automation in the Paper, Rubber Plastics and Polmerisation Industries*, Antwerp, Belgium, pp 162-174. 1983.

34. M. B. Zarrop and X.. Troyas. Model Based Predictive Control for Two Dimensional Dynamical Processes, in *Advances in Model - Based Predictive Control*, ed. D. Clarke, Oxford Science Publications. 1994.

11

Verification of Logic Controllers for Continuous Plants

S. Kowalewski, S. Engell and O. Stursberg

Process Control Laboratory, Chemical Engineering Department,
University of Dortmund, D-44221 Dortmund, Germany.
E-mail: {s.kowalewski | s.engell | o.stursberg}@ct.uni-dortmund.de

Summary. This contribution deals with the formal verification of logic controllers for processes with switched continuous dynamics. We present a consistent method for treating this problem in a rather general formulation and describe the corresponding computer support. The approach is characterized by the integration of available analysis tools for real-time and hybrid systems from computer science and a signal-flow oriented modular modeling framework for mixed discrete-continuous systems. The formal verification is based on determining the reachable discrete states of the resulting model and comparing it to a set of undesired states. The approach is illustrated by the modeling and analysis of a process engineering example.

1 Introduction

1.1 Motivation

Constellations in which continuous processes are controlled by logic controllers can be found in most industrial application areas. For example, in the process industries they arise when switching logic realized by programmable logic controllers (PLCs) or distributed control systems (DCSs) is used to perform control functions, e.g., process supervision to avoid unwanted or even dangerous states of the process or damage to the equipment, sequence control, startup and regular or emergency shutdown procedures, switching the mode of operation of quasi-continuous controllers, or supervision of sensor inputs and outputs to actuators. A significant part of this discrete control logic is critical for the safe operation of the process. Both with regard to the economic operation of the plant and from the point of view of safety and protection of the environment, it is highly desirable, that the enormous amount of control soft- and hardware which processes discrete signals and performs logical operations works correctly. However, whereas continuous controllers can be and in some cases even really are designed using sophisticated methods and extensive simulation studies, the discrete control logic is produced in a much less systematic manner. Usually, incomplete specifications lead to a first design and an implementation which is then improved by testing on-site. This situation creates an incentive to switch over, at least for the most critical parts of the control logic, to currently available rigorous, model-based methods which guarantee provably correct discrete control software by so-called *formal verification*. The aim of this contribution is to give an introduction to one representative of these methods.

The notion of formal verification originates from computer science where, in general terms, it means a mathematical proof that a model of an algorithm fulfills given formal properties. In the last decades, different representations and methods have been developed and, in the recent years, some of them have been applied successfully in the area of hardware and communication protocol design (see [5] for a survey). However, in the field of logic controllers for processing systems, in particular when continuous processes are involved, formal verification is currently not applied in practice and only a few research projects are described in the literature (see next subsection). There are several apparent reasons for this situation. The first one is that PLC or DCS software is usually developed by engineers and not by computer scientists. Thus, the developers are not familiar with the available formal methods and tools. Another reason is that most control software in this area is written by the PLC or DCS user for one particular application or a rather small number of similar plants. This is a different situation to hardware or protocol design where the software is part of a mass product and the verification effort pays off more easily.

The approach presented in this contribution aims at making formal verification of logic controllers for continuous processes practically more easily applicable. This objective is achieved by providing a modeling environment based on representations familiar to engineers (block diagrams and signals) and by the integration of recently developed analysis tools from computer science. The core of the verification procedure is to create a model of all possible, technically relevant behaviors of the plant without any assumption on the controller. This means, that any disturbance, failure and even undesired operator interaction and their consequences which cannot reasonably be excluded a priori have to be considered in this description. The model is then connected to a compatible model of the control program which is to be verified. This results in a model of the controlled (closed-loop) system which can be checked against specifications of the kind that certain discrete states or regions of the continuous state space must not be reached under all circumstances. Since the plant models which are needed for this purpose usually are much more complex than those of the desired behavior only, it is necessary to build them in a structured manner. The method presented here makes use of the modeling paradigm of Condition/Event systems (C/E systems) according to Sreenivas and Krogh [32] by which complex systems can be composed from simpler blocks by signal interconnection. C/E systems also make it possible to describe the dynamics of the modules on different levels of abstraction: discrete event, timed-discrete, or (switched) continuous. As automatic analysis is only possible for the first two kind of models, (switched) continuous systems have to be approximated by discrete event or timed-discrete systems for which tools are available.

The approach is an a-posteriori analysis. In the development process of logic control software, it can be applied when the logic control program was designed and the complete control system consisting of the controller and the given plant has to be checked against specifications for the behavior of the controlled process. Thus, it complements the final validation testing. It should not be seen as a substitute for testing because in this approach, all possible reactions of the continuous plant in all conceivable situations including equipment failures have to be included in the model, and in practice, this is can only be achieved up to a some level of certainty and complexity. But if the procedure can be applied successfully, it guarantees that all possible combinations and sequences of events which could possibly lead to an unwanted process state were checked completely.

1.2 Related Work

The pioneering work in the verification of logic controllers for industrial and in particular, chemical processing systems was done by to Moon et al. [26] for control programs represented in Relay Ladder Logic. They applied the symbolic model checking method from [4] in which the system which is to be verified is modeled as a finite state machine and the specifications of the desired behavior are represented by temporal logic expressions. Their approach has been extended to include plant models and recent work shows that the formal verification of logic controllers for processes of moderate size is possible [30]. The continuous dynamics of subsystems of the plant are discretized in an elementary manner: The range of values of each continuous variable is partitioned into intervals and the model simply describes the possible transitions between these intervals. Timers are incorporated in the same qualitative fashion by neglecting the timer value and keeping only two states, running and elapsed. This suffices if checking real-time constraints is not required, e.g., whether a controller response to a plant event is fast enough to avoid unwanted process behavior.

The same basic problem, i.e. model checking of Relay Ladder Logic diagrams, has been treated by Park and Barton [27] using completely different techniques. Here, the problem of checking whether a temporal logic formula holds for the model of the logic controller is mapped into the feasibility problem for a system of equalities and inequalities for binary variables. The latter is then solved by Integer Programming. Application results show that there exist examples for which this approach generates solutions faster than classical model checking. The problem of abstracting continuous or real-time information is solved in the same way as in the first approach.

A verification method based on a switched continuous model and, in principle, avoiding any abstraction has been suggested by Dimitriadis et al. [6 and 7]. The reachability problem is reformulated as an optimization problem in the discrete time domain which can be solved by Mixed Integer Programming. Basically, the optimization determines the worst possible behavior meaning that the system is most often in an undesired region of the continuous state space. The approach is general in the sense that it can be applied to hybrid systems as well as to purely discrete or purely continuous systems. Its strength lies in the ability to take advantage of well tested and efficient optimization procedures. A limitation is given by the fact that the size of the Mixed Integer Program grows with the product of the number of discrete time steps and the number of equations describing the plant and logical expressions describing the controller. Recently, a similar approach has been suggested by Bemporad and Morari [3]. Here, an iterative scheme is used to perform conventional reachability analysis.

In contrast to the work mentioned above, the approach presented here builds on recent extensions of model checking to include real-time and hybrid dynamics. These methods are based on the modeling framework of *Hybrid Automata* [2]. In this context, the first application to an industrial example in engineering was reported in [33]. Our approach introduces a modular, block-oriented framework for modeling hybrid systems based on Hybrid Automata connected by certain discrete valued signals. It aims at applications where formal verification requires a quantitative analysis of the interaction between timers or threshold values in the logic control program and the plant dynamics.

1.3 Outline

The remainder of the paper is organized as follows. To provide a motivation for the practical need for formal verification of logic control programs and to give an impression of the type of problems encountered in practice, we start by presenting a simple process engineering example. It is used later in the paper to illustrate important concepts and the application of the method. In Sec. 3, the basic methodology of the approach is sketched. Section 4 discusses important modeling issues. Based on this, the applied modeling framework is introduced in Sec. 5. Section 6 presents an approximation procedure for representing continuous systems by timed-discrete models. Section 7 is devoted to the translation of control code into formal models. Aspects of the analysis and necessary model transformations are described in Sec. 8. In Sec. 9, we present a computer tool which is being developed to support the described approach. A discussion concludes the paper.

1.4 Acknowledgement of Contributions

This presentation is a survey on a collaborative effort in which many people participated. The authors therefore most gratefully acknowledge the following contributions to the method and the computer tool presented in this paper. Jörg Preußig (Dortmund) developed the extension of the C/E systems to Timed C/E systems and implemented the translation of such models to *HyTech*. Heinz Treseler (Dortmund) realized the user interface and the data structures for C/E systems. Nanette Bauer (Dortmund) worked on the translation from PLC code to timed discrete models. André Deparade, Oliver Jenneskens and Tobias Severin (all Dortmund) contributed further to the development of the tool. R. Huuck, Y. Lakhnech, B. Lukoschus, and L. Urbina (Kiel, Germany) cooperated in the comparison of Timed C/E systems and Timed Automata and developed the translator from the first to the latter model.

2 Example

We consider the chemical batch-reactor illustrated by Fig. 1. The plant consists of a liquid-cooled reactor vessel with measuring, stirring and feed supply devices, and a control unit. The sensor *TI* measures the temperature T_R of the vessel content, and *LIS* is a limit switch indicating whether the level has reached an upper threshold. The process specification is as follows.

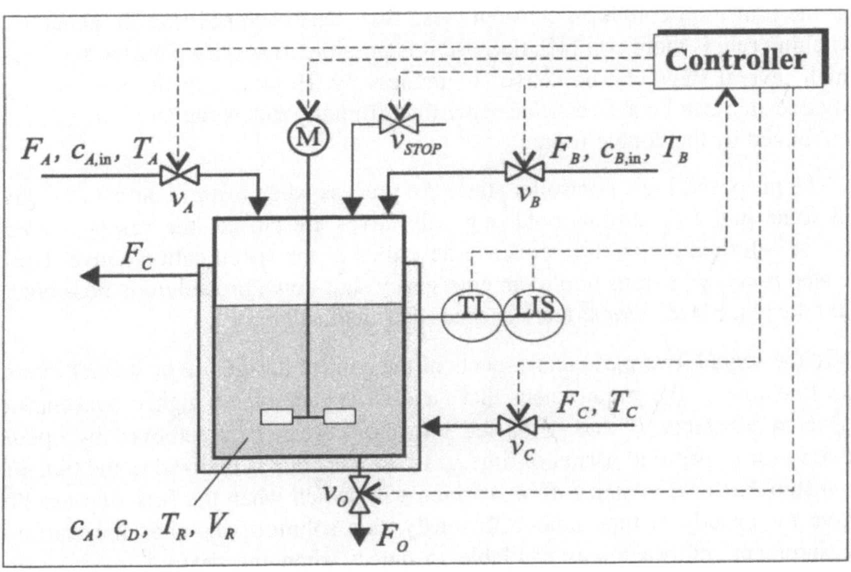

Fig. 1. Process diagram of the batch reactor.

First, a certain amount of highly concentrated, dissolved substance B *is fed into the reactor by opening valve* v_B. *After switching the stirrer motor on, a solution of substance* A *is added through valve* v_A *until the total reactor fluid volume* V_R *has reached an upper limit. To ensure that the chemical reaction* $2A + B \rightarrow D$ *will start properly, the cooling must not be operating for a certain period after* v_A *was opened to fill in substance* A. *However, the reaction is exothermic and will lead to a rise of the liquid temperature* T_R. *Thus, to avoid a dangerous situation, the cooling (liquid with flow* F_C *and Temperature* T_C*) has to be switched on at an appropriate instance during the starting phase of the reaction. When the product concentration has reached the desired value, the reactor is drained by opening valve* v_O *and the stirrer is switched off. The production sequence ends by closing the valves* v_C *and* v_O *when the vessel is empty. In the case of an emergency due to excessive temperature increase during the reaction, a stopper liquid can be discharged into the vessel by opening valve* v_{STOP}.

To run this process in the given plant, a logic controller is needed which realizes the specified sequence under consideration of all conceivable failures and disturbances. An example of such a controller is given by Fig. 2 in the form of a so-called *sequential function chart* (SFC), a specification and programming language for sequence control programs [18]. The SFC language consists of three main elements: steps (represented by labeled rectangles), transitions (short vertical lines with a logical condition), and actions (control commands attached to a step). The control sequence is defined by flowlines connecting steps and transitions. Steps can be active or inactive. A step becomes active when its preceding step has been active and the transition condition between these two steps becomes true. In general, the switching rule is more complex because it is possible to specify parallel branches in which several steps can be active simultaneously. However, in the SFC in Fig. 2 only one step can be active at the same time. Initially this is the step *idle* (which is symbolized by the double frame).

The proposed logic controller starts the process when a start command is given and some initial conditions hold (e.g., all valves are closed, the vessel is empty, etc.). It will then realize the sequence described in the specification above. During the step *reaction*, a transition to an emergency shut-down procedure is possible and will take place if the temperature exceeds a critical value.

In the sequel, two particular aspects of the control design are of special interest. The first one is the requirement that "a *certain amount* of highly concentrated, dissolved substance *B*" has to be fed into the reactor. It is achieved by opening valve v_B for a specified period of time t_B. In the SFC this is realized in the transition from step *Fill in B* to step *Fill in A* which will switch when the first step has been active for exactly t_B time units. Obviously, this solution was chosen because no measurement information is available to detect when the desired level for *B* is reached. (The sensor *LIS* only indicates the crossing of the maximal value of the level).

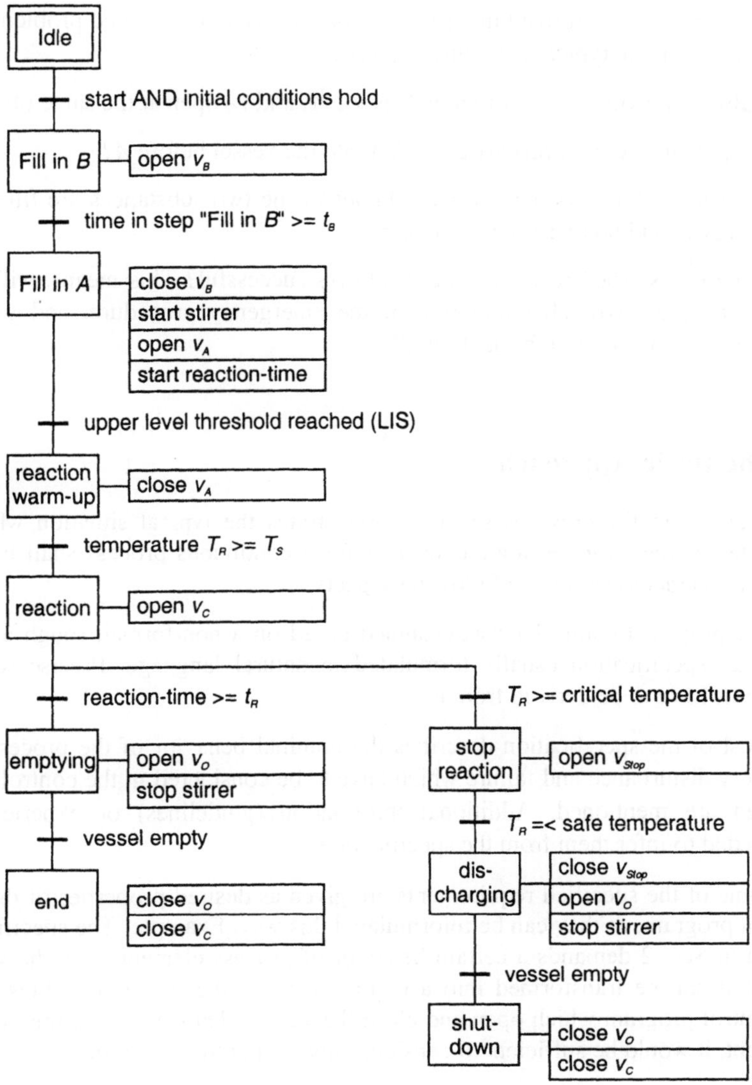

Fig. 2. Logic control program for the reactor example.

The second interesting requirement is that the "cooling has to be switched on at an *appropriate instance* during the starting phase of the reaction". This is realized by introducing a switching temperature threshold T_S at which the cooling valve v_C will be opened. Here, the necessary measurement signal for this control loop is available. It is generated by the sensor *TI*.

When the correctness of the control program from Fig. 2 is checked, the general question is whether the controlled process fulfills the given specification. Usually, this question has to be refined into several more precisely formulated problems. For the reactor example typical correctness questions are:

(1) Is always maximally one of the valves v_A, v_B, and v_O open at the same time?

(2) Is it guaranteed, that B is always filled into the vessel before A?

(3) Is it true that always the correct volumes of the two substances are filled into the vessel and no overflow will occur?

(4) Is the undisturbed reaction process always successful? This means, will there never be an avoidable initiation of the emergency procedure, and will the product concentration be as desired?

3 The Basic Approach

The example in the previous section demonstrates the typical situation which is encountered when a given logic controller for a continuous process shall be verified. It is characterized by the following aspects.

- The proposed controller was designed based on a non-formal, rough and abstract specification usually formulated in natural language. Precise requirements have to be derived from it.
- Most of the specification describes the nominal behavior of the process. Not every disturbance and failure which have to be considered in the controller design are mentioned. Additional information (guidelines) or experience is needed to infer them from the specification.
- Some of the specified requirements are given as desired properties of the control program, or they can be reformulated this way. Example: The specification (1) in Sec. 2 demands a certain behavior of process elements (i.e. the valves) but it can be transformed into a requirement for the output variables of the control program which open and close the valves. Thus, to check this requirement, it would be sufficient to consider only the control program.
- However, many requirements are not of this type. They are formulated in terms of the desired process behavior and cannot be mapped into specifications for the controller. The specifications (3) and (4) from Sec. 2 are examples for this. They describe properties of process variables (volume, concentration) which have no direct correspondence to the variables of the control program. Thus, for such specifications, information about the plant is needed. (Note, that this is true for designing the controller as well as for verifying it.) It can be obtained from semi-formal descriptions (e. g. flowcharts), general knowledge about the physical behavior, and engineering experience. Usually such information is available, no matter whether the plant already exists or is in the planning phase.

Considering the last aspect, it is a straightforward idea to build the verification not only on a formal model of the controller but also on a formal model of the plant. Fig. 3 illustrates the resulting situation which is the starting point of the proposed verification procedure: The control program already exists and was designed based on a non-formal specification of the closed loop behavior. This controller is supposed to be connected to a particular plant. Based on this setting, Fig. 4 shows the main necessary steps of the verification procedure in the form of a Petri net. The gray ovals represent the initial state already shown in from Fig. 3. From there, the following transitions have to be taken to reach the target which is the verification result in the top right corner: The major effort required is to derive a model of the plant. This model has to capture the *uncontrolled* behavior of the plant, i.e., all possible undesired situations, disturbances, faults and even obviously wrong operator actions have to be included. Otherwise, it is not possible to prove that a critical situation is reliably prevented by the controller. Then, the control program has to be translated into the same modeling framework such that a composition of the plant model and the controller becomes possible. The composition of the models of the plant and the controller constitutes the model of the controlled plant. This model can then be analyzed and compared to a formal description of the desired plant behavior. In the approach presented here, this will be an evaluation whether a set of undesired discrete states is reachable. In the following sections, the main steps are described and discussed in detail.

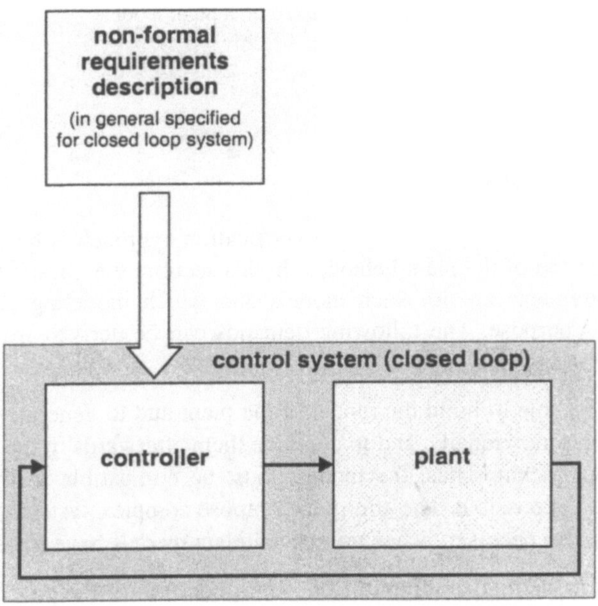

Fig. 3. Starting point of the approach.

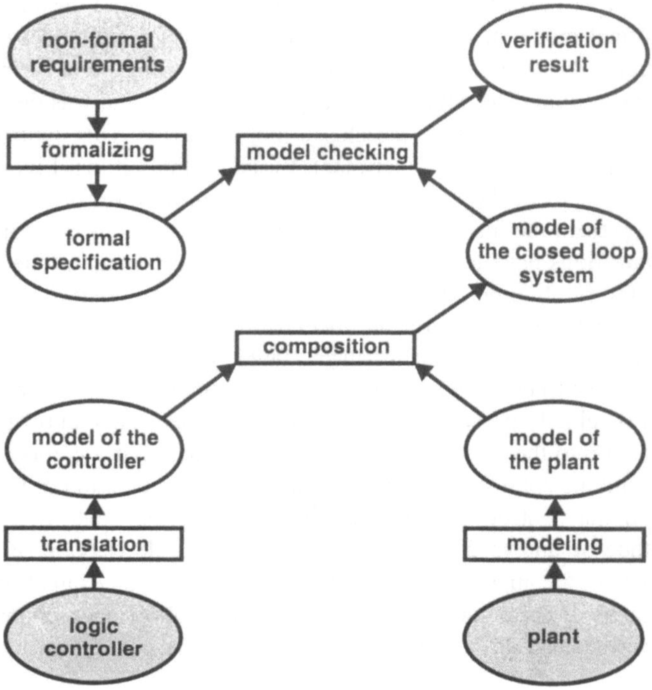

Fig. 4. Verification based on a plant model.

4 Modeling Issues

As described in the previous section, the verification approach is based on *models* of the controller and of the plant behavior. In this section, the question is discussed what the requirements are for such models and which modeling paradigms are suitable for this purpose. The following demands can be derived from the procedure illustrated in Fig. 4.

(a) One must be able to build the model of the plant and to generate the model of the controller individually, and to combine them afterwards in the model of the closed-loop system. Thus, the models must be compatible in the sense that composition is possible. The ability to compose complex systems from simpler modules is also necessary when non-trivial plant models have to be created.

(b) A uniform modeling paradigm for the controller and for the plant is needed. It must be able to describe the predominately discrete dynamics of the controller as well as the switched continuous dynamics of the plant, if necessary.

(c) It must be possible to include sufficient quantitative information about the system's dynamics in the models to check the specification requirements (e.g., if the requirement is that a value must be reached within a certain time, it must at least be possible to include clocks in the model). To reduce the modeling effort, it is desirable to be able to adjust the degree of abstraction of each model to the corresponding requirements.

(d) Finally, the resulting model must be analyzable in the sense that an algorithm exists which is able to determine the correctness of the specified properties automatically in finite time.

4.1 Different Levels of Abstraction of the Plant Dynamics

The requirements (b), (c), and (d) affect the abstraction level of the plant dynamics (purely discrete, discrete with real-time constraints, or continuous). This is illustrated in the sequel.

We first consider (c), i.e. whether the modeling paradigm is expressive enough to check the specification requirements, and revisit specification (2) from Sec. 2, i.e., whether B is always filled into the vessel before A. Obviously, in this case a purely discrete model which represents the possible contents of the vessel is sufficient. Fig. 5 shows a finite automaton for this task. Although this may sound uncommon to engineers usually dealing with transfer functions and differential equations, this automaton is a model of the plant. It only captures one particular aspect (the sequence of substances or mixtures being in the vessel - this is why the closing of the valves is not of interest), but it describes all possible behaviors.

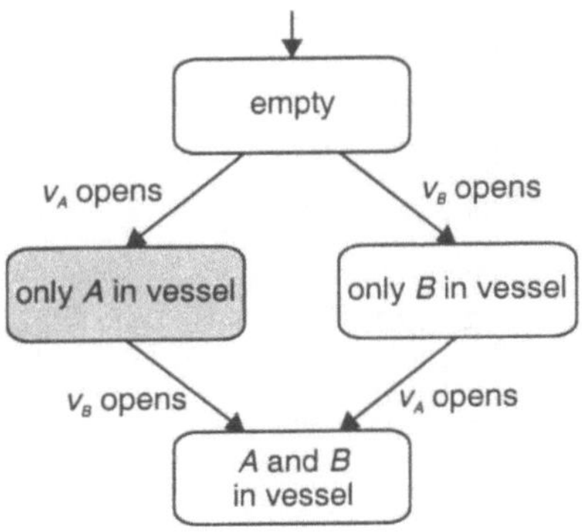

Fig. 5. Discrete model for the filling sequence in the reactor example.

Fig. 5 also illustrates how the specification can be expressed in these models. The most simple and most common way is to classify the discrete states of the automaton into allowed and forbidden ones. Of course, this is not possible for every specification. But for most practically interesting properties, a given automaton for the plant behavior can be modified such that the desired property is expressible by forbidden states. In our example, the state *only A in vessel* is forbidden which is symbolized by the gray shading. The property that B is always filled in before A can now be checked by connecting the model of Fig. 5 to an automaton model of the control program from Fig. 2 (which would be so similar to the SFC representation that we omit it here) and analyzing whether the state *only A in vessel* is unreachable.

For the specification (3), i.e. the correct volumes of the substances and no occurrence of overflow, the situation is different. Since a timer is used in the control program to determine the appropriate instance when to close valve v_B, the model of the filling process has to provide quantitative information about how fast the level rises when B is filled in. This can either be a time span or a continuous model of the level dynamics. The same problem arises for the overflow problem. Each controller has a reaction delay. So the question is not only whether the proposed control program will close valve v_A at all after it receives the threshold crossing signal from the sensor LIS, but also whether it will do this in a short enough time. Therefore at least timing information is needed in the plant model.

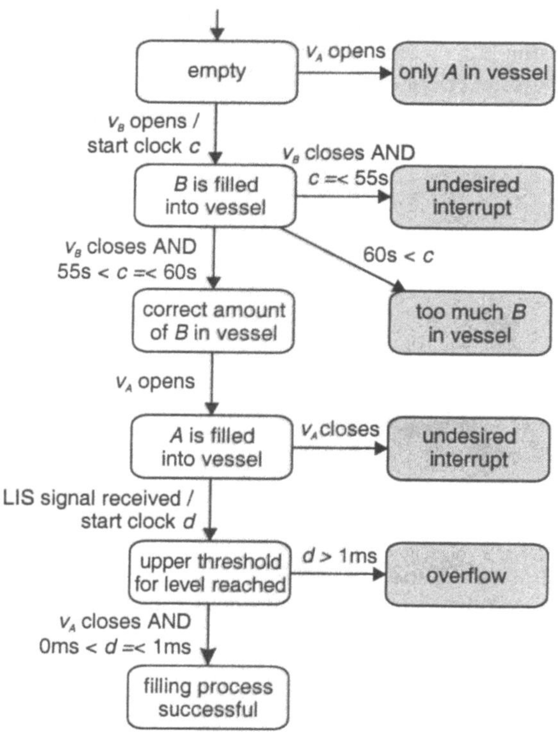

Fig. 6. Timed discrete model of the filling process for the reactor example.

Fig. 6 shows such a model for the filling process of the reactor. The difference to the automaton in Fig. 5 is the existence of two clocks which can be started by a transition and the values of which enter into conditions for other transitions. The clock c is started when the filling of B begins. It is then used to make the transitions out of this state depending on the time when valve v_B closes. If it closes before c is running for more than 55s, this means that the desired level has not been reached. If the filling time is between 55s and 60s, it can be assumed that the correct amount has been filled into the vessel. If 60s are exceeded, the desired level is exceeded as well. Clock d is used to determine whether the reaction on the LIS threshold signal is quick enough to avoid overflow. It is started when the signal comes in, and if v_A is not closed within 1 ms the transition to *filling process successful* is no longer possible and *overflow* becomes reachable.

The modeling paradigm from Fig. 6 is known as *Timed Automata* [1]. Tools are available for the composition and reachability analysis (see Sections 8.1 and 8.2). Thus, it also fulfills requirement (d) (and, obviously, requirement (b)).

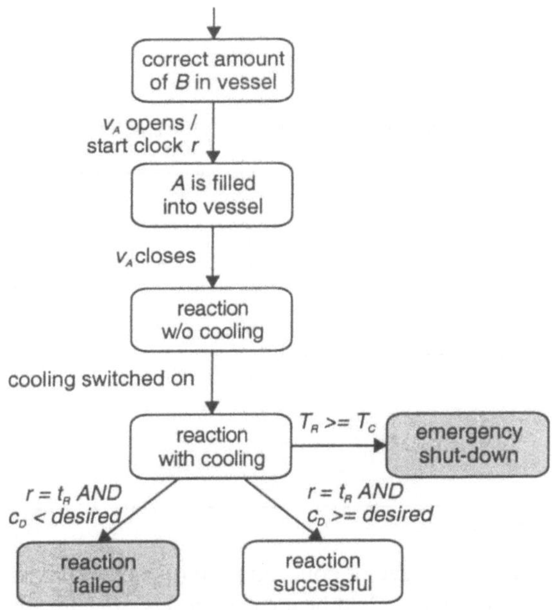

Fig. 7. Possible transition sequence for reactor example.

In some cases, timing information as it is used in the model in Fig. 6 is not available. This could be the case for example when the plant has not yet been erected or measurements are impossible for other reasons. An example for this situation is given by specification (4) which can be phrased also as: Is the switching temperature T_S (when the cooling is started) on the one hand side chosen sufficiently low to avoid reaching a dangerous temperature $T_{R,crit}$, and on the other hand side chosen sufficiently high to imply $c_D \geq c_{D,des}$ within the time t_R? Fig. 7 shows the corresponding model. The initial state is the one in which the reactor is already filled with substance B. The requirement is mapped into the two forbidden states *emergency shut-down* and *reaction failed.* Whether one of these two is reached or not depends on the time when the transition *cooling switched on* takes place (which is triggered by the controller when the temperature reaches the switching value T_S). But to infer this from the model, it must tell us first, whether the temperature T_R exceeds the critical value T_C before the reaction time t_R has elapsed, and secondly, what the concentration will be at this instance. This is necessary to decide whether the transitions to the forbidden states will occur or not. But obviously, this information is not part of the Timed Automaton in Fig. 7. We need an additional description of the continuous dynamics of the reaction in terms of the concentration and the temperature.

Table 1 presents a hybrid model of the reactor dynamics in the form of switched differential equations and algebraic equations. The closing of v_A, for example, is expressed by setting F_A to zero, and a Boolean Variable s is used to model the switching of the cooling.

Table 1. Hybrid model for the reactor.

DAE-System:

$$\frac{\partial c_A}{\partial t} = \frac{F_A}{V_R} \cdot \left(c_{A,\text{in}} - c_A\right) - 2 \cdot r, \quad \frac{\partial c_D}{\partial t} = r - \frac{c_D \cdot F_A}{V_R}, \quad r = c_A^2 \cdot k_0 \cdot \exp\left(\frac{-E_a}{R_m \cdot T_R}\right),$$

$$\frac{\partial T_R}{\partial t} = \frac{F_A}{V_R} \cdot \left(T_A - T_R\right) - \frac{s \cdot k_c \cdot A_c}{\rho \cdot c_p \cdot V_R} \cdot \left(T_R - T_c\right) + \frac{H_r \cdot r}{\rho \cdot c_p}$$

$$\frac{\partial V_R}{\partial t} = F_A, \quad A_c = \frac{\pi}{4} D_R^2 + \frac{4}{D_R} \cdot V_R$$

Variables and parameters:

A_c - heat transfer area; c_A – reactor concentration of A, $c_{A,\text{in}}$ - inlet concentration of A; c_D – reactor conc. of D, c_p - fluid heat capacity D_R - reactor diameter; E_a – activation energy; F_A - inlet flow of A -solution (= 0 if v_A is closed); H_r - molar reaction heat; k_c - heat transfer coefficient; k_0 – reaction rate constant; r - reaction rate; ρ - fluid density; R_m - gas constant; s – cooling switch (= 1 if v_C opened, and = 0 else); t - time; T_A – temperature of inlet flow; T_C - temperature of cooling fluid; T_R - temperature of reactor content

In principle, the model in Table 1 together with the model in Fig. 7 and a model of the control program provides enough information to check the specification. In this particular case, the resulting model can be simulated by integrating the differential equations from Table 1 with discrete switchings depending on the transitions in the automata. However, in the general case this is not possible because non-determinism in the discrete part (e.g., caused by the uncertainty which failure will occur next or at which point in time an operator action will take place) would require to consider a combinatorially exploding number of different switching sequences or even infinitely many switching instances in a continuous time interval. The only remedy is to apply symbolic methods which map an infinite state space into a finite one. However, the class of hybrid systems for which this is possible is restricted [2, 11]. Timed Automata are one of these classes, and tools to analyze them exist. Therefore, we decided to use them in our approach as the basic model for analysis. If models with more complex dynamics are necessary to capture relevant requirements, they have to be approximated by Timed Automata or equivalent models. A corresponding procedure will be presented in Sec. 6.

This opens two paths from the real system to the model, as it is illustrated by Fig. 8 (which can be regarded as a more detailed representation of the modeling step for the plant in Fig. 4): Either a plant module is directly modeled as a timed discrete event system using available knowledge of the process, or a (switched) continuous model is built using first principles and then approximated automatically. The first option is appropriate if the natural level of abstraction for the subsystem model is discrete and if only coarse knowledge of the plant dynamics is available. If the system under consideration has non-trivial continuous dynamics for which a model can be built with reasonable effort, the second way usually is less error-prone.

4.2 Interactions Between the Modules

The requirement (a) from Sec. 4 demands a modular modeling framework which offers the possibility to build structured models of large and complex systems from small modules. Therefore, a concept of interaction between subsystems is necessary. In the case of Timed Automata this is realized by synchronization of events using so-called synchronization labels. Two or more transitions in different automata can only be taken at the same time, if all of them are possible. If a transition takes place, all transitions with the same label participate. This concept has two major disadvantages: First, it is possible that the interconnection of two automata can lead to paradoxical behavior. This is the case when one automaton is forced to make a transition by its timing constraints but the other automaton blocks this transition because it contains a transition with the same label which is not possible at the moment. This phenomenon is called a *time-deadlock*. The second problem is that synchronization is a non-causal interconnection. There is neither a sender nor a receiver of an event. This makes it unintuitive, for example, to model that a controller forces switching of a valve.

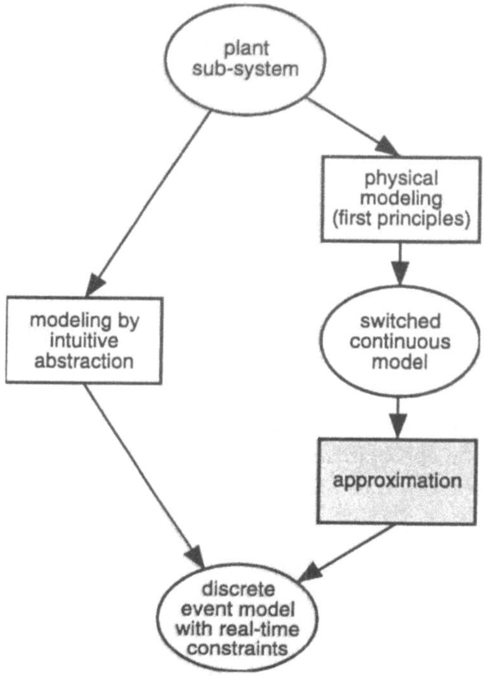

Fig. 8. Refinement of the modeling step in Fig. 4.

For these reasons we apply a different framework to build the models than to analyze them, namely Condition/Event (C/E) systems according to Sreenivas and Krogh [32]. They provide more powerful interconnection operators than Timed Automata do as will be explained in the next section. However, since we use analysis tools for Timed Automata, C/E systems are transformed into such a model for the purpose of analysis. This will be presented in Sec. 8.3.

5 Modeling with Condition/Event Systems

5.1 Preliminaries: Condition/Event Signals and Models

C/E systems were introduced to model interconnected discrete event systems (DES) and hybrid systems in a modular, block diagram and signal flow oriented fashion similar to continuous dynamical systems. When developing the model, Sreenivas and Krogh were motivated by the observation that in existing DES models the interaction of systems is either based on the synchronization of events or the conditioning of event occurrence depending on state information. This distinction is also well known from logic controller programming languages and from the field of switching circuits (e.g. edge-triggered function block inputs in contrast to regularly scanned inputs in the IEC 1131 standard [18]).

To incorporate both concepts into a single, unified representation of interconnected DES, two classes of signals are defined over the continuous time axis which can be input or output signals of a system: *condition signals* and *event signals* (cf. Fig. 9). Formally, a condition signal $c(\cdot)$ is a right continuous function c: $[t_0, \infty) \rightarrow C$ with limits from the left, with C being a non-empty and finite set of conditions. So, condition signals are piecewise constant and their values correspond to discrete states. An event signal $e(\cdot)$ is a function e: $[t_0, \infty) \rightarrow E_0 = E \cup \{0\}$ for which $e(t_0) = 0$ (the *null* or *zero event*) and $\{t \in [t_1, t_2] \mid e(t) \neq 0\}$ is finite for all finite intervals $[t_1, t_2] \in [t_0, \infty)$ and E is a nonempty and finite set of events. Event signals are point-wise nonzero and carry information about currently occurring state transitions. The set of all condition signals and all event signals on $[t_0, \infty)$ is written as $\boldsymbol{C}(C, t_0)$ and $\boldsymbol{E}(E, t_0)$, respectively.

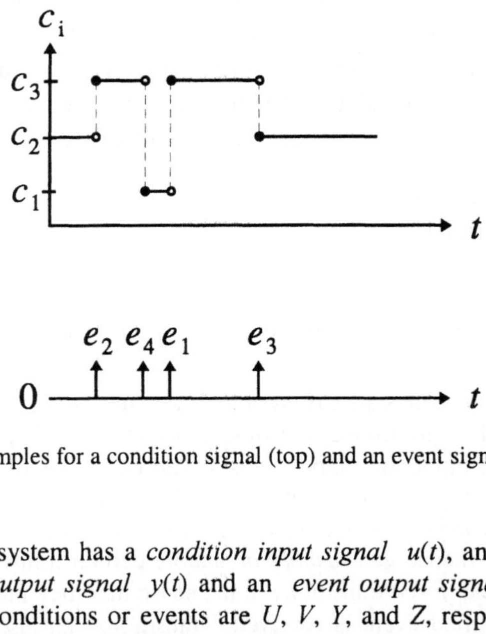

Fig. 9. Examples for a condition signal (top) and an event signal (bottom).

In general, a C/E system has a *condition input signal* $u(t)$, an *event input signal* $v(t)$, a *condition output signal* $y(t)$ and an *event output signal* $z(t)$. The corresponding sets of conditions or events are U, V, Y, and Z, respectively. The basic definition of a C/E system defines its input/output behavior by a mapping which gives the set of admissible output signal trajectories for each possible input signal trajectory: $\Sigma : \boldsymbol{C}(U, t_0) \times \boldsymbol{E}(V, t_0) \rightarrow 2^{\boldsymbol{C}(Y, t_0) \times \boldsymbol{E}(Z, t_0)}$, such that for any input $(u(\cdot), v(\cdot)) \in \boldsymbol{C}(U, t_0) \times \boldsymbol{E}(V, t_0)$ there exists at least one output $(y(\cdot), z(\cdot)) \in \boldsymbol{C}(Y, t_0) \times \boldsymbol{E}(Z, t_0)$ fulfilling $(y(\cdot), z(\cdot)) \in \Sigma(u(\cdot), v(\cdot))$.

This definition only characterizes the input/output behavior and provides no specification or restriction of the model of the internal behavior of a C/E system. In [32] any formal representation which describes an input/output relation according to the definition above is called a *C/E model*. Note that a C/E model does not necessarily have to be discrete. In particular, even continuous dynamics are possible as long as the system interacts with its environment by condition and event signals. Examples for C/E models based on Petri nets were presented by Sreenivas and Krogh [31] and Hanisch et al. [12]. In our verification approach, we apply two different C/E models, *discrete state C/E systems* and *C/E timers*.

5.2 Discrete State Condition/Event Systems

Discrete state C/E systems were introduced by Sreenivas and Krogh to model untimed DES. They are defined as follows.

Definition 1 (Discrete state C/E system). A discrete state C/E system is a 9-tupel $S=(U, V, \Sigma, Y, Z, f, g, h, \sigma_0)$ where U, V, Y, Z are the same as above, Σ is a finite set of *discrete states*; f, g and h are functions defined as: $f: \Sigma \times U \times V_0 \rightarrow 2^{\Sigma} - \emptyset$, the *state transition function* satisfying $\forall\ \sigma \in \Sigma,\ u \in U: \sigma \in f(\sigma, u, 0)$, $g: \Sigma \times U \rightarrow Y$, the *condition output function*, and $h: \Sigma \times \Sigma \times V_0 \rightarrow Z_0$, the *event output function* satisfying $\forall\ \sigma \in \Sigma: 0 = h(\sigma, \sigma, 0)$. V_0 and Z_0 symbolize that 0 is included. $\sigma_0 \in \Sigma$ is the *initial state*. Given a discrete state C/E system S and input signals $u(\cdot)$, $v(\cdot)$ defined on $[t_0,\infty)$, the set of valid of state trajectories and output signals consists of all triples of signals $(\sigma(\cdot), y(\cdot), z(\cdot))$ defined on $[t_0,\infty)$ satisfying

$$\sigma(t) \in f(\sigma(t^-), u(t^-), v(t)), \tag{1}$$

$$y(t) = g(\sigma(t), u(t)), \tag{2}$$

$$z(t) = h(\sigma(t^-), \sigma(t), v(t)), \tag{3}$$

where $\sigma(t^-)$ and $u(t^-)$ are abbreviations of $lim_{\Delta \to 0}\ \sigma(t-\Delta)$ and $lim_{\Delta \to 0}\ u(t-\Delta)$. ♦

The semantics of a discrete state C/E system can be interpreted as an untimed finite state machine which is embedded in a time-dependent signal space formed by the condition and event input and output signals. The two kinds of signals realize the above mentioned interaction concepts of conditioning and synchronization in a specific way: The condition input signal constitutes conditions for changes of the state of the system (hence can *disable or enable* certain transitions) whereas the event input signal can *force* transitions in the sense that they must occur at the instance when a specified event is received. Transitions can also take place spontaneously and can be non-deterministic. The two properties $\forall\ \sigma \in \Sigma,\ u \in U$: $\sigma \in f(\sigma, u, 0)$ and $\forall\ \sigma \in \Sigma: 0 = h(\sigma, \sigma, 0)$ of f and h guarantee that transitions and output events will not be forced by condition input changes.

To provide a flavor of how technological systems are modeled by means of discrete state C/E systems we present a small toy example. Figure 10 shows a tank for liquid material which can be drained via an outlet valve. In Fig. 11, a very simple C/E model for this system is illustrated. The model consist of two blocks, one for the level and one for the outlet valve. The two blocks are connected by one condition signal and one event signal. According to the nomenclature introduced in [24], condition signal flow lines are characterized by two black rectangles whereas event signal flow lines can be recognized by a lightning symbol. The discrete dynamics of each subsystem is represented as a state graph. For a complete discrete state C/E model of this block diagram, we have to specify the possible successor states and output signal values for all combinations of states and input signal values in both systems. This information is represented for each system separately by the three functions *f*, *g* and *h* which were introduced in Def. 1. For this example, we will not discuss the complete model but only some aspects of it in order to illustrate the important concepts of forcing and enabling/disabling.

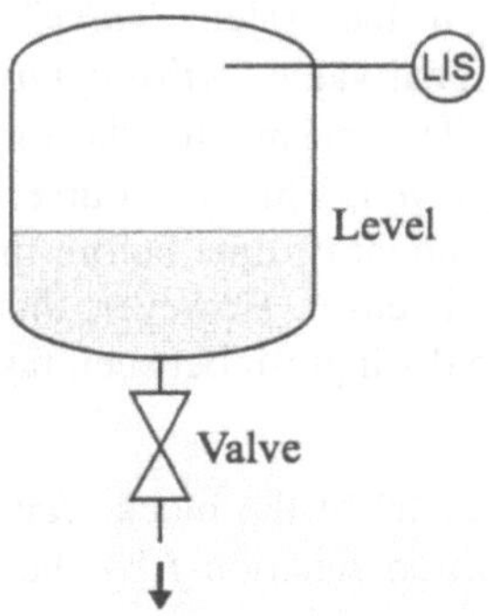

Fig. 10. Example: draining tank.

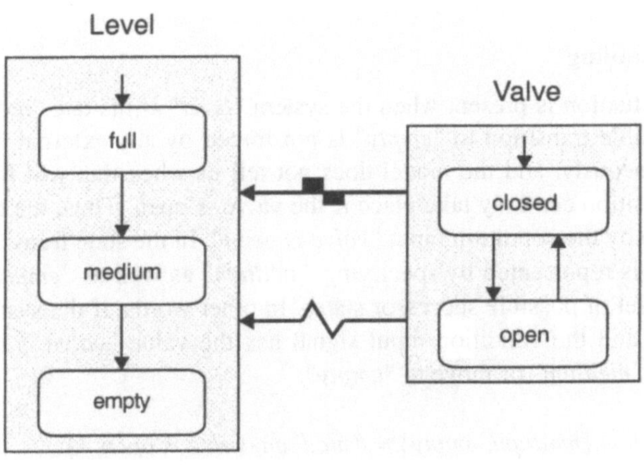

Fig. 11. Discrete state C/E system model of the draining tank example.

First, we define the condition and event signals. For the condition signal in Fig. 11, we have $C = \{valve\ is\ open,\ valve\ is\ closed\}$. The value set of the event signal is given by $E = \{valve\ opens,\ valve\ closes\}$. Thus, the value of the condition signal simply represents the state of the block "*valve*", while the event signal indicates the two possible transitions. Whereas the block "*valve*" is an autonomous system, the behavior of the block "*level*" is effected by its input signals. In the following we will use these effects to illustrate the different mechanisms of interaction which can be expressed by condition and event signals.

Forcing

Consider the initial situation when the tank is full and the valve is closed. A transition from "*open*" to "*closed*" in the system "*valve*" will then be indicated to the system "*level*" by the event signal value "*valve opens*" and it will *force* the transition from "*full*" to "*medium*". This means that the forced transition will take place exactly when the event is received. This, of course, is an abstraction because in reality it would take a finite amount of time before the effect of opening the valve could be measured be the level sensor. However, this simplification is very useful and common in engineering to distinguish between fast and slow phenomena and to obtain simpler models.

In the discrete state C/E model of the block "*level*", the described behavior is represented in the state transition function f by the fact that if the event "*valve opens*" is received, the set of possible successor states of "*full*" is the singleton "*medium*" (The "-" indicates that the value of the condition input is irrelevant for this transition):

$$\{medium\} = f(full,\ -,\ valve\ opens\}. \tag{4}$$

Enabling/disabling

A different situation is present when the system "*level*" is in state "*medium*". Now, the only possible transition to "*empty*" is not forced by any external input but will occur spontaneously, and the model does not tell us when this will happen. However, the transition can only take place if the valve is open. Thus, the transition has to be *enabled* by the condition input "*valve is open*". In the state transition function, this situation is represented by specifying "*medium*" as well as "*empty*" as the elements of the set of possible successor states. In other words, if the system "*level*" is in "*medium*" and the condition input signal has the value "*valve is open*", it can either stay in "*medium*" or move to "*empty*":

$$\{medium,\ empty\} = f(medium,\ valve\ is\ open,\ -\}. \tag{5}$$

In the same fashion, the *disabling* of the transition from *medium*" to "*empty*" by the condition signal value "*valve is closed*" is expressed by restricting the set of possible successor states to the current state. Thus, if the system "*level*" is in "*medium*" and the condition input signal has the value "*valve is closed*", it has to stay in "*medium*" and cannot move to "*empty*":

$$\{medium\} = f(medium,\ valve\ is\ closed,\ -\}. \tag{6}$$

5.3 Timed Condition/Event Systems

In addition to the discrete dynamics, quantitative timing information has to be incorporated in this framework in order to make it possible to formally verify control programs with timers against real-time constraints and to represent the plant dynamics at least in an approximate fashion . This can be done within the same conceptual framework by introducing *C/E timers* as a new class of C/E systems. C/E timers can again be coupled to the untimed subsystems by condition and event signals. Thus, the basic C/E system definition applies also to timers. The advantage is obvious: Describing the timing is part of the modular concept. The untimed dynamics and the timing conditions are separated and can be modeled independently. Timing information can be added to an already existing model without changing any block of the C/E block diagram by simply adding the necessary timer blocks and connecting them to the appropriate discrete blocks. Such configurations consisting of at least one discrete state C/E system and a C/E timer are called *timed C/E systems* (*TCESs* for short) [21]. C/E timers were introduced in [10] and are defined as follows.

Definition 2 (C/E timer). Given an initial time $t_0 \in I\!R$ and a *threshold time* $T_\theta \in I\!R^+-\{0\}$, a *C/E timer* θ on $[t_0, \infty)$ is a mapping θ: $\boldsymbol{E}(V_\theta, t_0) \to \boldsymbol{C}(Y_\theta, t_0) \times \boldsymbol{E}(Z_\theta, t_0)$, with $V_\theta = \{\text{“}t_\theta := 0\text{“}\}$, $Y_\theta = \{\text{“}0 < t_\theta < T_\theta\text{“}, \text{“}t_\theta \geq T_\theta\text{“}\}$, and $Z_\theta = \{\text{“}t_\theta = T_\theta\text{“}\}$, such that for any event input signal $v_\theta(t) \in \boldsymbol{E}(V_\theta, t_0)$, the output $(y_\theta(t), z_\theta(t))$ is determined by

$$y_\theta(t) = \begin{cases} "0 < t_\theta < T_\theta" & \text{if } \tau(t) < T_\theta, \\ "t_\theta \geq T_\theta" & \text{if } \tau(t) \geq T_\theta, \end{cases} \tag{7}$$

and $$z_\theta(t) = \begin{cases} "t_\theta = T_\theta" & \text{if } \tau(t) = T_\theta, \\ 0 & \text{else} \end{cases} \tag{8}$$

in which τ: $[t_0, \infty) \to [t_0, \infty)$ is the *clock function* given by $\tau(t_0) = T_\theta + \varepsilon$ with ε arbitrary but fixed and $\varepsilon > 0$, and for all $t \in [t_0, \infty)$:

$$\dot{\tau}(t) = \begin{cases} 1 & \text{if } v_\theta(t) = 0 \\ \text{undefined} & \text{else,} \end{cases} \tag{9}$$

and $$\tau(t) := 0 \quad \text{if } v_\theta(t) \neq 0\,. \quad \blacklozenge \tag{10}$$

A C/E timer can be interpreted as an alarm clock which is reset and started by the input event "$t_\theta := 0$" and which indicates that it has reached its threshold time T_θ by sending out an event "$t_\theta = T_\theta$". The condition outputs "$0 < t_\theta < T_\theta$" and "$t_\theta \geq T_\theta$" provide the information that the threshold has not yet been reached or has been crossed, respectively.

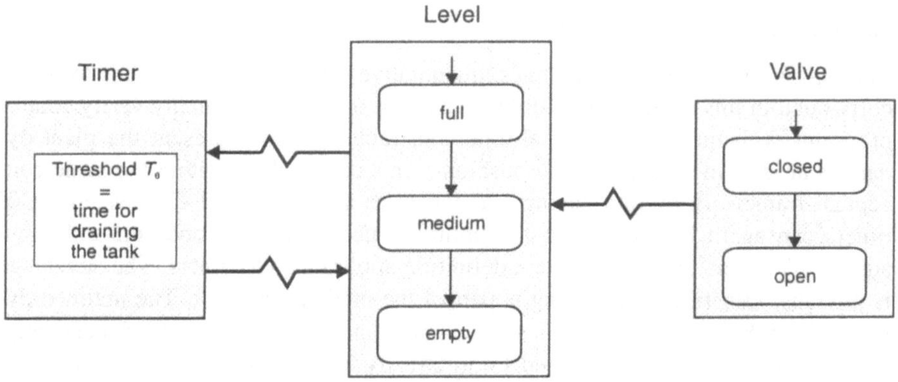

Fig. 12. Incorporating timing information to the C/E model of the draining tank.

Figure 12 illustrates the use of a C/E timer in the context of the draining tank example. We now assume that the outlet valve is opened only once and remains open then. Furthermore, the time needed to drain the full tank completely is known. This information can be incorporated into the model by adding a C/E timer into the block diagram as shown in Fig. 12. The timer is reset by the event "*level starts to fall*" which is generated by the transition from "*full*" to "*medium*" in the system "*level*" (which itself is forced by the opening of the valve). The clock function of the timer will then increase until it reaches the threshold representing the time for draining the tank. This threshold-crossing event is sent from the timer to the block "*level*" where it forces the transition from "*medium*" to "*empty*":

$$\{empty\} = f(medium,\ valve\ is\ open,\ t_\theta = T_\theta\}. \tag{11}$$

This model differs from the purely discrete one in that the transition from "*medium*" to "*empty*" is no longer spontaneous, but forced by the event "*draining time has elapsed*". Its time of occurrence is determined by the occurrence of the transition from "*full*" to "*medium*" and the pre-defined time for draining.

Figure 13 shows a sample trajectory of the clock function $\tau(t)$. It is easy to see that T_θ is the only free parameter for the C/E timer behavior, while it is not affected by the value of the initial offset ε. The internal behavior of a C/E timer is comparable to an integrator with constant input which is reset by an event input (here by "*level starts to fall*" at time t_1) and sends out an event when a threshold is reached (here "*draining time has elapsed*" at time t_2).

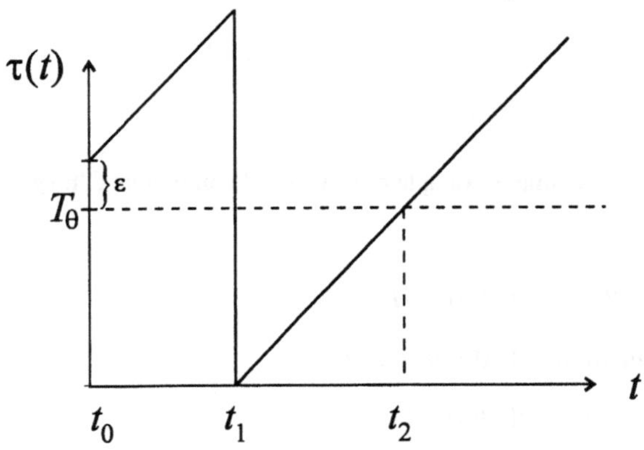

Fig. 13. Example of a clock function trajectory.

5.4 Switched Continuous Condition/Event Systems

The example in the previous section shows that the C/E system framework makes it possible to couple C/E systems and continuous systems in a convenient manner. It is therefore straightforward to generalize the concept of the C/E timer to *switched continuous C/E systems*: Instead of only integrators we allow arbitrary ordinary differential equations for the continuous state variables and additionally to resets by event inputs, the currently valid dynamics can be switched by the condition input. Fig. 14 illustrates the idea with the help of the draining tank example from Fig. 11. The discrete state block is replaced by a switched continuous C/E system with one continuous state variable, the level h. The differential equation governing the state dynamics now depends on the condition input: If the valve is open (and there is still liquid in the tank), the level decreases according to Torricelli's law; if the valve is closed (or the tank is empty), the level is constant.

The condition and event outputs are determined in the same fashion as in a C/E timer: The condition output signal provides the information in which pre-defined region of the continuous state space the system state currently is. The event output signal indicates the crossing of pre-defined threshold manifolds. In the example from Fig. 14, this could be realized as shown by (12).

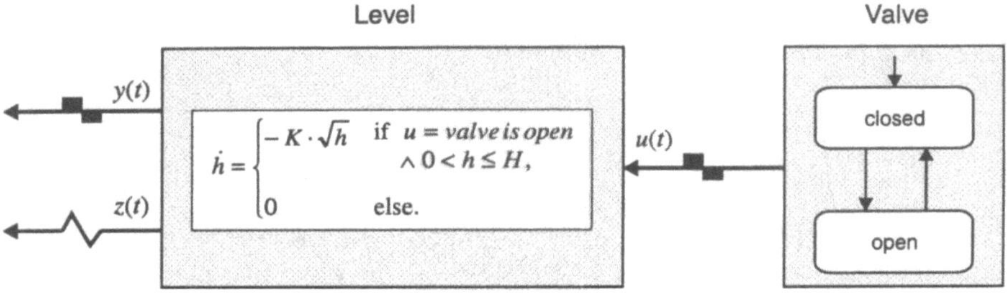

Fig. 14. A simple example of a switched continuous C/E system.

$$y(t) = \begin{cases} \textit{full} & \text{if } h(t) = H, \\ \textit{medium} & \text{if } 0 < h(t) \leq H, \\ \textit{empty} & \text{if } h(t) = 0. \end{cases}$$

$$z(t) = \begin{cases} \textit{becomes empty} & \text{if } h(t^-) < 0 \wedge h(t) = 0, \\ 0 & \text{else.} \end{cases} \tag{12}$$

In [24, 9, 8] a general modular modeling paradigm for hybrid systems was proposed in which C/E systems are connected to switched DAE-systems which include threshold functions to generate condition and event outputs from continuous variables. This framework is useful for, e.g., simulation of hybrid systems. For the purpose of verification, this model is too general. If continuous dynamics exceeding the complexity of resetable integrators are part of the model, they have to be approximated by TCESs. This is described in the next section.

6 Approximation of Switched Continuous Systems by Timed Condition/Event Systems

This section presents a method to derive analyzable, timed discrete models from given switched continuous systems. The procedure is implemented in the tool *Verdict* (see Sec. 9). A more detailed description and a comparison to other algorithms for timed discrete approximation of continuous systems appeared in [34].

6.1 Preliminaries

The starting point is a switched continuous C/E system as it was introduced in Sec. 5.4. The continuous dynamics which has to be approximated is given as a system of switched ODEs of the form

$$\dot{x} = f_u(x), \quad x \in X \subseteq I\!R^n, \quad u \in U \tag{13}$$

where X denotes the n-dimensional continuous state space and U the set of discrete inputs. We assume that the possibly non-linear functions f_u have a unique and continuous solution between the switching points of $u(t)$.

We introduce a rectangular partition of the continuous state space. The corresponding thresholds for the state variables are derived from technological constraints like variable boundaries, limit switches or safety margins. The controller receives the information whether the values of the state variables are below or above certain *landmarks* and computes an appropriate command depending on this discrete information. The threshold detection corresponds to a partition of the range of each state variable x_j into a finite number of discrete intervals (qualitative states), e. g. "low", "normal", "high" and "critical" for the temperature in a chemical reactor. Formally, we define a mapping D_X: $X \rightarrow \{1, ..., p_1)\} \times ... \times \{1, ..., p_n\}$ which divides the state space X into a finite set of n-dimensional rectangular partition elements. D_X is characterized by an ordered set of landmarks:

$$L_j = \{l_{j,0}, ..., l_{j,p_j}\}, \; j \in \{1, ..., n\}, \tag{14}$$

for each x_j where a landmark $l_{j,k}$, $k \in \{1, ..., p_j\}$ corresponds to one of the p_j thresholds. For each partition element, the mapping D_X generates an *index vector* $\boldsymbol{d} = (d_1, ..., d_n)$, which specifies the number of the actual discrete interval (defined by two consecutive landmarks) for each x_j:

$$d_j = \begin{cases} k & \text{if } x_j \in [l_{j,k-1}, l_{j,k}[, k \in \{1, ..., p_j - 1\} \\ p_j & \text{if } x_j \in [l_{j,p_j-1}, l_{j,p_j}] \end{cases} . \tag{15}$$

We assume that X is a bounded region such that the physically admissible range of x_j is limited by $l_{j,0}$ and l_{j,p_j}. The symbol ${}^{D}\boldsymbol{x}_d$ is used to denote the vector of intervals which corresponds to a particular index vector $\boldsymbol{d}$. As illustrated in Fig. 15, each ${}^{D}\boldsymbol{x}_d$ represents a rectangular partition element. The union of all ${}^{D}\boldsymbol{x}_d$ forms a partitioned state space ${}^{D}X$:

$${}^{D}X = \{{}^{D}\boldsymbol{x}_1, ..., {}^{D}\boldsymbol{x}_\pi\}, \; \pi = \prod_j p_j . \tag{16}$$

In what follows we assume that the discrete input $\boldsymbol{u}$ can only change when ${}^{D}\boldsymbol{x}$ changes. This corresponds to the usual closed-loop setting in which the logic controller changes its output as a reaction to a threshold crossing of the process.

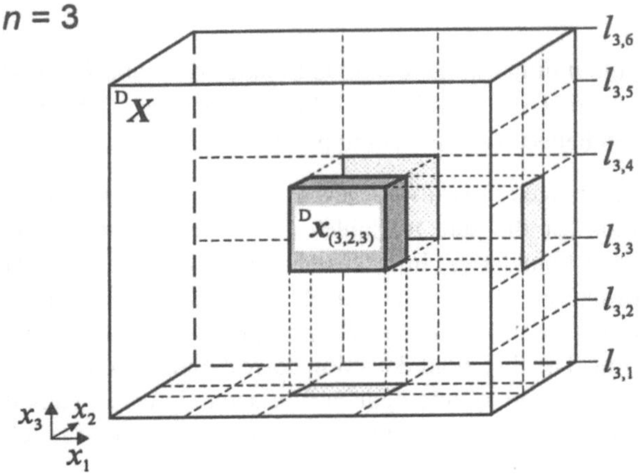

Fig. 15. A rectangular state space partition

6.2 Approximation of the Continuous Dynamics

The idea is to interpret the region between adjacent grid-points on the boundaries of the cells as discrete states and to determine transitions by calculating continuous trajectories between different states. This is different to earlier approaches [34, 35] in which the cells were directly mapped into the discrete states.

We introduce a grid on the boundary of each cell ${}^{D}\boldsymbol{x}$, i. e. assigning $(n\text{-}1)\cdot n_g$ grid-points equally spaced to all sides of the hyperboxes yields a grid with meshsize:

$$\varepsilon_{j,k} = \left| l_{j,k} - l_{j,k+1} \right| / \left(n_g - 1 \right). \tag{17}$$

The grid-points are labeled by $\gamma_{d,g}$, where $\boldsymbol{d}$ is again the cell identifier. The n-dimensional vector $\boldsymbol{g}$ indicates component-wise the number of the grid-point (starting with zero on the cell's corner with the lowest landmark indices), except of the invariant coordinate of the boundary section which gets a '-' or a '+' for the lower or upper side, respectively, of the hyperbox. Next, a set Σ_d of discrete states $\sigma_{d,s}$ is introduced where $\sigma_{d,s}$ represents the region of the cell boundary that is enclosed by $2^{(n-1)}$ grid-points lying adjacent on a boundary section (i. e. one side of the hyperbox ${}^{D}\boldsymbol{x}$). As shown in Fig. 16, the state identifier $\boldsymbol{s}$ is identical to the highest indexed grid-point identifier $\boldsymbol{g}$ being involved.

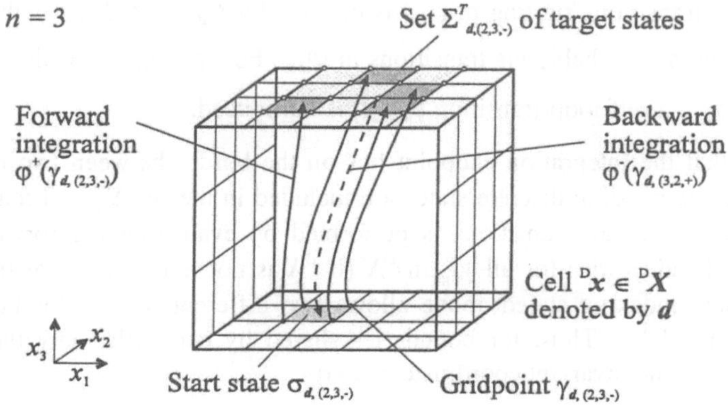

Fig. 16. Determination of transitions by forward/backward integration

The behavior within ${}^D\boldsymbol{x}$, described by Eq. 13, is mapped into a transition system by determining the set of feasible successor states for each $\sigma_{d,s} \in \Sigma_d$. First, we define classes of continuous trajectories in a cell ${}^D\boldsymbol{x}$:

Definition 3 (Forward/Backward cell trajectories). Let ${}^D\boldsymbol{x} \in {}^D\boldsymbol{X}$ be a cell indexed by $\boldsymbol{d}$, and $\partial({}^D\boldsymbol{x})$ its boundary. Assume that $[t_0, t_e]$ is a time interval on which a continuous solution of Eq. 13 exists where $\boldsymbol{x}(t_0) \in \partial({}^D\boldsymbol{x})$, $\boldsymbol{x}(t_e) \in \partial({}^D\boldsymbol{x})$ but $\boldsymbol{x}(]t_0, t_e[) \notin \partial({}^D\boldsymbol{x})$. Then, the solution of Eq. 13 in $[t_0, t_e]$ with $\boldsymbol{x}(t_0) = \gamma_{d,g}$ is called the *forward cell trajectory*, denoted by $\varphi^+(\gamma_{d,g})$, and the solution with $\boldsymbol{x}(t_e) = \gamma_{d,g}$ is the *backward cell trajectory* $\varphi^-(\gamma_{d,g})$. Additionally, the solution of Eq. 13 in $[t_0, \infty]$ with $\boldsymbol{x}(t_0) = \gamma_{d,g}$ and $\boldsymbol{x}(]t_0, \infty[) \notin \partial({}^D\boldsymbol{x})$ is a *forward cell trajectory* of $\gamma_{d,g}$ denoted by $\varphi^+_\infty(\gamma_{d,g})$. ♦

Transitions are assigned to those pairs of discrete states in ${}^D\boldsymbol{x}$ between which at least one cell trajectory exists:

Definition 4 (Cell-internal transition (CIT)). For two discrete states $\sigma_{d,s1}$, $\sigma_{d,s2}$ in Σ_d, a *cell-internal transition*, denoted by $\phi_{d,s1\to s2}$, is introduced, if for one $\gamma_{d,g}$ constituting $\sigma_{d,s1}$ $\exists$ $\varphi^+(\gamma_{d,g})$ where $\boldsymbol{x}(t_e)$ lies in the region of $\sigma_{d,s2}$, or for one $\gamma_{d,g}$ assigned to $\sigma_{d,s2}$ $\exists$ $\varphi^-(\gamma_{d,g})$ with $\boldsymbol{x}(t_0)$ in the region of $\sigma_{d,s1}$. ♦

The set of all transitions starting in $\sigma_{d,s1}$ is denoted by $\Phi_{d,s1}$ and $\Sigma^T_{d,s1}$ is the set of *target states* being reachable via transitions in $\Phi_{d,s1}$. For $\varphi^+_\infty(\gamma_{d,g})$, with $\gamma_{d,g}$ being assigned to $\sigma_{d,s1}$, a self-loop-transition $\phi_{d,s1\to s1}$ is introduced.

In the case that the integration endpoint lies on the border between two or more regions all corresponding discrete states are included in the set $\Sigma^T_{d,s1}$. Technically, the determination of all transitions is performed by evaluating the forward and backward cell trajectories for all $\gamma_{d,g}$ in ${}^D\boldsymbol{X}$ (if ${}^D\boldsymbol{X}$ is not left). Note that the grid-point and state indexing stated above allows two different indices for boundary sections inside of ${}^D\boldsymbol{X}$. Thus, for boundaries shared by two cells, only the index carrying a '-' for the invariant coordinate is used.

For numerical integration, the 4/5th-order Runge-Kutta-method with step-size adaptation was used. Besides leaving a cell region, the stopping criteria to capture cycling attractors as well as asymptotic approaches to steady states were an upper bound for the residence time in a cell and a lower bound for the progress of the integration.

The Definitions 3 and 4 lead to a time valuation of transitions in a straightforward manner if an interval enclosing the integration times of all cell trajectories connecting two cells is determined:

Definition 5 (Transition time interval (TTI)). For $t_0 = 0$ in the definitions of $\varphi^+(\gamma_{d,g})$ and $\varphi^-(\gamma_{d,g})$, t_e is the duration for which the cell trajectory is defined. If T_e is the set of durations of all forward and backward cell trajectories corresponding to a transition $\phi_{d,s1\to s2}$, its *transition time interval* is given by $\Delta t_{d,s1\to s2} = [\min(T_e), \max(T_e)]$, respectively by $\Delta t_{d,s1\to s1} = [0, \infty]$ for a self-loop transition. ♦

When all CITs and the corresponding TTIs are computed, the representation as a TCES is straightforward. A discrete state C/E system is introduced with a state space equal to the set of the cell boundary regions, $X = \Sigma_d$, and with a discrete transition for any CIT. The timing information is given by the TTIs. They can be regarded as the lower and upper bound for the instances at which a transition takes place. Thus, for any discrete transition two thresholds define its timing: the lower threshold T_l gives the earliest time at which the transition can happen, the upper threshold T_u gives the time at which the transition has to take place eventually. In other words, the condition "*time is beyond* T_l" enables the transition while the condition "*time is below* T_l" disables it. The event "*time reaches* T_u" will force the transition if it has not happened already in the meantime. It is obvious that this behavior can nicely be modeled by means of TCESs. Two C/E timers, one for T_l and one for T_u, are needed to realize the described interaction as it is shown in Fig. 17. Since this configuration is necessary for any TTI, we combine both timers to one system and call it a *C/E Interval timer*.

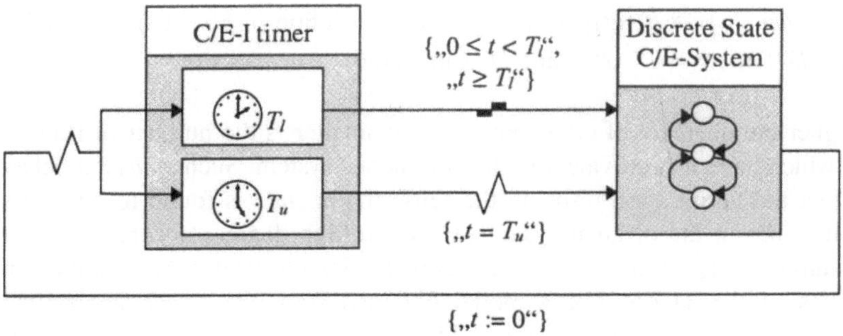

Fig. 17. C/E interval timer.

6.3 Properties of the Approximation Procedure

To assess whether the transformation method is suitable within a verification procedure, we define the properties *completeness* and *consistency*.

Definition 6 (Completeness). Assume that $D: X \rightarrow \Sigma$ denotes the mapping of the continuous state space X into the set Σ of discrete states of the TCES. The latter is called *state complete* iff $\forall\, x \in X$, there exists a state $\sigma \in \Sigma$ with $\sigma = D(x)$. Let $\varphi[t_1, t_2] \rightarrow X$, $\varphi(t_1) = x_1$, $\varphi(t_2) = x_2$ be a continuous trajectory on a time interval $[t_1, t_2]$ as a solution of Eq. (13). Then, the TCES with a state transition function f is called *dynamically complete* if for all continuous trajectories there exists a finite sequence of discrete states $(\sigma_1, \sigma_2, ..., \sigma_m)$ with $\sigma_2 = f(\sigma_1), ..., \sigma_m = f(\sigma_{m-1})$ and $\sigma_1 = D(x_1)$, $\sigma_m = D(x_2)$. The TCES is a *complete* abstraction of the switched continuous system if it is state complete and dynamically complete. ♦

Completeness should be understood as an important property of the generated TCES since a model showing all possible behavior of the original system is a precondition for the verification approach (see Sec. 3). In the approximation procedure, a cell trajectory constituting a discrete transition $\phi_{d,s1 \rightarrow s2}$ is interpreted as a representative for all continuous trajectories connecting the states $\sigma_{d,s1}$, $\sigma_{d,s2}$. Obviously, these representatives can only estimate the set of all actually possible transitions and do not necessarily form a strict inclusion of all feasible behaviors. Hence, the TCES obtained from this method is not always complete according to the above definition. To use the transformation within a formal verification approach, the discretization defined by Eq. (17) must be chosen fine enough to ensure a complete approximation.

Definition 7 (Consistency). A TCES is called a *r-consistent* approximation of a continuous system given by Eq. (13) iff for all finite sequences of states $(\sigma_1, \sigma_2, ...,$

σ_r) with length r and $\sigma_2 = f(\sigma_1)$, ..., $\sigma_r = f(\sigma_{r-1})$, there exists at least one continuous trajectory $\varphi[t_1, t_2] \rightarrow X$, $\varphi(t_1) = x_1$, $\varphi(t_2) = x_2$ as solution of Eq. (13) in $[t_1, t_2]$, such that $x_1 = D^{-1}(\sigma_1)$, $x_2 = D^{-1}(\sigma_m)$, and D as in Def. 6. ♦

The higher the degree r of consistency is, the smaller is the number of runs of the TCES which have no equivalent in the continuous system. Such *spurious* behaviors can effect the verification result in the sense that a state is found to be reachable while it is not in the original system. Hence, a high degree of consistency and a small number of spurious state sequences with a length greater than r are desirable properties of the TCES. For a model obtained from the above approximation procedure, 2-consistency is provable. This follows immediately from the definition of cell-internal transitions (Def. 4). The amount of state sequences of length $r > 2$ which constitute spurious behaviour depends also on the discretization defined by Eq. (17). In Sec. 9, these properties are discussed more detailled by the applying approximation method to the example system sketched in Sec. 2.

6.4 Application to the Reactor Example

Table 2 presents numerical results (Pentium 200MHz) from the application of the approximation procedure to the reactor model from Table 1. For the state space partitioning, a set of 2 landmarks for c_D, 3 landmarks each for c_A and V_R, and 4 landmarks for T_R were introduced in a first trial, where the medium T_R-landmark corresponds to T_S. This is the roughest partition which allows to answer the verification question in principle. The parameterization and landmark values were chosen such that they represent a pilot plant reactor. The relevant 3 input combinations are $\boldsymbol{u} = (v_A, v_C) \in \{$(*'open'*, *'closed'*), (*'closed'*, *'closed'*), (*'closed'*, *'open'*)$\}$. Setting the number of grid-points per cell and per dimension to $n_g = 2$, we obtain a set Σ of 76 discrete states for the reactor. Table 2 shows that the number of discrete states and transitions increases rapidly with n_g.

Table 2: Numerical results for the approximation of the reactor example.

Number of grid-points	Number of discrete states	Number of transitions	Size of the TCES file	CPU time (approx.)
2	76	407	50 kB	3 min
3	607	10916	1.3 MB	36 min
4	2052	20977	2.52 MB	180 min

7 Translation of Control Code into TCES

The input of control software to the verification tool is realized by a translator which maps *Instruction List* (*IL*) programs to TCES. The class of algorithmically verifiable IL programs is restricted. In our case, only IL code which can be represented by TCES can be analyzed. This excludes programs with arbitrary manipulation of real variables, e.g. timers which can be stopped and restarted. To support the user in developing verifiable control code, an IL editor is provided which allows the user to write only IL programs which are transformable to TCES.

The translation procedure is based on a mapping of all possible combinations of values of the variables in the IL program into states of the resulting C/E system. An exhaustive search is performed then to determine all possible state transitions, i.e., for each reachable state the program is simulated using all possible values of the input variables. The resulting C/E system is then minimized using standard techniques. For each timer function block being called in the program, a corresponding C/E timer is introduced.

When translating the control program into a C/E system representation, it has to be taken into account that PLCs operate in a so-called scanning mode and that the communication with the process is based on process data buffers. This means, the PLC executes the control program in a cyclic manner. In each cycle, first, the current measurements values are written into the input buffer. Then the program is executed and, after that, the contents of the output buffer are sent to the PLC output connections. During the execution of the program, inputs are read from the input buffer and outputs written to the output buffer. As a consequence of this operating mode, input signal changes do not effect the output immediately but after a certain time depending on the cycle time of the PLC. To capture this behavior, the input and output buffers are modeled as additional C/E systems which are driven by a C/E timer. Another C/E timer is used to trigger the execution of the program. By letting each of the two timers start the other one as soon as it elapses, an alternating sequence of writing data into both buffers at one time and reading data out of the buffers and execution of the program at the other time is achieved. The thresholds of the timers are set to the maximal time needed to update both buffers and to execute the program. Fig. 18 illustrates the model. A similar approach to model PLC behavior by C/E systems is described in [13].

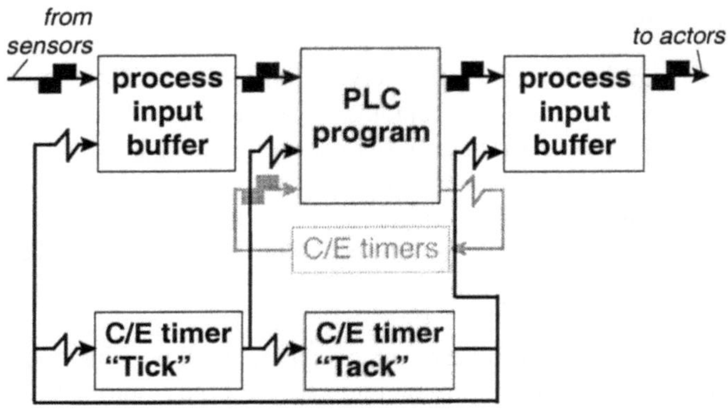

Fig. 18: C/E model of a PLC in scanning mode

8 Analysis

8.1 Timed Automata

As already mentioned in Sec. 4, for the analysis step our approach uses model checking techniques for Timed Automata (TA) [1 and 15]. To apply these tools we take advantage of the fact that TCESs are equivalent in expressive power to a subclass of TAs called *Safety TAs* [16]. This allows us to transform TCES models to TA models and use existing model checkers for TAs, e.g. *KRONOS* [36], *UPPAAL* [25] or *HyTech* [14].

TAs are a generalization of *Büchi Automata* by adding real-valued variables, called *clocks*, the value of which increases by a rate of one and which can be reset by transitions. Transitions can be made dependent on the values of the clocks by clock constraints, i.e. Boolean combinations of formulas of the form $c\#n$ with c being a clock, $\# \in \{=, <, \leq, >, \geq\}$ and n a non-negative integer. A TA is then defined over an alphabet Σ as a 4-tuple $A = (Q, q_0, C, E)$ with: Q, a finite set of locations (discrete states), $q_0 \in Q$, the initial location, C, a finite set of clocks, $E \subseteq Q \times Q \times \Sigma \times \Delta(C) \times 2^C$, the set of transitions, and an acceptance condition. A transition $(q, q', a, \delta, \rho) \in E$ represents a transition from q to q' on the input symbol a conditioned by the clock constraint $\delta \in \Delta(C)$ ($\Delta(C)$ is the set of constraints over C) resulting in a reset of the clocks $\rho \in 2^C$. A *Safety TA* is a TA whose acceptance condition is trivial (i.e. every state is an accepting state). The behavior of a TA is defined by *accepting runs* which correspond to *timed words* of the TA. Timed words are infinite sequences of pairs consisting of a symbol σ from Σ and a time stamp representing the point in time when a transition labeled with σ was made. The set of all timed words of a TA forms the language of the TA.

8.2 Reachability Analysis of Timed Automata

The *state* of a TA comprises the location and the values of the clocks. The reachability analysis of TAs is an iterative procedure determining the set of states which are reachable from an initial state, i.e. an initial location and initial values for the clock variables. In the symbolic variant, which is used by *HyTech*, the algorithm is based on a set representation of polyhedral subspaces in the clock variable space and operations on polyhedral sets as, e.g., intersection. The iteration roughly works as follows. Each time a location is reached with a certain polyhedron representing the possible clock values, it is checked whether the location has been reached before. If so, it is checked whether the current polyhedron is completely contained in the union of the polyhedra with which the location has been visited before. If this is the case, this branch of the search tree is left. If not, the polyhedron is "stretched" by letting time pass which increases the clock values. The resulting polyhedron can now be used to determine whether the transitions from the current location are possible. For each transition, it is intersected with the polyhedron representing the corresponding clock constraint. If the intersection is empty, the timing conditions prevent the transition. If not, the transition is possible. For each possible transition the iteration continues. It can be proved that the algorithm terminates for rational values of the thresholds [1].

8.3 Translation of Timed C/E Systems into Timed Automata

TCESs and TAs are based on different theoretical concepts: A TCES is specified as a relation associating a set of output signal trajectories to input signal trajectories, whereas the behavior of a TA is defined as a set of timed words. However, a translation from TCESs to TAs can be realized by a two-step procedure. First, the underlying transition system of the TCES is mapped into an isomorphic state transition structure for the TA. Then the relevant conditions and events for each transition of the TCES are mapped into transition constraints for the TA. An algorithm for automatic translation of a TCES into a TA is given in [17]. Technical problems of this procedure arise from subtle details which the following examples may illustrate.

- The interconnection of TA must be realized by synchronization labels (see Sec. 4.2). This makes it necessary to introduce self-loops in the resulting TA when the forcing and enabling mechanisms of TCES are transformed. Fig. 19 illustrates this solution. Apart from the synchronization of the forcing and the forced event, additional self-loops are needed to express that the forcing event can also take place when the receiving system is not ready to take part. (If the forced transition could also occur spontaneously, α-labeled self-loops would be necessary for each state in the sending system. This case is not illustrated in Fig. 19.) In the case of enabling, the enabled transition has to be synchronized with a new self-loop at the enabling state of the sending system.

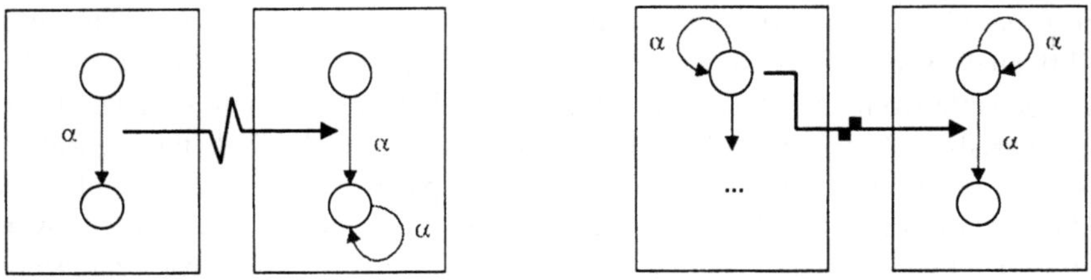

Fig. 19. Translation of forcing by events (left) and enabling by conditions (right) into TA synchronization.

- In TCESs, an input event "$t = T$" from a timer can *force* a transition to occur. In TAs, a guard of the kind "$t = T$" is only a *necessary* condition for the occurrence of the transition at time T. This problem can be solved by *invariants*, which are logical propositions on the clock variables and have to be true in the assigned location. The location must be left, if the invariant becomes false. However, a location cannot be entered if the invariant is false. So, if the above behavior of the TCES is modeled by an invariant "$t \leq T$" in the corresponding location of the TA, an additional location with the invariant "$t>T$" has to be introduced to express that the TCES state may be reached again after the clock has crossed its threshold. Another reason for this translation rule is that invariants in TA must have convex value sets. Fig. 20 illustrates the translation for a simple example.

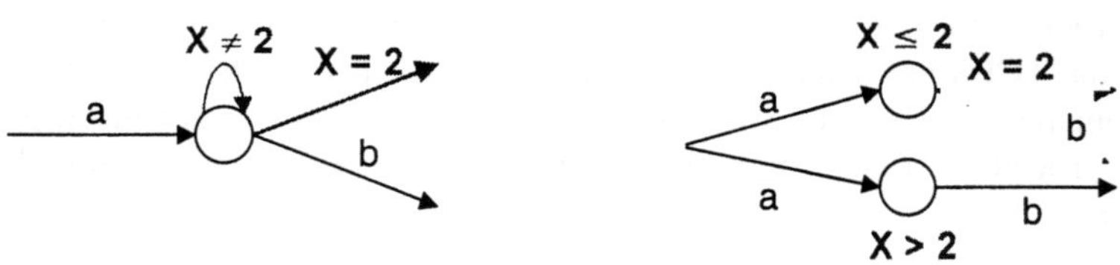

Fig. 20. Translation of non-convex self-loop conditions into convex invariants.

- For a transition in a TCES at time t, the condition input signal from a C/E timer is evaluated at time t^-, whereas the constraints in TAs are relevant for the time of the transition. Consequently, "<"-relations in TCESs have to be mapped into $\leq$-relations in TAs.

9 Tool Support

This section describes the tool *Verdict* which supports the presented verification approach. The basic idea is to combine an appropriate modeling framework with efficient analysis engines by introducing easily accessible user interfaces [20]. This approach allows the process engineer to apply model-based verification techniques without having to get into the theory of formal methods more deeply than necessary and to specify the problem using representations which are common in process engineering. This section describes how TCES models of the plant and of the controller can be set up by using *Verdict*. Depending on the function, the different features of the framework can be divided into four parts: a hierarchical set of plant editors, an approximation facility to convert subsystems with continuous dynamics into timed discrete models, an editor and translator to implement controller programs, and a compiler to translate the TCES representation of the controlled plant into the input format of appropriate analysis engines. The overall architecture of *Verdict* is depicted in Fig. 21. The components of the tool are described in the sequel.

The user interface for plant modeling consists of three hierarchically ordered graphical editors: On the top level (the *plant editor*), the overall plant can be modeled using flow charts, i. e. the plant is structured in terms of logical units, and both units and interconnecting mass streams are represented graphically. On an intermediate level, a *block-diagram editor* allows to specify each plant unit in more detail by introducing a set of modules. Each of these modules is a TCES and the connections between the modules are instantiated by condition signals and event signals.

On the third (and lowest) editor level, the internal dynamics of a module is specified by a so-called *system editor*: Two different types of module descriptions are possible, *timed discrete subsystems* and *continuous subsystems*. For the timed discrete type, the system editor allows to specify the behavior either in terms of transition diagrams or textually. Using transition diagrams, a module is described graphically by drawing a set of discrete states and labeled edges. For textual modeling, a modeling language, called *Celeste*, has been introduced [19]. It is a textual representation of TCES based on a syntax very close to *Structured Text* (specified in the IEC 1131-3 standard [18]) For continuous subsystems, the dynamics in terms of ordinary differential and algebraic equations is specified textually.

Libraries of standard elements for each modeling level support the modeling task by supplying templates for components which frequently occur in chemical plants, as e. g. storage tanks, pumps and evaporators. The controller logic can be specified by one or more subsystems of the timed discrete type as sketched above or edited in PLC code which is automatically translated into *Celeste* code (see Sec. 7). From this representation, translation algorithms are available which map the TCES model into input files for *HyTech* and *KRONOS*, as described in Sec. 8.3.

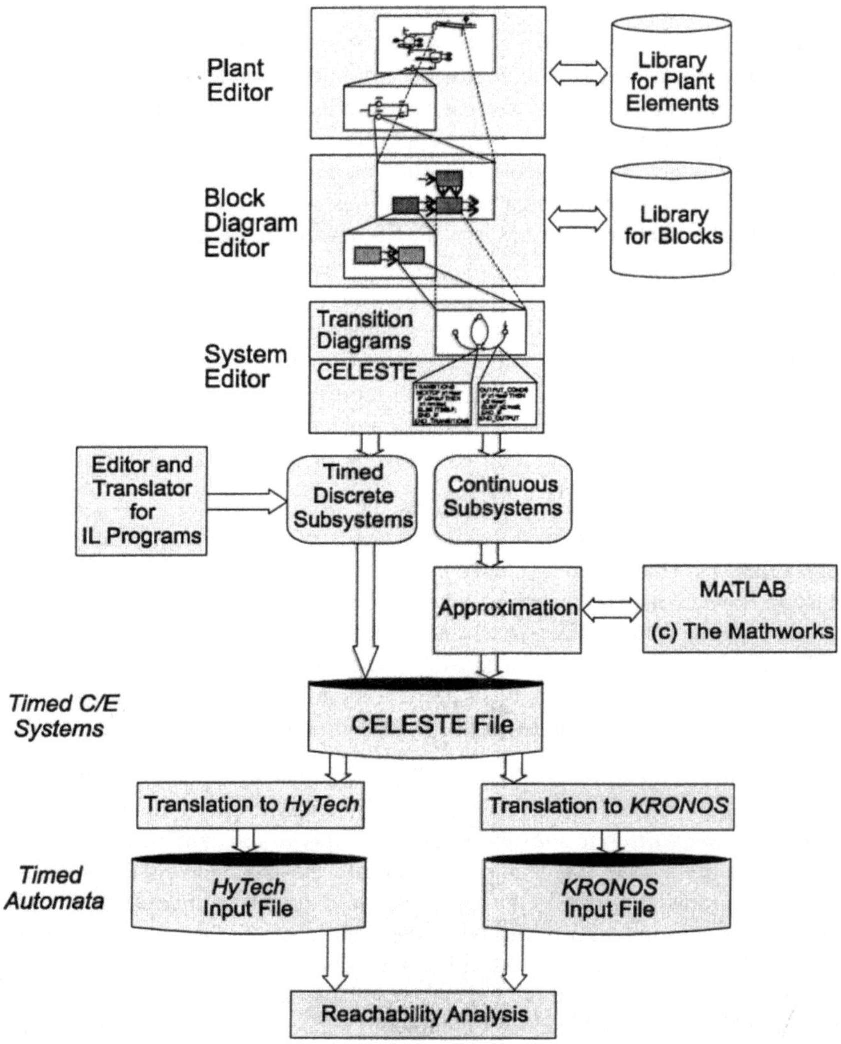

Fig. 21. Architecture of *Verdict*

To provide an impression of the use of *Verdict*, we sketch how it can be applied to the reactor example from Sec. 2. To create an appropriate model of this system in *Verdict*, one block each is introduced for the controller and the reactor on the block-diagram editor level. (If the reactor was part of a plant consisting of several modules, this structure would be realized in the plant editor.) As shown in Fig. 22, the two blocks are connected by condition signals carrying information about the setting of the valves v_A and v_C from the controller to the reactor. In the opposite direction, event signals are emitted when the liquid level, the temperature, or the product concentration cross thresholds at which controller actions must occur.

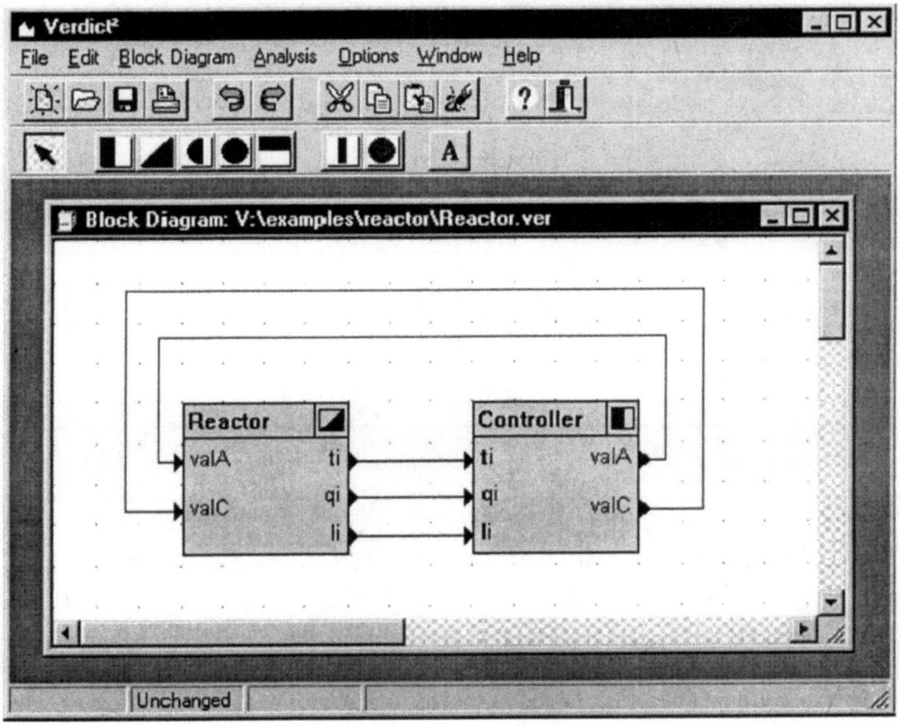

Fig. 22. Example modeled by means of the *Verdict* block-diagram editor.

The behaviors of both blocks are specified next on the system editor level: Instead of using the IL editor and translator, we here model the controller by means of the state diagram editor as depicted (in part) in Fig. 23 (left window). The controller model is straightforwardly obtained from implementing the transition sequence in Fig. 7. For each state, the name (and possibly an invariant), and for each transition the conditions (and possibly an action) have to be specified. As shown in Fig. 23 (right window), the user must additionally choose one of the transition types *'MUST_TO'* or *'MAY_TO'* which refer to enforced or enabled transitions, respectively. For the given example, the controller must switch from state *'react_with_cooling'* to *'critical_temperature'* when it receives the event signal *'above_*T_{crit}*'* from the reactor.

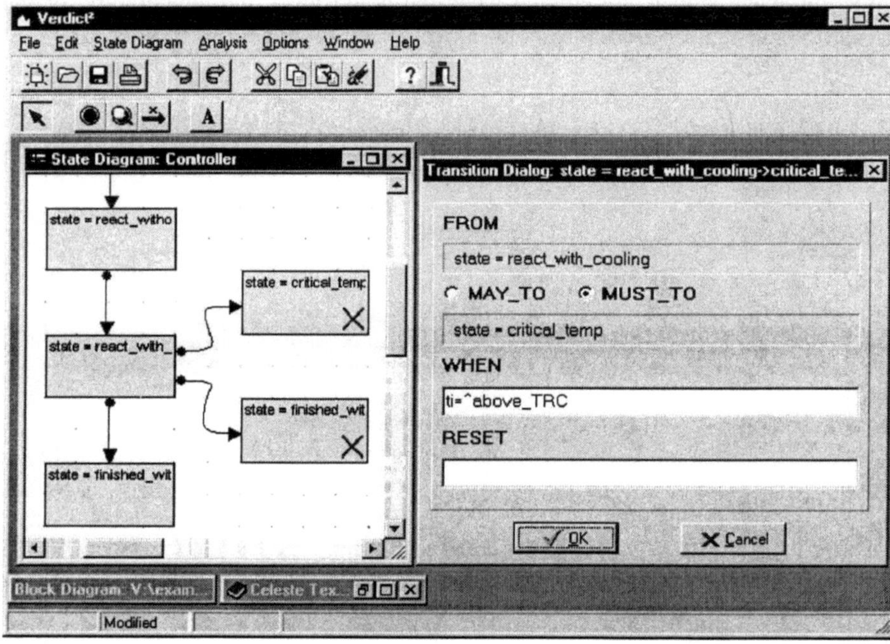

Fig. 23. Controller modeled by means of the *Verdict* state-diagram editor.

The continuous subsystem is specified textually in the *Celeste* language which provides the necessary constructs to implement switched continuous systems in a straightforward manner. Fig. 24 shows this for the reactor example consisting of the system of differential and algebraic equations given in Tab. 1 and the transition structure in Fig. 7. Additionally, the appropriate discretization parameters are specified (see Sec. 4.1). The different parts of the *Celeste* file are explained in the comments given between the "(*" and the "*)" symbols.

```
SYSTEM Reactor: CONTINUOUS

  VAR_INPUT
      (* Declaration of condition input signals. *)
    valA: CONDITION(openA, closedA);
    valC: CONDITION(openC, closedC);
  END_VAR

  VAR_INTERNAL
      (* Declaration of continuous state variables and *)
      (* algebraic variables with landmarks and units. *)
    ca:    STATE(0,0.01,2), 'kmole/^3';  (*concentration c_A*)
    cd:    STATE(0,0.8), 'kmole/m^3';  (*concentration c_D*)
    tr:    STATE(290,293,313,333), 'K';  (*temperature T_R*)
    vr:    STATE(1,1.02,1.5), 'm^3';  (*volume V_R*)
    fin:   ALGEBRAIC, 'm^3/s';  (*low into reactor*)
    ac:    ALGEBRAIC, 'm^2'; (*heat transfer area*)
  END_VAR
```

```
VAR_OUTPUT
    (* Declaration of event output signals. *)
    (* The values represent threshold crossings. *)
  ti:   EVENT(^above_TRS, ^above_TRC);
  qi:   EVENT(^above_CDS);
  li:   EVENT(^above_FULL);
END_VAR

APPROX_PAR
    (*Parameters necessary for the approximation routine.*)
  number_of_gridpoints=2;
    (*Initial values for state vector (ca,cd,tr,vr):*)
  init_state=([0, 0.01], 0, [290, 293], [1, 1.02]);
  input_combination=[openA,closedC;closedA,closedC;
      closedA,openC];
  time_scale=1;
  int_tolerance=1E-5;
END_PAR

PHYSIC_PAR
    (* Physical parameters, cf. Table 1 *)
  flow=8.333E-4, 'm^3/s';
  cin=5, 'kmole/m^3';
  tin=308, 'K';
  d1=1.2, 'm';
  ko=1.39E17, 'm^3/kmole/s';
  ea=1.22E5, 'kJ/kmole';
  gc=8.314, 'kJ/K/mole';
  rho=1.0E3, 'kg/m^3';
  cp=3, 'kJ/kg/K';
  dHr=-1.97E5, 'kJ/kmole';
  tc=283, 'K';
  kc=0.5, 'kW/m^2/K';
END_PAR


DYNAMICS
    (* System of switched differential equations. *)
    (* DER(x) = time derivative dx/dt *)
  IF valA=openA THEN
    fin:=flow;
  ELSE
    fin:=0;
  END_IF

  DER(vr):=fin;
  ac:=3.1414/4*d1^2+4/d1*vr;
  DER(ca):=fin/vr*(cin-ca)-2*ca^2*ko*exp(-ea/(gc*tr));
  DER(cd):=ko*exp(-ea/(gc*tr))*ca^2-cd*fin/vr;

  IF valC=closedC THEN
    DER(tr):=fin/vr*(tin-tr)-
             dHr/(rho*cp)*ca^2*ko*exp(-ea/(gc*tr));
  ELSE
    DER(tr):=fin/vr*(tin-tr)-
             dHr/(rho*cp)*ca^2*ko*exp(-ea/(gc*tr))
            -kc*ac/(rho*cp*vr)*(tr-tc);
  END_IF
END_DYNAMICS
```

```
 OUTPUT
    (* Event output.  *)
    (* The symbol ->| means crossing from below. *)

   IF tr->|313 THEN
      ti := ^above_TRS;
   ELSIF tr ->| 333 THEN
      ti := ^above_TRC;
   END_IF

   IF cd ->| 0.8 THEN
     qi := ^above_CDS;
   END_IF

   IF vr ->| 1.5 THEN
     li := ^above_FULL;
   END_IF
 END_OUTPUT

END_SYSTEM
```

Fig. 24. The switched continuous model of the reactor in *Celeste* representation.

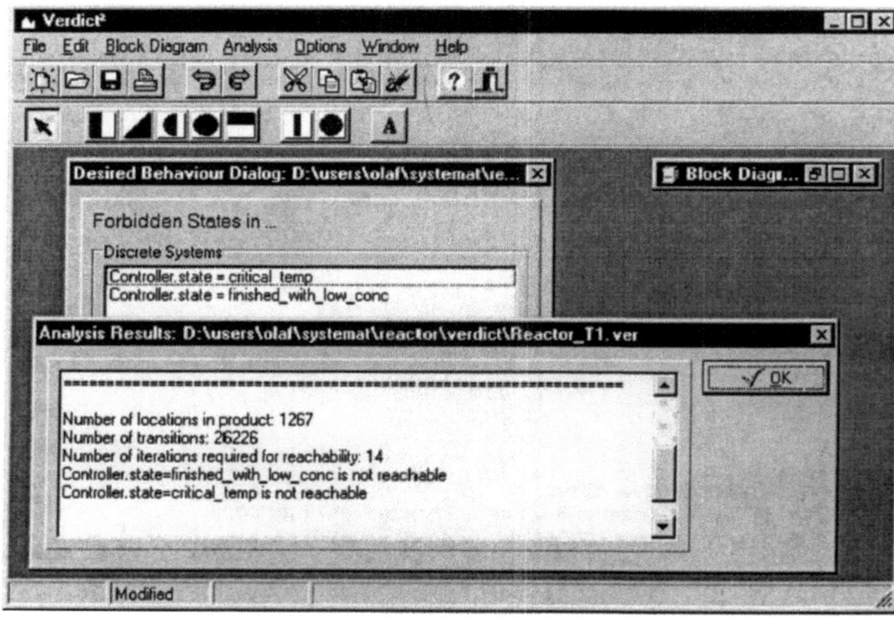

Fig. 25. Behavior specification and *HyTech* analysis result in *Verdict*.

Finally, the set of initial states for both blocks and the desired behavior of the overall system have to be specified. For the latter, this is just a list of the two forbidden controller states (see Fig. 25). Concerning the definition of forbidden states, the following has to be noted: In contrast to Fig. 7, here, we map the critical reactor states into forbidden controller states by taking advantage of the fact that there exist event output signals in the plant model indicating when undesired regions of the continuous state space are reached. A physical danger to the system, of course, would rather occur in the reactor than in the controller. However, since the discrete states of the reactor model are generated automatically, projecting the undesired situations into the controller model makes the behavior specification considerably easier.

Having the system modeled this way, the analysis is performed by the following steps: First, the syntax of the *Celeste* description of all subsystems and the consistency of the subsystems' interconnections are checked. If no errors are detected, the overall *Celeste*-representation is composed from the subsystem specifications and it is translated into the input code of the analysis tool – for this example, we chose *HyTech*. Then the reachability analysis is carried out.

For the reactor example, we started by checking the following three values for the switching temperature T_S: 303 K, 313 K, and 323 K. Choosing the setting given in the *Celeste* file in Fig. 24 (in particular, $n_g = 2$ and $c_A \in$ [0 kmole/m^3, 0.01 kmole/m^3], $c_D = 0$ kmole/m^3, $T_R \in$ [290 K, 293 K], $V_R \in$ [1 m^3, 1.02 m^3] as the initial state region), the approximation routine generated a discrete model containing 76 states and around 400 transitions (CPU-time for approximation on a Pentium, 200 MHz: ~7 minutes). *HyTech* found that for all three values for T_S neither of the two forbidden states is reachable (see the lower window in Fig. 25). However, from this result one cannot infer that the specification is fulfilled for these values. Since the approximation procedure is not guaranteed to be complete (cf. Sec. 6.3), it is possible that trajectories exist in the continuous system which are not represented by the approximating model. Therefore, to gain more confidence in the verification results, the partitioning has to be refined by introducing additional landmarks or by increasing n_g. We chose the second option and repeated the procedure for $n_g = 3$ which generated a discrete model with 608 states and more than 5500 transitions (CPU-time: around 30 minutes). With this model the following results were obtained. For T_S =303 K, the state "*finished_with_low_conc*" (finished with low concentration) is found to be reachable, whereas the state "*critical_temperature*" is not. For T_S =323 K, it is just the other way round. And for T_S =313 K, both forbidden states are not reachable.

Since $n_g = 3$ still is a rather coarse granularity, we checked whether the last result also holds for $n_g = 4$. The approximation took around 3 hours and the resulting discrete model has more than 2000 states and 21000 transitions. The analysis confirmed that both forbidden states cannot be reached for T_S =313 K. Based on this result, it is reasonable to assume that T_S =313 K fulfills the specification, because the partitioning for $n_g = 4$ is reasonably fine with respect to the gradient field variation and the naturally limited accuracy of the original switched continuous model. This was also confirmed by simulations of the reactor model. However, it cannot be excluded completely that there exist trajectories in the switched continuous system which would lead to the forbidden region and which are not represented by the timed-discrete model. Therefore, our current research is also concerned with approximation methods which guarantee completeness (cf. Sec. 10 and [28]).

It also has to be noted that with 4-dimensional systems like the reactor, a granularity of $n_g = 4$, and a number of thresholds in the same order of magnitude, we have reached the system size up to which the applicability of our approach is limited so far. For $n_g = 5$ and greater, the approximation time and the temporarily allocated memory during the *HyTech* analysis (which already exceeded 70 MB for $n_g = 4$ in the reactor example) becomes the limiting factor.

10 Conclusions

An approach to the formal verification of logic controllers for continuous processes was presented which addresses several problems of practical relevance. It is based on a clear distinction between the plant and the controller which permits a transparent, process specific formulation of the requirements. The underlying framework is modular and signal oriented which helps to build complex models. It permits the formulation of real-time constraints which addresses the fact that most industrial logic control programs contain timers and that dynamical processes can often be described sufficiently well by timed-discrete models. Continuous models are approximated automatically by timed discrete event models. This provides a systematic procedure to develop verifiable models of continuous systems. The approach takes advantage of existing techniques and tools for the verification of real-time and hybrid systems.

On the other hand, it has to be noted that the application scope of the method is limited by the computational cost of the analysis and the necessary effort to build the models. Whereas the complexity of reachability analysis in purely discrete systems grows exponentially with the number of interacting subsystems, in the case of TCESs and TAs the number of timers and clocks, respectively, is critical, too. Additionally, the approximation procedure is exponential in the number of discrete values of $\boldsymbol{u}(t)$ and the granularity of the discretization. A further source of complexity is the transformation step from TCES to TA because it is necessary to introduce additional transitions and states to represent the TCES semantics in TAs. Finally, the numerical robustness of *HyTech* is unsatisfactory. Overflow errors occur already for quite simple examples. For these reasons, the described approach up to now has only been applied to problems of a size comparable to the reactor from Sec. 2 (for a different example refer cf. [22]).

Therefore, our current research is concerned with techniques which will help us to treat more complex systems by this method. One approach is to utilize the structure of the underlying model, e.g. by compositional analysis [23]. Further work is about using Rectangular Automata as approximating systems [35, 28]. This class of automata makes it possible to retain more information of the continuous state while being amenable to a robust approximative analysis [29].

Acknowledgement

The work was partially supported by the German Research Council (DFG) in the special program *Analysis and Synthesis of Technical Systems with Continuous-Discrete Dynamics (KONDISK)* and by the temporary graduate school ("Graduiertenkolleg") *Modeling and Model-Based Design of Complex Technical Systems.*

References

1. R. Alur and D. L. Dill. A Theory of Timed Automata. *Theoretical Comp. Science* **126**, 1994, pp. 183-235.
2. R. Alur, C. Courcoubetis, N. Halbwachs, T.A. Henzinger, P.-H. Ho, X. Nicollin. A. Olivero, J. Sifakis, and S. Yovine. The Algorithmic Analysis of Hybrid Systems. *Theoretical Comp. Science* 138, 1995, pp. 3-34.
3. A. Bemporad and M. Morari. Verification of hybrid systems via mathematical programming. *2nd Int. Workshop on Hybrid Systems - Computation and Control*, March 1999, Nijmegen, Netherlands. Will appear in *Lecture Notes in Computer Science* 1569, Springer, 1999.
4. J.R. Burch, E.M. Clarke, K.L. McMillan, D.L. Dill and L.J. Hwang. Symbolic Model Checking: 10^{20} states and beyond. *Information and Comp.*, 98(2), 1992, pp. 142-170.
5. E.M. Clarke and R.P. Kurshan. Computer-aided Verification. IEEE Spectrum, June 1996, pp. 61-67.
6. V.D. Dimitriadis, N. Shah and C.C. Pantelides. A Case Study in Hybrid Process Safety Verification. *Computers and Chemical Eng.*, Vol. 20, Suppl., 1996, pp. S503-S508.
7. V.D. Dimitriadis, N. Shah and C.C. Pantelides. Modelling and Safety Verification of Discrete/Continuous Processing Systems. *AIChE Journal* 43(4), 1997, pp. 1041-1059.
8. S. Engell and I. Hoffmann. Modular Hierarchical Models of Hybrid Systems. Proc. 35th IEEE Conference on Decision and Control (CDC), Kobe, 1996, 142-143.
9. S. Engell, S. Kowalewski and B. H. Krogh. Discrete Events and Hybrid Systems in Process Control. In: J.C. Kantor, C.E. Garcia, and B. Carnahan (Eds.) *Chemical Process Control V: Assessment and New Directions for Research.* AIChE Symposium Series, Vol. 93, No. 316, 1997, pp. 165-176.
10. S. Engell, S. Kowalewski, B. H. Krogh and J. Preußig. Condition/event systems: a powerful paradigm for timed and untimed discrete models of technical systems. In *F. Breitenecker u. I. Husinsky (Eds.): EUROSIM 95*, Vienna, Austria, , Elsevier, Amsterdam, 1995, pp. 421-426.
11. G. Fafferriere, G.J. Pappas, and S. Yovine. Decidable Hybrid Systems. Technical Report UCB/ERL M98/39, University of California at Berkeley, June 1998. Presented at *School on Computational Aspects and Applications of Hybrid Systems*, Grenoble, France, October 1998.

12. H.-M. Hanisch, A. Lüder and M. Rausch. Controller Synthesis for Net Condition/Event Systems with a Solution to Incomplete State Observation. *European Journal of Control* 3, 1997, pp. 280-291.
13. H.-M. Hanisch, J. Thieme, A. Lüder and O. Wienhold. Modeling of PLC behavior by means of Timed Net Condition/Event Systems. Proc. *6th IEEE Conf. Emerging Technologies and Factory Automation* (*ETFA'97*), Los Angeles, 1997, pp. 391-396.
14. T.A. Henzinger, P.S. Ho and H. Wong-Toi. *HyTech*: A model checker for hybrid systems. *Software Tools for Technology Transfer*, 1(1,2), Springer, 1997, pp. 110-122.
15. T.A. Henzinger, X. Nicollin, J. Sifakis and S. Yovine. Symbolic Model Checking for Real-Time Systems. *Information and Computation* 111, 1994, pp. 193-244.
16. R. Huuck, Y. Lakhnech, L. Urbina, S. Engell, S. Kowalewski and J. Preußig. Comparing Timed C/E Systems with Timed Automata. Proc. *Int. Workshop on Hybrid and Real-Time Systems* (*HART'97*), Grenoble, France, *Lecture Notes in Computer Science* 1201, Springer, 1997, pp. 81-86.
17. R. Huuck, Y. Lakhnech, L. Urbina, S. Engell, S. Kowalewski and J. Preußig. Integrating Timed Condition/Event Systems and Timed Automata for the Verification of Hybrid Systems. Will appear in *Journal of Parallel and Distributed Computing Practices*, 1999.
18. IEC. Int. Standard 1131: *Programmable Logic Controllers*, Part 3: *Languages*. International Electrotechnical Commission, 1992.
19. S. Kowalewski and H. Treseler. *CELESTE* - eine Sprache zur modellbasierten Verifikation von diskreten Steuerungen. In: E. Schnieder, D. Abel (Hrsg.), *5. Fachtagung Entwurf komplexer Automatisierungssysteme* (EKA'97), Braunschweig, Germany, May 1997, Vol. 2, pp. 568-587. (In German)
20. S. Kowalewski and H. Treseler. *VERDICT* - A Tool for Model-Based Verification of Real-Time Logic Process Controllers. Proc. *5th Int. Workshop on Parallel and Distributed Real-Time Systems* (*WPDRTS'97*), Geneva, Switzerland, 1997. IEEE Comp. Soc. Press.
21. S. Kowalewski and J. Preußig. Timed Condition/Event systems: a framework for modular discrete models of chemical plants and verification of their real-time discrete control. Proc. *2nd Int. Workshop Tools and Algorithms for the Construction of Systems* (*TACAS*), Passau, Germany, *Lecture Notes in Computer Science* 1066, Springer, 1996, pp. 225-240.
22. S. Kowalewski, S. Engell, J. Preußig, and O. Stursberg. Verification of Logic Controllers for Continuous Plants Using Timed Condition/Event-System Models. Will appear in *Automatica - Special Issue on Hybrid Systems*, March 1999.
23. S. Kowalewski, Y. Lakhnech, B. Lukoschus, and L. Urbina. On the composition of condition/event systems. Proc. *4th Workshop on Discrete Event Systems* (WODES'98), Cagliari, Italy, IEE Press, London, 1998, pp. 349-354.
24. B.H. Krogh. Condition/Event Signal Interfaces for Block Diagram Modeling and Analysis of Hybrid Systems. Proc. *8th International Symposium on Intelligent Control Systems*, Chicago, USA, 1993, pp. 180-185.

25. K.G. Larsen, P. Pettersson, and W. Yi. UPPAAL in a nutshell. *Software Tools for Technology Transfer*, 1(1,2), Springer, 1997, pp. 134-152.
26. I. Moon, G.J. Powers, J.R. Burch and E.M. Clarke. Automatic Verification of Sequential Control Systems Using Temporal Logic. *AIChE Journal*, 38(1), 1992, pp. 67-75.
27. T. Park and P.I. Barton. Implicit Model Checking of Logic Based Control Systems, AIChE Journal, 43(9), 1997, pp. 2246-2260.
28. J. Preußig, O. Stursberg, and S. Kowalewski. Reachability Analysis of a Class of Switched Continuous Systems by Integrating Rectangular Approximation and Rectangular Analysis. *2nd Int. Workshop on Hybrid Systems - Computation and Control*, March 1999, Nijmegen, Netherlands. Will appear in *Lecture Notes in Computer Science* 1569, Springer, 1999.
29. J. Preußig, S. Kowalewski, T.A. Henzinger, and H. Wong-Toi. An algorithm for the approximate analysis of simple rectangular automata. Proc. *5th Int. School and Symposium on Formal Techniques in Fault Tolerant and Real Time Systems (FTRTFT'98)*, Lyngby, Denmark, *Lecture Notes in Computer Science* 1486, Springer, 1998, pp. 228-240.
30. S.T. Probst, G.J. Powers, D.E. Long, and I. Moon. Verification of a logically controlled solids transport system using symbolic model checking. *Computers and Chemical Engineering* 21(4), 1997, pp. 417-429.
31. R.S. Sreenivas and B. H. Krogh. Petri net based models for Condition/Event Systems. Proc. *American Control Conference*, Boston, USA, 1991.
32. R.S. Sreenivas and B.H. Krogh. On condition/event systems with discrete state realizations. *Discrete Event Dynamic Systems* 1(2), 1991, pp. 209-236.
33. T. Stauner, O. Müller, and M. Fuchs. Using *HyTech* to verify an automotive control system. Proc. *Int. Workshop on Hybrid and Real-Time Systems (HART'97)*, Grenoble, France, *Lecture Notes in Computer Science* 1201, Springer, 1997, pp. 139-155.
34. O. Stursberg, S. Kowalewski and S. Engell. Generating Timed Discrete Models of Continuous Systems. Proc. *2nd IMACS Symposium on Mathematical Modelling of Systems (MATHMOD)*, Vienna, Austria, 1997, pp. 203-210.
35. O. Stursberg, S. Kowalewski, I. Hoffmann and J. Preußig. Comparing Timed and Hybrid Automata as Approximations of Continuous Systems. *Hybrid Systems IV, Lecture Notes in Computer Science* 1273, Springer, 1997, pp. 361-377.
36. S. Yovine. *KRONOS*: A Verification Tool for Real-Time Systems. *Software Tools for Technology Transfer*, 1(1,2), Springer, 1997, pp. 123-133.

12

Nonlinear Predictive Control and Moving Horizon Estimation – An Introductory Overview

F. Allgöwer[1], *T. A. Badgwell*[2], *J. S. Qin*[3], *J. B. Rawlings*[4], *and S. J. Wright*[5]

[1] ETH Zürich, Institut für Automatik, 8092 Zürich, Switzerland, e-mail: allgower@aut.ee.ethz.ch
[2] Rice University, Chemical Engineering Department, Houston, TX 77005, USA, e-mail: tab@rice.edu
[3] The University of Texas at Austin, Department of Chemical Engineering, Austin, TX 78712-1062, USA, e-mail: qin@che.utexas.edu
[4] University of Wisconsin at Madison, Department of Chemical Engineering, Madison, WI 53706, USA, e-mail: jbraw@che.wisc.edu
[5] Argonne National Laboratory, Mathematics and Computer Science Division, Argonne, IL 60439, USA, e-mail: wright@mcs.anl.gov

Summary. In the past decade model predictive control (MPC) has become a preferred control strategy for a large number of processes. The main reasons for this preference include the ability to handle constraints in an optimal way and the flexible formulation in the time domain. Linear MPC schemes, i.e. MPC schemes for which the prediction is based on a linear description of the plant, are by now routinely used in a number of industrial sectors and the underlying control theoretic problems, like stability, are well studied. Nonlinear model predictive control (NMPC), i.e. MPC based on a nonlinear plant description, has only emerged in the past decade and the number of reported industrial applications is still fairly low. Because of its additional ability to take process nonlinearities into account, expectations on this control methodology are high.

In this article we give an introduction and an overview of the field of MPC with a special emphasis on nonlinear model predictive control. Also a fresh look on the fairly new area of moving horizon estimation, which is the dual of MPC, is given. For both areas the problem formulation, system theoretical properties and computational aspects are discussed together with condensed accounts of their respective histories and issues pertaining to their industrial application.

1 Introduction

The essence of model predictive control (MPC) is to optimize, over the manipulatable inputs, forecasts of process behavior. The forecasting is accomplished with a process model over a finite time interval, usually called *prediction horizon*. The first input of the optimal input sequence is injected into the plant and the problem is solved again at the next time interval using updated process measurements and a shifted horizon. This methodology is what is now referred to as model predictive control. Often the synonymous terms *receding horizon control*, or *moving horizon control* are used in reference to the way in which the time window considered in the calculation shifts forward from one time step to the next.

The model is the essential element of an MPC controller. Historically, the models of choice in early industrial MPC applications were time domain, input/output, step or impulse response models [95, 26, 81]. Part of the early appeal of MPC for practitioners in the process industries was undoubtedly the ease of understanding provided by this model form. It has become more common for MPC researchers, however, to discuss models in state-space form

$$x_{k+1} = f(x_k, u_k) \tag{1}$$

$$y_k = h(x_k) \tag{2}$$

$$u_k \in \mathbb{U} \tag{3}$$

$$x_k \in \mathbb{X} \tag{4}$$

in which x is the n-vector of states, y is the p-vector of (measurable) outputs and u is the m-vector of (manipulatable) inputs and k is the discrete time sample number. Most of our subsequent discussion focuses on discrete time even though the continuous time case is equally well developed. The discussion of MPC in state-space form has several advantages including easy generalization to multi-variable systems, ease of analysis of closed-loop properties, and on-line computation. Furthermore, the wealth of linear systems theory: the linear quadratic regulator theory, Kalman filtering theory, internal model principle, etc., is immediately accessible for use in MPC starting with this model form.

In general the objective function minimized at each time instance can be

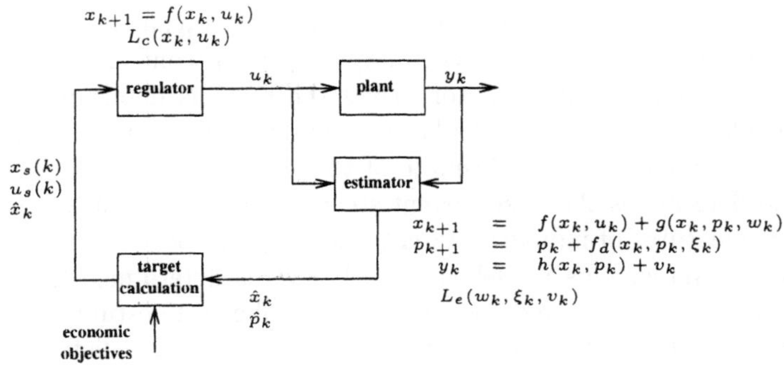

Figure 1: MPC controller consisting of: receding horizon regulator, state estimator, and target calculator.

chosen fairly arbitrary. However, in most cases cost functions of the form

$$J(x_k, \mathbf{u}_k^N) = \Phi(x_N) + \sum_{i=0}^{N} L(x_i, u_i) , \tag{5}$$

with the so-called stage cost L, terminal penalty term Φ, the input vector $\mathbf{u}_k^N$ and prediction horizon N are used.

From a theoretical perspective, the significant shift in problem formulation came from the MPC practitioners who insisted on maintaining constraints, particularly input constraints in the problem formulation. Optimization over inputs subject to hard constraints leads immediately to nonlinear control, and that departure from the well understood and well tested linear control theory provided practitioners with an important, new control technology and motivated researchers to understand better this new framework. Certainly optimal control with constraints was not a new concept in the 1970s, but the implementation in a moving horizon fashion of these open-loop optimal control solutions subject to constraints at each sample time was the new twist that had not been fully investigated.

The use of nonlinear models in MPC is motivated by the possibility to improve control by improving the quality of the forecasting. In which settings use of nonlinear models for forecasting delivers improved control performance is an open issue, however. For continuous processes maintained at nominal operating conditions and subject to small disturbances, the potential improvement would appear small. For processes operated over large regions of the state space — semi-batch reactors, frequent product grade changes, processes subject to large disturbances, for example — the advantages of nonlinear models appear larger.

Consider Figure 1, which shows some of the issues that must be addressed in a practical implementation of MPC. The regulator is discussed in Section 3. Its function is to optimize over a sequence of future inputs a multi-stage objective function with stage cost L measuring regulation error and size of inputs.

This piece is often called the model predictive controller, but as shown in Figure 1, it is only one piece of the complete control system. The steady state to which the regulator drives the state is determined by the target calculator. The target calculator evaluates the process and economic objectives, the current best estimate of the state plus any modeled disturbances and determines the best current steady operating point, x^s_k, u^s_k. We discuss some features of industrial target calculators in Section 6. The current state of the system plus modelled disturbances is determined from the sequence of inputs and observed measurements by the state estimator. Industrial implementations often include integrator models for the disturbances to obtain offset free control in the presence of nonzero disturbances. In Section 4 we discuss the theory of using online optimization to construct the state estimator. This dual to the regulation problem optimizes over the initial state and disturbance sequence a multi-stage cost function with stage cost L_e measuring data fitting error and size of disturbances. This formulation is known as moving horizon estimation.

We do not mean to imply that Figure 1 depicts the only way or even an industrially implemented way to organize the functionality of predictive controllers. We merely find it a useful construct for the purposes of organizing this paper. For example, we are interested in the issue of moving horizon estimation, independent of whether a separation principle holds or one plans to use separation to address the output feedback problem. Full dynamic tracking problems rather than repeated optimization of steady-state operating points are also important in applications, and we do not treat that problem here.

Because the implementations we discuss are online optimizations, computational efficiency is vital to success in applications, and an overview of computational issues is presented in Section 5. Much of the motivation for MPC research stems from its importance in applications, and Section 6 provides a review of applications from the process industries. We present briefly the future outlook and conclusions in Sections 7 and 8.

2 Historical Development of MPC and MHE

The economic operating point of a typical process unit often lies at the intersection of constraints [81]. A successful industrial controller must therefore maintain the system as close as possible to constraints without violating them. An adequate theory for the control of *constrained* dynamic systems, linear or nonlinear, still has to be constructed although encouraging advances are now being made. Model Predictive Control (MPC) is one of the few methods available for handling hard constraints, and this fact has made it an important tool for the control engineer, particularly in the process industries where plants being controlled are sufficiently 'slow' to permit its implementation.

The MPC research literature is by now large and diverse, but review articles have appeared at regular intervals. [37] provide an excellent review of the literature up to the late 1980s in which researchers studied the effect of control and cost horizons and cost parameters on stability when the system is linear, the cost quadratic, and hard constraints are absent. For further reading, the three MPC papers presented at the CPC V conference in 1996 are an excellent starting point [59, 66, 83]. Morari and Lee provide another recent review [74]. Kwon provides a very extensive list of references [54]. Moreover, several excellent books have appeared recently [75, 103, 10].

The origins of predictive control can be traced back to Lee and Markus who anticipated in 1967 [58] this methodology. Propoi in 1963 [82] was the first to explicitly propose a form of MPC using a linear programming formulation. Thomas [107] appeared to be the first to use the term *receding horizon* estimation and control in an LQ formulation.

The focus of this overview is nonlinear model predictive control with constraints, and we are concerned primarily with state regulation and output feedback via state estimation and separation. For that reason we cannot cover other important versions of model predictive control. Generalized Predictive Control (GPC)[20, 21] deserves special mention because it has found widespread success in applications. GPC is built around a linear input/output model, and researchers have focused on many issues not addressed here, such as tuning disturbance models to meet performance and robustness specifications, and trajectory tracking [19, 12].

The early major contribution to receding horizon (model predictive) control for nonlinear systems was the demonstration by Keerthi and Gilbert [50] that, for time-varying, *constrained, nonlinear*, discrete-time systems, the addition of a terminal *stability constraint* $x_{k+N} = 0$ to the open-loop optimal control problem ensures that, under mild conditions, the resultant receding horizon controller is stabilizing. This result is a significant generalization of the earlier linear results. The discussion of this literature continues in Section 3.

Next we discuss some of the history of the industrial developments.

2.1 Industrial MPC Development

2.1.1 IDCOM

The first description of MPC control applications was presented by Richalet et al. in 1976 [95, 96]. They described their approach as Model Predictive Heuristic Control (MPHC). The commercial software was referred to as IDCOM, an acronym for Identification and Command. The distinguishing features of IDCOM are (i) finite impulse response (FIR) model for the plant; (ii) quadratic performance objective over a finite prediction horizon; (iii) future output behavior specified by a reference trajectory; (iv) input and output constraints included in the formulation; and (v) optimal inputs computed using a heuristic iterative algorithm.

The MPHC algorithm drives the predicted future output as closely as possible to a reference trajectory, defined as a first order path from the current output value to the desired setpoint. The speed of the desired closed loop response is set by the time constant of the reference trajectory.

Richalet et al. make the important point that dynamic control must be embedded in a hierarchy of plant control functions in order to be effective. They describe four levels of control.

- Level 3 - Time and space scheduling of production
- Level 2 - Optimization of setpoints to minimize costs and ensure quality and quantity of production
- Level 1 - Dynamic multivariable control of the plant
- Level 0 - Control of ancillary systems; PID control of valves.

They point out that significant benefits do not come from simply reducing the variations of a controlled variable through better dynamic control at level 1. The real economic benefits come at level 2 where better dynamic control allows the controlled variable setpoint to be moved closer to a constraint without violating it. This argument provides the basic economic motivation for using MPC technology. This concept of a hierarchy of control functions is fundamental to advanced control applications and seems to have been followed by many practitioners.

2.1.2 DMC

Engineers at Shell Oil developed MPC technology independently in the early 1970's, with an initial application in 1973. Cutler and Ramaker presented details of an unconstrained multivariable control algorithm which they named Dynamic Matrix Control (DMC) at the 1979 National AIChE Meeting [25] and at the 1980 Joint Automatic Control Conference [26]. In a companion paper Prett and Gillette [81] described an application of DMC technology to an FCCU reactor/regenerator in which the algorithm was modified to handle nonlinearities and constraints. Key features of the DMC control algorithm include: (i) linear step response model for the plant; (ii) quadratic performance objective with move suppression over a finite prediction horizon; (iii) future output behavior specified by following the setpoint; and (iv) optimal inputs computed as the solution to a least-squares problem.

By using the step response model one can write predicted future output changes as a linear combination of future input moves. The matrix that ties the two together is the so-called *Dynamic Matrix*. Using this representation allows the optimal move vector to be computed as the solution to a least-squares problem. Feedforward control is readily included in this formulation by modifying predicted future outputs.

The objective of a DMC controller is to drive the output as close to the setpoint as possible in a least-squares sense with a penalty term on the input moves. This results in smaller computed input moves and a less aggressive output response. As with the IDCOM reference trajectory, this technique provides a degree of robustness to model error. Move suppression factors also provide an important numerical benefit in that they can be used to directly improve the conditioning of the numerical solution.

The initial IDCOM and DMC algorithms represent the *first generation* of MPC technology; they had an enormous impact on industrial process control and served to define the industrial MPC paradigm.

2.1.3 QDMC

The original IDCOM and DMC algorithms provided excellent control of unconstrained multivariable processes. Constraint handling, however, was still somewhat ad-hoc. Engineers at Shell Oil addressed this weakness by posing the DMC algorithm as a Quadratic Program (QP) in which input and output constraints appear explicitly [24, 36]. A distinguishing feature of QDMC over DMC is that the control moves are solved as a QP with hard constraints rather than least squares without constraints.

The default QDMC algorithm requires strict enforcement of input and output constraints at each point of the prediction horizon. Constraints that are strictly enforced are referred to as *hard constraints.* Constraints for which violations are allowed are referred to as *soft constraints.* In practice García and Morshedi [36] report that hard output constraints are typically required to be satisfied over only a portion of the horizon which they refer to as the *constraint window.* The constraint window generally starts at some point in the future and continues on to steady state. They report that if nonminimum phase dynamics are present, performance is improved by pushing the constraint window farther into the future. This amounts to ignoring hard output constraints during the initial portion of the closed loop response.

The QDMC algorithm can be regarded as representing a *second generation* of MPC technology, comprised of algorithms which provide a systematic way to implement input and output constraints. This was accomplished by posing the MPC problem as a QP.

2.1.4 IDCOM-M, SMOC, and PCT

As MPC technology gained wider acceptance, and problems tackled by MPC technology grew larger and more complex, control engineers implementing second generation MPC technology ran into other practical problems. The QDMC algorithm provided a systematic approach to incorporate hard input and output constraints, but there were several obvious limitations: (i) There was no clear way to handle an infeasible solution. For example it is possible for a feedforward disturbance to lead to an infeasible QP; what should the control do to recover from infeasibility? (ii) The soft constraint formulation is not completely satisfactory because it means that all constraints will be violated to some extent, as determined by the relative weights. Clearly some output constraints are more important than others, however, and should never be violated. (iii) Process models are still limited to step response models which are incapable of representing unstable processes. (iv) The constant-disturbance output feedback is sub-optimal and unrealistic especially when low-level PID controllers are in place.

Fault tolerance is also an important practical issue. Rather than simply turning itself off as signals are lost, a practical MPC controller should remain online and try to make the best of the sub-plant under its control. It also became increasingly difficult to translate control requirements into relative weights for a single objective function. Including all the required trade-offs in a single objective function means that relative weights have to be assigned to the value of output setpoint violations, output soft constraint violations, inputs moves, and optimal input target violations.

These issues motivated engineers at Shell (France), Adersa and Setpoint, Inc. to develop new versions of MPC algorithms. The version marketed by Setpoint was called IDCOM-M, while the Adersa version was referred to as HIECON (Hierarchical Constraint Control). The IDCOM-M controller was first described in a paper by Grosdidier et al. [43]. Distinguishing features of the IDCOM-M algorithm include: (i) controllability supervisor to screen out ill-conditioned plant subsets; (ii) multi-objective function formulation; quadratic output objective followed by a quadratic input objective; (iii) controls a single future point in time for each output, called the coincidence point, chosen from a reference trajectory; (iv) constraints can be hard or soft, with hard constraints ranked in order of priority.

The SMOC algorithm [65] developed at Shell France is very close to the "modern" development of MPC with the following distinguishing features: (i) State space models are used which can represent both stable and unstable processes. (ii) Full state estimation known as *extended close-loop observer* is used for output feedback; constant output disturbance is simply a special case. (iii) A distinction is introduced between *controlled variables* that are in the control objective and *feedback variables* that are used for state estimation, resolving the difficulty for interdependent output variables. (iv) Input and output constraints are observed via a QP formulation.

The SMOC and IDCOM-M algorithms are two of several that represent a *third generation* of MPC technology; others include the PCT algorithm sold by Profimatics, the RMPCT controller developed by Honeywell, and the PFC algorithm developed by Adersa. This generation distinguishes between several levels of constraints (hard, soft, ranked), provides some mechanism to recover from an infeasible solution, addresses the issues resulting from a control structure that changes in real time, uses state estimation as optimal output feedback, and allows for a wider range of process dynamics (stable and unstable) and controller specifications.

While MPC theory has advanced to the point where nominal stability can be guaranteed with several schemes, the majority of the current industrial MPC algorithms is based on ideas inherited from the original DMC and IDCOM algorithms. Products sold by Adersa, Aspen Technology, Continental Controls, Honeywell, and Pavilion Technology share such features as a finite prediction horizon and options for impulse or step response models. The main emphasis in recent years has been to allow for a wider range of model types, including nonlinear models, state space models (e.g., SMOC from Shell) [65], and better integration of the controller interface into existing distributed control hardware [83, 85].

2.2 Historical Notes of Moving Horizon Estimation

While MPC control technology had its origins in industry, moving horizon estimation (MHE) was developed primarily by researchers in academia. An early application of MHE for nonlinear process systems was the work of Jang and coworkers [46]. Their strategy ignores disturbances and attempts only to estimate the initial state of the system. Thomas [107] and Kwon and coworkers [55] discussed earlier moving horizon strategies for linear systems. Limited memory and adaptive filters for linear systems are analogous to MHE, because only a fixed window of data is considered (see [47] for a discussion of limited memory filters). Many researchers in the process systems extended the work of Jang and coworkers. Bequette and coworkers [6, 87] investigated moving horizon strategies for state estimation as a logical extension of model predictive control. Edgar and coworkers [51, 61] investigated moving horizon strategies for nonlinear data reconciliation. Because MHE is formulated as an optimization problem, it is possible to handle explicitly inequality constraints. Robertson and Lee [97, 98] have investigated the probabilistic interpretation of constraints in estimation. Muske and Rawlings [77, 76] derived some preliminary conditions for the stability of state estimation with inequality constraints. Tyler and Morari [108] demonstrated how constraints may result in instability for non-minimum phase systems.

In parallel to the research done in process systems, MHE was also investigated by researchers in other areas of automatic control. Zimmer [116] investigated an MHE strategy similar to the approach of Jang and coworkers [46] and derived conditions for stability using fixed point theorems. Moraal and Grizzle [73] also derived conditions for stability using fixed point theorems, but they formulated the estimation problem as the solution of a set of algebraic equations. Michalska and Mayne [72] investigated an MHE strategy similar to the approach of Jang and coworkers and derived conditions for stability using Lyapunov arguments.

3 Nonlinear Model Predictive Control Theory

In the previous sections we discussed the general setup of model predictive control and gave a brief account on its history and development. In this section we will focus our attention on the system theoretic properties of NMPC like stability, robustness etc. It will turn out that standard finite horizon NMPC with arbitrary horizons and arbitrary cost function does, in general, not lead to a stable closed loop, not even in the nominal undisturbed case. We will elaborate on this in the next section, followed by an overview of NMPC approaches that guarantee stability of the closed loop.

3.1 Finite Horizon NMPC

We consider the class of nonlinear systems described by the following set of difference equations

$$x_{k+1} = f(x_k, u_k) \tag{6}$$

subject to the input and state constraints

$$u_k \in \mathbb{U}\,, \qquad x_k \in \mathbb{X} \tag{7}$$

where $x_k \in \mathbb{R}^n$ represents the state and $u_k \in \mathbb{R}^m$ the input. In its simplest form, $\mathbb{U}$ and $\mathbb{X}$ are given by

$$\mathbb{U} := \{u_k \in \mathbb{R}^m | u_{min} \leq u_k \leq u_{max}\} \tag{8}$$

$$\mathbb{X} := \{x_k \in \mathbb{R}^n | x_{min} \leq x_k \leq x_{max}\}\,. \tag{9}$$

For the moment we restrict our attention to the simplest case where we only consider the regulation problem (i.e. bringing the system state from an initial condition x_0 back to a fixed operating point). The NMPC open loop optimization problem that we have to solve at each time instance is given by:

Problem 3.1 (NMPC Control Problem)
Solve

$$\min_{\mathbf{u}_k^N} J(x_k, \mathbf{u}_k^N) \tag{10}$$

subject to

$$x_{l+1|k} = f(x_{l|k}, u_{l|k}) \qquad x_{0|k} = x_k \tag{11}$$

$$u_{l|k} \in \mathbb{U}\,, \quad l \in [0, N-1] \tag{12}$$

$$x_{l|k} \in \mathbb{X}\,, \quad l \in [0, N]\,, \tag{13}$$

with

$$J(x_k, \mathbf{u}_k^N) = \Phi(x_{N|k}) + \sum_{i=0}^{N-1} L(x_{i|k}, u_{i|k})\,. \tag{14}$$

Internal controller variables predicted at time instance k are denoted by a double index separated by a vertical line where the second argument denotes the time instance at which the prediction is made. The first variable is the prediction time in the controller, starting at $l = 0$ for each real time instance k. The decision variable is $\mathbf{u}_k^N = \{u_{0|k}, u_{1|k}, \ldots, u_{N-1|k}\}$, where the upper index N specifies the length of the control horizon. The respective lengths of the prediction and control horizons are assumed to be equal. At each time instance k we apply the first element of the solution sequence $\mathbf{u}_k^{N\star}$ [1] of the optimal open loop control Problem 3.1, i.e.

$$u_k = u_{0|k}^{\star} \tag{15}$$

and refer to the so controlled plant as the *closed loop under NMPC control.*

In the next sections we frequently refer to the following assumptions:

(A_c1) $f : \mathbb{R}^n \times \mathbb{R}^m \to \mathbb{R}^n$ is twice continuously differentiable on $\mathbb{X} \times \mathbb{U}$.

(A_c2) W.l.o.g. we assume $0 = f(0,0)$. Furthermore this steady state is considered as the desired operating point to which the regulator should drive the state.

(A_c3) The set $\mathbb{U} \subset \mathbb{R}^m$ is compact and convex, $\mathbb{X} \subseteq \mathbb{R}^n$ is closed and simply connected with $(0,0)$ in the interior of $\mathbb{X} \times \mathbb{U}$.

(A_c4) At each time instance k the full state x_k is available for measurement (state-feedback).

(A_c5) No disturbances (persistent or not) act on the system.

(A_c6) There is no mismatch between the model used to predict the dynamic behavior and the real plant.

(A_c7) The stage cost L in (14) is a quadratic function in x and u

$$L(x_{i|k}, u_{i|k}) = x_{i|k}^T Q x_{i|k} + u_{i|k}^T R u_{i|k} \,, \tag{16}$$

with Q, R positive definite.

Some of these assumptions are only needed for some specific NMPC schemes (see the sections below). Others, like for example (A_c7), can be relaxed and are only included to simplify the exposition of some results. In Sections 3.6 and 7 we will make some comments about the more realistic and practically meaningful case were the above assumptions are not satisfied.

When an *infinite* prediction and control horizon is used, i.e. N is set to ∞, and Assumptions (A_c4)-(A_c6) hold, it follows directly from Bellmann's Principle of Optimality [4] that the open loop (predicted) input and state trajectories computed at a specific instance in time as the solution of the

[1] Optimal values are denoted by a star.

NMPC optimization Problem 3.1, are in fact equal to the closed loop trajectories of the nonlinear system (actual trajectories). This fact implies closed loop stability, as any feasible predicted trajectory goes to the origin.

The most intuitive way to achieve stability in NMPC is thus the use of an infinite horizon. This method is generally referred to as *infinite horizon NMPC* and the following result holds:

Theorem 3.2 (Infinite Horizon NMPC) *[10, 50, 70]*
Suppose a) Assumptions (A_c1)-(A_c6) *are satisfied, b) the open loop optimal control Problem 3.1 with* $N = \infty$, $\Phi = 0$ *is feasible at* $k = 0$. *Then the origin of the closed loop system under NMPC control is asymptotically stable and the region of attraction contains the set of states for which b) holds.*

In linear MPC, infinite horizons (approximated by very large horizons) are in many cases a practical route to achieving stability as there exist very efficient ways to solve the corresponding huge quadratic programming problems (see Section 5). For nonlinear problems, on the other hand, the solution of such large optimization problems is extremely difficult if not impossible to obtain. Therefore finite horizons are indispensable in NMPC and infinite horizon NMPC only plays a role as a conceptual theoretical method.

When a *finite* prediction and control horizon is used, the actual closed loop input and state trajectories will in general differ from the predicted open loop trajectories even if no model plant mismatch and no disturbances are present. This obvious fact has two immediate consequences. Firstly, our actual goal is to compute a feedback such that a performance objective determined by the cost functional as in (14) *over the infinite horizon* is minimized. It is by no means clear that repeatedly minimizing a *finite horizon objective* (with the same stage cost L) in a receding horizon manner leads to an optimal solution for the infinite horizon problem. In fact, the two solutions will in general differ significantly, if a short horizon is chosen. Hence, when using finite horizons in standard NMPC, the stage cost cannot be chosen straightforwardly based on the desired physical objectives. Secondly, if the predicted and the actual trajectories differ, there is no guarantee that the closed loop system will be stable. It is indeed easy to construct examples for which the closed loop becomes unstable if a (small) finite horizon is chosen. It is therefore essential to study the stability properties of NMPC and to develop methods that assure that the closed loop is stable.

In principle, it would be desirable to have a "controller design procedure" that would allow to determine stabilizing prediction and control horizons for an NMPC setup based on the plant model and the chosen stage cost. This problem is, however, very difficult and, to the best of the authors knowledge, has not yet been solved. There are even no analysis methods available that permit to analyze closed loop stability based on knowledge of the plant model, the objective functional and the horizon lengths!

Nevertheless, there are possibilities to achieve closed loop stability despite the fact that this property cannot be analyzed. The idea behind these approaches is to modify the NMPC setup such that stability of the closed loop can be guaranteed independent of the choice of the horizon length, independent of the choice of the stage cost and for any plant to be controlled. This is usually achieved by adding suitable equality or inequality constraints (and possibly suitable additional terms to the cost functional) to the setup. These additional constraints are not motivated by physical restrictions or desired performance requirements but have the sole purpose of enforcing stability of the closed loop. Therefore, they are usually termed *stability constraints* [66, 67]. NMPC formulations with a modified setup to achieve closed loop stability independent of the choice of the performance parameters in the cost functional are usually labeled NMPC approaches *with guaranteed stability.* A well-known control method with guaranteed stability is the standard linear quadratic regulator (LQR) that also achieves closed loop stability independent of the choice of the positive definite weighting matrices in the quadratic cost functional [48].

In the following sections we will discuss several NMPC formulation with guaranteed stability, also commenting on their respective merits and drawbacks.

3.2 NMPC with Zero State Terminal Equality Constraint

The most widely suggested NMPC scheme with guaranteed stability utilizes a stability constraint in the form of a *zero state terminal equality constraint* [50, 52, 68]

$$x_{N|k} = 0\,, \tag{17}$$

that forces the state to be zero at the end of the finite horizon. In the context of LQ control, such a terminal equality constraint is quite intuitive, since it corresponds to an infinite initial condition of the backwards Riccati difference equation, whose solution gives rise to a stabilizing state feedback controller for linear controllable systems [52]. In fact, if the terminal equality constraint is enforced, i.e. the predicted system state $x_{N|k}$ at the end of the finite horizon is at the origin and the input thereafter is set to be zero, then the unperturbed nominal system stays at the origin forever. This way, the prediction horizon can be thought of as expanding to infinity.

Keerthi and Gilbert [50] were the first to show that feasibility of an NMPC formulation with zero state terminal equality constraint implies stabilization of a class of nonlinear constrained systems. Later, this result was further expanded in [93, 69]. Continuous time versions can be found in [68]. Under reasonable conditions, asymptotic stability for the closed-loop system can be proven.

Theorem 3.3 (NMPC w/Zero State Terminal Equality Constraint) *[69, 50, 68]*
Suppose a) Assumptions (A_c1)-(A_c6) are satisfied, b) the open loop optimal control Problem 3.1 together with the additional stability constraint (17) is feasible at $k = 0$. Then the origin of the closed loop system under NMPC

control is asymptotically stable and the region of attraction contains the set of states for which b) holds.

Guaranteeing stability by imposing a zero state terminal equality constraint is by far the most popular technique at present. For one, this is certainly due to the clear theoretical framework, but also due to the fact that no off-line computation or tuning is needed. On the other hand, a terminal equality constraint is an artificial additional burden that may require significant extra on-line computation cost (see [16, 14] for a comparison to other approaches) and even more importantly, leads in many cases to a severely restricted region of operation due to feasibility problems. In addition, if the value function is not continuous at the origin, it can be shown [93] that if the horizon length in NMPC with zero state terminal equality constraint is increased to infinity, the resulting trajectories do not necessarily approach the same ones as for infinite horizon NMPC. Thus, as pointed out in the previous section, the desired closed loop performance can not easily be adjusted by a straightforward choice of the stage cost in the optimization problem. For the reasons stated, this approach can only be recommended with some reservations for practical applications.

3.3 Dual-mode NMPC

From a computational point of view, an exact numerical satisfaction of a zero state terminal equality constraint is impossible. If only an approximated satisfaction of the terminal equality constrained is enforced, i.e. we only require the terminal state to lie in an "small" region around the origin, then the guaranteed stability is in general lost. In order to relax the equality stability constraint while not compromising the closed loop asymptotic stability of the origin, the so-called *dual-mode NMPC scheme* was introduced in [71]. The term dual-mode refers to two different controllers that are applied in different regions of the state space depending on the state being inside or outside of some *terminal region* that contains the origin. If the state is outside this terminal region, an NMPC controller with a variable horizon is applied. If the current state lies inside the terminal region, a linear state feedback $u_k = Kx_k$ is applied. Closed-loop control with this scheme is implemented by switching between the two controllers, i.e.

$$u_k = \begin{cases} u^{\star}_{0|k}, & \text{if} \quad x_k \notin \Omega \\ Kx_k, & \text{if} \quad x_k \in \Omega\,. \end{cases} \tag{18}$$

The terminal region Ω and the state feedback are determined off-line such that the terminal region is a positive invariant region of attraction for the nonlinear system controlled by the linear state feedback law and that the input and state constraints are both satisfied with this linear controller in Ω.

In the NMPC optimization problem that is solved if the state is outside Ω, the length N_k of the horizons is taken as an additional minimizer, i.e. (10)

is replaced by

$$\min_{\mathbf{u}_k^N, N_k} J\left(x_k, \mathbf{u}_k^N, N_k\right), \tag{19}$$

and an additional terminal inequality stability constraint

$$x_{N_k|k} \in \Omega_B \tag{20}$$

is introduced such that the state at the end of the horizon has to lie on the boundary Ω_B of the terminal region Ω.

Under the assumption that

(A_c8) the Jacobian linearization $x_{k+1} = Ax_k + Bu_k$ of (6) with $A := \frac{\partial f}{\partial x}(0,0)$ and $B := \frac{\partial f}{\partial u}(0,0)$ is stabilizable

a simple procedure to determine a suitable terminal region Ω and a linear feedback matrix K that satisfy the above mentioned requirements in the continuous time case is proposed in [68]. In discrete time a similar scheme is summarized in the following procedure [32]:

Procedure 3.4

Step 1 : *Find a locally stabilizing linear state feedback $u_k = Kx_k$, based on the Jacobian linearization of (6). Choose a constant $\kappa \in (0,1)$.*

Step 2 : *Solve the Lyapunov equation $A_K^T \tilde{P} A_K - \tilde{P} = -Q_K$ with $Q_K = Q + K^T R K$ for the positive definite and symmetric matrix $\tilde{P}$.*

Step 3 : *Find the largest possible α_1 such that $\Omega_1 \subseteq \mathbb{X}$ and $Kx_k \in \mathbb{U}, \forall x_k \in \Omega_1$ with $\Omega_1 := \{x_k \in \mathbb{R}^n | x_k^T \tilde{P} x_k \leq \alpha_1\}$.*

Step 4 : *Determine $\Omega = \{x_k \in \mathbb{R}^n | x_k^T \tilde{P} x_k \leq \alpha\}$ via iterations of*

$$\max_{x_k}\{\phi(x_k)^T \tilde{P} \phi(x_k) + 2\phi(x_k)^T \tilde{P} x_k + (1-\kappa) x_k^T Q_K x_k | x_k^T \tilde{P} x_k \leq \alpha\} \tag{21}$$

by reducing α from α_1, until the optimal value of (21) is non-positive. $\phi(x_k) = f(x_k, Kx_k) - A_K x_k$ is the difference between the linearized and the nonlinear system both controlled with feedback $u_k = Kx_k$.

A similar procedure is also used in the quasi-infinite horizon NMPC approach as discussed in Section 3.5.

Under fairly weak conditions, it can be shown that starting from outside the terminal region, the nonlinear system with the predictive controller will reach the terminal region in a finite time. Closed-loop stability follows from the use of the stabilizing local linear feedback law thereafter. This is made precise in the following theorem:

Theorem 3.5 (Dual-mode NMPC) *[71]*
Suppose a) Assumptions (A_c1)-(A_c8) are satisfied, b) the open loop optimal control Problem 3.1 with (10) replaced by (19) together with the additional stability constraint (20) is feasible at $k = 0$. Then the origin of the nominal closed loop system under feedback (18) is asymptotically stable and the region of attraction contains the set of states for which b) holds.

Discrete time versions of dual-mode NMPC with fixed horizons are described in [56, 100].

Computationally this approach is more attractive than the one imposing a terminal equality constraint, as inequality constraints can be handled more effectively during optimization than equality constraints. In addition, less feasibility problems have to be expected and hence the region of attraction will be larger. It can also be shown [71] that the feasibility of the optimization problem, and not necessarily the optimality, is needed for closed-loop stability. A drawback of this approach is the involved implementation due to the required switching between control strategies, the need to determine a local stabilizing state feedback gain K and the terminal region Ω, and the fact that the predicted open loop and the actual closed loop trajectories will in general be different.

3.4 Contractive NMPC

In contractive NMPC as suggested in [112, 31], a stability constraint of the form

$$\|x_{N_k|k}\|^2 \leq \alpha^2 \|x_k\|^2 \,, \qquad \alpha \in (0,1) \tag{22}$$

is added to Problem 3.1. This constraint directly forces the magnitude of the state vector to contract by a pre-specified factor each time a new input is calculated. This constraint is referred to as *contraction stability constraint.* Different from standard NMPC, the *entire* input function $\mathbf{u}_k^{N_k\star}$ over the interval $[k, k+N_k-1]$ is applied to the nonlinear system. The next optimization problem is only solved at time instance $k+N_k$. In the formulation suggested by [31], an additional second stability constraint of the form

$$\max_{j=0,\ldots,N_k} \|x_{j|k}\|^2 \leq \beta^2 \|x_k\|^2 \,, \qquad \beta \in [1,\infty) \tag{23}$$

is included in the formulation and, like in the dual-mode NMPC case, the horizon length $N_k \leq N_{max}$ is used as an additional minimizer. Closed-loop asymptotic stability is proven by showing that $\|x_{k+j}\|^2 \leq \alpha^{2j}\|x_k\|^2 \to 0$ as $j \in \mathbb{N} \to \infty$ holds.

Theorem 3.6 (Contractive NMPC) *[31]*
Suppose a) Assumptions (A_c1)-(A_c8) are satisfied, b) the open loop optimal control Problem 3.1 with (10) replaced by (19) subject to the additional stability constraints (22), (23) is feasible at $k = \sum_{l=0}^{j} N_l$ for $j = 0,1,2,\cdots$. Then the origin of the nominal closed loop system under NMPC control is asymptotically stable.

Assumption b) in this theorem is much stronger than the corresponding assumptions in Theorems 3.2, 3.3, 3.5, 3.7 as feasibility has to be *assumed* for all time instances for which an optimization is performed. Hence, this approach is not very attractive for applications, since the feasibility of the optimization problem at subsequent sampling instances is in general not guaranteed.

For the other methods suggested, feasibility at $k = 0$ *implies* feasibility at all later times. Choosing large values for $\alpha \in (0,1)$ might obtain the required feasibility, but a large α implies a slow contraction of the state vector. Moreover, despite the free specification of the desired control performance by the objective functional, the achieved control performance will be influenced directly by the choice of design parameters α and β.

As a generalization of this approach, the contraction constraint (22) on the state can be replaced by a contraction constraint on some positive definite function $V(x)$, i.e. $V(x_{k|N_k}) \leq \alpha^2 V(x_k)$, $\alpha \in (0,1)$. The enforced contraction guarantees the monotonic decay of V, thus permitting a Lyapunov based proof of closed loop stability using V as Lyapunov function [66].

3.5 Quasi-infinite Horizon NMPC

In the quasi-infinite horizon NMPC scheme [13, 16] an inequality stability constraint

$$x_{N|k} \in \Omega \tag{24}$$

and a quadratic *terminal penalty term*

$$\Phi(x_{N|k}) = x_{N|k}^T P x_{N|k} \tag{25}$$

are added to the standard setup. The basic idea behind this scheme is that the terminal penalty term is not a performance specification that can be chosen freely, but rather that matrix P in (25) is determined off-line according to a specific procedure such that (14) with Φ chosen as in (25) approximates an infinite horizon prediction cost functional. This way closed-loop stability can be achieved, while only an optimization problem over a finite horizon must be solved numerically.

Consider an infinite horizon cost functional defined by

$$J^{\infty}\left(x_k, \mathbf{u}_k^{\infty}\right) = \sum_{i=0}^{\infty} L\left(x_{i|k}, u_{i|k}\right)$$

with $\mathbf{u}_k^{\infty}$ a control sequence of infinite length. This cost functional can be split up into two parts

$$\min_{\mathbf{u}_k^{\infty}} J^{\infty}\left(x_k, \mathbf{u}_k^{\infty}\right) = \min_{\mathbf{u}_k^{\infty}} \left\{ \sum_{i=0}^{N-1} L\left(x_{i|k}, u_{i|k}\right) + \sum_{i=N}^{\infty} L\left(x_{i|k}, u_{i|k}\right) \right\} . \tag{26}$$

The goal is to approximate the second term on the right hand side of (26) by a terminal penalty term Φ. Without further restrictions, this is usually not possible for general nonlinear systems. However, if we ensure that the trajectories of the closed-loop system remain within some neighborhood of the origin (terminal region) for the time interval $[k+N, \infty)$, then, an upper bound on this second term can be found. The terminal region Ω is constructed such that a local state feedback law

$$u_{i|k} = K x_{i|k} \tag{27}$$

asymptotically stabilizes the nonlinear system in Ω and renders Ω positively invariant for the closed loop. Because of the invariance property of Ω, it suffices to impose an additional terminal (point) inequality constraint $x_{N|k} \in \Omega$ (see (24)) in order to ensure that the trajectories of the nonlinear system with u_k given by (27) indeed remain in Ω beyond time instance $k+N$. We now have

$$\min_{\mathbf{u}_k^\infty} J^\infty\left(x_k, \mathbf{u}_k^\infty\right) \le \min_{\mathbf{u}_k^N} \left\{ \sum_{i=0}^{N-1} L\left(x_{i|k}, u_{i|k}\right) + \sum_{i=N}^{\infty} L\left(x_{i|k}, K x_{i|k}\right) \right\} . \tag{28}$$

If the terminal region Ω and the terminal penalty matrix P are chosen according to Procedure 3.4 and $P = \frac{\tilde{P}}{\kappa}$, then it can be shown [16] that

$$\sum_{i=N}^{\infty} L\left(x_{i|k}, K x_{i|k}\right) \le x_{N|k}^T P x_{N|k} \tag{29}$$

holds. Substituting (29) into (28) leads to

$$\min_{\mathbf{u}_k^\infty} J^\infty\left(x_k, \mathbf{u}_k^\infty\right) \le \min_{\mathbf{u}_k^N} J(x_k, \mathbf{u}_k^N) \, , \tag{30}$$

with J being the finite horizon objective functional as in (14). This implies that the optimal value of the finite horizon optimization problem bounds that of the corresponding infinite horizon optimization problem. In this sense, the prediction horizon in the discussed nonlinear MPC scheme can be thought of as expanding quasi to infinity. Property (30) can be exploited to prove asymptotic stability for the closed-loop system independent of the specification of the desired control performance.

Theorem 3.7 (Quasi-infinite Horizon NMPC) *[16]*
Suppose a) Assumptions (A_c1)-(A_c8) are satisfied, b) the terminal region Ω is chosen according to Procedure 3.4 and the terminal penalty matrix P is chosen as $P = \frac{\tilde{P}}{\kappa}$, c) the open loop optimal control Problem 3.1 with $\Phi(x_{N|k}) = x_{N|k}^T P x_{N|k}$ and with the additional stability constraint (24) is feasible at $k = 0$. Then the origin of the nominal closed loop system under NMPC control is asymptotically stable and the region of attraction contains the set of states for which c) holds.

The idea of using a local state feedback law and a terminal inequality constraint is inspired by the dual-mode nonlinear MPC scheme described in Section 3.3. However, dual-mode nonlinear MPC scheme requires switching between the nonlinear predictive controller outside the terminal region and the linear feedback controller inside the terminal region. In the quasi-infinite horizon nonlinear MPC scheme, closed-loop control is determined by solving an NMPC problem no matter whether the system is inside or outside the terminal region.

Similar versions for discrete time nonlinear systems can be found in [33, 28]. In [28], the terminal state penalty term is non-quadratic and taken as the value of the infinite remaining cost when the local linear feedback (27) is applied:

$$\Phi(x_{N|k}) = \sum_{i=N}^{\infty} L\left(x_{i|k}, Kx_{i|k}\right) . \tag{31}$$

Clearly, this is in general not implementable. Thus, it is suggested to approximate the infinite sum in (31) by a sum of *sufficient length.* This means that the terminal penalty term will be evaluated by "integrating" $x_{i+1|k} = f(x_{i|k}, Kx_{i|k})$ on-line for a sufficiently long time horizon. In addition to the high on-line computational effort, it remains open how long the on-line integrating time horizon has to be in order to guarantee stability?

Like in the dual-mode approach, the use of the terminal inequality constraint gives the quasi-infinite horizon nonlinear MPC scheme computational advantages. The implementation is simpler than the dual-mode approach, because no switching between control strategies is needed. Moreover, the additional terminal penalty term is introduced to approximate an infinite horizon objective functional and thus predicted open loop and actual closed loop trajectories will be at least similar and control performance can be adjusted via the stage cost L. Like for dual-mode NMPC, it is not necessary to find optimal solutions of Problem 3.1 in order to guarantee stability. Feasibility also implies stability here [16].

3.6 Robust NMPC

The NMPC schemes discussed in the previous sections do require that the system being controlled is identical to the model used for prediction, i.e. that no model/plant mismatch or unknown disturbances are present. Clearly this is a very unrealistic assumption for practical applications and the development of an NMPC framework to address robustness issues is of paramount importance. Like in the nominal stability and performance case, the resulting difference between the predicted open loop and actual closed-loop trajectory is the main obstacle.

Despite this difficulty a number of results have been published for *linear* predictive control schemes (see for example [35, 113, 38, 114, 53, 1, 60, 18]). Even though the analysis of robustness properties in *nonlinear* NMPC must still be considered an unsolved problem in general, some preliminary results are available. Firstly, some of the schemes discussed above (including zero state terminal equality, dual-mode, contractive, quasi-infinite horizon NMPC) have some inherent robustness properties, or can be made robust by simple changes (for example the use of conservative terminal inequality constraints in the dual-mode approach [71]). However, these results essentially only state that sufficiently "small" model uncertainties do not jeopardize closed loop stability. But they do not permit to derive controllers that guarantee stability for a given uncertainty description with given bounds on the size of this uncertainty. Typically, for all these approaches no explicit quantitative nonlinear uncertainty model or only very simple ones, like for example gain and additive perturbations in [27], are used. This is not surprising as the definition of meaningful uncertainty descriptions for nonlinear systems is an open problem not only in the NMPC context, but also in other areas of control. First NMPC approaches that make use of quantitative uncertainty models are described in [39, 2, 17, 102, 64, 115]. More detailed discussions of this issue can be found in [66, 67, 15].

4 MHE theory

Motivation A state feedback regulation problem has been presented in Section 3 as part of the model predictive controller. These regulators require the state, and, particularly if using nonlinear process models, we would like to use a consistent model in the regulator and estimator. This requirement invites us to assess the theory for nonlinear state estimation.

Moreover, the state estimation problem occurs in a much wider context than as part of feedback control. Drawing conclusions from data and models is a fundamental issue and state estimation may be part of process monitoring and diagnosis, inferential sensor development, and product property estimation.

4.1 Introduction

In this work we consider online optimization as a means for designing state estimators. This formulation is known as moving horizon state estimation (MHE). Much of this discussion is motivated by the success of online optimization as a means of designing state regulators as discussed in the previous section. Along the way, we discuss some of the similarities and differences in these two problems.

With imperfect models, one can sometimes augment the models with other physical information, such as constraints on states, to improve the estimation procedure. Although one can enforce constraints with any unconstrained estimator through projection, for example, that is similar to clipping in control and it may not provide the sharpest estimator properties. Moving horizon estimation allows constraints on estimates as part of the problem formulations, and we would like to better understand the implications of these constraints.

The remainder of this section presents the optimization problem statements for full information estimation and moving horizon estimation for both linear and nonlinear models subject to constraints. Then we present results on (deterministic) estimator stability. Section 4.3 summarizes the status of several aspects of this approach.It should be emphasized throughout that while the general notion of estimating the state via optimization is a mainstream approach and was present in the earliest estimation work, it has only recently become possible to consider treating realistic industrial problems as *online* optimizations. The possible benefits we can derive from this extra computational capability is what we would like to explore further.

Many of these results are recent [88]. The majority of the results for linear constrained estimation are based on [90, 89], and the nonlinear results are based on [91], which also considers time-varying systems.

Problem Statement. In this section we investigate optimization strategies for estimating the state of systems modeled by nonlinear difference equation of the form

$$x_{k+1} = f(x_k) + g(x_k, w_k) \tag{32}$$

$$y_k = h(x_k) + v_k \tag{33}$$

in which we impose the following constraints on the state and disturbances:

$$x_k \in \mathbb{X}, \qquad w_k \in \mathbb{W}.$$

in which the functions $f : \mathbb{R}^n \to \mathbb{R}^n$, $g : \mathbb{R}^n \times \mathbb{R}^m \to \mathbb{R}^n$ with $g(\cdot, 0) = 0$, and $h : \mathbb{R}^n \to \mathbb{R}^p$. We assume the sets $\mathbb{X}$ and $\mathbb{W}$ are closed with $0 \in \mathbb{W}$.

We formulate the optimal constrained estimation problem as the solution to the following optimal control problem

$$(P1): \qquad \Phi_T^* = \min_{x_0, \{w_k\}_{k=0}^{T-1}} \Phi_T(x_0, \{w_k\}_{k=0}^{T-1}),$$

subject to the constraints

$$x_k \in \mathbb{X} \text{ for } k = 0, \dots T, \qquad w_k \in \mathbb{W} \text{ for } k = 0, \dots, (T-1), \tag{34}$$

in which the objective function is of the form

$$\Phi_T(x_0, \{w_k\}_{k=0}^{T-1}) \triangleq \sum_{k=0}^{T-1} L(w_k, v_k) + \Gamma(x_0). \tag{35}$$

The dynamics are specified by the model (32). We assume the stage cost function $L : \mathbb{R}^m \times \mathbb{R}^p \to \mathbb{R}_+$ and the initial penalty $\Gamma : \mathbb{R}^n \to \mathbb{R}_+$. The solution to $(P1)$ at time T is the state sequence $\{\hat{x}_{k|T-1}\}_{k=0}^{T}$ and disturbance sequence $\{\hat{w}_{k|T-1}\}_{k=0}^{T-1}$. Recall $v_k = y_k - h(x_k)$, so the decision variable x_k is implicit in the objective function through the definition of (32). We define $\hat{v}_{k|T-1} = y_k - h(\hat{x}_{k|T-1})$.

We refer to the formulation $(P1)$ as the full information estimator; the problem setup is depicted in Figure 2. The size of the full information estimation problem increases with time as more data are processed. The problem $(P1)$ has T stages, so the computational complexity scales at least linearly with T. Unless the process is linear, unconstrained, and the cost functions are quadratic, in which case the optimal estimator is the Kalman filter, the online solution of $(P1)$ is impractical due to the increasing computational burden. To make the problem tractable, we need to bound the problem size. One strategy to reduce $(P1)$ to a fixed dimension optimal control problem is

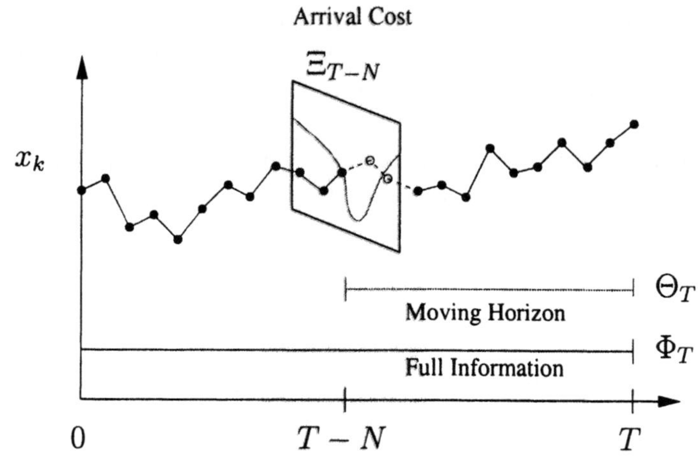

Figure 2: Full information and equivalent moving horizon state estimation problem and arrival cost.

to employ a moving horizon approximation. Unlike the full information estimator, the MHE does not determine the full state sequence $\{\hat{x}_k\}_{k=0}^{T}$. Rather, the MHE determines the truncated sequence $\{\hat{x}_k\}_{k=T-N}^{T}$. The key to preserving stability with the moving horizon approximation is the construction of the initial penalty that approximately summarizes the past data. Consider again the problem $(P1)$. We can arrange the objective function by breaking the time interval into two pieces as follows.

$$\begin{aligned}\Phi_T(x_0, \{w_k\}_{k=0}^{T-1}) &= \sum_{k=0}^{T-1} L(w_k, v_k) + \Gamma(x_0) \\ &= \sum_{k=0}^{T-N-1} L(w_k, v_k) + \sum_{k=T-N}^{T-1} L(w_k, v_k) + \Gamma(x_0)\end{aligned}$$

Because we use a state-variable description of the system, the quantity

$$\sum_{k=T-N}^{T-1} L(w_k, v_k)$$

depends only on the state x_{T-N} and the decision variables w_k and v_k for $k \in [(T-N), \ldots, (T-1)]$. Hence, we can reformulate $(P1)$ as the following equivalent optimal control problem

$$(P1') \qquad \min_{x_{T-N}, \{w_k\}_{T-N}^{T-1}} \sum_{k=T-N}^{T-1} L(w_k, v_k) + \Xi_{T-N}(x_{T-N}),$$

in which

$$\Xi_{T-N}(\bar{x}) = \min_{x_0, \{w_k\}_{k=0}^{T-N-1}} \left\{ \Phi_{T-N}(x_0, \{w_k\}_{k=0}^{T-N-1}) \ : \ x_{T-N} = \bar{x} \right\},$$

subject to the constraints

$$x_k \in \mathbb{X} \text{ for } k = 0, \ldots, (T-N), \qquad w_k \in \mathbb{W} \text{ for } k = 0, \ldots, (T-N-1).$$

This problem is also depicted in Figure 2. It is relatively straightforward to demonstrate the equivalence of the solutions to $(P1)$ and $(P1')$ using forward dynamic programming (c.f. [7]). We refer to the initial penalty Ξ_j as the *arrival cost.*[2] Arrival cost is a fundamental concept in MHE, because it allows us to recast the full information problem $(P1)$ as a fixed horizon $(P1')$ problem. We can view the arrival cost as an equivalent statistic [106] for summarizing the past data $\{y_k\}_{k=0}^{T-N}$ not accounted for explicitly in the objective function of $(P1')$. The arrival cost serves as an equivalent statistic by penalizing deviation of x_{T-N} away from $\hat{x}_{T-N|T-N-1}$. In particular, if we have high confidence in the optimal estimate $\hat{x}_{T-N|T-N-1}$, then the cost of choosing $x_{T-N|T-1}$ far away from $\hat{x}_{T-N|T-N-1}$ is large. Likewise, if we have little confidence in the optimal estimate $\hat{x}_{T-N|T-N-1}$, then the cost of choosing $x_{T-N|T-1}$ far away from $\hat{x}_{T-N|T-N-1}$ is small.

For the vast majority of systems, we do not possess an algebraic expression for the arrival cost. Notable exceptions are unconstrained linear systems with quadratic objectives. Assume

$$f(x_k) = Ax_k, \qquad g(x_k, w_k) = Gw_k, \qquad h(x_k) = Cx_k,$$

in which $A \in \mathbb{R}^{n\times n}$, $G \in \mathbb{R}^{m\times n}$, and $C \in \mathbb{R}^{p\times n}$. If

$$L(w_k, v_k) = w_k^T Q^{-1} w_k + v_k^T R^{-1} v_k,$$

and

$$\Gamma(x_0) = (x_0 - \bar{x})^T \Pi_0^{-1} (x_0 - \bar{x}),$$

in which Q, R, Π_0 are symmetric positive definite matrices and $\bar{x} \in \mathbb{R}^n$, then the arrival cost is given by

$$\Xi_j(x) = (x_j - \hat{x}_{j|j-1})^T \Pi_j^{-1} (x_j - \hat{x}_{j|j-1}).$$

The matrix Π_j is obtained by iterating the discrete-time Riccati equation

$$\Pi_{j+1} = GQG^T + A^T \Pi_j A - A\Pi_j C^T (R + C\Pi_j C^T)^{-1} C\Pi_j A^T.$$

This result is obtained by deriving the deterministic Kalman filter using forward dynamic programming (c.f. [23]).

When the system is nonlinear or constrained, a general algebraic expression for the arrival cost does not exist. Yet to implement the estimator online, we require an algebraic expression. Ideally, we want the moving horizon estimate to be as close as possible to the full information estimate. One solution is to attempt to formulate MHE as the solution to a numerically tractable, though approximate, version of $(P1')$. Using an algebraic approximation for

[2]Other researchers have used the term *cost to come* (c.f [3]) or *cost to arrive* (c.f. [109]).

the arrival cost, the MHE accounts for the data not included in the estimation window. The past data are indirectly accounted for through the initial penalty by penalizing deviation away from the past estimate in accordance with our confidence in the estimate. Because this choice is an approximation, we need to ensure that MHE does not improperly weight the old data. Estimator divergence may result if the initial penalty biases the old data by too strongly weighting the past estimates, while performance may suffer if the initial penalty neglects the old data by not sufficiently weighting them. In Section 4.2 we discuss stability and performance implications of approximate representations of the arrival cost.

We formulate the moving horizon approximation to the full information estimation problem, or MHE, as the following optimal control problem

$$(P2) \qquad \Theta_T^* = \min_{x_{T-N}, \{w_k\}_{k=T-N}^{T-1}} \Theta_T(x_{T-N}, \{w_k\}_{k=T-N}^{T-1}),$$

subject to the constraints

$$x_k \in \mathbb{X} \text{ for } k = (T-N), \dots, T, \qquad w_k \in \mathbb{W} \text{ for } k = (T-N), \dots, (T-1),$$

in which the objective function is of the form

$$\Theta_T(x_{T-N}, \{w_k\}_{k=T-N}^{T-1}) \triangleq \sum_{k=T-N}^{T-1} L(w_k, v_k) + \Gamma_{T-N}(x_{T-N}).$$

We define the stage cost functions L equivalently to $(P1)$ and $\Gamma_k : \mathbb{R}^n \to \mathbb{R}_+$ for $k \geq 0$. The solution to $(P2)$ at time T is the state sequence $\{\hat{x}_{k|T-1}^{\text{mh}}\}_{k=T-N}^{T}$ and disturbance sequence $\{\hat{w}_{k|T-1}^{\text{mh}}\}_{k=T-N}^{T-1}$. Let $\hat{v}_{k|T-1}^{\text{mh}} = y_k - h(\hat{x}_{k|T-1}^{\text{mh}})$. We drop the superscript when it is obvious from context that we are referring to the solution of $(P2)$. We use $\hat{x}_T \triangleq \hat{x}_{T|T-1}$ as shorthand for the current estimate. The initial penalty Γ_j represents the approximate arrival cost.

4.2 Stability of Estimators

We define stability subsequently, but the simplest idea is the following: if the model is correct, the noise sequences are zero, and the only error entering the state estimation problem is that the initial state is not correct, does the state estimate converge to the true state as more output measurements become available. If yes, then we say the estimator is (asymptotically) stable (in the sense of an observer).

Given this extremely idealistic question, one might naturally wonder why researchers would exert any effort trying to answer it. After all, the estimators are *applied* in an environment in which the model is not correct, the noise sequences are not zero, and we do not expect the state estimate to converge to the true state. This statement, while true, misses the main point of stability analysis. A *failure* to have this simple stability property tells us a lot. It is unlikely that estimators without this property have any other useful theoretical properties.[3] We hasten to point out that it is also easy to construct trivially stable estimators that have no chance of working in application; just set the arrival cost to zero, for example. The simple notion of stability described above follows immediately because the optimization of cost immediately moves the initial state to the correct value. This stable estimator would have, however, few other practically important properties such as insensitivity to measurement noise and model error.

So the current motivation underlying stability analysis of MHE is the following. First, include in the optimization problem of the estimator the features that have been found useful in practice such as tuning parameters to address the tradeoff between speed of estimator response and filtering of measurement noise and sensitivity to model error. Add the new features allowed by the online optimization framework, such as constraints and model nonlinearity. Then, impose the requirement that this practically motivated estimator has at least this simple stability property. The hope is that this exercise may produce candidate estimators that are both useful in practice and fruitful for further theoretical development.

Notation, Definitions, and Basic Assumptions

Definition 4.1 *A function* $\alpha : \mathbb{R}_+ \to \mathbb{R}_+ : x \mapsto \alpha(x)$ *is a* K_+-function *if it is continuous, strictly monotone increasing,* $\alpha(x) > 0$ *for* $x \neq 0$*, and* $\alpha(0) = 0$.

For any $\bar{x} \in \mathbb{R}^n$ and $\{w_k\}_{k=0}^{T-1} \in \mathbb{W}^T$, let

$$x(T, \bar{x}, \{w_k\}_{k=0}^{T-1})$$

denote the solution to the difference equation (32), or the T-step state transition map, with the initial condition $x_0 = \bar{x}$ and disturbance sequence $\{w_k\}_{k=0}^{T-1}$. When we consider the disturbance free response of the system, i.e. $w_k = 0$, we use the following simplification,

$$x(T, \bar{x}),$$

for the solution to the difference equation (32) with the initial condition $x_0 = \bar{x}$. We define the T-step output response map given the initial condition $x_0 = \bar{x}$ and disturbance sequence $\{w_k\}_{k=0}^{T-1}$ as

$$y(T, \bar{x}, \{w_k\}_{k=0}^{T-1}) \left\{ \triangleq h\left(x(T, \bar{x}, \{w_k\}_{k=0}^{T-1})\right) \right\},$$

with the similar simplification,

$$y(T, \bar{x}) \left\{ \triangleq h\left(x(T, \bar{x})\right) \right\},$$

[3] Consider, for example, the extended Kalman filter; the EKF has been shown useful in practice, but it has proven difficult to establish a strong supporting theory for its use. See [94] and the references therein for an excellent overview of available EKF stability results.

for the disturbance free response of the system. Note the difference between y_k and $y(k,\cdot,\cdot)$. The vector y_k denotes the observation of the output at time index k and the vector $y(k,\cdot,\cdot)$ denotes the predicted output at time index k.

Definition 4.2 *A system is* observable *if there exists a positive integer T_o and a K_+-function φ such that for any two states x_1 and x_2,*

$$\varphi(\|x_1 - x_2\|) \leq \sum_{j=0}^{T-1} \|y(j,x_1) - y(j,x_2)\|$$

for all $T \geq T_o$.

The observability condition states that if the prediction residuals are small, then the estimation error is small. Mathematically this condition requires that the mapping from the state $\bar{x}$ to the sequence $\{y(j,\bar{x})\}_{j=0}^{T_o-1}$ is one-to-one. When the system is linear (i.e. $f(x) = Ax$ and $h(x) = Cx$), then the observability condition is satisfied when the observability grammian

$$V_T \triangleq \sum_{j=0}^{T-1} A^{jT} C^T C A^j$$

is positive definite for all $T \geq n$.

In order to guarantee the problems ($P1$) and ($P2$) are well posed, we impose the following conditions on the model (32), stage cost functions L, initial penalty Γ, and feasible region.

A_e1 The functions f, g, and h are Lipschitz continuous.

A_e2 L and Γ are continuous with $L(0,0) = 0$ and $\Gamma(0) = 0$.

A_e3 There exist K_+-functions η and γ such that

$$\eta(\|(w,v)\|) < L(w,v) < \gamma(\|(w,v)\|),$$
$$\eta(\|x - \tilde{x}\|) < \Gamma(x) - \Gamma(\tilde{x}) \leq \gamma(\|x - \tilde{x}\|),$$

for $(w,v) \in (\mathbb{W} \times \mathbb{R}^p) \setminus (0 \times 0)$ and $x \in \mathbb{X}$. The vector $\tilde{x}$ denotes the reference state or initial guess for the state. Often we take $\Gamma(\tilde{x}) = 0$.

The vector $\tilde{x}$ in assumption (A_e3) denotes the initial condition, or initial guess, for the estimator. Because the choice of $\tilde{x}$ influences the estimate, we use the notation $\hat{x}_T(\tilde{x})$ when we wish to emphasize the effect of a particular initial condition.

To guarantee that a solution exists to ($P1$) and ($P2$), we require that the feasible region is nonempty.

A_e4 For all T, there exists x_0 and $\{w_k\}_{k=0}^{T-1}$ such that

1. $x\left(k, x_0, \{w_j\}_{j=0}^{k-1}\right) \in \mathbb{X}$ for $k = 0,\ldots,T$,
2. $w_k \in \mathbb{W}$ for $k = 0,\ldots,(T-1)$,

Observer Stability The following discussion of observer stability is premised on classical Lyapunov stability theory for dynamical systems. The concepts are completely analogous to their classical counterpart. To account for constraints, we have modified the definition of stability in an analogous manner to [50].

Before proceeding with the abstract definition of observer stability, we first introduce the concept of an observer. When the state is not directly available from measurements, it is necessary to infer the current state of the system. The observer addresses the problem of inferring the current state of the following deterministic (i.e. $(w_k, v_k) \equiv 0$) system

$$x_{k+1} = f(x_k) \tag{36a}$$

$$\hat{y}_k = h(x_k). \tag{36b}$$

When the functions f and h are linear and time invariant, the Luenberger observer [62, 63] solves the observer problem. The basic strategy of the Luenberger observer is to design a linear output feedback law that stabilizes the error dynamics of the system (36). Note the observer problem is solved if the initial condition x_0 is known exactly, because the model then tracks the system perfectly. Luenberger emphasized any system is an observer for another system. Whether the observer is effective is a different question. However, many observers are directly available without the need for rational design. This concept is demonstrated by Soroush [105], in which the author shows many models of chemical reactors are stable observers without modification.

One strategy for the observer problem is to solve directly for x_0. If the system (36) is observable and there are sufficient data, the initial condition can be uniquely determined from the data. This strategy is employed iteratively using moving horizon observers in [46], [116], [73], and [72]. The terminology moving horizon observer is used, because only the state of a deterministic system is reconstructed. This strategy is limited for several reasons. The primary reason is the problem statement represents an idealization, because our knowledge of the true system is not complete. In addition, state and measurement noise prevent a consistent state estimate being recovered from a fixed window of data. By assuming a deterministic model structure, the moving horizon observer strategies may amplify rather than filter the effect of noise on the estimate. Finally, the iterative strategies may break down, because cost decreases are not always possible. Even though the problem statement is an idealization, it is prudent to ask whether the estimation strategy proposed in this work can address this *simple* problem.

In addition to representing a degree of basic competency for the estimator, observer stability has the additional property that small numerical errors are bounded and asymptotically rejected. This property is also important because the calculations are performed with finite precision arithmetic.

We begin our analysis by making precise the notion of observer stability. The concept of observer stability and asymptotic observer stability is equivalent to stating that the error $\bar{e}\left\{\triangleq x - \hat{x}\right\} = 0$ is an asymptotically stable equilibrium point for the error dynamics of the system (36) combined with the observer.

Definition 4.3 *The estimator is a* stable observer *for the system (36) if for any $\epsilon > 0$ there corresponds a number $\delta(\epsilon) > 0$ and a positive integer $\bar{T}$ such that if $\|\bar{x} - \tilde{x}\| \leq \delta(\epsilon)$ and $\tilde{x} \in \mathbb{X}_0$, then $\|x(T, \bar{x}) - \hat{x}_T(\tilde{x})\| \leq \epsilon$ for all $T \geq \bar{T}$.*

Definition 4.4 *The estimator is an* asymptotically stable observer *for the system (36) if it is stable and there is a fixed number $\vartheta > 0$ such that if $\|\bar{x} - \tilde{x}\| \leq \vartheta$ and $\tilde{x} \in \mathbb{X}_0$, then for all $\mu > 0$ there exists a positive integer $\tilde{T}(\mu)$ such that $\|x(T, \bar{x}) - \hat{x}_T(\tilde{x})\| \leq \mu$ for all $T \geq \tilde{T}$.*

Full Information Estimator The full information problem provides a stable estimator by design. Although this paper contains no proofs of results, it may be useful to show the structure of the arguments to see where and how the assumptions are used. We show the essential arguments below for the full information case. Readers interested in more complete arguments and the MHE results should see [88].

Proposition 4.5 *If we assume (A_e1)–(A_e4), (32) is observable, and the evolution of (36) respects the constraints $\mathbb{X}$, then the full information estimator is an asymptotically stable observer for the system (36).*

We assume throughout the argument that $T \geq T_o$, and demonstrate estimator convergence. Showing Lyapunov stability is not difficult but not as interesting. If we assume (A_e1)–(A_e4), then a solution exists to $(P1)$ for all T. Hence, by optimality and (A_e4), we have that $\Gamma(\bar{x}) \geq \Phi_k^*$ for all k in which $\bar{x}$ is the initial state generating the data. Writing out the cost function explicitly, we regroup the optimal cost as follows

$$\begin{aligned}
\Phi_T^* &= \sum_{k=0}^{T-1} L(\hat{w}_{k|T-1}, \hat{v}_{k|T-1}) + \Gamma(\hat{x}_{0|T-1}), \\
&= L(\hat{w}_{T-1|T-1}, \hat{v}_{T-1|T-1}) + \sum_{k=0}^{T-2} L(\hat{w}_{k|T-1}, \hat{v}_{k|T-1}) + \Gamma(\hat{x}_{0|T-1}).
\end{aligned}$$

Because $\{\hat{w}_{k|T-1}\}_{k=0}^{T-2}$ and $\hat{x}_{0|T-1}$ are feasible at time index $k = T-1$, we obtain the inequality

$$\sum_{k=0}^{T-2} L(\hat{w}_{k|T-1}, \hat{v}_{k|T-1}) + \Gamma(\hat{x}_{0|T-1}) \geq \Phi_{T-1}^*$$

by optimality. This inequality implies

$$\Phi_T^* - \Phi_{T-1}^* \geq L(\hat{w}_{T-1|T-1}, \hat{v}_{T-1|T-1}).$$

Since Φ_T^* is nondecreasing and bounded above by $\Gamma(\bar{x})$, the sequence of optimal costs Φ_T^* converges to $\Phi_\infty^* < \infty$. Convergence implies the partial sum

$$\sum_{k=T-N}^{T-1} L(\hat{w}_{k|T-1}, \hat{v}_{k|T-1}) \to 0,$$

as $T \to \infty$. By the choice of function L, (A_e3), the sequences $\{\hat{w}_{k|T-1}, \hat{v}_{k|T-1}\}$ go to zero as $T \to \infty$, and, by Lipschitz continuity of the model (A_e1) and observability, the estimation error $\|\hat{x}_T - x(T, \bar{x})\| \to 0$ as claimed.

Linear, constrained moving horizon estimator We assume the system generating the data sequence $\{y_k\}$ is modeled by the following linear, time-invariant, discrete-time system:

$$x_{k+1} = Ax_k + Gw_k, \tag{37a}$$

$$y_k = Cx_k + v_k, \tag{37b}$$

in which $x_k \in \mathbb{R}^n$, $y_k \in \mathbb{R}^p$, and $w_k \in \mathbb{R}^m$. We also assume $\{w_k\}$ and $\{v_k\}$ are sequences of independent, random variables with zero mean and covariances Q and R respectively. The random variable w_k typically models unmeasured disturbances and model inaccuracies, while the random variable v_k models sensor noise. We assume the initial state of the system x_0 is normally distributed with mean $\bar{x}$ and covariance Π_0. If the matrices Q, R, and Π_0 are symmetric positive definite, then the optimal state estimate at time T given observations $\{y_k\}_{k=1}^{T-1}$ is obtained from the solution of the following mathematical program

$$\Phi_T^*(\bar{x}) = \min_{x_0, \{w_k\}_{k=0}^{T-1}} \Phi_T(x_0, \{w_k\}_{k=0}^{T-1}) \tag{38}$$

subject to Equations 37 in which

$$\Phi_T(x_0, \{w_k\}_{k=0}^{T-1}) = \sum_{k=0}^{T-1} v_k^T R^{-1} v_k + w_k^T Q^{-1} w_k + (x_0 - \bar{x})^T \Pi_0^{-1} (x_0 - \bar{x}).$$

We denote the optimal state and disturbance estimates at time k given observations $\{y_k\}_{k=1}^{T-1}$ as $\{\hat{x}_{k|T-1}\}_{k=0}^{T}$ and $\{\hat{w}_{k|T-1}\}_{k=0}^{T-1}$. In particular, if

$$(\hat{x}_{0|T-1}, \{\hat{w}_{k|T-1}\}_{k=0}^{T-1}) \in \arg \min_{x_0, \{w_k\}_{k=0}^{T-1}} \Phi_T(x_0, \{w_k\}_{k=0}^{T-1}),$$

then $\{\hat{x}_{k|T-1}\}$ denotes the solution to (38) at time index $k \geq 0$ with the initial condition $\hat{x}_0$ and disturbance sequence $\{\hat{w}_{j|T-1}\}_{j=0}^{k-1}$. Again this problem ins the full information estimator, because it uses all of the available measurements.

With an eye towards maintaining a quadratic program for efficient online implementation, we are interested in inequality constraints of the form

$$w_k \in \mathbb{W} \triangleq \{w : Dw \leq d\}, \tag{39a}$$
$$x_k \in \mathbb{X} \triangleq \{x : Mx \leq m\}, \tag{39b}$$

in which $D \in \mathbb{R}^{n_w \times m}$, $M \in \mathbb{R}^{n_x \times n}$, $d \in \mathbb{R}^{n_w}$ and $m \in \mathbb{R}^{n_x}$. In probabilistic terms, constraints may be used to model random variables with truncated normal densities. Constraints on the unmeasured disturbances w_k are used typically to model truncated densities. Constraints on the state may also be used to model truncated densities. However, state constraints are used typically to complete the conservation equations forming the model. It is important to note that constraints change the statistical properties of the variables. See [98] and [97] for further discussion regarding the role of constraints in estimation.

The largest effect of model nonlinearity enters at this point. If we restrict ourselves to linear models, even in the presence of constraints, we can under-bound the arrival cost. In fact, we can use the covariance of the unconstrained, linear Kalman filter for this purpose. That appears to be an ideal solution to the problem, because if the constraints remain inactive, we are using the true arrival cost and MHE reduces to the Kalman filter, which is the optimal as well as a stable solution to the estimation problem. This result is summarized in the following.

Proposition 4.6 *Suppose Q, R, and Π_0 are positive definite, (C, A) is observable, $N \geq n$ and the evolution of (40) respects the constraint $\mathbb{X}$ (i.e. $A^k \bar{x} \in \mathbb{X}$ for $k \geq 0$). Then the constrained moving horizon estimator is an asymptotically stable observer for the system (40).*

$$\begin{aligned} x_{k+1} &= Ax_k \\ y_k &= Cx_k \end{aligned} \tag{40}$$

Nonlinear moving horizon estimator The problem now becomes more challenging. We have no obvious candidate to under-bound the arrival cost. If we resort to linearization to produce an approximate covariance, then the approach suffers the same shortcomings as the EKF. Although this linearized covariance may be a good approximation of arrival cost for many systems and may work well in practice, it does not necessarily under-bound arrival cost and stability does not necessarily follow. Because practitioners are making increasing use of nonlinear models in estimation, however, some effort to address the nonlinear problem would seem warranted.

One proposal that guarantees stability is to generate *online* a sequence of MHE initial penalties as scaled EKF covariances [91]. The basic idea is to construct online a comparison cost by solving the estimator optimization with zero initial penalty. This estimator is known to be a stable estimator,

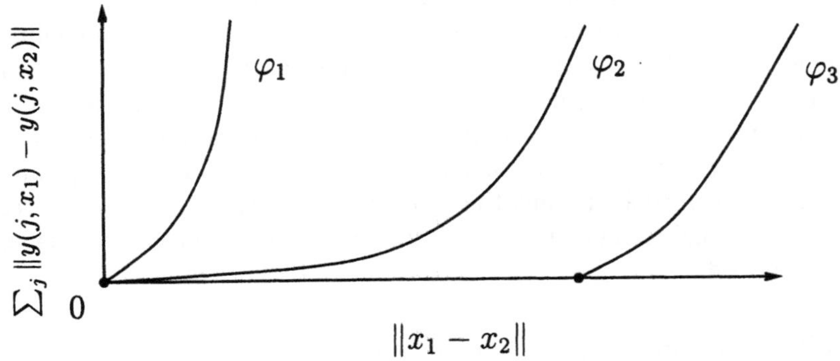

Figure 3: Well-conditioned (φ_1), ill-conditioned (φ_2), and unobservable (φ_3) systems. See also Definition 4.2

but not a very flexible one from a practical viewpoint. Then one scales the EKF covariances such that the initial penalty under-bounds this comparison cost, rather than the arrival cost, and stability follows. A serious remaining challenge is the online computation of the comparison cost, however. A naive implementation would require global optimization of a nonconvex objective, which is not generally suitable for online application. This situation is much the same as in nonlinear MPC regulation in which generating suitable terminal regions online, as required in quasi-infinite horizon approaches, raises the same optimization issues.

As proposed in the regulation literature, suboptimal strategies are also possible for MHE. From the practical perspective, even the suboptimal strategies are still out of reach as a guaranteed online computation. It is often true, however, that state of the art optimizers produce perfectly adequate solutions in reasonable computing time, and we discuss some of these examples in the next sections. Further understanding of the interplay between what is computable and what is required for the estimator theory may prove valuable. Even if the theoretical requirements for nonlinear MHE remain out of reach in the online implementation, the understanding and insight generated by the research effort may pay off. We may expect at least to understand more clearly the influence of initial state penalties and horizon lengths on estimator stability and performance.

4.3 Remarks

Observability. Observability is a serious problem in any realistic industrial setting. Of course, one evaluates the *conditioning* of the estimation problem, in preference to the yes/no question of observability as given in Definition 4.2. As displayed in Figure 3, a poorly conditioned estimation problem is as bad as an unobservable system in any practical setting.

Given the dearth of on-line sensors for product properties, application of MHE to chemical processes will likely prove challenging. One modelling challenge is to formulate low-order state space models or partition the larger state space models and determine the essential states to be estimated from the available measurements. Moreover, one often adds integrator states to the outputs to account for unmodelled disturbances that are not typically zero mean as assumed in formulating problem ($P1$). Adding these states may worsen the conditioning of the estimation problem, so the designer faces a difficult tradeoff.

Constraints. Estimator constraints are the main new feature that separates MHE from traditional linear estimators. The constrained estimator is nonlinear, even for linear models, which may provide new capabilities. Again, the designer faces tradeoffs in using constraints. Often a more complete and nonlinear model enforces constraints on states that are lost in a reduced order or linearized model. MHE allows one to add these known physical constraints back to the estimation problem without requiring the full nonlinear model, which may be computationally intractable for online use. However, in this situation, the reduced or linearized model and physical constraints may not be consistent and the estimator may encounter online infeasibilities. Further research and understanding of this issue may be required before implementation of constrained estimators can become reliable and predictable.

Full information problem and moving horizon approximation. Optimization of the estimates in the full information setting is preferable but not always practical. MHE may be regarded as an approximation of this full information problem. Treating the approximation of arrival cost, which separates full information into effects of neglected and considered data, may prove to be the key to MHE design. These results show that underbounding the arrival cost is sufficient to maintain stability, but it is not clear what design procedures are best for performance and robustness to modelling error.

Comparison cost. For nonlinear models, even underbounding arrival cost seems difficult, in general. An alternative is to generate online a comparison cost to ensure estimator stability. These ideas are, of course, closely related to the use of forgetting factors in adaptive filtering and control [41]. The best means to resolve this issue is far from clear at present, and more research is required.

Probabilistic setting. In the unconstrained, linear state estimation problem, a probabilistic treatment is tractable and provides further insight [49]. The disturbances are modelled as Gaussian distributed and zero mean, and one maximizes the conditional density of the state given the output measurements. In this setting, one also readily computes the covariances of the estimates, which are valuable measures of the uncertainty of the estimates [104, 47].

In the constrained, linear problem, a probabilistic setting is still tractable. The full information estimator maximizes the conditional density assuming the disturbances are truncated Gaussian distributed with zero mean [99]. For the nonlinear model, computing the conditional densities becomes markedly more complex [80].

5 Computational Issues

MPC strategies can be designed and implemented by means of algorithms from the fields of numerical optimization and numerical linear algebra. The interplay between MPC and optimization theory is a field of both historical and current interest; significant advances continue to be reported. In this section, we outline the connection between the these disciplines and review some of the results obtained to date. We then venture into areas of current research, discussing a variety of optimization techniques and their possible relevance to issues in MPC.

From an optimization point of view, the structure of MHE problems is similar to that of MPC problems, so the techniques we describe below can be applied to MHE problems as well.

We start with a discussion of algorithms for linear MPC. Such methods are interesting in their own right, and also as solvers for subproblems that arise in algorithms for nonlinear MPC.

5.1 Formulation Issues in Linear MPC

To describe algorithms for linear MPC, we consider the following formulation of the problem:

$$\min_{u,x,\epsilon} \Phi(u,x,\epsilon) = \sum_{k=0}^{N-1} \frac{1}{2}(x_k^T Q x_k + u_k^T R u_k + 2x_k^T M u_k) + z^T x_k + \frac{1}{2} x_N^T \bar{Q}_N x_N + \bar{z}^T x_N, \tag{41}$$

subject to

$$\begin{aligned} x_0 &= \hat{x}_j, \quad \text{(fixed)} && \text{(42a)}\\ x_{k+1} &= Ax_k + Bu_k, \quad k=0,1,\ldots,N-1, && \text{(42b)}\\ Du_k + Gx_k &\le d, \quad k=0,1,\ldots,N-1, && \text{(42c)}\\ Fx_N &\le f, && \text{(42d)} \end{aligned}$$

where the matrices Q and R are positive semidefinite. We note that this formulation is quite general. The quantities D, G and d in (42c) can be chosen to represent constraints on the inputs u_k and states x_k alone, as well as mixed constraints. Bounds on the change in input $\Delta u_k = u_{k+1} - u_k$ can be expressed in the form (41), (42) by redefining the state vector x_k to be the quantity $(x_k^T, u_{k-1}^T)^T$; see for example [92]. Equality constraints can be handled by replacing them with a pair of inequalities, since $c_i(w) = 0$ is equivalent to the pair $c_i(w) \leq 0$, $-c_i(w) \leq 0$. When we have access only to a state vector y_k defined as $y_k = Cx_k$, and when the objective function and output constraints are defined in terms of y_k rather than x_k, we can simply substitute for y_k in the relevant functions. We also need to replace the initial constraint $x_0 = \hat{x}_j$ by the pair $y_0 = \hat{y}_j$, $y_0 = Cx_0$, since we have access only to the output vector $\hat{y}_j$. The result of these substitutions is a form very similar to (41), (42).

The formulation (41), (42) also permits the use of "soft constraints." We could, for instance, penalize violations of the inequality $Hx_k \leq h$ by including weighted sum-of-absolute-values and weighted sum-of-squares terms in the objective, as follows: By introducing ϵ_k to represent the violations, we obtain the constraints

$$\epsilon_k = Hx_k - h, \quad \epsilon_k \geq 0, \quad k = 0, 1, \ldots, N-1. \tag{43}$$

We then include in the objective the term

$$\gamma_1 \mathbf{1}^T \epsilon_k + \gamma_2 \epsilon_k^T \epsilon_k, \tag{44}$$

where $\mathbf{1} = (1, 1, \ldots, 1)^T$ and γ_1 and γ_2 are nonnegative weighting terms. By replacing the state vector x_k by the augmented state vector $(x_k^T, \epsilon_k^T)^T$, we can easily redefine the data in (41) and (42c) to accommodate (43), (44).

Note that the quadratic program defined by (41), (42) is quite sparse if N is medium to large in size. The unknowns u_k, $k = 0, 1, \ldots, N-1$ and x_k, $k = 1, 2, \ldots, N$ each appear in only a small number of objective and constraint terms. Both the number of variables and the number of nonzeros in the problem data depend linearly on N, whereas in a dense problem the number of nonzeros in the problem data would depend cubically on N. One possible technique for solving (41), (42) would be to eliminate the state variables x_k, $k = 1, 2, \ldots, N$ using the equations (42b). Though it reduces the number of unknowns, this procedure has the unfortunate effect of making the problem completely dense, and experience shows that solution times scale like N^3.

5.2 Algorithms for Linear MPC: Interior-Point Methods

It is possible to devise better approaches for solving (41), (42) that are capable of exploiting the sparse, stagewise structure of the problem. These approaches can use either *interior-point* or *active-set* methodology. We outline interior-point methods in this section, in the context of the general convex quadratic program

$$\min_w \frac{1}{2} w^T H w + h^T w \text{ subject to } Aw = a, \; Gw \leq g, \tag{45}$$

where H is a positive semidefinite matrix. Because of convexity, the first-order necessary conditions are sufficient for optimality. By introducing a slack vector s for the inequality constraints $Gw \leq g$, and Lagrange multipliers ζ and λ for the equality and inequality constraints, respectively, we have that w is a solution of (45) whenever there are s, ζ, and λ such that

$$\begin{aligned} Hw + h + A^T\zeta + G^T\lambda &= 0, && \text{(46a)} \\ Aw &= a, && \text{(46b)} \\ Gw + s &= g, && \text{(46c)} \\ s \geq 0, \;\; \lambda \geq 0, \;\; s^T\lambda &= 0. && \text{(46d)} \end{aligned}$$

Primal-dual interior-point methods are essentially modified Newton methods applied to the system of equations formed by

- the first three (linear) equations in (46), and
- the nonlinear equation
$$\Lambda S \mathbf{1} = 0, \tag{47}$$
where $\Lambda \stackrel{\text{def}}{=} \text{diag}\,(\lambda_1, \lambda_2, \ldots)$, $S = \text{diag}\,(s_1, s_2, \ldots)$, and, as before $\mathbf{1} = (1, 1, \ldots, 1)^T$.

A sequence of iterates $(w^\ell, \zeta^\ell, \lambda^\ell, s^\ell)$, $\ell = 0, 1, 2, \ldots$, is generated by setting

$$(w^{\ell+1}, \zeta^{\ell+1}, \lambda^{\ell+1}, s^{\ell+1}) = (w^\ell, \zeta^\ell, \lambda^\ell, s^\ell) + \alpha_\ell(\Delta w^\ell, \Delta\zeta^\ell, \Delta\lambda^\ell, \Delta s^\ell). \tag{48}$$

The initial point $(w^0, \zeta^0, \lambda^0, s^0)$ is chosen so that λ^0 and s^0 have all components strictly positive. By choice of α_ℓ, we ensure that this condition holds at all subsequent iterates as well. (In fact, this condition is the origin of the term "interior point.") The search direction $(\Delta w^\ell, \Delta\lambda^\ell, \Delta\zeta^\ell, \Delta s^\ell)$ is obtained by solving the following modified Newton equations:

$$\begin{bmatrix} H & A^T & G^T & 0 \\ A & 0 & 0 & 0 \\ G & 0 & 0 & I \\ 0 & 0 & S^\ell & \Lambda^\ell \end{bmatrix} \begin{bmatrix} \Delta w^\ell \\ \Delta\zeta^\ell \\ \Delta\lambda^\ell \\ \Delta s^\ell \end{bmatrix} = \begin{bmatrix} h - Hw^\ell - A^T\zeta^\ell - G^T\lambda^\ell \\ a - Aw^\ell \\ g - Gw^\ell - s^\ell \\ -\Lambda^\ell S^\ell e + r^\ell_{\lambda s} \end{bmatrix}. \tag{49}$$

By eliminating Δs^ℓ from this system we obtain the more compact system

$$\begin{bmatrix} H & A^T & G^T \\ A & 0 & 0 \\ G & 0 & -(\Lambda^\ell)^{-1}S^\ell \end{bmatrix} \begin{bmatrix} \Delta w^\ell \\ \Delta\zeta^\ell \\ \Delta\lambda^\ell \end{bmatrix} = \begin{bmatrix} h - Hw^\ell - A^T\zeta^\ell - G^T\lambda^\ell \\ a - Aw^\ell \\ g - Gw^\ell - (\Lambda^\ell)^{-1}r^\ell_{\lambda s} \end{bmatrix}. \tag{50}$$

Elimination of the dual variables λ^ℓ leads to an even more compact system with coefficient matrix

$$\begin{bmatrix} H + G^T\Lambda^\ell(S^\ell)^{-1}G & A^T \\ A & 0 \end{bmatrix} \tag{51}$$

Further reduction of this system may be possible when H is diagonal or otherwise has special structure. In general, however, the step is calculated by factoring one of the forms (50) or (51).

If the perturbation $r^\ell_{\lambda s}$ in the final term of the right-hand side were zero, the system (49) is simply Newton's method applied to (46a), (46b), (46c), and (47). Inclusion of the term $r^\ell_{\lambda s}$ is essential, however, to the performance of the algorithm, both from the theoretical viewpoint of being able to prove convergence and low computational complexity, and from the viewpoint of good practical performance. This term typically includes "centering" information that stabilizes the iterates by forcing all the pairwise products $\lambda^\ell_i s^\ell_i$ to go to zero at roughly the same rate, and also higher-order information to accelerate the convergence. The choice of line search parameter α_ℓ in (48) is also important from both theoretical and practical standpoints; straightforward formulae are available that have proved to be quite effective.

For details of both theory and implementation of interior-point methods, see the book of Wright [111]. Most of this book concerns linear programming, but Chapter 8 discusses the fairly straightforward extensions of these methods to convex quadratic programming.

The chief advantages of the interior-point approach when applied to the problem (41), (42) concern its robustness, its ability to exploit the structure of the problem in a transparent way, and (not least) the relative ease with which it can be implemented. To expose the structure, we note that the first-order conditions (46) when specialized to (41), (42) yield the following system of equalities and bounds:

$$
\begin{aligned}
Qx_k + Mu_k + z - A^T p_{k+1} + p_k + G^T \lambda_k &= 0, \; k = 0, \ldots, N-1, && \text{(52a)}\\
(p_0 &= 0), &&\\
\bar{Q}x_N + \bar{z} + p_N + F^T \lambda_N &= 0, && \text{(52b)}\\
M^T x_k + Ru_k - B^T p_{k+1} + D^T \lambda_k &= 0, \; k = 0, \ldots, N-1, && \text{(52c)}\\
x_{k+1} - Ax_k - Bu_k &= 0, \; k = 0, \ldots, N-1, && \text{(52d)}\\
(x_0 &= \hat{x}_j), &&\\
Du_k + Gx_k + s_k - d &= 0, \; k = 0, \ldots, N-1, && \text{(52e)}\\
Fx_N + s_N - f &= 0, && \text{(52f)}\\
\lambda_k \geq 0, \;\; s_k \geq 0 \;\; \lambda_k^T s_k &= 0, \; k = 0, \ldots, N. && \text{(52g)}
\end{aligned}
$$

(We have introduced slack vectors s_k, $k = 0, 1, \ldots, N-1$ and Lagrange multipliers for the constraints (42c), slack vector s_N and Lagrange multiplier λ_N for (42d), and adjoint variables—that is, Lagrange multipliers—p_{k+1}, $k = 0, 1, \ldots, N-1$ for the state equation (42b).) The linear system (50) when adapted to this special case has as its solution the steps in the variables x_{k+1}, u_k, and p_{k+1}, $k = 0, 1, \ldots, N-1$, together with λ_k, $k = 0, 1, \ldots, N$. By reordering this system so that all the primal and dual variables are grouped together according to stage index k (rather than partitioning the primals from the duals, as in (49), (50), and (51)), we obtain a linear system at each interior-point iteration whose coefficient matrix has its nonzero blocks clustered about the diagonal. Gaussian elimination algorithms that are specialized for banded systems can therefrom be applied to calculate a solution in $O(N)$ operations.

We can eliminate the λ_k variables from (52) to obtain a matrix with more compact form, corresponding to (51). The resulting linear system has a special form: It can be viewed as the optimality conditions for a time-varying linear MPC problem in which auxiliary constraints (42c) do not appear. Hence, it can be solved by using standard techniques for problems of this type, for example, Riccati substitution. The latter method can also be viewed as block Gaussian elimination, with a particular elimination strategy for the blocks. For details, see [92].

In MPC, the quadratic programming problem (41) is not solved once in isolation, but rather is re-solved from a sequence of starting points $\hat{x}_j$ in succession. If the model is adequate, and if there have been no significant upsets, the solution to (41) obtained from one starting point should, after an appropriate time shift, provide a good starting point for the next MPC problem, which starts from $\hat{x}_{j+1}$. The use of *warm starts* is made more complicated for interior-point methods by the additional requirements on the iterates. These include the requirement that, in the notation of (46), the iterates must remain strictly positive in their λ and s components, must be well-centered (that is, the pairwise products $\lambda_i s_i$ have similar values for each i), and must be such that the infeasibilities in the linear equations (46a)–(46c) are not too much greater than the complementarity gap $\lambda^T s$. These conditions are not satisfied at the solution of the current problem—in particular, some of the components of λ and s will be zero—so some adjustment of the point is required before it can be used to restart the interior-point method from the new starting point. The adjustment techniques are fairly ad-hoc in practice; they usually consist of choosing a target value μ in some heuristic fashion and increasing λ_i and/or s_i so that $\lambda_i s_i = \mu$ is satisfied for each i. A good warm start can decrease the number of interior-point iterates required by the solver by a factor of about 3.

5.3 Algorithms for Linear MPC: Active-Set Methods

A second class of methods for (41) makes a sequence of guesses about which of the inequality constraints (42c), (42d) are active (that is, satisfied as equalities at the solution) and which are inactive (that is, the left-hand side is strictly less than the right). For each guess of the active set, we find the solution of the *reduced* quadratic program in which the constraints in the postulated active set are enforced as equality constraints, while the constraints outside the active set are ignored. By inspecting this solution, we can detect whether or not our guess of the active constraints is correct and, if not, we change the active set, usually by adding or deleting a single constraint, and repeat the process. The efficiency of the method comes from the fact that the active set changes only slightly from one iteration to the next. Therefore, it is possible to find the solution of each reduced quadratic program economically by using the information from the solution of the previous reduced quadratic program.

We give some specifics of the active-set procedure for the general quadratic program (45). Given an active set $\mathcal{A}_\ell \subset \{1, 2, \ldots, m\}$, where m is the number of rows in G, the reduced QP at each iteration has the form

$$\min_w \frac{1}{2} w^T H w + h^T w \text{ s.t. } Aw = a, \; \bar{G}^\ell w = \bar{g}^\ell, \tag{53}$$

where

$$\bar{G}^\ell = [G_{i\cdot}]_{i \in \mathcal{A}_\ell}, \qquad \bar{g}^\ell = [g_i]_{i \in \mathcal{A}_\ell}.$$

Here, $\bar{G}^\ell$ is the row submatrix of G corresponding to the current active set $\mathcal{A}_\ell$, and $\bar{g}^\ell$ is the corresponding right-hand side subvector. If the current iterate is w^ℓ, we can perform the change of variables $w = w^\ell + d$ in (53), where d is the step vector, to obtain the equivalent problem

$$\min_d \frac{1}{2} d^T H d + (h + H w^\ell)^T d \text{ subject to } Ad = a - Aw^\ell, \; \bar{G}^\ell d = \bar{g}^\ell - \bar{G}^\ell w^\ell. \tag{54}$$

The solution d satisfies the first-order conditions for this problem, which can be written as the following linear system:

$$\begin{bmatrix} H & A^T & (\bar{G}^\ell)^T \\ A & 0 & 0 \\ \bar{G}^\ell & 0 & 0 \end{bmatrix} \begin{bmatrix} d^\ell \\ \zeta^\ell \\ \lambda^\ell \end{bmatrix} = \begin{bmatrix} -h - Hw^\ell \\ a - Aw^\ell \\ \bar{g}^\ell - \bar{G}^\ell w^\ell \end{bmatrix}. \tag{55}$$

Similarly to (46), we have introduced multipliers ζ and λ^ℓ, but because there are no inequality constraints (53), the complementarity conditions (46d) do not appear.

If the solution of (55) has $d^\ell = 0$, we check the Lagrange multipliers for optimality. If all components of λ^ℓ are nonnegative, we conclude that the feasible vector w^ℓ is a solution of (45). If not, we choose one of the negative components from λ^ℓ and delete the corresponding index from the active set $\mathcal{A}_\ell$ to obtain the active set $\mathcal{A}_{\ell+1}$ for the next iteration.

If the solution d^ℓ is nonzero, we step from w^ℓ in the direction d^ℓ. To be specific, we set

$$w^{\ell+1} = w^\ell + \alpha_\ell d^\ell,$$

where α_ℓ is the largest value of α in the range $[0, 1]$ such that $w^\ell + \alpha d^\ell$ remains feasible with respect to the constraints $G_{i\cdot} w \leq g_i$, $i \notin \mathcal{A}_\ell$, that is, the constraints outside the active set. If one of the previously inactive constraints becomes active, we add it to $\mathcal{A}_\ell$ to produce $\mathcal{A}_{\ell+1}$.

Addition or deletion of a constraint from the active set corresponds to addition or deletion of a row and column from the coefficient matrix in (55). The solution at iterate $\ell + 1$ can therefore be obtained by updating the factorization of the matrix to account for this minor change—usually a much more efficient operation that recomputing a new factorization from scratch. For general dense matrices, strategies for updating factorizations after addition or deletion of a column are well known; see for example Fletcher [34, Chapter 10] or Nocedal and Wright [79, Chapter 16]. These strategies can be applied blindly if we formulate the linear MPC subproblem by using the state equation (42b) to eliminate the variables x_k, $k = 0, 1, \ldots, N$, since the resulting problem in the variables u_k, $k = 0, 1, \ldots, N-1$ has a dense Hessian matrix H and constraint matrix G. As we have noted already, however, this reformulation is unappealing because its operation count scales poorly with problem size. The factorization of the matrix in (55) at the initial point alone requires $O(N^3)$ operations, while the update process at each iteration would require $O(N^2)$ operations.

By considering the linear MPC problem in its original form (41) (as for the interior-point approach), we find that the matrix in (55) can be reordered to form a banded structure at each iteration, similar to the structure that is encountered at each iteration of the interior-point method. As in the latter case, the bandwidth of this matrix is independent of N, so a complete factorization can be performed at a cost of $O(N)$ operations. The difficulty with applying the active-set approach to this formulation is however the strategy for updating the factors after addition/deletion of a row and column. The cost of updating is more expensive, *relative to the cost of a full factorization of the matrix in (41)*, than is the case for general quadratic programming problems. Moreover, the linear algebra processes for performing and maintaining these factorizations are more difficult to program in an efficient and stable way.

We partially explain this point by considering the modifications that are needed for the factorization when columns are added to the matrix in (55). Suppose that we have computed an LU factorization of this matrix, after performing permutations to the rows and columns to transform it to a more banded structure and also to promote stability. That is,

$$P \begin{bmatrix} H & A^T & (\bar{G}^\ell)^T \\ A & 0 & 0 \\ \bar{G}^\ell & 0 & 0 \end{bmatrix} Q = PWQ = LU,$$

where P and Q are permutation matrices and L is lower triangular and U upper triangular Suppose that we add q constraints to the active set, so that a matrix with q rows (which we denote by C) is added to $\bar{G}_\ell$. We seek to augment L and U to account for the additional rows and columns in the matrix. Specifically, we seek submatrices L_1 and L_2 (with q rows each) and U_1 and U_2 (with q columns each) such that

$$\begin{bmatrix} PWQ & C^T \\ C & 0 \end{bmatrix} = \begin{bmatrix} L & 0 \\ L_1 & L_2 \end{bmatrix} \begin{bmatrix} U & U_1 \\ 0 & U_2 \end{bmatrix}.$$

By matching the left- and right-hand sides of this expression, we find that the submatrices can be computed by performing the following operations in turn:

perform forward-substitution to compute U_1 from $LU_1 = C^T$;
perform forward-substitution to compute L_1 from $U^T L_1^T = C^T$;
compute the product $-L_1U_1$, and perform an LU factorization to obtain L_2 and U_2 from $L_2U_2 = -L_1U_1$.

If P and Q were chosen appropriately, the number of nonzeros in L and U should be linear in N, so the total number of operations in this updating procedure $O(N)$ multiplied by the number of rows q in C. Rather than allow q to become too large (and cause the cost of the factorization L_2U_2 to increase rapidly), it is best to perform a fresh refactorization of the matrix in (55) periodically with the current active constraint matrix $\bar{G}^\ell$.

The new submatrices L_1 and U_1 are in general fairly dense. On the other hand, if the rows and columns represented in C had been included in the original matrix W, rather than added at a later step of the optimization algorithm, the corresponding rows and columns in the L and U factors would have been quite sparse. Hence, this situation contrasts with the situation in which H and G are general dense matrices, where the cost of updating the factorization as new rows are added to G incrementally is not much higher than the cost of including these rows in G a priori.

Constraint deletion is a little more difficult to explain, and we give just an outline of one possible scheme here. Suppose that the last row of $\bar{G}^\ell$ is discarded, so that the system to be solved at iteration $\ell+1$ has the following form:

$$\begin{bmatrix} H & A^T & (\bar{G}^{\ell+1})^T \\ A & 0 & 0 \\ \bar{G}^{\ell+1} & 0 & 0 \end{bmatrix} \begin{bmatrix} d^{\ell+1} \\ \zeta^{\ell+1} \\ \lambda^{\ell+1} \end{bmatrix} = \begin{bmatrix} -h - Hw^{\ell+1} \\ a - Aw^{\ell+1} \\ \bar{g}^{\ell+1} - \bar{G}^{\ell+1}w^{\ell+1} \end{bmatrix}, \tag{56}$$

where

$$\bar{G}^\ell = \begin{bmatrix} \bar{G}^{\ell+1} \\ y^T \end{bmatrix},$$

where y^T represents the discarded constraint. By introducing a scalar dummy variable η, we can rewrite the system (56) as follows:

$$\begin{bmatrix} H & A^T & (\bar{G}^{\ell+1})^T & 0 \\ A & 0 & 0 & 0 \\ \bar{G}^{\ell+1} & 0 & 0 & 0 \\ 0 & 0 & 0 & -2 \end{bmatrix} \begin{bmatrix} d^{\ell+1} \\ \zeta^{\ell+1} \\ \lambda^{\ell+1} \\ \eta \end{bmatrix} = \begin{bmatrix} -h - Hw^{\ell+1} \\ a - Aw^{\ell+1} \\ \bar{g}^{\ell+1} - \bar{G}^{\ell+1}w^{\ell+1} \\ 0 \end{bmatrix}. \tag{57}$$

This system clearly has the same solution as the one in (56), with the addition of the dummy variable η whose value is zero. The coefficient matrix in this system is also nonsingular if and only if the matrix in (56) is nonsingular. Note now that we can express the matrix in (57) as the sum of the matrix of (55) and two rank-one terms, that is,

$$\begin{bmatrix} H & A^T & (\bar{G}^{\ell+1})^T & 0 \\ A & 0 & 0 & 0 \\ \bar{G}^{\ell+1} & 0 & 0 & 0 \\ 0 & 0 & 0 & -2 \end{bmatrix} = \begin{bmatrix} H & A^T & (\bar{G}^{\ell})^T \\ A & 0 & 0 \\ \bar{G}^{\ell} & 0 & 0 \end{bmatrix} \tag{58}$$

$$-\begin{bmatrix} y \\ 0 \\ 0 \\ e_q \end{bmatrix} \begin{bmatrix} 0 & 0 & 0 & e_q^T \end{bmatrix} - \begin{bmatrix} 0 \\ 0 \\ 0 \\ e_q \end{bmatrix} \begin{bmatrix} y^T & 0 & 0 & e_q^T \end{bmatrix},$$

where e_q is the vector $(0, 0, \ldots, 0, 1)^T$. Since we know the factorization of the matrix in (55), we can apply the Sherman-Morrison-Woodbury formula (see Golub and Van Loan [40, p. 50]) using these factors and the rank-one terms in (58) to obtain a solution of (57). we omit the details of this process, but remark that its cost is similar to the cost of performing a constraint addition, and the same comments regarding efficiency apply.

All of our discussions of updating above have been framed in the language of linear algebra operations on the matrix in (55) arising from the KKT conditions. As for the interior-point case, where we mentioned solution of the linear system via a Riccati substitution, we can reformulate the updating operations in terms of algorithms that are more familiar to control practitioners. Graham [42] explores the process of adding/deleting constraints from the active set in the context of the differential dynamic programming (DDP) algorithm, and shows that the update operations can also be interpreted in terms of modifications to this method, as well as in the terms we described above.

One recent implementation of active-set methods for problems of the form (41) is the package QDTOC, implemented by Ralph and Xiao [86] (private communication) and available at URL

`http://www.ms.unimelb.edu.au/~danny/DTOC/`.

This package also contains an interior-point solver.

The relative efficiency of interior-point and active-set approaches depends on many factors. When N is large relative to the dimensions of the vectors x_k and u_k, the interior-point approach generally is faster, but when N is relatively small, active-set approaches have the advantage. When a "warm start" is available, the active-set approach often is able to use it to good advantage and converge to the updated solution in a small number of steps. The savings from a warm start are generally less significant in the case of an interior-point method. From a programming point of view, the interior-point approach is significantly easier to implement.

5.4 Algorithms for Nonlinear MPC

We discuss algorithms for nonlinear MPC with respect to the plant model (1), (2), the objective (14), and general constraints. We restate the problem here as

$$\min_{u,x,y} \Phi(u, x, y) = \phi(y_N) + \sum_{k=0}^{N-1} L(y_k, u_k), \tag{59}$$

subject to

$$y_0 = \hat{y}_j, \qquad \text{(fixed)} \tag{60a}$$

$$\begin{aligned} x_{k+1} &= f(x_k, u_k), \quad & k = 0, 1, \ldots, N-1, & \qquad (60b) \\ y_k &= g(x_k), & k = 0, 1, \ldots, N, & \qquad (60c) \\ h(u_k, y_k) &\leq 0, & k = 0, 1, \ldots, N-1, & \qquad (60d) \\ \bar{h}(y_N) &\leq 0. & & \qquad (60e) \end{aligned}$$

Note that the constraints are generalizations of those in (42) that explicitly account for the possible discrepancy between state vector x_k and output vector y_k. As in the linear case, we could eliminate the output vector by substituting from the equation (60c), but derivation of algorithms generally is cleaner with the formulation above.

From the optimization viewpoint, the natural approach is to apply standard algorithms for nonlinear programming to (59), (60), exploiting the structure to attain effiency. The sequential quadratic programming (SQP) algorithm is an iterative technique in which each step is obtained by solving an approximation to (59), (60), in which the objective is replaced by a quadratic approximation and the constraints by linear approximations. Application of SQP to problems of the form (59), (60), including variants that enhance efficiency, are discussed by Biegler [9]; see also the reference list in this paper. We give details of this approach with respect to a general formulation of the nonlinear programming problem, namely,

$$\min_{w} \psi(w) \text{ subject to } c(w) = 0, \; h(w) \leq 0, \qquad (61)$$

where the equality constraint vector c has n_e components and the inequality constraint vector h has n_i components. The Lagrangian function for this problem is given by

$$\mathcal{L}(w, \zeta, \lambda) \stackrel{\text{def}}{=} \psi(w) + \zeta^T c(w) + \lambda^T h(w), \qquad (62)$$

and the first-order necessary conditions for optimality of w are that these exist vectors $\zeta \in \mathbb{R}^{n_e}$ and $\lambda \in \mathbb{R}^{n_i}$ such that

$$\begin{aligned} \nabla_w \mathcal{L}(w, \zeta, \lambda) &= 0, & (63a) \\ c(w) &= 0, & (63b) \\ h(w) + s &\leq 0, & (63c) \\ s \geq 0, \; \lambda \geq 0, \; s^T \lambda &= 0. & (63d) \end{aligned}$$

Given a current iterate w, and possibly also estimates of the Lagrange multipliers ζ and λ, the problem to be solved for the step Δw at each iteration of SQP is as follows:

$$\min_{\Delta w} \nabla \psi(w)^T \Delta w + \tfrac{1}{2} \Delta w^T B \Delta w, \qquad (64)$$

subject to

$$c(w) + \nabla c(w)^T \Delta w = 0, \quad h(w) + \nabla h(w)^T \Delta w \leq 0, \qquad (65)$$

where B is an approximation to the Hessian of the Lagrangian, that is,

$$B \approx \nabla^2_{ww}\mathcal{L}(w,\zeta,\lambda) = \nabla^2\psi(w) + \sum_{i=1}^{n_e} \zeta_i \nabla^2 c_i(w) + \sum_{i=1}^{n_i} \lambda_i \nabla^2 h_i(w). \tag{66}$$

When exact second derivatives of the objective and constraint functions are available, and when second-order sufficient conditions are satisfied at the solution, use of the exact Lagrangian Hessian $B = \nabla^2_{ww}\mathcal{L}(w,\zeta,\lambda)$ can lead to superlinear local convergence, when $\nabla^2_{ww}\mathcal{L}(w,\zeta,\lambda)$ is positive definite on the subspace defined by the null space of the active constraint gradients at the solution. Besides the fact that second derivatives may be difficult or inconvenient to calculate, this variant requires additional algorithmic machinery in order to guarantee convergence even to a point the satisfies the first-order conditions (63). Techniques based on the use of trust-regions and line searches are used to modify the length and/or the direction of the computed SQP step with the aim of ensuring that the steps produce a decrease in a certain merit function, whose minima to coincide with the first-order points for (61).

When applied to the nonlinear MPC problem (59), (60), the Lagrangian Hessian $B = \nabla^2_{ww}\mathcal{L}(w,\zeta,\lambda)$ has a block-diagonal form, so that the quadratic subproblem (64), (65) has a similar structure to that of the linear MPC problem (41), (42). When the entire Lagrangian Hessian is in fact positive semidefinite (not just its projection onto the null space mentioned above), we can apply the active-set or interior-point techniques described above for the linear MPC case, to calculate the step. Warm start capabilities become significant on later iterations, when w nears the solution w^*, since the active set is likely to remain much the same from one quadratic subproblem (64), (65) to the next.

In general, we cannot expect the Lagrangian Hessian to be positive definite even at the solution, let alone at early iterates. Therefore, quasi-Newton techniques have been proposed that maintain an approximation B that mimicks the behavior of $\nabla^2_{ww}\mathcal{L}$ while retaining positive definiteness. Superlinear convergence results are also possible for such algorithms, but they still require the application of globalization techniques to yield convergence to first-order points for the problem (61). In the context of the nonlinear MPC problem (59), (60), this approach has the additioanl disadvantage that the standard techniques for updating B yield a matrix that is completely dense. Its structure does not mirror that of the exact Lagrangian Hessian, and the cost of solving the subproblem may be considerable.

Another approach, discussed by Biegler [9], does yield quadratic subproblems that are convex and structured, while sacrificing local superlinear convergence. This approach takes note of the fact that the objective function components ϕ and L in (59) are usually convex (and usually convex quadratic, in fact), and therefore sets B to be the Hessian of the objective—that is, $B = \nabla^2\psi(w)$ in the notation of (66)—ignoring the contributions from the constraint terms of $\nabla^2_{ww}\mathcal{L}(w,\zeta,\lambda)$. This so-called Gauss-Newton approach still requires additional machinery to ensure global convergence.

More detailed discussions of SQP approaches can be found in Nocedal and Wright [79, Chapter 18] and Fletcher [34], and Bertsekas [8].

5.5 Robust MPC

Robustness to various types of uncertainty and model error is an active research area in MPC as outlined in Section 3.6.

Linear matrix inequalities (LMI) and semidefinite programming (SDP) are two optimization paradigms that provide important tools for controller design, including robust design. A linear matrix inequality problem has the form

$$\text{find } x \in \mathbb{R}^m \text{ such that } F_0 + \sum_{i=1}^{m} x_i F_i > 0,$$

where "> 0" denotes positive definiteness and F_i, $i = 0, 1, \ldots, m$, are given symmetric matrices. (In practice, many LMIs are formulated with a matrix as the unknown, leading to more efficient computation.) LMI applications to control are discussed extensive by Boyd et al [11]. In particular, [11, Chapter 7] shows the formulation of unconstrained robust state feedback control as an LMI.

In SDP, we seek the vector $x \in \mathbb{R}^m$ that minimizes $b^T x$ subject to the constraint

$$F_0 + \sum_{i=1}^{m} x_i F_i \geq 0,$$

where "≥ 0" denotes positive semidefiniteness. Mathematically, the difference between LMI and SDP is the difference between an optimization problem and a feasibility problem, and algorithms for SDP can obviously be applied to LMI by choosing any objective function and iterating until the constraints are satisfied (or until it is determined that no feasible point exists). Interior-point algorithms for SDP have been the subject of intense research in the optimization community during the past five years. See the seminal book of Nesterov and Nemirovskii [78] for a description of the fundamental principles of these algorithms, and consult the following URL for many papers on theory, algorithms, and software for SDPs:

`http://www.mcs.anl.gov/otc/InteriorPoint/`

6 Industrial Applications

Since its initial development in the 1970's, MPC technology has spread throughout the process industries. Table 1 summarizes data obtained by Qin and Badgwell from two MPC technology vendor surveys up to early 1998 [83, 85]. Note that this is a count of MPC applications performed by the vendors themselves. As such, these results do not include in-house applications of licensed technology by end-user companies, nor does it include

Table 1: Summary of vendor-implemented MPC applications

Application	Adersa	Aspen	CCI[1]	Honeywell	Pavilion	Total
Refining	280	930	-	290	13	1513
Chemicals	5	430	15	50	20	520
Food	41	-	-	-	9	50
Pulp & Paper	-	15	-	30	1	46
Air and Gas	-	5	18	5	-	28
Polymers	-	1	-	-	15	16
Utilities	-	5	2	2	-	9
Other	45	30	1	-	-	76
Total	371	1416	36	377	58	2258

[1] CCI: Continental Controls, Inc.

applications of technology developed in-house by companies such as Shell and Eastman Chemical. For these reasons one should regard these data as providing a conservative lower bound on the true number of MPC applications.

Over 2200 MPC applications are reported in Table 1. The majority of applications (67 %) were in the refining area, where MPC technology has established a solid track record of success. A significant number of applications can also be found in chemicals (this includes petrochemicals), food processing, and the pulp and paper industries. The number of polymer processing applications has grown rapidly in the past few years due to the availability of MPC products using nonlinear models. The *Other* category includes a wide variety of interesting applications by Adersa such as river dam level control, ship auto-pilots, automotive engine and transmission controls, and robot motion control.

In this section we describe the control algorithms used in several commercial NMPC products [85]: Aspen Target from Aspen Technology, Multivariable Control (MVC) from Continental Controls, NOVA Nonlinear Controller (NOVA-NLC) from DOT Products, and Process Perfecter from Pavilion Technology. The industrial implementations of NMPC is almost the same as their linear counterparts, except for the use of a nonlinear model, which, however, brings up the challenges in model derivation, output feedback, optimization, and stability and performance analysis.

6.1 A General MPC Calculation

MPC controllers usually receive an optimal steady-state operating point from an overlying optimizer, or they may compute an optimal operating point using an internal steady-state optimizer. The general objectives of an MPC controller, in order of importance, are:

1. prevent violation of input and output constraints

2. drive the controlled variables (CV's) to their steady-state optimal values
3. drive the manipulated variables (MV's) to their steady-state optimal values using remaining degrees of freedom
4. prevent excessive movement of MV's
5. when signals and actuators fail, control as much of the plant as possible.

The translation of these objectives into a mathematical problem statement involves a number of approximations and trade-offs that define the basic character of the controller.

6.2 Output Feedback

All of the industrial MPC algorithms use the same form of feedback for stable processes, based on comparing the current measured process output y_k^m to the current predicted output y_k:

$$p_k = y_k^m - y_k. \tag{67}$$

The bias p_k term is added to the model for use in subsequent predictions:

$$y_{j|k} = h\left(x_{j|k}\right) + p_k. \tag{68}$$

This form of feedback is equivalent to assuming an output disturbance that remains constant for all future time [65]. In some cases the CV measurement may not be available at each control execution; this may happen, for example, when the CV measurement is provided by an analyzer. In this case one can skip the bias update for the affected CV for a number of control intervals.

The constant output disturbance treatment, which extends to integrated white noise disturbance in a stochastic framework, is equivalent to a state observer design with a specifically chosen feedback gain. While by no means optimal, this method of feedback removes steady-state offset [93], which provides theoretical support for its use. An optimal feedback strategy calls for the MHE material presented earlier in Section 4. The constant disturbance treatment, however, results in an overall unobservable system if the process includes integrating variables. In all MPC practice integrating variables are modeled in differential forms.

6.3 Model Formulation

The first issue encountered in NMPC implementation is the derivation of a dynamic nonlinear model suitable for model predictive control. In the general practice of linear MPC, the majority of dynamic models are derived from plant testing and system identification. For NMPC, however, the issue of plant testing and system identification becomes much more complicated.

Because step response and impulse response models are non-parsimonic, a class of state-space model is adopted in the Aspen Target product, which has a linear dynamic state equation and a nonlinear output relation. The output nonlinearity is modeled with a linear relation superimposed with a nonlinear neural network. This type of nonlinear model can approximate any discrete time nonlinear processes with fading memory.

Input-output models are found in the MVC and the Process Perfecter products. To simplify the system identification task, both products use a static nonlinear model superimposed upon a linear dynamic model. The identification of the linear dynamic model is based on plant test data from pulse tests, while the nonlinear static model is a neural network built from historical data. This model makes the assumption that the process dynamics remain linear over the entire range of operation. Asymmetric dynamics (e.g., different local time constants), as a result, cannot be represented by this model.

Since empirical modeling approaches can be unreliable and require tremendous amount of experimental data, two of the vendors provide the option to use first principles models. These products usually ask the user to provide the first principles models with some kind of open equation editor, then the control algorithms can use the user-supplied models to calculate future control moves. The model parameters are typically estimated from plant data.

Hybrid models that combine first principles knowledge with empirical modeling are also found in the commercial packages. The first principles can be steady-state balance equations, a nonlinear function of physical variables that generates another physically meaning variable, such as production rate, or simply gain directions to validate empirical models.

6.4 Steady-State Optimization

Most industrial NMPC products split the control calculation into a local ste-ady-state optimization followed by a dynamic optimization. Optimal ste-ady-state targets are computed for each input and output; these are then passed to a dynamic optimization to compute the optimal input sequence required to move toward these targets. These calculations involve optimizing a quadratic objective that includes input and output contributions. The exception is the NOVA-NLC controller that performs the dynamic and steady-state optimizations simultaneously.

6.5 Dynamic Optimization

At the dynamic optimization level, an MPC controller must compute a set of MV adjustments that will drive the process to the steady-state operating point without violating constraints. All of the algorithms described here use a specific form of the following objective:

$$J = \sum\nolimits_{j=1}^{N} \|y_{j|k} - y_j^{\text{ref}}\|_{Q_j}^2 + \|\Delta u_{j|k}\|_{S_j}^2 + \|u_{j|k} - u_j^{\text{ref}}\|_{R_j}^2,$$

where y_j^{ref} and u_j^{ref} are reference trajectories for y and u, respectively. Specific treatments are: (i) only the process input and output terms using constant weight matrices ($Q_j = Q$, $R_j = R$, $S_j = 0$) and a reference trajectory; (ii) all three terms with constant weights ($Q_j = Q$, $R_j = R$, $S_j = S$); (iii) all three terms with the option of one or two norms ($Q_j = Q$, $R_j = R$, $S_j = S$); and (iv) one term only ($R_j = 0$, $S_j = 0$) with *trajectory weighting* that makes Q_j gradually increase over the horizon N. With this type of weighting, control errors at the beginning of the horizon are less important than those towards the end of the horizon, thus allowing a smoother control action.

There are basically two types of constraints used in industrial MPC technology; hard and soft. All of the NMPC algorithms allow hard MV maximum, minimum, and rate of change constraints to be defined. Most products perform rigorous optimizations subject to the hard input constraints. The PFC algorithm, however, enforces input hard constraints by clipping input values that exceed the input constraints. It should be noted that this method does not, in general, result in an optimal solution in the sense of satisfying the Karush-Kuhn-Tucker (KKT) conditions for optimality. Most control products enforce *soft* output constraints as part of the dynamic optimization. The exclusive use of hard output constraints is generally avoided in MPC technology because a disturbance can cause such a controller to lose feasibility.

Industrial NMPC controllers use four basic options to specify future CV behavior; a setpoint, zone, reference trajectory or funnel. All of the NMPC controllers described here provide the option to drive the CV's to a fixed setpoint, with deviations on both sides penalized in the objective function. In addition, all of the controllers provide some way to detune the controller using move suppression, a reference trajectory, or time-dependent weights. All of the controllers also provide a CV zone control option, designed to keep the CV within a zone defined by upper and lower boundaries.

Industrial NMPC controllers generally evaluate future CV behavior over a finite *prediction horizon*. The length of the horizon P is a basic tuning parameter for these controllers, and is generally set long enough to capture the steady-state effects of all computed future MV moves. This is an approximation of the infinite horizon solution for closed loop stability discussed earlier, and may explain why none of the industrial NMPC algorithms considered here include a terminal state constraint.

There are three different methods to parameterize the MV profile: (i) a single move; (ii) multiple MV moves; and (iii) basis functions such as polynomials, suitable for batch process control [83, 85]. Hicks and Ray [44] discussed the basis function parameterization in optimal control. The usual multiple move parameterization can be regarded as using the columns of an identity matrix as basis functions.

6.6 Computational Methods

The Aspen Target product uses a multi-step Newton-type algorithm [30, 29], and makes use of analytical model derivatives. The Newton's algorithm makes use of the QPKWIK solver which has the advantage that intermediate solutions, although not optimal, are guaranteed feasible. This permits early termination of the optimization algorithm if the optimum is not found within the sampling time. Aspen Target uses the same QPKWIK engine for local steady-state optimization and the dynamic MV calculation.

The MVC and Process Perfecter products use a generalized reduced gradient (GRG) code called GRG2 [57]. The NOVA-NLC product uses the NOVA optimization package, a proprietary mixed complementarity nonlinear programming code developed by DOT Products.

6.7 Control Design and Tuning

The NMPC control design and tuning procedure is generally described as follows [22, 45, 101]:

1. From the stated control objectives, define the size of the problem, and determine the relevant CV's, MV's, and measured disturbance variables (DV's)
2. Test the plant systematically by varying MV's and DV's; capture and store the real-time data showing how the CV's respond
3. Derive a dynamic model for the plant
4. Configure the NMPC controller and enter initial tuning parameters
5. Test the controller off-line using closed loop simulation to verify the controller performance.
6. Download the configured controller to the destination machine and test the model predictions in *open-loop* mode
7. Commission the controller and refine the tuning as needed.

Controller tuning is always required not only for stability, but also for trade-offs between performance and robustness. It is true that a high performance controller will not be robust with respect to model mismatch, but a low performance controller is not necessarily robust. Most MPC products provide tuning "knobs" to de-tune the controller: (i) move suppression factors, which are weights on Δu; (ii) MV horizons; (iii) time constant of the reference trajectory; and (iv) time interval of the funnel or frustum.

7 Future Outlook

By now, *linear* MPC has to be considered a quite mature field. Since a quarter of a century this technique has been successfully applied to a variety of industrial control problems as outlined in Sect. 6 and many of the issues affecting industrial practice have been addressed. Current developments with respect to application issues are mainly concerned with the improvement of existing solutions rather than addressing new issues (see [84, 74] for a discussion of some notable exceptions). Linear MPC *theory* has also caught up considerably with industrial developments during the last decade. The control theoretic properties of nominal MPC for constrained linear systems is reasonably well understood and computational procedures for solving the on-line optimization problem are available (see for example [59, 110] for an overview and discussion). At present active areas of research include the consideration of robustness to various types of uncertainty formulations plus the further exploration of the connections between industrial I/O MPC formulations as used by most vendors and the state space formulations that are popular in academia.

Although many academic research and industrial application results are also available for *nonlinear model predictive control*, many important issues related to industrial practice, control theoretical understanding and computational realization are as yet unresolved here.

Open Problems — Industrial Applications of NMPC

Most importantly there are no systematic modelling approaches for building nonlinear dynamic models for NMPC. In the case of empirical approaches, guidelines for plant tests are needed to build a reliable model.

Furthermore most current NMPC implementations use the traditional bias correction to the model prediction based on current measurements. While this approach is meaningful for linear MPC because of the principle of superposition, it is questionable how general this approach is to problems of disturbance attenuation and output feedback for nonlinear processes.

Criteria on where NMPC is needed and how much can be gained over linear control techniques including linear MPC are desirable but difficult to obtain. Benchmarks on the justification of NMPC are needed on a array of industrial processes.

Other issues that are applicable to linear MPC technology [83] should also be of the same level of concern for NMPC, if not more. These issues include multiple prioritized objective functions, determining controllable subprocesses, tuning, ill-conditioning, fault tolerance, monitoring, and links to higher automation levels. The inclusion of integer decision variables in MPC [5] may also prove very beneficial in industrial applications.

Open Problems — NMPC Theory

As stressed in Sect. 3.6, merely the surface has been scratched in robust NMPC, despite a few known results on robustness analysis and robust synthesis. A much clearer understanding of this issue is needed and current research is focusing on this.

In most existing nonlinear MPC schemes, the tracking problem can only be dealt with, when the required state trajectory can be calculated from the desired (output) trajectory. Trajectory generation (for nonlinear systems) is an old but still challenging problem and the presence of constraints imposes additional difficulties in this respect.

In practice the whole state vector is usually not available for feedback. Therefore, the problem of output feedback control is essential from an application point of view. Only few preliminary stability results are known to date, and all of these results incorporate a nonlinear observer, which is very difficult to design by itself. Future advances must directly aim at deriving robustly stable output feedback schemes.

At present, disturbances are only considered in a very rudimentary way in most nonlinear MPC schemes. One main goal of control is however the suppression of the effect of unknown and unmeasured disturbances, and thus nonlinear MPC schemes are needed for which disturbance attenuation requirements can be included in the performance objective.

Open Problems — NMHE Theory

Among other things, future MHE research should focus on what key issues of practice can be addressed in this framework that are out of reach with previous approaches. Because moving horizon estimation with linear models produces simple, positive definite quadratic programs, on-line implementation is possible today for many process applications. The use of constraints on states or state disturbances presents intriguing opportunities, but it is not clear what applications benefit from using the extra physical knowledge in the form of constraints. Nonlinear, fundamental models coupled with moving horizon state estimation may start to play a larger role in process operations. State estimation for unmeasured product properties based on fundamental, nonlinear models may have more impact in the short term than closed-loop regulation with these models. State estimation using empirical, nonlinear models is already being used in commercial process monitoring software. Moreover, state estimation is a wide ranging technique for addressing many issues of process operations besides feedback control, such as process monitoring, fault detection and diagnosis.

Open Problems — On-line optimization in NMPC

Speed and the assurance of a reliable solution in real-time are major limiting factors in existing applications. However, the field of optimization offers many tools which, appropriately specialized, provide a means to solve many of the current and future issues posed in MPC formulations. Further development is needed in the structure-exploiting quadratic programming and nonlinear programming solvers discussed in Section 5, especially in the areas of improved global convergence properties, more efficient implementations of active-set strategies, and warm starting. In the MPC context, further investigation is needed to incorporate real-time imperatives into the algorithms, that is, strategies that produce a feasible though possibly suboptimal solution within a specified time, or that use "dead time" between stages to enhance the global convergence behavior. Semdefinite programming and second-order cone programming, for which efficient algorithms have recently become available, is an valuable tool for tackling many problems in the related area of controller design. Exploitation of structure is once again crucial to efficient implementation of these algorithms, and is a topic of current investigation.

8 Conclusions

In this paper we have given a tutorial introduction to the areas of nonlinear model predictive control and moving horizon estimation that have both attracted considerable industrial and academic interest over recent years.

Model predictive control for linear constrained systems has been shown to provide an excellent control solution both theoretically and practically. The incorporation of nonlinear models poses a much more challenging problem mainly because of computational and control theoretical difficulties. We have discussed in some detail several NMPC formulations that guarantee closed loop stability and pointed out the state of the art with respect to robustness considerations.

As summarized in this paper, the theory of linear moving horizon estimation is in about the same shape as the theory of receding horizon regulation. As MHE is applied by practitioners, it is not clear what problem formulations if any will become dominant. As one considers the widely varying purposes of estimators, such as inferential sensing and product property monitoring, model nonlinearity may no longer be negligible. From both the research and applications perspectives, this area seems ripe for further attention.

The rapid pace of developments in model predictive control and moving horizon estimation has given rise to a host of new problems where optimization is of key relevance. Concurrently, developments in optimization, and especially in interior-point and active set-methods, have produced a new set of algorithms that may be especially helpful in this context. In this paper we reviewed both algorithms for linear MPC and extensions for the nonlinear case.

The state of industrial application of NMPC is growing rapidly and seems to follow academically available results more closely than linear MPC did. Specifically, the past three years have seen rapid progress in the development and application of NMPC algorithms, with a total of 88 applications reported by the vendors included in a recent survey discussed here. However, none of the NMPC algorithms provided by vendors and used in the applications of this survey include stability constraints as required by control theory for nominal stability; instead they rely implicitly upon setting the prediction horizon long enough to effectively approximate an infinite horizon. Future developments in NMPC control theory will hopefully contribute to making the gap between academic and industrial developments even smaller.

Acknowledgments

FA wants to thank R. Findeisen and H. Chen for many stimulating discussions over the past years.
SJQ would like to acknowledge the financial support from National Science Foundation and members of TWMCC.
JBR would like to acknowledge C.V. Rao, D.Q. Mayne and W.H. Marquardt for many helpful discussions of this material. The financial support of the National Science Foundation, through grant CTS–9708497, and the industrial members of the Texas-Wisconsin Modeling and Control Consortium is gratefully acknowledged.
SJW acknowledges support from the Mathematics, Information, and Computational Sciences Division subprogram of the Office of Computational and Technology Research, U.S. Department of Energy, under Contract W-31-109-Eng-38.
All authors are indebted to P. Menold and R. Findeisen for their active and professional support during the writing of this paper.

References

1. T. A. Badgwell. A robust model predictive control algorithm for stable linear plants. In *Proc. Amer. Contr. Conf.*, pages 1618–1622, Albuquerque, 1997.
2. T.A. Badgwell. A robust model predictive control algorithm for stable nonlinear plants. In *Proc. Int. Symp. Adv. Control of Chemical Processes, ADCHEM*, pages 477–481, Banff, Canada, 1997.
3. T. Başar and P. Bernhard. *H_∞ Optimal Control and Related Minimax Design Problems: A Dynamic Game Approach.* Systems & Control: Foundations & Applications. Birkhäuser Verlag, Boston, Basel, 1991.
4. R.E. Bellman and S.E. Dreyfus. *Applied Dynamic Programming.* Princeton University Press, Princeton, New Jersey, 1962.
5. A. Bemporad and M. Morari. Predictive control of constrained hybrid systems. In F. Allgöwer and A. Zheng, editors, *Nonlinear Predictive Control.* Birkhäuser, Basel, 1999. (In press).
6. B.W. Bequette. Nonlinear predictive control using multi-rate sampling. *Can. J. Chem. Eng.*, 69:136–143, February 1991.
7. D.P. Bertsekas. *Dynamic Programming and Optimal Control*, volume 1. Athena Scientific Press, Belmont, MA, 1995.
8. D.P. Bertsekas. *Nonlinear Programming.* Athena Scientific Press, Belmont, MA, 1995.
9. L. Biegler. Efficient solution of dynamic optimization and NMPC problems. In F. Allgöwer and A. Zheng, editors, *Nonlinear Predictive Control*, Basel, 1999. Birkhäuser. (In press).
10. R. R. Bitmead, M. Gevers, and V. Wertz. *Adaptive Optimal Control – The Thinking Man's GPC.* Prentice Hall, New York, 1990.
11. S. Boyd, L. El Ghaoui, E. Feron, and V. Balakishnan. *Linear Matrix Inequalities in System and Control Theory.* SIAM, Philadelphia, 1994.

12. E.F. Camacho and C. Bordons. *Model Predictive Control.* Springer, Berlin, 1998.
13. H. Chen and F. Allgöwer. A quasi-infinite horizon predictive control scheme for constrained nonlinear systems. In *Proc. 16th Chinese Control Conference*, pages 309–316, Qindao, 1996.
14. H. Chen and F. Allgöwer. A quasi-infinite horizon nonlinear predictive control scheme for stable systems: Application to a CSTR. In *Proc. Int. Symp. Adv. Control of Chemical Processes, ADCHEM*, pages 471–476, Banff, Canada, 1997.
15. H. Chen and F. Allgöwer. Nonlinear model predictive control schemes with guaranteed stability. In C. Berber, R. und Kravaris, editor, *Nonlinear Model Based Process Control*, pages 465–494. Kluwer Academic Publishers, Dodrecht, 1998.
16. H. Chen and F. Allgöwer. A quasi-infinite horizon nonlinear model predictive control scheme with guaranteed stability. *Automatica*, 34(10):1205–1218, 1998.
17. H. Chen, C.W. Scherer, and F. Allgöwer. A game theoretic approach to nonlinear robust receding horizon control of constrained systems. In *Proc. Amer. Contr. Conf.*, pages 3073–3077, Albuquerque, 1997.
18. H. Chen, C.W. Scherer, and F. Allgöwer. A robust model predictive control scheme for constrained linear systems. In *5th IFAC Symposium on Dynamics and Control of Process Systems, DYCOPS-5*, pages 60–65, Korfu, 1998.
19. D. W. Clark. Advances in model-based predictive control. In D. W. Clarke, editor, *Advances in Model-Based Predictive Control*, pages 3–21. Oxford University Press, 1994.
20. D. W. Clarke, C. Mohtadi, and P. S. Tuffs. Generalized predictive control – Part I. The basic algorithm. *Automatica*, 23(2):137–148, 1987.
21. D. W. Clarke, C. Mohtadi, and P. S. Tuffs. Generalized predictive control – Part II. Extensions and interpretations. *Automatica*, 23(2):149–160, 1987.
22. DMC Corp. DMC: Technology Overview. Product literature from DMC Corp., July 1994.
23. H. Cox. On the estimation of state variables and parameters for noisy dynamic systems. *IEEE Trans. Automat. Contr.*, 9(1):5–12, 1964.
24. C. Cutler, A. Morshedi, and J. Haydel. An industrial perspective on advanced control. In *AIChE Annual Meeting*, Washington, D.C., 1983.
25. C.R. Cutler and B.L. Ramaker. Dynamic Matrix Control – A computer control algorithm. In *AIChE National Meeting*, Houston, 1979.
26. C.R. Cutler and B.L. Ramaker. Dynamic Matrix Control – A computer control algorithm. In *Proc. Joint Automatic Control Conference*, volume 1, San Francisco, CA, 1980. Paper No. WP5-B.
27. G. De Nicolao, L. Magni, and R. Scattolini. On the robustness of receding horizon control with terminal constraints. *IEEE Trans. Automat. Contr.*, 41(3):451–453, 1996.
28. G. De Nicolao, L. Magni, and R. Scattolini. Stabilizing receding-horizon control of nonlinear time-varying systems. In *Proc. 4rd European Control Conference ECC'97*, Brussels, 1997.

29. N. M.C. de Oliveira and L. T. Biegler. An extension of Newton-type algorithms for nonlinear process control. *Automatica*, 31(2):281–286, 1995.

30. N.M.C. de Oliveira and L.T. Biegler. Constraint handling and stability properties of model-predictive control. *AIChE J.*, 40(2):1138–1155, 1994.

31. S.L. De Oliveira. *Model predictive control for constrained nonlinear systems.* PhD thesis, Swiss Federal Institute of Technology (ETH), Zürich, Switzerland, 1996.

32. R.H. Findeisen and F. Allgöwer. A procedure to determine a region of attraction for discrete time nonlinear systems. Technical Report # 99.14, Automatic Control Laboratory, Swiss Federal Institute of Technology (ETH), Zürich, Switzerland, April 1999.

33. R.H. Findeisen and J.B. Rawlings. Suboptimal infinite horizon nonlinear model predictive control for discrete time systems. Technical Report # 97.13, Automatic Control Laboratory, Swiss Federal Institute of Technology (ETH), Zürich, Switzerland, 1997. *Presented at the NATO Advanced Study Institute on Nonlinear Model Based Process Control.*

34. R. Fletcher. *Practical Methods of Optimization.* John Wiley & Sons, New York, 1987.

35. C.E. García and M. Morari. Internal Model Control. 1. A unifying review and some new results. *Ind. Eng. Chem. Process Des. Dev.*, 21:308–323, 1982.

36. C.E. García and A.M. Morshedi. Quadratic programming solution of dynamic matrix control (QDMC). *Chem. Eng. Commun.*, 46:73–87, 1986.

37. C.E. García, D.M. Prett, and M. Morari. Model Predictive Control: Theory and practice - A survey. *Automatica*, 25(3):335–347, 1989.

38. H. Genceli and M. Nikolaou. Robust stability analysis of constrained L_1-norm model predictive control. *AIChE J.*, 39(12):1954–1965, 1993.

39. H. Genceli and M. Nikolaou. Design of robust constrained model-predictive controllers with Volterra series. *AIChE J.*, 41(9):2098–2107, 1995.

40. G.H. Golub and C.F. Van Loan. *Matrix Computations.* The Johns Hopkins University Press, Baltimore, 1996.

41. G.C. Goodwin and K.S. Sin. *Adaptive Filtering Prediction and Control.* Prentice-Hall, Englewood Cliffs, New Jersey, 1984.

42. A.J. Graham. Active-set methods for quadratic discrete-time optimal control problems. Master's thesis, Department of Mathematics and Statistics, University of Melbourne, Parkville, Victoria, 1998.

43. P. Grosdidier, B. Froisy, and M. Hammann. The IDCOM-M controller. In T.J. McAvoy, Y. Arkun, and E. Zafiriou, editors, *Proceedings of the 1988 IFAC Workshop on Model Based Process Control*, pages 31–36, Oxford, 1988. Pergamon Press.

44. G.A. Hicks and W.H. Ray. Approximation methods for optimal control synthesis. Can. J. Chem. Eng., 49:522–528, 1971.

45. Inc. Honeywell. RMPCT concepts reference. Product literature from Honeywell, Inc., October 1995.

46. S.-S. Jang, B. Joseph, and H. Mukai. Comparison of two approaches to on-line parameter and state estimation of nonlinear systems. *Ind. Eng. Chem. Process Des. Dev.*, 25:809–814, 1986.

47. A.H. Jazwinski. *Stochastic Processes and Filtering Theory.* Academic Press, New York, 1970.

48. R.E. Kalman. Contributions to the theory of optimal control. *Bull. Soc. Math. Mex.*, 5:102–119, 1960.

49. R.E. Kalman. A new approach to linear filtering and prediction problems. *Trans. ASME, J. Basic Engineering*, pages 35–45, March 1960.

50. S.S. Keerthi and E.G. Gilbert. Optimal infinite-horizon feedback laws for a general class of constrained discrete-time systems: Stability and moving-horizon approximations. *J. Opt. Theory and Appl.*, 57(2):265–293, 1988.

51. I.W. Kim, M.J. Liebman, and T.F. Edgar. A sequential error-in-variables method for nonlinear dynamic systems. *Comp. & Chem. Eng.*, 15(9):663–670, 1991.

52. D.L. Kleinman. An easy way to stabilize a linear constant system. *IEEE Trans. Automat. Contr.*, 15:692, 1970.

53. M. V. Kothare, V. Balakrishnan, and M. Morari. Robust constrained model predictive control using linear matrix inequalities. *Automatica*, 32(10):1361–1379, 1996.

54. W.H. Kwon. Advances in predictive control: Theory and applications. In *1st Asian Control Conference*, Tokyo, 1994.

55. W.H. Kwon, A.M. Bruckstein, and T. Kailath. Stabilizing state-feedback design via the moving horizon method. *Int. J. Contr.*, 37(3):631–643, 1983.

56. L. L. Chisci, A. Lombardi, and E. Mosca. Dual-receding horizon control of constrained discrete time systems. *European Journal of Control*, 2:278–285, 1996.

57. L.S. Lasdon and A.D. Warren. GRG2 user's guide. Cleveland State University, Cleveland, OH, 1986.

58. E. B. Lee and L. Makus. *Foundations of Optimal Control Theory.* John Wiley & Sons, 1967.

59. J.H. Lee and B. Cooley. Recent advances in model predictive control and other related areas. In J.C. Kantor, C.E. Garcia, and B. Carnahan, editors, *Fifth International Conference on Chemical Process Control - CPC V*, pages 201–216. American Institute of Chemical Engineers, 1996.

60. J.H. Lee and Z.H. Yu. Worst-case formulations of model predicitve control for systems with bounded parameters. *Automatica*, 33(5):763–781, 1997.

61. M.J. Liebman, T.F. Edgar, and L.S. Lasdon. Efficient data reconciliation and estimation for dynamic processes using nonlinear programming techniques. *Comp. & Chem. Eng.*, 16(10/11):963–986, 1992.

62. D.G. Luenberger. Observers for multivariable systems. *IEEE Trans. Automat. Contr.*, 11(2):190–197, 1966.

63. D.J. Luenberger. An introduction to observers. *IEEE Trans. Automat. Contr.*, 16:596–602, 1971.

64. L. Magni. *Nonlinear Receding Horizon Control: Theory and Application.* PhD thesis, Università degli Studi di Pavia, 1998.

65. P. Marquis and J. P. Broustail. SMOC, a bridge between state space and model predictive controllers: Application to the automation of a hydrotreating unit. In T. J. McAvoy, Y. Arkun, and E. Zafiriou, editors, *Proceedings of the 1988 IFAC Workshop on Model Based Process Control*, pages 37–43, Oxford, 1988. Pergamon Press.

66. D.Q. Mayne. Nonlinear model predictive control: An assessment. In J.C. Kantor, C.E. Garcia, and B. Carnahan, editors, *Fifth International Conference on Chemical Process Control - CPC V*, pages 217–231. American Institute of Chemical Engineers, 1996.

67. D.Q. Mayne. Nonlinear model predictive control: Challenges and opportunities. In F. Allgöwer and A. Zheng, editors, *Nonlinear Predictive Control.* Birkhäuser, Basel, 1999. (In press).

68. D.Q. Mayne and H. Michalska. Receding horizon control of nonlinear systems. *IEEE Trans. Automat. Contr.*, 35(7):814–824, 1990.

69. E.S. Meadows, M.A. Henson, J.W. Eaton, and J.B. Rawlings. Receding horizon control and discontinuous state feedback stabilization. *Int. J. Contr.*, 62(5):1217–1229, 1995.

70. E.S. Meadows and J.B. Rawlings. Receding horizon control with an infinite horizon. In *Proc. Amer. Contr. Conf.*, pages 2926–2930, San Francisco, 1993.

71. H. Michalska and D.Q. Mayne. Robust receding horizon control of constrained nonlinear systems. *IEEE Trans. Automat. Contr.*, AC-38(11):1623–1633, 1993.

72. H. Michalska and D.Q. Mayne. Moving horizon observers and observer-based control. *IEEE Trans. Automat. Contr.*, 40(6):995–1006, 1995.

73. P.E. Moraal and J.W. Grizzle. Observer design for nonlinear systems with discrete-time measurements. *IEEE Trans. Automat. Contr.*, 40(3):395–404, 1995.

74. M. Morari and J.H. Lee. Model predictive control: Past, present and future. In *Proc. PSE'97-ESCAPE-7 Symposium*, Trondheim, 1997.

75. E. Mosca. *Optimal, predictive and adaptive control.* Information and System Science Series. Prentice Hall, Englewoods Cliffs, New Jersey, 1994.

76. K.R. Muske and J.B. Rawlings. Nonlinear moving horizon state estimation. In R. Berber, editor, *Methods of Model Based Process Control*, pages 349–365, Netherlands, 1995. Kluwer Academic Publishers.

77. K.R. Muske, J.B. Rawlings, and J.H. Lee. Receding horizon recursive state estimation. In *Proc. Amer. Contr. Conf.*, pages 900–904, San Francisco, 1993.

78. Y. Nesterov and A. Nemirovsky. *Interior point polynomial methods in convex programming: Theory and applications.* SIAM, Philadelphia, 1994.

79. J. Nocedal and S.J. Wright. *Numerical Optimization.* Springer, New York, 1999. (In press).

80. D. Ocone. Probability densities for conditional statistics in the cubic sensor problem. *Mathematics of Control, Signals, and Systems*, 1:183–202, 1988.

81. D.M. Prett and R.D. Gillette. Optimization and constrained multivariable control of a catalytic cracking unit. In *Proceedings of the Joint Automatic Control Conference*, volume 1, 1980. Paper No. WP5-C.

82. A. I. Propoi. Use of linear programming methods for synthesizing sampled-data automatic systems. *Automn. Remote Control*, 24(7):837–844, July 1963.

83. S.J. Qin and T.A. Badgwell. An overview of industrial model predictive control technology. In Kantor et al. [84], pages 232–256.

84. S.J. Qin and T.A. Badgwell. An overview of industrial model predictive control technology. In J.C. Kantor, C.E. Garcia, and B. Carnahan, editors, *Fifth International Conference on Chemical Process Control - CPC V*, pages 232–256. American Institute of Chemical Engineers, 1996.

85. S.J. Qin and T.A. Badgwell. An overview of nonlinear model predictive control applications. In F. Allgöwer and A. Zheng, editors, *Nonlinear Predictive Control*. Birkhäuser, Basel, 1999. (In press).

86. D. Ralph and Y. Xiao. Private communication, March 1999.

87. Y. Ramamurthi, P.B. Sistu, and B.W. Bequette. Control-relevant dynamic data reconciliation and parameter estimation. *Comp. & Chem. Eng.*, 17(1):41–59, 1993.

88. C.V. Rao. *Moving Horizon Estimation of Constrained and Nonlinear Systems*. PhD thesis, University of Wisconsin–Madison, 1999.

89. C.V. Rao, J.B. Rawlings, and J.H. Lee. Stability of constrained linear moving horizon estimation. In *Proc. Amer. Contr. Conf.*, San Diego, 1999. to appear.

90. C.V. Rao, J.B. Rawlings, and J.H. Lee. Constrained linear state estimation - a moving horizon approach. Submitted for publication in *Automatica*, January 1999.

91. C.V. Rao, J.B. Rawlings, and D.Q. Mayne. Constrained state estimation for nonlinear discrete-time systems: Stability and moving horizon approximations. Submitted for publication in *IEEE Transactions on Automatic Control*, 1999.

92. C.V. Rao, S.J. Wright, and J.B. Rawlings. Application of interior-point methods to model predictive control. *J. Opt. Theory and Appl.*, 99:723–757, 1998.

93. J.B. Rawlings, E.S. Meadows, and K.R. Muske. Nonlinear model predictive control: A tutorial and survey. In *Proc. Int. Symp. Adv. Control of Chemical Processes, ADCHEM*, Kyoto, Japan, 1994.

94. K. Reif, S. Günther, E. Yaz, and R. Unbehauen. Stochastic stability of the discrete-time extended Kalman filter. *IEEE Trans. Automat. Contr.*, 44(4):714–728, 1999.

95. J. Richalet, A. Rault, J.L. Testud, and J. Papon. Algorithmic control of industrial processes. In *Proc. 4th IFAC Symposium on Identification and System Parameter Estimation*, pages 1119–1167, Tbilisi, 1976.

96. J. Richalet, A. Rault, J.L. Testud, and J. Papon. Model predictive heuristic control: Application to industrial processes. *Automatica*, 14:413–428, 1978.

97. D.G. Robertson and J.H. Lee. A least squares formulation for state estimation. *J. Proc. Contr.*, 5(4):291–299, 1995.

98. D.G. Robertson, J.H. Lee, and J.B. Rawlings. A moving horizon-based approach for least-squares estimation. *AIChE J.*, 42(8):2209–2224, 1996.

99. Douglas Robertson. *Development and Statistical Interpretation of Tools for Nonlinear Estimation.* PhD thesis, Auburn University, 1996.

100. P.O.M. Scokaert, D.Q. Mayne, and J.B. Rawlings. Suboptimal model predictive control (feasibility implies stability). *IEEE Trans. Automat. Contr.*, 44(3):648–654, 1999.

101. Inc. Setpoint. SMC-Idcom: A state-of-the-art multivariable predictive controller. Product literature from Setpoint, Inc., October 1993.

102. O. Slupphaug and B.A. Foss. Bilinear matrix inequalities and robust stability of nonlinear multi-model MPC. In *Proc. Amer. Contr. Conf.*, pages 1689–94, Philadelphia, 1998.

103. R. Soeterboek. *Predictive control – a unified approach.* Prentice Hall, New York, 1992.

104. E.D. Sontag. *Mathematical Control Theory.* Springer-Verlag, New York, 1990.

105. M. Soroush. Nonlinear state-observer design with application to reactors. *Chem. Eng. Sci.*, 52(3):387–404, 1997.

106. C. Striebel. Sufficient statistics in the optimum control of stochastic systems. *J. Math. Anal. Appl.*, 12:576–592, 1965.

107. Y. A. Thomas. Linear quadratic optimal estimation and control with receding horizon. *Electron. Lett.*, 11:19–21, 1975.

108. M.L. Tyler and M. Morari. Stability of constrained moving horizon estimation schemes. Technical Report # 96.18, Automatic Control Laboratory, Swiss Federal Institute of Technology (ETH), Zürich, Switzerland, April 1996.

109. S. Verdu and H.V. Poor. Abstract dynamic programming models under commutativity conditions. *SIAM J. Contr. Optim.*, 25:990–1006, 1987.

110. S. J. Wright. Applying new optimization algorithms to model predictive control. In J.C. Kantor, C.E. Garcia, and B. Carnahan, editors, *Fifth International Conference on Chemical Process Control – CPC V*, pages 147–155. American Institute of Chemical Engineers, 1996.

111. S.J. Wright. *Primal-Dual Interior-Point Methods.* SIAM Publications, Philadelphia, 1997.

112. T. H. Yang and E. Polak. Moving horizon control of nonlinear systems with input saturation, disturbances and plant uncertainty. *Int. J. Contr.*, 58(4):875–903, 1993.

113. E. Zafiriou. Robust model predictive control of processes with hard constraints. *Comp. & Chem. Eng.*, 14(4/5):359–371, 1990.

114. A. Zheng and M. Morari. Robust stability of constrained model predictive control. In *Proc. Amer. Contr. Conf.*, pages 379–383, 1993.

115. A. Zheng and W.H. Zhang. Robust model predictive control of large-scale nonlinear constrained systems. To appear in *Proc. 14th IFAC World Congress*, paper N-7a-02-2, 1999.

116. G. Zimmer. State observation by on-line minimization. *Int. J. Contr.*, 60(4):595–606, 1994.

MIX
Papier aus verantwortungsvollen Quellen
Paper from responsible sources
FSC® C105338

If you have any concerns about our products,
you can contact us on
ProductSafety@springernature.com

In case Publisher is established outside the EU,
the EU authorized representative is:
Springer Nature Customer Service Center GmbH
Europaplatz 3, 69115 Heidelberg, Germany

Printed by Libri Plureos GmbH
in Hamburg, Germany